HANDBOOK OF FIBER OPTIC
DATA COMMUNICATION

HANDBOOK OF
FIBER OPTIC
DATA COMMUNICATION

CASIMER DeCUSATIS
Editor-in-Chief
IBM Corporation
Poughkeepsie, New York

ERIC MAASS
Associate Editor
Motorola, Incorporated
Chandler, Arizona

DARRIN P. CLEMENT
Associate Editor
Lasertron, Incorporated
Bedford, Massachusetts

RONALD C. LASKY
Associate Editor
Cookson America, Incorporated
Providence, Rhode Island

 **ACADEMIC PRESS**

San Diego   London   Boston
New York   Sydney   Tokyo   Toronto

This book is printed on acid-free paper. ∞

ACADEMIC PRESS
525 B Street, Suite 1900, San Diego, CA 92101-4495, USA
1300 Boylston Street, Chestnut Hill, MA 02167
http://www.apnet.com

ACADEMIC PRESS LIMITED
24–28 Oval Road, London NW1 7DX, UK
http://www.hbuk.co.uk/ap/

Library of Congress Cataloging-in-Publication Data

Handbook of fiber optic data communication / Casimer DeCusatis . . . [et al.].
    p.   cm.
    Includes bibliographical references and index.
    ISBN 0-12-437162-0 (alk. paper)
    1. Optical communications. 2. Fiber optics. 3. Data transmission
systems.   I. DeCusatis, Casimer.
TK5103.59.H3515   1997
621.39'81--dc21   97-29071
                                                             CIP

Printed in the United States of America
97 98 99 00 IP 9 8 7 6 5 4 3 2 1

*To the people who give meaning to my life and
taught me to look for wonder in the world:
my wife Carolyn, my daughters, Anne and Rebecca, my parents,
my godmother, Isabel, and her mother, Mrs. Crease.* — CD

*To my wife and family for all of their encouragement
and support.* — RL

*To my wife Cathee, who is radiantly pregnant with our
first child together as this goes to press! I would also
like to thank Professor Ulf Österberg and Ronald C. Lasky for
introducing me to the world of fiber optic communications
and Casimer DeCusatis for his leadership on this book.* — DPC

*To my wife, Joan, and our children, David and Pamela, for
their support; and to my mother, Lillian, and my late father, Roger.
I would also like to thank Casimer DeCusatis for his leadership,
and Darrin Clement for helping me become increasingly involved
in this project.* — EM

Contents

Part 4 The Manufacturing Technology

Chapter 15 Semiconductor Laser and Light-Emitting Diode Fabrication 535

Wenbin Jiang and Michael S. Lebby

Chapter 16 Receiver, Laser Driver, and Phase-Locked Loop Design Issues 587

Dave Siljenberg

Chapter 17 Packaging Assembly Techniques 608

Glenn Raskin

Chapter **22** Manufacturing Challenges 783

Eric Maass

Contributors

Numbers in parentheses indicate the pages on which the authors' contributions begin.

Kurt Aretz (147), Siemens Corporation, Semiconductor Division-HL, Berlin, Germany

Daniel Baldwin (633), Manufacturing Research Center, School of Mechanical Engineering, Georgia Institute of Technology, 813 Ferst Drive Northwest, Atlanta, Georgia 30332

Carl Beckmann (385), Thayer School of Engineering, 8000 Cummings Hall, Dartmouth College, Hanover, New Hampshire 03755

Alan F. Benner (331), IBM RS6000SP Power Parallel Systems, Poughkeepsie, New York 12601

Darrin P. Clement* (633), Lasertron, Incorporated, 9 Oak Park, Bedford, Massachusetts 01803

John D. Crow (331), IBM Thomas J. Watson Research Center, Yorktown Heights, New York 10598

Carolyn J. Sher DeCusatis (87), Lighting Research Center, Rensselaer Polytechnic Institute, Troy, New York 12180

Casimer DeCusatis (191, 415), IBM Corporation, Poughkeepsie, New York 12601

George DeMario (131, 147), Siemens Corporation, Semiconductor Division-HL, Wappingers Falls, New York 12590

Jim Hayes (304), Fotec, Incorporated, 151 Mystic Avenue, Suite 7, Medford, Massachusetts 02155

* Present address: Banyan Systems, 120 Flanders Road, Westboro, Massachusetts 01581.

R. T. Hudson (262), Siecor Corporation, 800 17th Street Northwest, Hickory, North Carolina 28601

Georg Jeiter (131), Siemens Corporation, Semiconductor Division-HL, Berlin, Germany

Ching-Long (John) Jiang (87), Amp, Inc., Lytel Division, MS 300-001, 61 Chubb Way, Post Office Box 1300, Somerville, New Jersey 08876

Wenbin Jiang (38, 535), Phoenix Applied Research Center, Motorola, Incorporated, 2100 East Elliot Road, MS EL703, Tempe, Arizona 85284

D. R. King (262), Siecor Corporation, Hickory, North Carolina 28601

Michael Langenwalter (131, 147), Siemens Corporation, Semiconductor Division-HL, Berlin, Germany

Ronald C. Lasky (633, 673), Cookson America, Incorporated, One Cookson Place, Providence, Rhode Island 02903

Michael S. Lebby (38, 535), Phoenix Applied Research Center, Motorola, Incorporated, 2100 East Elliot Road, MS EL703, Tempe, Arizona 85284

Greg LeCheminant (304), Hewlett-Packard, Santa Rosa Systems Division, Santa Rosa, California 95403

Chung-Sheng Li (759), IBM Thomas J. Watson Research Center, Yorktown Heights, New York 10598

Daniel M. Litynski (704), Photonics Research Center, Department of Electrical Engineering and Computer Science, Thayer Road, United States Military Academy, West Point, New York 10996

Eric Maass (115, 783), Motorola, Incorporated, 2100 East Elliot Road, Tempe, Arizona 85284

Yann Y. Morvan (673), MPM Corporation, Medway, Massachusetts 02035

Ulf L. Österberg (3), Thayer School of Engineering, Dartmouth College, Hanover, New Hampshire 03755

Glenn Raskin (608), Motorola, Incorporated, 2501 South Price Road, M/D G651, Chandler, Arizona 85248

T. R. Rhyne (262), Siecor Corporation, Hickory, North Carolina 28601

Andre H. Sayles (704), Photonics Research Center, Department of Electrical Engineering and Computer Science, Thayer Road, United States Military Academy, West Point, New York 10996

Carsten Schwantes (147), Siemens Corporation, Semiconductor Division-HL, Berlin, Germany

Barry L. Shoop (704), Photonics Research Center, Department of Electrical Engineering and Computer Science, Thayer Road, United States Military Academy, West Point, New York 10996

Dave Siljenberg (587), IBM Corporation, 3605 Highway 52 North, Rochester, Minnesota 55901

Herwig Stange (147), Siemens Corporation, Semiconductor Division-HL, Berlin, Germany

Daniel J. Stigliani, Jr. (439), IBM Corporation, Poughkeepsie, New York 12601

Ray D. Sundstrom (115), Motorola, Incorporated, 2501 South Price Road, M/D G235, Chandler, Arizona 85248

Rakesh Thapar** (496), InterOperability Laboratory, University of New Hampshire, Durham, New Hampshire 03824

Frank Tong (759), IBM Thomas J. Watson Research Center, Yorktown Heights, New York 10598

T. A. Torchia (262), Siecor Corporation, Hickory, North Carolina 28601

Scholto Van Doren (415), Siemens Corporation, Santa Clara, California 95054

** Present address: FORE Systems, 501 Ashley Place, Wexford, Pennsylvania 15090.

Introduction

And God said, "Let there be light," and there was light.
God saw that the light was good, and separated the light from the darkness.
God called the light "ON" and the darkness He called "OFF."
And there was digital communication—the first day.

And God said, "Let there be an expanse between the light and dark to separate them."
And it was so.
God called the expanse "the modulation region."
And there was analog communication—the second day.

And in the beginning, there was the Internet, and the Net showed academic users and federal contractors a small amount of text, static information, and email.
And the Net was good.
And the Net begat the Web, which added color images, sound, and movies; and the corporate users discovered that the Web could show prospective customers their wares, and competition increased to see who could have the coolest Web page.
And the Web begat intranets that let the corporate users do business on-line, and begat browser software that made everyone a publisher in a global library with no index system. These all went forth, were fruitful, and multiplied . . . and multiplied . . . and multiplied. And the Web was very good.

—After Genesis, the New Optics Version (Anonymous) and Ingrid Meyer,
Commun. Week, *p. 61, September 1996*

This handbook deals with the use of optoelectronics for fiber optic data communication systems. Although the use of this technology in telecommunications is well established, and many good books are available in this area, data communications has only recently emerged as a distinct subfield. This book is a result of the need for a new text addressing the specific needs of the data communications industry.

Given the interest in and importance of this technology, we have attempted to assemble a handbook that covers all the fundamental aspects of fiber optics and optoelectronics for data communications. The handbook is divided into five sections, each dealing with a key area of optoelectronics and fiber optics in datacom: the technology, the links, applications, manufacturing, and the future. Within each section, leading experts from industry, academia, and government describe the technology and its implications for the novice while providing extensive tabulated information and up-to-date references for the practicing professional. The book is also suitable as a text for an advanced undergraduate or graduate course.

We, the editors, hope that this handbook proves useful to all these groups, and that it will contribute in some small measure to our growing understanding of this new discipline known as fiber optic datacom.

In order to fully appreciate the depth and breadth of this field, we provide a brief (and by no means comprehensive) look backward to its roots in the following section.

A Brief History of Optical Data Communications

Communicating using light is a very basic concept (the word "light" appears in the Bible 232 times). Visual communication systems have been used for centuries, including such concepts as smoke signals, beacon fires, and semaphores. (Written references to signal fires appear as early as the 5th century B.C. in the play "Agamemnon" by Aeschylus.) Even more modern concepts, such as optical communications for telephone systems, are far from being new ideas; in 1880, Alexander Graham Bell developed the Photophone, capable of transmitting speech over several hundred meters using visible light beams. (For comparison, Marconi demonstrated wireless radio for the first time in 1895. The Photophone was criticized as a fanciful idea by some, particularly in editorial cartoons showing huge lanterns strung between telephone poles.) Although Bell's early system was crude by today's standards and proved to be impractical, it nevertheless set the stage for exploration of optical frequencies for communications. Over the years, progress in fiber optics has been the result of an interdisciplinary collaboration involving electrical engineers, physicists, materials scientists, and others (this method of development has also characterized semiconductor electronics, as we will discuss shortly). As a result, many of the key enabling technologies, such as semiconductor lasers, low-loss optical fiber, and integrated electronics, were developed at about the same time. We can trace the development of modern optical fiber communications to a combination of incremental innovations in the existing art and scientific breakthroughs (such as the invention of the laser) that led to entirely new technologies.

As communication engineers sought to investigate higher and higher frequencies for transmission, eventually leading to microwave systems in the early 1940s, speculation on the use of optical communications began in the years following World War II. This background of theory was put to use when the laser was first described by Townes and Schawlow in 1958 and subsequently demonstrated for the first time by Maiman in 1960. (It has been said that AT&T was at first hesitant to file a patent application for the laser because it could not see what it had to do with telephones.) Thus began a prolific period of development in the technology that would make optical communications practical. Following a proposal in 1966 by Kao, a British engineer with the Standard Telecommunications Laboratory, that loss in glass fibers could be significantly reduced to very low levels, Corning Glass Works proceeded to develop the first practical optical fiber (with loss below 20 dB/km) in 1970. Advances in the following years led to losses of well below 1 dB/km, at about the same time that semiconductor lasers that were capable of continuous operation at room temperature became available. By the mid-1970s demonstrations of optical fiber communications systems were well under way, which ultimately led to the proliferation of fiber systems in the telecommunications infrastructure.

At a meeting of the American Mathematical Society in 1940, George Stibitz conducted an obscure experimental demonstration of the information superhighway. Stibitz controlled a digital-relay computer at Bell Laboratories in New York from a remote site in Hanover, New Hampshire, by transmitting data over ordinary telephone lines. At the time, the phone company had little interest in the emerging computer business; digital computers at the time used vacuum tube components as switching elements, which were perhaps 1 in. in diameter, 3 in. high, and consumed enough power while operating that they became too hot to touch. (The first vacuum tube diode was developed by Fleming in 1904. Two years later, in 1906, the first semiconductor version was demonstrated by Pickard; however, this proved too unreliable and was soon abandoned in favor of the vacuum tube amplifiers that were developed in the same year.) In the late 1940s, transistors were invented at Bell Labs and the junction transistor was proposed by Shockley; it offered a 100 times reduction in size and energy consumption and had begun to replace tubes by the late 1950s. (The Nobel Prize in physics was awarded in 1956 to Shockley, Bardeen, and Brattain for their work, which was the first Nobel award ever given for the invention of an engineering device.) Subsequent dramatic reductions in size, coupled with the invention of the integrated circuit by Kilby of Texas Instruments (and independently by Noyce at Fairchild Semiconductor) around 1958 led ultimately to modern very large-scale integration devices in which feature size is measured in fractions of a micrometer or about a 100-million times reduction in size over vacuum tubes. Interestingly, this is close to the order of magnitude increase available in data transmission speed between voice-quality telephone and the available 20 THz bandwidth of optical fiber.

The development of these technologies spawned the nascent computer industry. Once again, this was not without precedent; around 1883, Babbage had proposed the first serious effort to build a mechanical calculator machine, and even his work was preceded by similar ideas from Pascal, Leibnitz, and Schickhard, dating as far back as 1633! The first working electromechanical calculator, however, was built by IBM engineers in 1930 under the direction of Professor Aiken of Harvard University (the IBM Automatic Sequence Controlled Calculator, Mark I). This saw operation for more than 15 years, until in 1946 the first electronic calculator, ENIAC, was built at the University of Pennsylvania by Eckert and Mauchly. (This "Electronic Numeric Integrator and Computer" contained 18,000 vacuum tubes and filled a room 10 × 13 m.). That same year, IBM built the Model 603, the precursor to small, general-purpose electronic computers. From these humble beginnings grew a multibillion dollar international industry, which has seen the introduction of low-cost computers in many businesses, homes, and schools.

With the advancement of modern, inexpensive digital computer technology came another paradigm shift, the emergence of networking (increasingly based on fiber optic networks). It is possible to view the evolution of computer systems in three stages; the first being large mainframes with centralized control of data, the second being low-cost personal computers, and the third (emerging) stage being the proliferation of networking. ("The data center of the future will be software and a network"—*Network Computing,* May 1995.) As with many so-called paradigm shifts, the networking trend did not emerge overnight; its roots go back to the late 1960s, when the Advanced Research Project Agency (ARPA) began its early efforts to link together computers at major federally funded research centers, both universities and private corporations. (Even this work was not without precedent; similar projects had previously been explored by the National Physics Laboratory in the United Kingdom and the Societe Internationale de Telecommunications Aeronotique in France.) In 1970, the first fruits of this effort resulted in four West Coast universities going on-line together as ARPAnet (after the newly renamed Advanced Research Project Agency). The pace of change was rapid, however. In 1972, the first International Conference on Computer Communications (ICCC) was held in Washington, DC, to discuss progress of these early efforts. The chairman of this conference was Vinton Cerf, who, along with Robert Kahn, would release the standard Internet protocol TCP/IP just four years later. It was also Cerf who proposed linking ARPAnet with the National Science Foundation's CSNet via a TCP/IP gateway in 1980, which some consider to be the birth of the modern Internet. It is unusual, to say the least, that a government-funded research project intended to protect national security during the Cold War grew into a global network with no central control, where anarchy rules and there are virtually no restrictions on the exchange of information. It could very easily have happened differently.

Predicting the future of data communications is a proposition fraught with uncertainty. (Tom Watson, Sr., founder of IBM Corporation, boldly forecast in 1943 that there would be a worldwide market for perhaps five computers.) As the computer industry has continued to evolve, many people have tried to speculate what the future will bring. In 1965, Gordon Moore (founder and current chairman of Intel Corporation) projected that computing power as measured by the logic density of silicon integrated circuits would grow exponentially, approximately doubling every 12–18 months. This has held true for about the past 30 years and has come to be known as Moore's Law. (Mr. Moore's partner, Andrew Grove, has often expressed frustration with the performance of the telecommunications industry compared with the computer industry by noting that telecommunications bandwidth appears to double only once every 100 years.) Recent presentations by Mr. Moore forecast that this trend will continue through the next decade or so as feature sizes approach 0.18 μm as the economic problem associated with fabricating such chips begin to be felt. Observing that the usefulness of this computer power depends on the number of networked users, in 1980 Bob Metcalfe argued that the value of a network can be measured by the square of the number of users. Metcalfe's Law is more fully described by George Gilder in his forthcoming book, *Telecosm;* Mr. Gilder is also the namesake for the Gilder Paradigm, which predicts in part that future communication system designs will be influenced by a key scarcity of bandwidth. In this same spirit, similar predictions can be made concerning the growth of bandwidth requirements in data communications, following the recent introduction of fiber optics (before the widespread use of fiber, data rates appear to have increased only incrementally over a relatively long period of time). During a recent technical conference, for example, one of the editors introduced a graph showing the extrapolated growth of input/

output bandwidth on mainframes and large servers as a measure of leading-edge application requirements. This bandwidth has been growing exponentially since about 1988, when optical fiber first became available as an option on the IBM System/390. This trend, labeled by one of the meeting attendees as DeCusatis' Law, is projected to continue for at least the next several generations of complementary metal oxide semiconductor-based large systems and perhaps beyond.

In the past 20 years, computer systems have become ubiquitous; by contrast, optical fiber is only now beginning to emerge as a serious alternative to copper wiring in computer interconnections. This area is experiencing rapid growth, however, estimated by some to be as much as 25% compounded annually through the turn of the century. As of this writing, there is enough fiber installed in the world today to stretch between the earth and the moon 28 times. Emerging technologies, such as vertical cavity lasers and "smart pixels," are expected to play an important role in the next generation of data communication systems. New computer applications that can take advantage of fiber's huge bandwidth should drive increased growth in this area; in fact, the technologies of computers, telecommunications, and multimedia are merging at an astounding rate, in a revolution comparable to the invention of the movable-type printing press by Guttenberg some 500 years ago. The growth rate of the Internet and World Wide Web is staggering; some recent estimates place the number of independent CPUs attached to the Internet at well over 20 million. By the time you finish reading this sentence, another new site will have been added to the World Wide Web. We are told that industry, government, entertainment, education, and even our very lifestyles are on the verge of radical new changes brought about by these technologies; in order to realize the true potential of these new ideas, we need to deal with considerable intellectual, engineering, political, and financial challenges. In 1960, Marshall McCluhan first described the "global village" in which the world is brought closer together by strong communication systems providing free access to everyone regardless of location and traditional barriers were abolished by the power of the network. As we move closer to this vision, we hope that the technology described in this book will contribute in a positive way to reshaping our world.

Acknowledgments

Preparation of this work would not have been possible without the combined efforts of many individuals. We thank the authors for their support and dedication in preparing the material in a concise, timely manner, and for incorporating our suggestions to make the final manuscript even stronger than the sum of its parts. Thanks also to our publisher, Academic Press.

Casimer DeCusatis

Darrin P. Clement

Eric Maass

Ronald C. Lasky

General Historical References

1. Special issue. 1977 (September). Microelectronics. *Sci. Am.*
2. C. Weiner. 1973 (January). How the transistor emerged. *IEEE Spectrum,* 24–33.
3. M. F. Wolff. 1776 (August). The genesis of the integrated circuit. *IEEE Spectrum,* 45–53.
4. S. E. Miller and A. G. Chynoweth. 1979. Forward. In *Optical fiber telecommunications.* New York: Academic Press.
5. J. Gowar. 1984. Historical perspective. In *Optical communication systems.* Section 1.1. Englewood Cliffs, NJ: Prentice Hall.
6. G. Gilder. 1996 (December). The Gilder Paradigm, *Wired,* 225–228. Reprinted from the *Gilder Technology Report,* published by G. Gilder, P.O. Box 660, Housatonic, MA 01236 (1996).
7. C. DeCusatis. 1996 (October 21). Requirements for fiber optic data communication. Paper presented at the Optical Society of American Technical Group Meeting on Fiber Optic Data Communication, Rochester, NY; also see "A tutorial on fiber optic data communications," C. DeCusatis, in press.
8. Repealing Moore's Law. 1996 (December). *OEM* 4(34): 82–83; also see Late News. 1996 (December 16). *Electrical Eng. Times,* 8.
9. G. Moore. 1995. Lithography and the future of Moore's Law. SPIE Proc. 2440 (Optical/Laser Microlithography VII), 2–17.

Part 1 | The Technology

Chapter 1 | Optical Fiber, Cable, and Connectors

Ulf L. Österberg

*Thayer School of Engineering, Dartmouth College, Hanover,
New Hampshire 03755*

1.1. Light Propagation

1.1.1. RAYS AND ELECTROMAGNETIC MODE THEORY

Light is most accurately described as a vectorial electromagnetic wave. Fortunately, this complex description of light is often not necessary for a satisfactory treatment of many important engineering applications.

In the case of optical fibers used for tele- and data communication it is sufficient to use a scalar wave approximation to describe light propagation in single-mode fibers and a ray approximation for light propagation in multimode fibers.

For the ray approximation to be valid the diameter of the light beam has to be much larger than the wavelength. In the wave picture we will assume a harmonically time-varying wave propagating in the z direction with phase constant β. The electric field can be expressed as

$$E = E_\mathrm{o}(x, y)\cos(\omega t - \beta z). \tag{1.1}$$

This is more conveniently expressed in the phasor formalism as

$$E = E_\mathrm{o}(x, y)e^{j(\omega t - \beta z)}, \tag{1.2}$$

where the real part of the right-hand side is assumed.

A wave's propagation in a medium is governed by the wave equation. For the particular wave in Eq. (1.2) the wave equation for the electric z component is

$$\nabla_t^2 E_z(x, y) + \beta_t^2 E_z(x, y) = 0, \tag{1.3}$$

3

HANDBOOK OF FIBER OPTIC
DATA COMMUNICATION

where we have introduced

$$\nabla_t^2 = \frac{\partial^2}{\partial_{x^2}} + \frac{\partial^2}{\partial_{y^2}} \qquad \text{[Transverse Laplacian]},$$

$$\beta_t^2 = k^2 n^2 - \beta^2 \qquad \text{[Transverse phase constant]},$$

$$k = \frac{2\pi}{\lambda} \qquad \text{[Free space wave vector]},$$

$$n(x, y) \qquad \text{[Refractive index]}.$$

The variable kn corresponds to the phase constant for a plane wave propagating in a medium with refractive index n. There is an equivalent wave equation to Eq. (1.3) for the H_z component. We have to solve only the wave equation for the longitudinal components E_z and H_z. The reason for this is that E_x and E_y can both be calculated from E_z and H_z using Maxwell's equations.

1.1.2. SINGLE-MODE FIBER

In an infinitely large isotropic and homogeneous medium a light wave can propagate as a plane wave and the phase constant for the plane wave can take on any value, limited only by the available frequencies of the light itself. When light is confined to a specific region in space, boundary conditions imposed on the light will restrict the phase constant β to a limited set of values. Each possible phase constant β represents a mode. In other words, when light is confined it can propagate only in a limited number of ways.

For an engineer it is important to find out how many modes can propagate in the fiber, what their phase constants are, and their spatial transverse profile. To do this we have to solve Eq. (1.3) for a typical fiber geometry (Fig. 1.1). Because of the inherent cylindrical geometry of an optical fiber, Eq. (1.3) is transformed into cylindrical coordinates and the modes of spatial dependence are described with the coordinates r, ϕ, and z. Because the solution is dependent on the specific refractive index profile, it has to be specified. In Fig. 1.2 the most common refractive index profiles are shown. For step-index profile in Fig. 1.2c, a complete analytical set of solutions can be given [3]. These solutions can be grouped into three different types of modes: TE, TM, and hybrid modes, of which the hybrid modes are further separated into EH and HE modes. It turns out that for typical

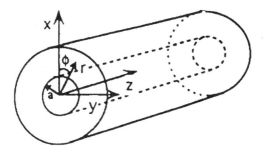

Fig. 1.1 Typical fiber geometry. Reprinted from Ref. [1], p. 12, courtesy of Academic Press.

fibers used in tele- and data communication the refractive index difference between core and cladding, $n_1 - n_2$, is so small (~0.002–0.008) that most of the TE, TM, and hybrid modes are degenerate and it is sufficient to use a single notation for all these modes—the LP notation. An LP mode is referred to as LP$_{\ell m}$, where the ℓ and m subscripts are related to the number of radial and azimuthal zeros of a particular mode. The fundamental mode, and the only one propagating in a single-mode fiber, is the LP$_{01}$ mode. This mode is shown in Fig. 1.3. To quickly figure out if a particular LP mode will propagate, it is very useful to define two dimensionless parameters, V and b.

$$V = ka\sqrt{n_1^2 - n_2^2} \approx \frac{2\pi}{\lambda} \cdot a \cdot n_1 \sqrt{2\Delta}, \qquad (1.4)$$

where a is the core radius, λ is the wavelength of light, and $\Delta \approx \dfrac{n_1 - n_2}{n_1}$.
The V number is sometimes called the normalized frequency.

The normalized propagation constant b is defined as

$$b = \frac{(\beta^2/k^2) - n_2^2}{n_1^2 - n_2^2}, \qquad (1.5)$$

where b is the phase constant of the particular LP mode, k is the propagation constant in vacuum, and n_1 and n_2 are the core and cladding refractive indexes, respectively.

Equation (1.5) is very cumbersome to use because b has to be calculated from Eq. (1.3). For LP modes Gloge *et al.* [4] have shown that to a very

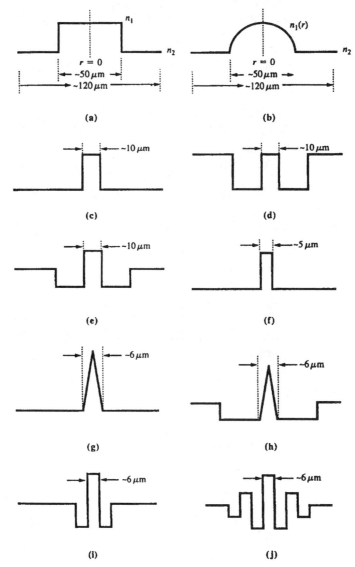

Fig. 1.2 Refractive index profiles of (a) step-index multimode fibers, (b) graded-index multimode fibers, (c) match-cladding single-mode fibers, (d, e) depressed-cladding, single-mode fibers, (f–h) dispersion-shifted fibers, and (i, j) dispersion-flattened fibers. Reprinted from Ref. [2], p. 125, courtesy of Irwin.

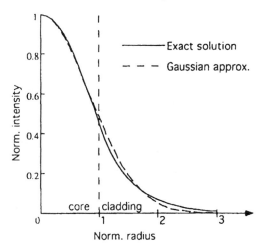

Fig. 1.3 Comparison between exact and Gaussian approximation of the radial mode field distribution for the lowest order mode. Reprinted from Ref. [1], p. 14, courtesy of Academic Press.

good accuracy the following formulas can be used to calculate b for different $LP_{\ell m}$ modes:

$$LP_{01}: b_{01} = 1 - \left[\frac{1 + \sqrt{2}}{1 + (4 + V^4)^{\frac{1}{4}}}\right]^2 \tag{1.6}$$

$$\begin{array}{c} LP_{\ell m} \\ {\scriptstyle \ell m \neq 01} \end{array} : b_{\ell m} = 1 - \frac{U_c^2}{V^2} \cdot exp\left[\frac{2}{S}\left(arcsin\left(\frac{S}{U_c}\right) - arcsin\left(\frac{S}{V}\right)\right)\right]$$

$$S = \sqrt{U_c^2 - \ell^2 - 1} \tag{1.7}$$

$$U_c = A - \frac{B - 1}{8A} - \frac{4(B - 1)(7B - 31)}{3(8A)^3}$$

$$A = \pi \cdot \left[m + \frac{1}{2}(\ell - 1) - \frac{1}{4}\right]; \ B = 4(\ell - 1)^2.$$

The graphs in Fig. 1.4 were generated using Eqs. (1.6) and (1.7). The

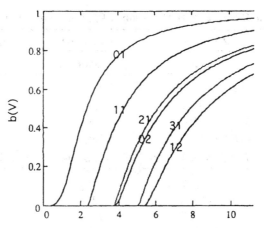

Fig. 1.4 Cutoff frequencies for the lowest order LP modes. Reprinted from Ref. [1], p. 15, courtesy of Academic Press.

normalized propagation constant b can vary only between 0 and 1 for guided modes; this corresponds to

$$n_2 k < \beta < n_1 k. \tag{1.8}$$

The wavelength for which b is zero is called the cutoff wavelength, i.e.,

$$b_{\ell m}(V_{co}) = 0 \Rightarrow \lambda_{co} = \frac{2\pi}{V_{co}} \cdot a \cdot n_1 \sqrt{2\Delta}. \tag{1.9}$$

Therefore, for wavelengths longer than the cutoff wavelength the mode cannot propagate in the optical fiber.

Cutoff values for the V number for a few LP modes are given in Table 1.1. The fundamental mode can, to better than 96% accuracy, be described using a Gaussian function

$$E(r) = E_o \exp\left[-\left(\frac{r}{w_g}\right)^2\right], \tag{1.10}$$

where E_o is the amplitude and $2w_g$ is the mode field diameter (MFD) (Fig. 1.3). The meaning of the MFD is shown in Fig. 1.5. The MFD for the fundamental mode is larger than the geometrical diameter in a single-mode

Table 1.1 **Cutoff Frequencies of Various LP$_{lm}$ Modes in a Step Index Fiber**[a]

$\ell = 0$ modes	$J_1(V_c) = 0$	$\ell = 1$ modes	$J_0(V_c) = 0$
Mode	V_c	**Mode**	V_c
LP$_{01}$	0	LP$_{11}$	2.4048
LP$_{02}$	3.8317	LP$_{12}$	5.5201
LP$_{03}$	7.0156	LP$_{13}$	8.6537
LP$_{04}$	10.1735	LP$_{14}$	11.7915
$\ell = 2$ modes	$J_1(V_c) = 0; V_c \neq 0$	$\ell = 3$ modes	$J_2(V_c) = 0_i; V_c \neq 0$
LP$_{21}$	3.8317	LP$_{31}$	5.1356
LP$_{22}$	7.0156	LP$_{32}$	8.4172
LP$_{23}$	10.1735	LP$_{33}$	11.6198
LP$_{24}$	13.3237	LP$_{34}$	14.7960

[a] Reprinted from Ref. [5], p. 380, courtesy of Cambridge University Press.

(SM) fiber and much smaller than the geometrical diameter in a multimode (MM) fiber. The optimum MFD is given by the following formula [7]:

$$\frac{w_g}{a} = 0.65 + 1.619 \cdot V^{-3/2} + 2.87 \cdot V^{-6}, \tag{1.11}$$

where a is the core radius. Equation (1.11) is valid for wavelengths between $0.8\ \lambda_{co}$ and $2\ \lambda_{co}$.

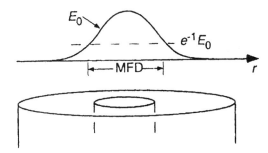

Fig. 1.5 The electric field of the HE$_{11}$ mode is transverse and approximately Gaussian. The mode field diameter is determined by the points where the power is down by e^{-2} or where the amplitude is down by e^{-1}. The MFD is not necessarily the same dimension as the core. Reprinted from Ref. [6], p. 144, courtesy of Irwin.

If the radial distribution for higher order modes is needed it is necessary to use the Bessel functions [3]. In Fig. 1.6 the radial intensity distribution is shown for five LP modes in a fiber with $V = 8$. Recommended specifications for a single-mode fiber are summarized in Table 1.2.

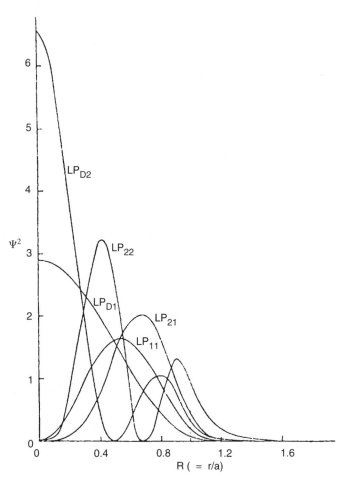

Fig. 1.6 Radial intensity distributions (normalized to the same power) of some low-order modes in a step index fiber for $V = 8$. Notice that the higher order modes have a greater fraction of power in the cladding. Reprinted from Ref. [5], p. 382, courtesy of Cambridge University Press.

Table 1.2 **CCITT Recommendation G.652[a]**

Parameters	Specifications
Cladding diameter	125 μm
Mode field diameter	9–10 μm
Cutoff wavelength λ_{co}	1100–1280 nm
1550 nm bend loss	$\leq$1 dB for 100 turns of 7.5 cm diameter
Dispersion	$\leq$3.5 ps/nm · km between 1285 and 1330 nm
	$\leq$6 ps/nm · km between 1270 and 1340 nm
	$\leq$20 ps/nm · km at 1550 nm
Dispersion slope	$\leq$0.095 ps/nm^2 · km

[a] Reprinted from Ref. [2], p. 126, courtesy of Irwin.

1.1.3. MULTIMODE FIBER

The previous discussion has in principle been for a step-index MM fiber. Because of the severe differences in propagation time between different modes in a step-index fiber, these are not commonly used in practice. Instead, a graded refractive index core is used for a MM fiber (Fig. 1.7). The various graded-index profiles are generated by

$$n^2(r) = \begin{cases} n_1^2\left(1 - 2\Delta\left(\dfrac{r}{a}\right)^q\right); & 0 \leq r \leq a \\ n_1^2(1 - 2\Delta) = n_2^2; & r \geq a, \end{cases} \tag{1.12}$$

for different q's, q is called the profile exponent. The optimum profile is the one that gives the minimum dispersion; this occurs for q at slightly less than 2.

The total number of modes that can propagate in a MM fiber is given by

$$N = \frac{1}{2}\frac{q}{q+2}V^2. \tag{1.13}$$

For a step-index fiber $q = \infty$ and $N = \dfrac{V^2}{2}$. Equation (1.13) is only valid for large V numbers.

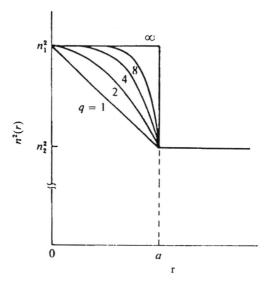

Fig. 1.7 The refractive index variation for a power law profile for different values of q. Reprinted from Ref. [5], p. 396, courtesy of Cambridge University Press.

One can calculate the different ray paths that are possible in a MM fiber using a geometrical optics approach [8]. A more accurate view of the different modes can be obtained by solving Eq. (1.3) using the so-called WKB approximation [3]. Using this analysis the phase constants for the different modes can be shown to obey the following relationship:

$$\beta_m = nk \sqrt{1 - 2\Delta\left(\frac{m}{n}\right)^{q/q+2}}, \, m = 1,2, \ldots , N. \qquad (1.14)$$

1.1.4. OPTICAL COUPLING

A first approach to estimate how much light can be coupled into an optical fiber is to use the ray picture. In this picture the light is confined within the core if it undergoes total internal reflection at the core–cladding boundary. This will occur only for light entering the fiber within an acceptance cone defined by the angle θ (Fig. 1.8). Rather than stating the angle θ for an optical fiber, it is the convention to give $\sin\theta$, which is called the numerical aperture (NA). The NA is defined as

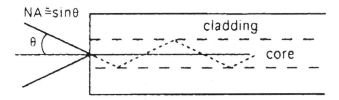

Fig. 1.8 Acceptance angle for an optical fiber. Reprinted from Ref. [1], p. 10, courtesy of Academic Press.

$$NA = n \cdot \sin\theta, \tag{1.15}$$

where n is the refractive index of the medium the light is coming from. In the case of coupling into an optical fiber, light is usually coming from air and subsequently $n \approx 1$. A more useful formula for the NA can be obtained if we use the dimensionless parameter Δ,

$$NA \approx n_1 \sqrt{2\Delta}. \tag{1.16}$$

For an incoherent light source such as a light emitting diode (LED) one can show that the total power accepted by the fiber is given by [9]

$$P \approx \pi \cdot B \cdot A_{\text{fiber}} \cdot (NA)^2, \tag{1.17}$$

where B is the LED's radiance (units for radiance is watts per area and steradian).

It is more common to give a coupling efficiency; thus, giving the total power accepted by the fiber, the efficiency is defined as [10]

$$\eta = \frac{P_{\ell m}}{P_{\text{in}}}, \tag{1.18}$$

where P_{in} is the power launched into the fiber and $P_{\ell m}$ the power accepted by the $LP_{\ell m}$ mode. For link budget analyses it is more convenient to deal with coupling losses in units of decibels α:

$$\alpha = 10 \cdot log\ \eta. \tag{1.19}$$

A more general definition of the coupling efficiency η for an optical fiber that obeys the weekly guiding approximation can be given as [10]

$$\eta = \left| \frac{n_2}{Z_0} \iint\limits_{A^\infty} E_i \cdot E^*_{\ell m} dA \right|^2, \tag{1.20}$$

where n_2 is the refractive index of the cladding, z_o is the free space wave impedance, and E_i and $E_{\ell m}$ are the electric field amplitudes for the incoming light and for the light propagating in mode $LP_{\ell m}$ in the fiber, respectively.

Coherent light from a laser can often be approximated with a Gaussian beam; furthermore, if we restrict ourselves to a SM fiber, so that the LP_{01} mode can also be approximated as a Gaussian field, it is possible to calculate η analytically [7, 11]:

$$\eta = \left(\frac{4D}{B}\right)e^{-AC/B}, \tag{1.21}$$

where

$$A = \frac{(k's_1)^2}{2}$$

$$B = G^2 + (D + 1)^2$$

$$C = (D + 1)F^2 + 2DFG\sin\theta + D(G^2 + D + 1)\sin^2\theta$$

$$D = \left(\frac{s_2}{s_1}\right)^2$$

$$F = \frac{2\Delta}{k's_1^2}$$

$$G = \frac{2\Delta Z}{k's_1^2}$$

$$k' = \frac{2\pi n_o}{\lambda}$$

where ΔZ is the separation distance between source and fiber, Δ is the lateral displacement, θ is the angular displacement, and s_1 and s_2 are the mode field radii or spot sizes of the source field and fiber field, respectively. The refractive index of the medium between source and fiber is denoted n_o.

Equation (1.21) takes into account four different coupling cases at once (Fig. 1.9). If only one of these different coupling cases is present at a time Eq. (1.21) can be simplified to:

Case 1.1.   Spot-size mismatch $s_1 \neq s_2$:

$$\eta = \left(\frac{2s_1s_2}{s_1^2 + s_2^2}\right)^2. \tag{1.22}$$

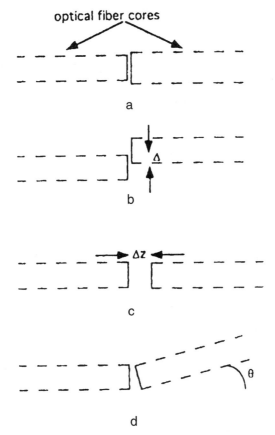

Fig. 1.9 Different coupling cases. Reprinted from Ref. [1], p. 18, courtesy of Academic Press.

Case 1.2. Transverse offset Δ:

$$\eta = \exp\left(-\left(\frac{\Delta}{s_2}\right)^2\right). \tag{1.23}$$

Case 1.3. Longitudinal offset ΔZ:

$$\eta = \frac{1}{1 + \left(\dfrac{\Delta Z}{2Z_0}\right)^2}, \tag{1.24}$$

where Z_0 is the Rayleigh range.

Case 1.4. Angular misalignment θ:

$$\eta = \exp\left(-\left(\frac{\theta}{\theta_0}\right)^2\right). \tag{1.25}$$

If a lens is used in between the emitter and fiber, some modifications to the previous formulas have to be done. What the lens can do for us is to match the output angle of the emitter to the acceptance angle of the receiving fiber. If properly done, the power coupled into the fiber is multiplied with the lens magnification factor M: $M = \dfrac{d_{\text{rec}}}{d_{\text{em}}}$, (see Fig. 1.10). All the preceding formulas need to be corrected for reflection losses. If the refractive index of the medium between the source and the fiber is denoted n_{o}, the coupled power into the fiber is reduced with a factor R,

$$R = \left(\frac{n_1 - n_{\text{o}}}{n_1 + n_{\text{o}}}\right)^2. \tag{1.26}$$

1.2. Optical Fiber Characterization

1.2.1. OPTICAL FIBER MATERIALS

The material is primarily chosen to provide the minimum attenuation. Table 1.3 shows order of magnitude attenuation at three different wavelengths for four common glass types.

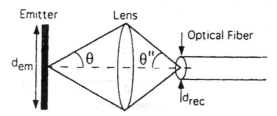

Fig. 1.10 Lens for improving coupling efficiency. Reprinted from Ref. [1], p. 20, courtesy of Academic Press.

Table 1.3 **Scattering Loss for Several Representative Glass Materials**[a]

Material	dB/km		
	633 nm	**800 nm**	**1060 nm**
Fused silica	4.8	1.9	0.6
Soda lime	8.5	3.3	1.1
Borosilicate crown	7.7	3.0	1.0
Lead silicate	47.5	18.6	6.0

[a] Reprinted from Ref. [1], p. 21, courtesy of Academic Press.

For tele- and data communication fibers, fused silica glass is the preferred material. To provide guiding of the light, the core of the fiber is doped with a few molar percentage of a substance that increases the refractive index. It is also possible to dope the cladding such that its refractive index becomes lower than the pure silica glass index in the core (Fig. 1.11).

1.2.2. ATTENUATION

Attenuation is a very important factor in designing effective long-distance fiber optic networks. Consequently, the fabrication methods have improved dramatically during the past 30 years so that attenuation is measured in a

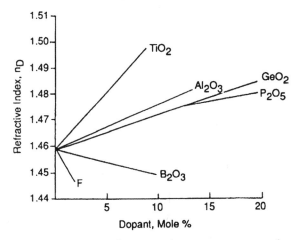

Fig. 1.11 Influence of different dopant mole percentages on refractive index of silica glass. Reprinted from Ref. [1], p. 22, courtesy of Academic Press.

few tens of a dB/km. The dB is defined in Eq. (1.19). The various factors affecting the attenuation, in the 0.8- to 1.6-μm wavelength region, are listed in Table 1.4. Figure 1.12 shows schematically how some of the factors contributing to the overall light attenuation vary with the wavelength in the near-infrared wavelength region. Typical total losses for an optical fiber, at the three different transmission windows at 800, 1300, and 1550 nm, are $\approx$2 or 3, $\leq$0.5, and $\leq$0.2 dB/km, respectively. These numbers are for SM fibers; from Fig. 1.13 it can be seen that MM fibers have slightly higher losses. The losses dealt with to date have been due to either intrinsic properties of the glass or extrinsic properties (such as OH and transition metal contents) that come from the particular fabrication method used. In addition, there are bending losses. If the fiber has been improperly cabled or installed these bending losses can be substantial. Bending losses are divided into micro- and macro-bending losses. Micro-bending losses are due to nanometer-size deviations in the fiber. Macro-bending losses are due to visible bends in the fiber. Figure 1.14 shows qualitatively how micro- and macrobends contribute to the overall loss in a SM and MM fiber.

1.2.3. DISPERSION

Dispersion is due to the fact that different wavelengths experience different propagation constants, $\beta(\lambda)$, and therefore travel with different velocities causing a longer temporal pulse at the end of the fiber. Dispersion does not alter the wavelength (frequency) content of the light pulse. From a communications point of view dispersion is a very important factor because

Table 1.4 **Factors Affecting Attenuation**[a]

Intrinsic loss mechanisms
    Tail of infrared absorption by Si–O coupling
    Tail of ultraviolet absorption due to electron transitions in defects
    Rayleigh scattering due to spatial fluctuations of the refractive index
Absorption by impurities
    Absorption by molecular vibration of OH
    Absorption by transition metals
Structural imperfections
    Geometrical nonuniformity at core–cladding boundary
    Imperfection at connection or splicing between fibers

[a] From Ref. [3].

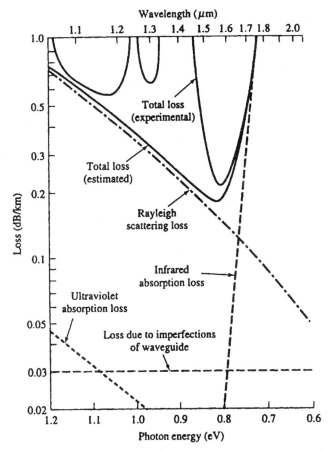

Fig. 1.12 Transmission loss in silica-based fibers. Reprinted from Ref. [12], p. 474, courtesy of Irwin.

it directly affects the bit rate. The following are three major components contributing to the dispersion:

- Material
- Waveguide
- Intermodal

The material and waveguide dispersion are often referred to as chromatic dispersion, measured in units ps/nm · km, which is defined as

$$D_{\text{chr}} = -\frac{1}{L}\frac{dt_{\text{g}}}{d\lambda}, \tag{1.27}$$

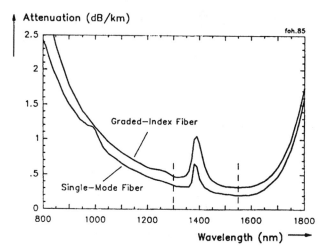

Fig. 1.13 Wavelength dependence of fiber attenuation. Reprinted from Ref. [14], p. 8. Copyright © 1989 by Hewlett Packard Company. Reproduced with permission.

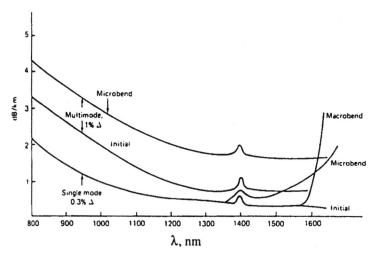

Fig. 1.14 Bend-induced losses of optical fibers. Reprinted from Ref. [15], p. 1.33, courtesy of McGraw-Hill.

where L is the fiber length and t_g is the time required to propagate a distance L. The subscript g refers to group velocity. When a pulse propagates in a dispersive medium it propagates with the group velocity,

$$V_g = \frac{d\omega}{d\beta}. \tag{1.28}$$

The phase travels with the phase velocity given by

$$V_p = \frac{\omega}{\beta} = \frac{c}{n}. \tag{1.29}$$

The dispersion properties are completely determined by the group velocity.

Material dispersion originates from the fact that the refractive index is a function of wavelength, $n(\lambda)$:

$$D_{\text{mat}} = -\frac{\lambda}{c}\frac{d^2 n_1}{d\lambda^2}. \tag{1.30}$$

The refractive index can be calculated from the Sellmeier equation:

$$n_1^2(\lambda) = 1 + \sum_{i=1}^{3} \frac{a_i \lambda^2}{\lambda^2 - b_i}, \tag{1.31}$$

where the constants a_i and b_i may vary with the particular fiber composition. These constants have been accurately measured and tabulated for different Ge, B and P concentrations [16].

Waveguide dispersion is due to the different propagation characteristics of the light in the core and cladding. Keep in mind that a large portion of the light (30–40%) travels in the cladding for the LP_{01} mode around the cutoff frequency for the next higher order mode. The waveguide dispersion can be approximated, using the normalized propagation constant b, as

$$D_{\text{wg}} \simeq -\frac{\lambda}{c} \cdot n_1 \cdot \Delta \cdot \frac{d^2 b}{d\lambda^2}. \tag{1.32}$$

The total chromatic dispersion as well as its constituents are plotted in Fig. 1.15. As shown in Fig. 1.2, an optical fiber can have many different refractive index profiles. How the refractive index profile can affect the dispersion properties of the fiber is shown in Fig. 1.16.

Intermodal dispersion arises from the different travel times for the different modes in the fiber. For modal dispersion the units are nsec/km and D_{mod} is defined as follows [2]:

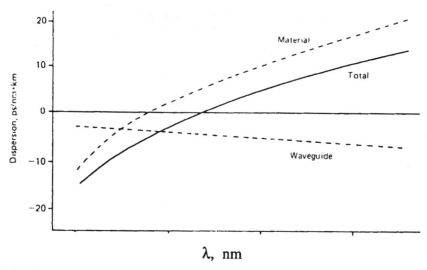

Fig. 1.15 Dispersion vs wavelength for single-mode fiber. Reprinted from Ref. [15], p. 1.38, courtesy of McGraw-Hill.

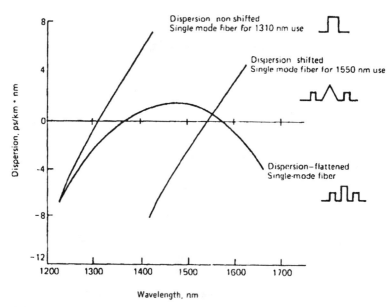

Fig. 1.16 Spectral disposition for various single-mode fiber profiles. Reprinted from Ref. [15], p. 1.38, courtesy of McGraw-Hill.

Table 1.5 **Physical Properties of Materials**[a]

Property	Silica glass	Copper	Aluminum	Steel
Chemical sign	SiO_2	Cu	Al	Fe
Specific gravity (g/cm³)	2.20	8.9	2.70	7.9
Tensile strength (kg/mm²)	500	25	10	120
Young's modulus (kg/mm²)	7200	12,000	6300	20,000
Elongation (%)	2–8	20–30	7–20	5–15
Coefficient of thermal expansion	5×10^{-7}	1.7×10^{-6}	2.3×10^{-5}	1×10^{-5}
Specific heat (cal/°C · g)	0.20	0.09	0.5	0.1

[a] Reprinted from Ref. [18], p. 53, courtesy of Dekker.

$$D_{mod} = t_g^{max} - t_g^{min}, \qquad (1.33)$$

where t_g^{max} and t_g^{min} are the propagation times for the modes traveling the longest and shortest distances, respectively. Using the group index N_1 [3],

$$N_1 = n_1 + k\frac{dn_1}{dk}, \qquad (1.34)$$

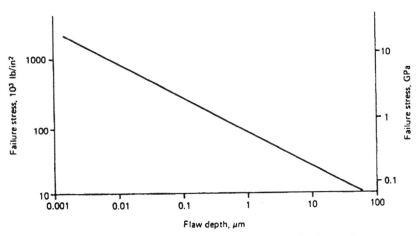

Fig. 1.17 Failure stress vs flaw size. Reprinted from Ref. [15], p. 1.40, courtesy of McGraw-Hill.

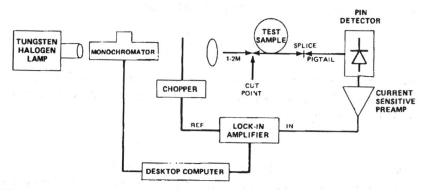

Fig. 1.18 Experimental setup for attenuation measurements by the cutback technique. This same apparatus can be used to measure cutoff wavelength. Reprinted from Ref. [20], p. 336, courtesy of Academic Press.

where the term $k\dfrac{dn_1}{dk}$ is the material dispersion for a plane wave traveling in the core region; we can approximately write the modal dispersion in a step-index fiber as follows [2]:

$$D_{\text{mod}} \simeq -\frac{N_1}{c}\Delta, \tag{1.35}$$

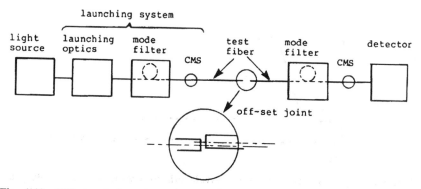

Fig. 1.19 Offset technique for mode field measurement—a block diagram; CMS, cladding mode stripper. Reprinted from Ref. [18], p. 426, courtesy of Marcel Dekker Inc.

and for a graded-index fiber, with $q = 2(1 - \Delta)$, as

$$D_{\text{mod}} \simeq -\frac{N_1}{c}\frac{\Delta^2}{8}. \tag{1.36}$$

Note that for most fibers $N_1 \simeq n_1$ is an excellent approximation [3].

The total fiber disposition in a MM fiber, units nesc/km, can be shown to be the following [2]:

$$D_{\text{tot}}^2 = D_{\text{chr}}^2 \cdot \Delta\lambda^2 + D_{\text{mod}}^2, \tag{1.37}$$

where $\Delta\lambda$ is the spectral bandwidth of the light source.

Knowledge of the total dispersion, D_{tot}, can be transmitted into bandwidth (BW), units MHz · km, as follows [17]:

$$BW \simeq \frac{350}{D_{\text{tot}}}. \tag{1.38}$$

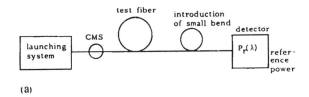

(a)

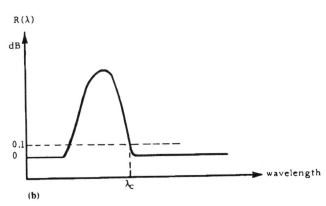

(b)

Fig. 1.20 Typical setup of cutoff wavelength measurement: (a) bending and (b) typical measurement of graph of the cutoff wavelength, single-mode reference (bending method). Reprinted from Ref. [18], (a) p. 430, and (b) p. 431, courtesy of Marcel Dekker Inc.

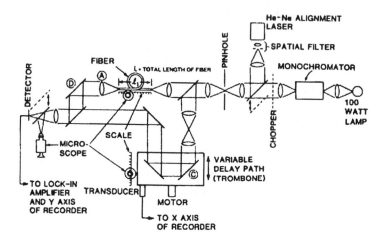

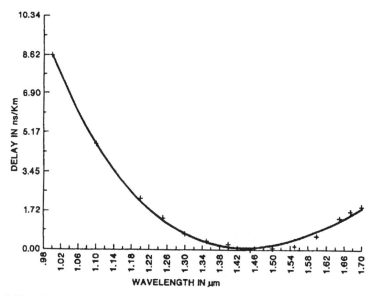

Fig. 1.21 Optical arrangement of dispersion measured by the interferrometric technique and the delay curve. Reprinted from Ref. [20], p. 351, courtesy of Academic Press.

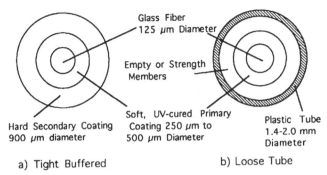

Glass Fiber
125 μm Diameter

Empty or Strength
Members

Hard Secondary Coating
900 μm diameter

Soft, UV-cured Primary
Coating 250 μm to
500 μm Diameter

Plastic Tube
1.4-2.0 mm
Diameter

a) Tight Buffered

b) Loose Tube

Fig. 1.22 Common single-fiber cable types. Reprinted from Ref. [1], p. 36, courtesy of Academic Press.

1.2.4. MECHANICAL PROPERTIES

Optical fibers have outstanding signal characteristics such as low attenuation and small dispersion. For optical fibers to be practically useful they must also have good mechanical properties to withstand environmental strains. In Table 1.5, physical properties for different materials used in communication cables are compared.

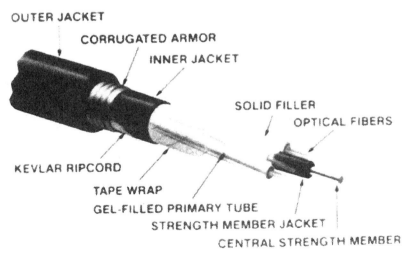

OUTER JACKET

CORRUGATED ARMOR

INNER JACKET

SOLID FILLER

OPTICAL FIBERS

KEVLAR RIPCORD

TAPE WRAP

GEL-FILLED PRIMARY TUBE

STRENGTH MEMBER JACKET

CENTRAL STRENGTH MEMBER

Fig. 1.23 Fiber optic cable-corrugated armor. Reprinted from Ref. [21], p. 247, courtesy of Academic Press.

Silica glass has a very high threshold for breaking when there are no flaws, either on the surface or inside the glass. The tensile strength has been estimated to be 20 GPa [19]. Because optical glass fibers have a large surface-to-volume ratio they are very prone to failure from surface flaws that propagate into the glass. The actual stress for which the fiber breaks as a function of the size of the surface flaw is shown in Fig. 1.17. When fibers are tested for their mechanical reliability it is common to use short

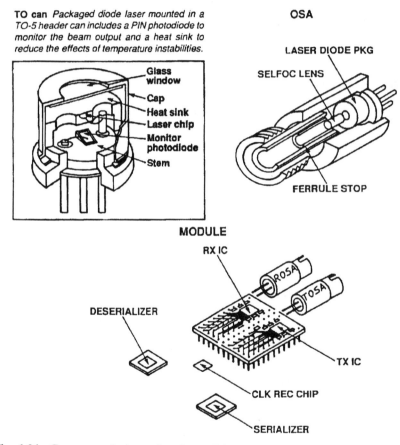

Fig. 1.24 Common solution today; Laser TO can → optical subassembly (OSA) → module. There are three levels of packaging and several active alignments. Reprinted from Ref. [22], p. 134, courtesy of Academic Press.

fibers for destructively testing the fibers' breaking threshold for static and dynamic loads. Additionally, a screen test is performed on long samples of the fibers to find the effective strength [19].

1.2.5.   *MEASUREMENT TECHNIQUES*

In this section I will briefly introduce one measurement method each for measuring attenuation, mode field diameter, cutoff wavelength, and chromatic dispersion.

Attenuation (the Cutback Method)

The transmission is measured as a function of wavelength for the full-length fiber, $P_L(\lambda)$. The fiber is then cut approximately 1 or 2 m away from the input end and a new transmission measurement is performed, $P_r(\lambda)$ (Fig. 1.18). The attenuation constant α is then calculated from as follows [20]:

$$\alpha = 10 \cdot \log \frac{P_L(\lambda)}{P_r(\lambda)}. \tag{1.39}$$

ST SERIES CONNECTOR

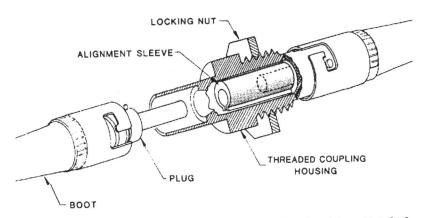

Fig. 1.25 Straight tip ferrule single fiber connector. Reprinted from Ref. [23], p. 303, courtesy of Academic Press.

Mode Field Diameter (the Near-Field Method)

A small area detector is used to scan the magnified image of the fiber end (Fig. 1.19). The experimental data are fitted to Eq. (1.10), from which the mode-field diameter can be calculated.

Cutoff Wavelength (the Single-Bend Attenuation Method)

In this method the power transmission is measured as a function of wavelength for two different fiber bending situations. In measurement 1 a prescribed bend diameter of 28 cm is used (CCITT and EIA standards). In the second measurement an additional loop with diameter 4–6 cm is used

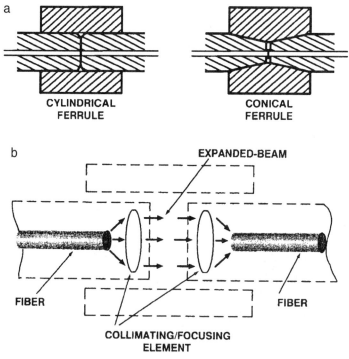

Fig. 1.26   (a) Example of two popular alignment mechanisms used in fiber optic connectors. (b) Schematic representation of a typical expanded-beam type connector. Reprinted from Ref. [23], p. 316 (a) and p. 318 (b), courtesy of Academic Press.

to, as much as possible, attenuate the higher order modes. The cutoff is then defined as the wavelength for which the normalized power becomes linear to within 0.1 dB (Fig. 1.20).

Chromatic Dispersion (Interferometric Method)

This method measures the phase delay for different wavelengths. The source is an incoherent high-pressure arc lamp and a typical delay curve is shown in Fig. 1.21.

1.3. Cable Designs

Because an optical fiber is very brittle and susceptible to chemical degradation it is important to protect the optical performance. Two common types of fiber optic cable construction are shown in Fig. 1.22. Common for all cable types is that the individual fibers are coated with some kind of organic material, examples of which include the following [16]:

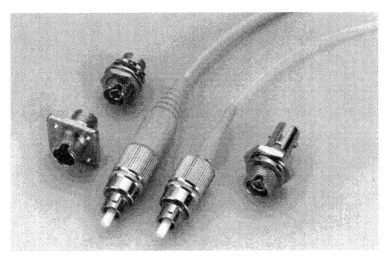

Fig. 1.27 FC connectors and adaptors. Reprinted from Ref. [24], p. 16, courtesy of Molex Fiber Optics, Inc.

- Polydimethyl siloxane
- Silicone oils
- Extrudates
- Acrylates

For ease of manufacturing and to prevent the glass surface from being scratched the coating is applied during the drawing process. Often, the first coating is not sufficient to protect the fiber and additional coatings are used, which are commonly referred to as buffering. The buffered fiber is

Fig. 1.28 (A, B) Spring latches. Reprinted from Ref. [24], p. 24 (A), and p. 34 (B), courtesy of Molex Fiber Optics, Inc.

then embedded in various strength members and additional protective jackets. These additional layers have as their primary goals to protect the fibers from water, rodents, and temperature fluctuations and to reduce stress in the fiber (Fig. 1.23).

1.4. Connectors

1.4.1. CONNECTOR TYPES

It is important to be able to easily reconfigure communication links for cost and performance optimization, to replace link subunits, and for relocating equipment [1]. To do this, connectors are a key component. There are two types of fiber optic connections:

- Fiber to transceiver
- Fiber to fiber

An example of a laser transceiver module is shown in Fig. 1.24. The fiber-to-fiber connectors are divided into simplex (one fiber to one fiber) and duplex (two fibers to two fibers) connectors (Fig. 1.25). The two basic connector designs are butt joints and expanded beam joints (Figs. 1.26a and 1.26b).

1.4.2. MECHANICAL PROPERTIES

Connectors are further differentiated depending on the particular mechanical method used to join the connectors.

FC connectors use threaded fasteners (Fig. 1.27). SC connectors use a spring latch (Fig. 1.28). These connectors can come as both simplex and duplex.

Variations on the SC connector, developed for specific standardized local area networks, are FDDI, ESCON, and FCS connectors (Figs. 1.29a–c). A common connector for MM data communication networks is the SMA connector (Fig. 1.30).

1.4.3. OPTICAL PROPERTIES

There is obviously some additional loss produced into a communications link with the incorporation of a connector (Fig. 1.31). As can be seen from Fig. 1.31 there is both a local and a distributed loss resulting from the

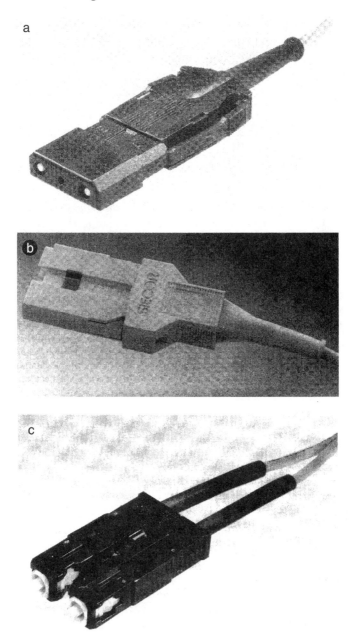

Fig. 1.29 Duplex connectors: (a) ESCON (R), (b) FDDI, (c) FCS duplex (fiber channel standard compliant). Reprinted from Ref. [1], p. 50, courtesy of Academic Press.

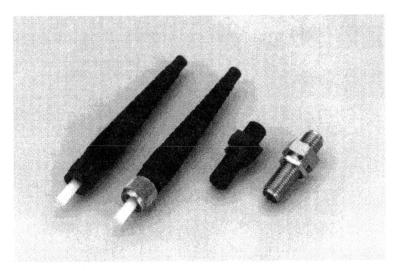

Fig. 1.30 SMA connectors and adapters. Reprinted from Ref. [24], p. 42, courtesy of Molex Fiber Optics, Inc.

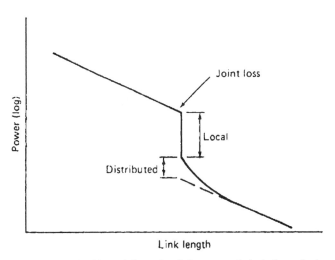

Fig. 1.31 Local and distributed (transient) losses and their launch dependence. Reprinted from Ref. [15], p. 4.40, courtesy of McGraw-Hill.

Table 1.6   Connector Loss Factors[a]

Extrinsic	Intrinsic
Lateral offset	Mode field diameter mismatch
Axial gap	Index profile mismatch
Tilt	Core eccentricity
End-face quality	Fiber and ferrule hole clearance
Reflections	Ferrule hole eccentricity

[a] Reprinted from Ref. [1], p. 55, courtesy of Academic Press.

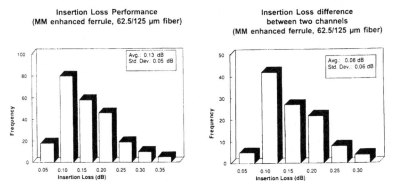

Fig. 1.32 Insertion loss performance for FDDI connectors. Reprinted from Ref. [24], p. 46, courtesy of Molex Fiber Optics, Inc.

connector joint. Contributing connector loss factors are shown in Table 1.6. The local loss factors can be calculated using the formulas in Section 1.1.4. The distributed losses are due to changes in the model launching conditions [15].

Connector losses can be as high as 0.5 dB (this corresponds to an additional km of fiber at 1.3 μm). With improved manufacturing technology, losses between 0.1 and 0.2 dB are obtained for new connectors. Loss data for a FDDI connector are shown in Fig. 1.32.

References

1. Webb, J. R., and U. L. Österberg. 1995. In *Optoelectronics for Datacommunication*. San Diego: Academic Press.

2. Liu, M. M. K. 1996. Fiber, cable and coupling. In *Principles and applications of optical communications,* eds. R. C. Lasky, U. L. Osterberg, and D. P. Stigliani. Chicago: Irwin.

3. Okoshi, T. 1982. *Optical fibers.* New York: Academic Press.

4. Marcuse, D., D. Gloge, and E. A. J. Marcatiti. 1979. *Guiding properties of fibers. Optical fiber telecommunications.* Orlando: Academic Press.

5. Ghatak, A., and K. Thyagarajan. 1989. *Optical electronics.* Cambridge, UK: Cambridge University Press.

6. Pollock, C. R. 1995. *Fundamentals of optoelectronics.* Chicago: Irwin.

7. Marcuse, D. 1977. *Loss analysis of single-mode finger splices. Bell System Tech. J.* 56: 703–718.

8. Cheo, P. K. 1990. *Fiber optics and optoelectronics.* 2nd. ed. Englewood Cliffs, NJ: Prentice-Hall.

9. Powers, J. P. 1993. *An introduction to fiber optic systems.* Homewood, IL: Aksen.

10. Neumann, E. G. 1988. *Single-mode fibers: Fundamentals.* Berlin: Springer-Verlag.

11. Nemoto, S., and T. Makimoto. 1979. Analysis of splice loss in single-mode fibers using a Gaussian field approximation. *Opt. Quantum Electron.* 11:447–457.

12. Maurer, R. D. 1973. Glass fibers for optical communication. Proc. IEEE 61: 452–462.

13. Chen, C. L. 1996. *Principles and applications of optical communications.* Chicago: Irwin.

14. Hentschel, C. 1989. *Fiber optics handbook.* Boeblingen, Germany: Hewlett Packard.

15. Allard, F. C. 1990. *Fiber optics handbook. For engineers and scientists.* New York: McGraw-Hill.

16. Shibata, N., and T. Edahiro. 1982. *Trans. Inst. Electron. Commun. Eng. Jpn.* 65:166–172.

17. Hecht, J. 1987. *Understanding fiber optics.* Indianapolis, IN: Sams.

18. Murata, H. 1996. *Handbook of optical fibers and cables.* New York: Dekker.

19. Kalish, D., Key, P. L., Kurkjian, C. R., Tanyal, B. K., and Wang, T. T. 1979. Fiber characterization—mechanical. In *Optical fiber telecommunications,* eds. S. E. Miller and C. G. Chynoweth. Boston: Academic Press.

20. Philen, D. L., and W. T. Anderson. 1988. Optical fiber transmission evaluation. In *Optical fiber telecommunications II,* eds. S. E. Miller and I. P. Kaminow. Boston: Academic Press.

21. Gartside, C. H., P. D. Patel, and M. R. Santana. 1988. Optical fiber cables. In *Optical fiber telecommunications II,* eds. S. E. Miller and I. P. Kaminow. Boston: Academic Press.

22. Radcliffe, J., C. Paddock, and R. Lasky. 1995. Integrated circuits, transceiver modules and packaging. In *Optoelectronics for datacommunication,* eds. R. C. Lasky, U. L. Österberg, and D. P. Stigliani. San Diego: Academic Press.

23. Young, W. C., and D. R. Frey. 1988. Fiber connectors. In *Optical fiber telecommunications II,* eds. S. E. Miller and I. P. Kaminow. Boston: Academic Press.

24. Molex Fiber Optics, Inc. 1996. *Fiber optic product catalog,* No. 1096.

Chapter 2 | Optical Sources: Light-Emitting Diodes and Laser Technology

Wenbin Jiang
Michael S. Lebby

Phoenix Applied Research Center, Motorola, Incorporated, Tempe, Arizona 85284

2.1. Introduction

In this chapter, we will present some fundamentals of optical transmitter design for fiber optic data communications. Specifically, we will consider the operating principles behind the three most common types of optical sources used for data communications:

- Light-emitting diodes (LEDs) — edge emitting (E-LEDs) and surface emitting
- Edge-emitting semiconductor laser diodes
- Vertical cavity surface-emitting lasers (VCSELs)

As discussed in Chapter 1, the sources for fiber optic communications most often operate at wavelengths near 1300 or 1550 nm, and VCSELS are currently limited to operation near 780–980 nm. A detailed discussion of laser and LED sources could easily fill a book by itself; for our purposes, we will present some of the fundamentals of this technology and refer the interested reader to some of the many good treatments in the literature [1–3]. Fabrication and manufacturing of these devices, with particular emphasis on VCSELs, will be discussed further in Chapter 15.

2.2. Technology Fundamentals

An optical source for fiber optic communications must have properties that are slightly different from those required for other applications. Specifically, it must exhibit a high radiance over a narrow band of wavelengths that

HANDBOOK OF FIBER OPTIC
DATA COMMUNICATION

coincide with the transmission window of the fiber, typically 0.8–1.55 μm. The emissive area should be no greater than the core diameter of the fiber, and the distribution of incident radiation should match the numerical aperture of the fiber (the acceptance cone). In addition, the source should be easily modulated at high frequencies (up to several gigahertz), have a modest cost compared with other components of the datacom system, and be at least as reliable as other large system computer components (several hundred thousand power-on hours with minimal variation in optical output power). Semiconductor sources satisfy these requirements by emission of light at a p–n junction through the process of injection luminescence. In this chapter, we will begin with a brief review of semiconductor properties, then describe the fundamentals of LEDs, edge-emitting laser diodes, and VCSELs.

For the range of wavelengths commonly used in data communications, the efficiency of both sources and detectors is determined by the bandgap energy. It is often desirable for the bandgap of the detectors to be slightly less than that of the source materials. To begin, then, we will consider some fundamentals of semiconductor physics (more information on this topic is provided in Chapter 3). Intrinsic semiconductor materials are characterized by an electrical conductivity much lower than that of pure metals, which increases rapidly with temperature. As described in Chapter 3, the Fermi level lies between the conduction and valence bands of the material. These materials may be doped with various impurities as either n type or p type so that the Fermi level moves either closer to the conduction or valence bands. Common room-temperature semiconductors include germanium and silicon, from group IV(b) of the periodic table, as well as several binary compounds from groups III(b) and V(b), such as gallium arsenide (GaAs) and indium phosphide(InP).

Silicon or germanium can be made n type by doping with donor impurities, such as P, As, or other group V elements, whereas they can be made p type by doping with acceptor impurities such as B, Ga, or other group III elements. These dopant concentrations are fairly small, approximately 1e-14 to 1e-21 per cubic centimeter in Si — or approximately from 1 dopant atom per billion silicon atoms to approximately 1 per thousand silicon. Typically, the Fermi level is uniform throughout the doped material; when an abrupt junction between p-type and n-type materials is formed, the conduction, valence, and Fermi bands bend to accommodate the difference. Electrons tend to accumulate on the n side of the junction, and holes develop on the p side. A thin depletion region is formed at the junction itself — the depletion region is effectively depleted of both hole and electron carriers.

An external voltage applied across the junction will have the effect of raising or lowering the potential barrier between the two materials, depending on its polarity. When the p side is connected to a positive voltage, we say that the junction is forward biased; the opposite condition is known as reverse bias. In a forward-biased junction, excess carriers are injected into either side of the depletion region, where they are the minority carriers. These excess minority carriers tend to diffuse away from either side of the depletion layer. Thickness of the depletion layer decreases under forward bias conditions. There is thus a balance of carrier flow; injection of carriers over the depletion region (which gives rise to an excess carrier concentration) is balanced by both diffusion away from the junction and carrier recombination.

Recombination is the process in which the electrons and holes combine. Several different kinds of recombination are possible, although we are only concerned with the radiative recombination processes that release photons; these are direct band-to-band transitions across the bandgap. Direct bandgap semiconductor materials (such as GaAs and other III–V compounds) have an energy–k relationship such that the conduction band minimum and the valence band maximum occur at the same value of k, as shown in Fig. 2.1. With direct bandgap materials, the holes and electrons can recombine directly; by contrast, in an indirect bandgap material such as silicon, the conduction band minimum does not correspond to the valence band maximum, and it is more complicated to recombine while satisfying the requirements of both conservation of energy and conservation of momentum. The nonradiative transitions, in which the energy is lost to thermal energy within the crystal, involve intermediate energy levels that trap carriers deep within the crystal lattice.

In direct bandgap materials, the radiative recombination process is proportional to the excess minority carrier concentration and gives rise to the creation of radiation from the junction. The process is known as injection luminescence and is the fundamental mechanism for the creation of optical radiation from semiconductor optical sources. We will not discuss this effect in great detail, other than to note that the operation of such sources is also influenced by the minority carrier lifetime, diffusion length near the junction, injection efficiency, and width of the depletion layer. Equivalent circuits for the p–n junction (discussed in Chapter 3) allow the modeling of junction capacitance and series resistance; these factors affect how rapidly

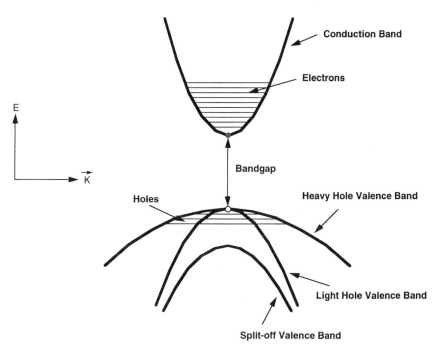

Fig. 2.1 Energy band structure (energy vs k) for direct bandgap semiconductor.

the optical source can be modulated or, conversely, how fast a semiconductor photodiode can respond to incident radiation.

The recombination of electrons and holes in a forward-biased $p-n$ junction can produce optical radiation at wavelengths of 0.8–1.7 μm, suitable for data communications. If recombination takes place in several stages, it is also possible for more than one photon of longer wavelength to be emitted. When only a single photon of energy (E) is produced, the wavelength may be determined from Planck's Law:

$$\lambda = hc/E = 1.24/E \ (\mu\text{m/eV}),\qquad(2.1)$$

where h is Planck's constant and c is the speed of light. The probability that an electron of energy E_2 will recombine with a hole of energy E_1 is proportional to the concentration of electrons at E_2 and the concentration of holes at E_1. The spectrum of the radiated energy may be determined by integrating the product of these carrier concentrations over all values of E_1 or E_2, subject to the constraint that the difference between E_1 and E_2 is the desired photon energy. Using a simple model for the behavior of carriers in a semiconductor, it is possible to derive the emission spectra due to injection luminescence; an LED will give rise to an approximately Gaussian spectra with a half-width of between 30 and 150 nm or more at room temperature.

It is desirable to design devices with high internal quantum efficiency, defined as the ratio of the rate of photon generation to the rate at which carriers are injected across the junction. A related parameter is the external quantum efficiency, defined as the ratio of the number of emitted photons to the number of carriers crossing the junction. External quantum efficiency is smaller than internal quantum efficiency for several reasons. For instance, some light will be reabsorbed before it can reach the emitting surface. Only light emitted toward the semiconductor–air interface is useful for coupling into the fiber, and a small percentage of this light is reflected back from the surface. Furthermore, only light reaching the surface at less than the critical angle will be coupled into an adjacent optical fiber. For a semiconductor material with an internal quantum efficiency of η_i, an injection current of I will give rise to an optical power, P, of

$$P = \eta_i \cdot (I/q) \cdot E, \tag{2.2}$$

where q is the charge on an electron. It is also possible to derive expressions for the behavior of these sources at high modulation frequencies; it can be shown that they exhibit a high frequency cutoff at modulation frequencies greater than

$$f = 1/\tau_p, \tag{2.3}$$

where τ_p is the mean lifetime of the excess minority carriers; it gives rise to a characteristic diffusion length, L,

$$L = \sqrt{\mu \cdot k \cdot T \cdot \tau_p/q} = \sqrt{(D \cdot \tau_p)}, \tag{2.4}$$

where μ is the electron mobility, k is Boltzmann constant, and the quantity D is known as the electron diffusion coefficient. From these expressions, we can determine that modulation frequencies above $f = 20$–25 MHz are difficult to obtain; however, bandwidths as high as several hundred MHz can be obtained from the double-heterostructure diodes that we will discuss in this chapter.

The structure of a typical surface-emitting LED is shown in Fig. 2.2. Note that for some structures, part of the substrate may be etched away to minimize the distance between the active area and the emitting surface. A thin layer of insulating oxide may be used to separate the positive contact

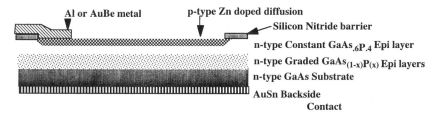

Fig. 2.2   Cross-section view of an LED.

from the p layer, except in the region of the active layer where the current flow is concentrated. The LED junction temperature is a critical parameter; the wavelength distribution of the LED changes with increasing temperature, internal quantum efficiency falls off, and the lifetime of the device is significantly decreased. In general, for GaAs and GaAlAs devices, the peak junction temperature should be kept below 50 or 100°C by the use of heat sinks or other devices. If the active layer is kept close to the heat sink, the thermal inpedance is small, and higher current densities may be used without an excessive rise in temperature.

In general, LEDs are diffuse (Lambertian) sources; for a typical GaAs LED forward biased at 100 mA and consuming 150 mW of power, the optical output is only on the order of 50 μW. To compensate for this rather low power, various optical coupling and packaging schemes are employed; these will be discussed in other chapters. One alternative is the edge-emitting LED, which offers small emissive area and higher radiance. The edge-emitting LED takes advantage of a slightly different approach — the double-heterojunction diode. Heterojunctions are produced at the boundary between two different types of materials whose physical lattice sizes match but that have different bandgap energies and other properties. Without going into details about the physics involved, we note that such structures offer higher injection efficiency, improved carrier confinement (and hence higher probability for radiative transitions to occur), and higher modulation bandwidths. The double heterostructure is simply a layered approach that further helps to confine excess minority carriers that are injected over the forward-biased p–n junction. For such devices, the modulation bandwidth increases as the square root of the current density and inversely as the square root of the active layer thickness (note that the modulation current must be a small fraction of the DC bias current, or nonlinear effects will occur that limit device performance). With current densities of 50 A/mm^2, devices such as this routinely offer 1 or 2 mW output power from a 50-μm aperture. These can also be made to emit at the longer wavelengths — in the 1.3- to 1.55-μm range.

The light output of an LED is based on spontaneous emission and tends to be incoherent, with typical spectral widths on the order of 100–150 nm. The two types of LEDs, surface-emitting LEDs and ELEDs, are characterized by high divergence of the output optical beams (120° or more), slow rise times (>1 ns) that generally limit the data communication applications of LEDs to less than 200 MHz, low temperature dependence, insensitivity to optical reflections, and output powers on the order of 0.1–3 mW.

By contrast, laser sources tend to exhibit much higher output powers (3–100 mW or more), narrower spectral widths (<10 nm), smaller beam divergence (5–10°), faster modulation rates (hundreds of MHz to several GHz), and higher sensitivity to both temperature fluctuations and optical reflections back into the laser cavity. This is because they are based on stimulated emission of coherent radiation — indeed, the word "laser" is an acronym for light amplification by stimulated emission of radiation. The radiative recombination process responsible for injection luminescence in LEDs is the result of spontaneous bandgap emissions. Transitions may also be stimulated by the presence of radiation of the proper wavelength; all photons produced by stimulated emission have the same frequency and are in phase with the stimulating emission, making this a source of coherent radiation. This is the type commonly produced by semiconductor edge-emitting lasers and vertical cavity lasers. Like all lasers, the semiconductor laser diode requires three essential operating elements:

1. A means for optical feedback, usually provided by a cleaved facet or multilayer Fresnel reflector
2. A gain medium, consisting of a properly doped semiconductor material
3. A pumping source, provided by current applied to the diode

For a semiconductor material with gain g and loss α, placed in a cavity of length L between two partially reflective mirrors of reflectivity R_1 and R_2, the minimum requirement for lasing action to occur is that the light intensity after one complete trip through the cavity must at least be equal to its starting intensity. This is called the threshold condition, given by the following relation:

$$R_1 R_2 \exp\left(2\left(g - \alpha\right)L\right) = 1. \tag{2.5}$$

Thus, the laser diode begins to operate when the internal gain exceeds a threshold value. Two key parameters associated with the laser are its efficiency of converting electrical current into laser light, known as the slope efficiency *dP/dI*, and the amount of current required before the laser begins stimulated emission, known as the threshold current I_{th}. This information is summarized in the power vs current (*P* vs *I*) characteristic curve of a laser diode, as shown in Fig. 2.3. Applying current, *I*, to a laser device initially gives rise to spontaneous emission of light, as in an LED; as the current is increased, the cavity and mirror losses are overcome, and the laser passes the threshold and begins to emit coherent light. The optical power emitted above threshold, *P*, is given by

$$P = (I - I_{th}) \, dP/dI. \tag{2.6}$$

Threshold current is found by extrapolating the linear region of the *P/I* curve. Low threshold currents and high efficiencies are desirable in laser

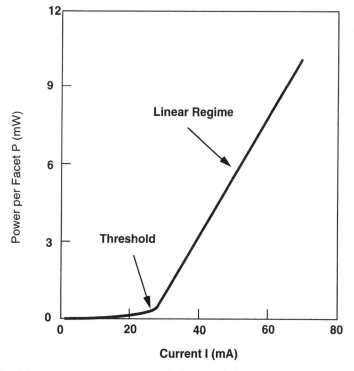

Fig. 2.3 Power vs current (*P* vs *I*) characteristic curve of a laser diode.

diodes. The slope efficiency, dP/dI, is related to the external differential quantum efficiency, η, by

$$\eta = (\lambda\ q)\ /\ (h\ c)\ (dP/dI), \tag{2.7}$$

where λ is the wavelength, q is the charge of an electron, h is the Plank constant, and c is the speed of light. The threshold current is a strong function of temperature, T, according to the relation

$$I_{th}\ (T) = I_{th}\ (T_1)\ \exp\ [(T - T_1)/T_0], \tag{2.8}$$

where T_0 is the laser's characteristic temperature, and T_1 is room temperature (300 K). A large T_0 is desirable because it translates into a reduced sensitivity of threshold current to operating temperature. Usually, it is only possible to measure the case temperature of a packaged laser diode. The junction temperature is always higher than the case temperature in typical applications. The junction temperature can be estimated by taking into account the device thermal resistance, the input electrical power, and the output optical power.

Increasing the diode current above threshold allows the gain coefficient to stabilize at a value near the threshold point, just large enough to overcome losses in the media:

$$g = \alpha - \frac{1}{2L}\ \ln\ (R_1 R_2). \tag{2.9}$$

Because of the high gain coefficients that can be obtained, the cavity of a double heterostructure laser can be made much smaller than other types of lasers; 1 mm or less is typical. For lasers biased above threshold, the slope of the power vs current characteristic curve in the spontaneous emission region corresponds approximately to the external quantum efficiency. The slope in the lasing region is related to the differential quantum efficiency, η_d, by

$$\eta_d = \frac{q}{E}\ dP/dI. \tag{2.10}$$

In practice, this ideal characteristic curve may be less well behaved; considerable research has been devoted to eliminating undesirable effects, such as "kinked" curves that correspond to unwanted fluctuations in the laser power. This has led to the development of many features, such as stripe geometry and buried heterostructure lasers, that are beyond the scope of this discussion. Compared with LED emissions, laser light has a narrower bandwidth (<10 nm, making it less susceptible to certain kinds of noise in the fiber link), it is more directional (so that the external quantum efficiency may be improved), and its modulation bandwidth is greater (up to several Ghz). Laser sources at short wavelengths (780–850 nm) are typically used with multimode fiber, whereas longer wavelengths (1.3 or 1.55 μm) are used with single-mode fiber.

2.3.   Device Structure — LED

In this and the following sections, device structures for LEDs, edge-emitting lasers, and VCSELSs will be introduced, with particular emphasis on the laser technologies. In Chapter 15, the manufacturing issues for each will be introduced, again with particular emphasis on the laser technologies.

LEDs use either direct bandgap or indirect bandgap materials for spontaneous light emission. When holes, in the valence band of Fig. 2.1, combine with electrons in the conduction band, photons of light with energy consistent with the bandgap are emitted. The material selection is dependent on the desired wavelength of light to be emitted. For example, whereas GaAs emits light at approximately 880 nm, GaP emits light at approximately 550 nm and InGaP emits light at approximately 670 nm. The light emission begins as soon as the LED is forward biased: Unlike the laser diodes discussed in the following section, there is no threshold current.

A surface-emitting LED consists of a p–n junction, with a controlled concentration profile between the p-typed and n-typed material. Figure 2.2 shows the cross section of an LED. The substrated in this example is GaAs; other III–V compounds, such as GaP, may be appropriate depending on the desired wavelength of light to be emitted. The substrate layer and the graded layer above are doped n type, and they are also doped to provide the bandgap for the desired wavelength of emitted light. A constant layer caps the graded layer and is capped in turn by a dielectric material such as silicon nitride. The p region of the p–n junction is formed by the diffusion of Zn doping into an opening in the dielectric. The p–n junction is contacted on the top by a metal such as gold, AuBe, or aluminum contacting the p-type, diffused Zn-doped region. The n-type region is contacted on the back side, with a metal such as AuSn or AuGe in contact with the n-doped substrate.

Device characteristic curves for LEDs include the forward current vs applied voltage curve and the light emission or luminous intensity vs forward current curve. Other useful characteristics in optoelectronic applications include the spectral distribution and the luminous intensity versus temperature and angle, the latter due to the need to couple into a medium such as optical fiber. An example of each is shown in Fig. 2.4. The spectral distribution of an LED is wider than that of a laser, incurring more chromatic dispersion. The lower frequency with which an LED can be switched, the wider angle of emission, and the larger spectral distribution are all disadvantages of LEDs compared to lasers. The primary advantage of LEDs over lasers for data communication is the low cost due to the significantly simpler processing.

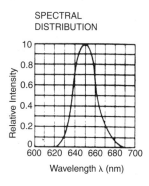

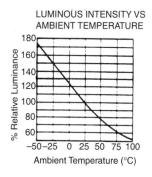

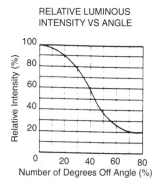

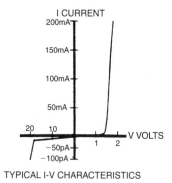

Fig. 2.4 Characteristics of an LED: spectral distribution, luminous intensity versus temperature and angle, and I vs V characteristics.

2.4. Device Structure — Lasers

Edge-emitting semiconductor lasers have been around for more than 30 years. Semiconductor lasers were first reported in 1962 [4–7]. Initial devices were based on forward-biased GaAs p–n junctions. Optical gain was provided by electron–hole recombination in the depletion region, and optical feedback was provided by polished facets perpendicular to the junction plane. This type of homojunction design meant that the carrier confinement of those lasers was poor, and the high laser threshold prohibited laser operation at room temperature. The concept of using wider bandgap material as one or both cladding layers to improve the laser carrier confinement and thus to reduce the leakage current was first proposed in 1963 [8]. Optical mode confinement was also expected to improve because a larger refractive index of the center active layer would provide a waveguide effect. Seven years later, a continuous wave (CW) GaAs/AlGaAs double-heterojunction (DH) semiconductor laser operating at room temperature was demonstrated using a liquid-phase epitaxial (LPE) growth technique [9, 10]. Commercial application of edge-emitting semiconductor lasers has since become practical. Today, the worldwide semiconductor laser annual sales revenue has exceeded $400 million [11].

In the late 1970s, Iga *et al.* [12] proposed a semiconductor laser oscillating perpendicular to the device surface plane to overcome the difficulties facing edge-emitting semiconductor lasers that oscillate in parallel to the device surface plane. This type of laser is termed VCSEL, pronounced "VIXEL." VCSELs have demonstrated many advantages over edge-emitting semiconductor lasers. First, the monolithic fabrication process and wafer-scale probe testing as per the silicon semiconductor industry substantially reduces the manufacturing cost because only known good devices are kept for further packaging [13, 14]. Second, a densely packed two-dimensional (2-D) laser array can be fabricated because the device occupies no larger of an area than a commonly used electronic device [15]. This is very important for applications in optoelectronic integrated circuits. Third, the microcavity length allows inherently single longitudinal cavity mode operation due to its large mode spacing. Temperature-insensitive devices can therefore be fabricated with an offset between the wavelength of the cavity mode and the active gain peak [16, 17]. Finally, the device can be designed with a low numerical aperture and a circular output beam to match the optical mode of an optical fiber, thereby permitting efficient coupling without additional optics [14, 18].

A conventional edge-emitting semiconductor laser utilizes its cleaved facets as laser cavity reflectors because the length of the active layer is usually several hundred micrometers. The long active length provides enough optical gain to overcome the cavity reflector loss even though the reflectivity of the facets is only ~30%. In comparison, a VCSEL needs both of its surfaces to be highly reflective to reduce the cavity mirror loss because its active layer is less than 1 μm thick. The first VCSEL was demonstrated with GaInAsP/InP in 1979, which operated pulsed at 77°K with annealed Au at both sides as reflectors [12]. A room-temperature pulsed-operating VCSEL was demonstrated with a GaAs active region in 1984 [19]. Room-temperature CW-operating GaAs VCSELs were achieved by improving both the mirror reflectivity and the current confinement [20].

Currently, an output power of more than 100 mW has been obtained from an InGaAs/GaAs VCSEL with GaAs/AlAs monolithic diffractive Bragg reflectors (DBR) [21, 22]. VCSELs with lasing threshold of sub-100 μA [23, 24] or wall-plug efficiency of more than 50% have been reported with lateral oxidized–Al confinement blocks [25, 26]. Room-temperature CW InGaAsP/InP VCSELs have met some difficulties primary due to a low index difference between GaInAsP and InP, which causes difficulty in preparing a highly reflective monolithic DBR [27]. Nevertheless, CW InGaAsP VCSELs at 1.5 μm have been reported using GaAs/AlAs DBR mirrors [28, 29].

Two-dimensional (2-D) arrayed VCSELs can find important applications in stacked planar optics, such as the simultaneous alignment of a tremendous number of optical components used in parallel multiplexing lightwave systems and parallel optical logic systems, free space optical interconnects, etc. High-power lasers can also be made with phase-locked 2-D arrayed VCSELs. Some 2-D arrayed devices have been demonstrated [15, 30, 31] and efforts have been made in coherently coupling these arrayed lasers [32, 33].

2.4.1. EDGE-EMITTING LASERS

Two types of lasers have been extensively studied. They include near-infrared $Al_y Ga_{1-y} As/Al_x Ga_{1-x} As$ ($x > y \geq 0$) and long wavelength $In_x Ga_{1-x} As_y P_{1-y}$/InP DH edge-emitting semiconductor lasers. The epitaxial structures for both types of lasers are similar. They are usually $n-p-p$ type, $n-i-p$ type, or $n-n-p$ type. LPE used to be the dominant epitaxial growth technique during the 1970s and early 1980s for high-quality semiconductor laser material growth, but it has gradually been taken over predominantly by metal organic chemical vapor deposition (MOCVD) techniques. Molecular beam epitaxy (MBE) is also a growth technique that has been used to demonstrate semiconductor lasers in research and development environments with a limited commercial success. A GaAs/Al_xGa_{1-x}As DH laser is shown in Fig. 2.5 consisting of a multiple of compound semiconductor layers grown on a n^+-type GaAs substrate. The p-type active layer of the GaAs semiconductor laser, in which stimulated emission is amplified, is made of GaAs doped by Be, C, or Zn at 1×10^{17} cm^{-3}. The active layer thickness ranges from 50 to 2000 Å. On top of the active layer is a p-type Al_xGa_{1-x}As cladding layer doped to 1×10^{18} cm^{-3} using Be, C, or Zn. Below the active layer is a n-type Al_xGa_{1-x}As cladding layer doped to 1×10^{18} cm^{-3} using Si, Sn, or Te. Cladding layer thickness usually ranges from 1500 Å to 1 μm. Within each Al_xGa_{1-x}As layer, x is the value of aluminum mole fraction. When x is 0.3, for example, the energy bandgap of Al_xGa_{1-x}As is approximately 1.8 eV — 0.4 eV wider than that of GaAs active layer. When the p–n junction is forward biased, electrons in the n-type cladding region are injected into the p-type active region. With a p-

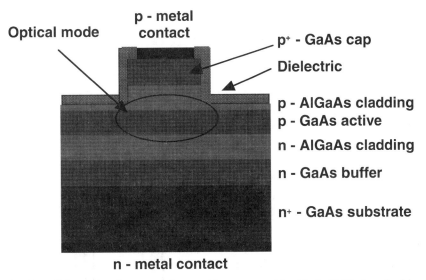

Fig. 2.5 Schematic diagram of an ridge waveguide GaAs/AlGaAs double-heterojunction (DH) laser.

type semiconductor of wide bandgap on the other side of the p–n junction, the injected minority carriers are mostly confined within the p-type active region. This carrier confinement allows population inversion to occur and optical gain to increase efficiently. In addition, the refractive index of GaAs is higher than that of $Al_xGa_{1-x}As$, and this acts like a waveguide to confine the majority of generated light within the GaAs active layer. The light that is not confined and penetrates into the $Al_xGa_{1-x}As$ cladding layers will not be absorbed by the cladding materials because of the wider bandgap and will therefore benefit laser action.

Although an epitaxy-ready GaAs substrate can be of good quality, there may exist some level of surface defects due to either grown-in crystal defects or mechanical polishing. Microscopic substrate surface flatness and native oxide on the substrate surface are also of great concern because they add a degree of difficulty to the epitaxial growth. To ensure a high-quality epitaxial crystal structure, the substrate usually first goes through the cleaning procedure, and a GaAs buffer layer with the same type of doping as that in the substrate is then deposited before the growth of any DH structure. There are also reports that high-quality epitaxial structures can be grown without buffer layers if substrates are thoroughly cleaned before the growth [34]. For the device structure shown in Fig. 2.5, the GaAs buffer layer is doped with a n-type dopant such as Si, Sn, or Te, typically having a concentration on the order of 1×10^{18} cm^{-3}. Its thickness ranges from 100 nm to several micrometers. The GaAs cap layer on top of the DH structure is very heavily doped with a p-type dopant such as Be, C, or Zn, typically at a concentration above 1×10^{19} cm^{-3}. This permits a low-resistivity metal Ohmic contact to be used for electrical conduction. Although a higher impurity doping concentration helps reduce the device series resistance, there is a limit at which the impurities can be incorporated into the host crystal structure. The adverse effect of overdoping is to form impurity clusters that behave as nonradiative recombination centers in the active region and the cladding layers, introducing internal optical loss or so-called free carrier absorption. The net effect is that the laser threshold current will increase. Therefore, the doping level at each layer of the device structure should be carefully adjusted to achieve the optimum designed laser performance.

The double-heterostructure semiconductor laser represents the single largest constituent of today's total semiconductor laser production because of its application in compact disk (CD) players and CD data storage. The current CD laser market volume is greater than 80 million units per year

[11]. The high volume of this market has driven the unit cost of a packaged laser to less than $1, with a large volume pricing of approximately $1. This has allowed businesses in the fiber optics market to use the low-cost CD laser in a historically high-cost environment. The CD laser has been used very successfully for some short-distance optical data links and has decreased laser wavelength specification from 850 nm for GaAs DH laser to 780 nm for CD laser [35, 36]. Because a CD laser operates at a wavelength of 780 nm, its active layer is made of AlGaAs with an Al mole fraction of approximately 15%. For example, a CD laser may have an epitaxial structure consisting of 0.1-μm-thick undoped $Al_{0.15}Ga_{0.85}As$ active layer sandwiched between a n-type $Al_{0.6}Ga_{0.4}As$ cladding layer and a p-type $Al_{0.6}Ga_{0.4}As$ cladding layer grown on a n-type GaAs substrate, as shown in Fig. 2.6 [37]. The device fabrication is similar to any other type of DH lasers. Typically, a CD laser has a threshold of 20–50 mA at room temperature and is operated at an output power of approximately 3–5 mW, as shown in Fig. 2.7. The laser wavelength is usually between 770 and 795 nm. The low-threshold CD laser is mostly used for portable consumer electronics powered by batteries. When used in a CD player, a CD laser is usually designed to operate at multimode or self-pulsation mode in GHz range in order to reduce the feedback noise due to light reflection from a disk [37]. This type of CD laser, however, cannot be used for multi-Gb data communications.

p - metal contact

p+ - GaAs cap

p - $Al_{0.6}Ga_{0.4}As$ cladding

$Al_{0.15}Ga_{0.85}As$ active (0.1 μm)

n - $Al_{0.6}Ga_{0.4}As$ cladding

n - GaAs buffer

n+ - GaAs substrate

n - metal contact

Fig. 2.6   Schematic diagram of an epitaxial structure for an edge-emitting semiconductor laser at 780 nm.

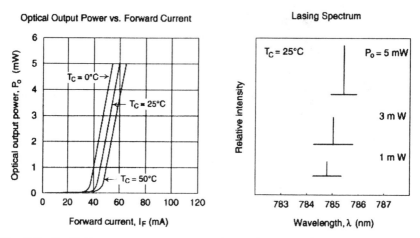

Fig. 2.7 (Left) Output power vs current of a CD laser, and (right) the correspondent laser spectrum.

When the active thickness of a DH laser is reduced to become comparable to the de Broglie wavelength [38], quantum mechanical effect starts to occur and the layer becomes a quantum well (QW). These QWs have been specifically used to design a new class of single quantum well (SQW) lasers. Two or more QWs can be placed between the two cladding layers to form a multiquantum well (MQW) laser. The layers separating the wells in the MQW laser are called barrier layers. Compared to a SQW laser, the MQW laser has a larger optical mode confinement factor, resulting in lower threshold carrier density and lower threshold current density. In comparison with a DH laser, the QW laser has a smaller active volume, a lower lasing threshold, and a higher differential gain, leading to increased relaxation oscillation frequency and reduced relative intensity noise. Quantum well lasers with small signal modulation frequencies above 20 GHz have been demonstrated [39, 40]. High-speed semiconductor lasers are important for large-bandwidth optical communications, as can be seen by the rapid deployment of optical fibers for transoceanic telecommunication cables and networking backbones throughout the Untied States and the world.

Shown in Fig. 2.8 is an energy band diagram of a GaAs/AlGaAs graded-index (GRIN) SQW laser structure [41]. The epitaxial layers are grown by MOCVD on a n^+-type GaAs substrate. The growth starts with a buffer layer of 2 μm, linear graded from GaAs to $Al_{0.6}Ga_{0.4}As$ and n-type doped at 2×10^{18} cm^{-3}. The n-type cladding layer of $Al_{0.6}Ga_{0.4}As$ has a thickness

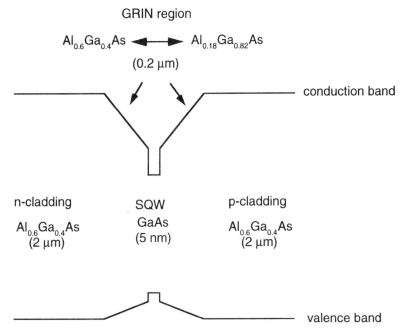

Fig. 2.8   Energy band diagram of a graded index (GRIN) single quantum well laser.

of 2 μm, doped at 2×10^{18} cm^{-3}. The *n*-side GRIN region is undoped with a thickness of 0.2 μm linear graded from Al$_{0.6}$Ga$_{0.4}$As to Al$_{0.18}$Ga$_{0.82}$As. The undoped GaAs QW is 50 Å thick. The *p*-side GRIN region is undoped with a thickness of 0.2 μm linear graded from Al$_{0.18}$Ga$_{0.82}$As to Al$_{0.6}$Ga$_{0.4}$As. The *p*-type cladding layer of Al$_{0.6}$Ga$_{0.4}$As has a thickness of 2 μm, doped at 1×10^{18} cm^{-3}. The *p*-type GaAs cap layer is heavily doped at 1×10^{19} cm^{-3}, with a thickness of 0.5 μm. The GRIN layers in this structure are effective in improving both the carrier and the optical mode confinement. A threshold current density of only 200 A/cm^2 has been demonstrated from such a broad-area laser with a length of 400 μm and a stripe width of 180 μm [41].

Further reduction of laser threshold current density can be achieved by using a strained QW active layer, such as In$_x$Ga$_{1-x}$As sandwiched between the GaAs barriers with $0 < x < 0.25$, or between the InP barriers with $x > 0.25$. The biaxial strain caused by the slight lattice mismatch between the two material systems alters the valence band edge by removing the degeneracy of the heavy hole and the light hole, resulting in reduced transparency carrier density and increased modal gain and thus reduced threshold current density [42]. The schematic energy band diagrams in k space for the strained gain medium are shown in Fig. 2.9. The strain in the active region also results in a larger differential gain, which helps improve the device operation bandwidth [40, 43]. Due to critical thickness constraints, the amount of strain in the active region and the active layer thickness, or the number of QWs with strained $In_xGa_{1-x}As$, is limited. To overcome the critical thickness barrier, strain-compensated active layers are used to increase the net total active thickness and the differential gain [44–47], thereby improving the optical mode confinement and reducing the laser threshold current density. The strain-compensated active structure consists of compressively strained quantum wells and tensile-strained barriers, or vice versa, so that the compressive strain and the tensile strain are mutually compensating for each other. Active layers exceeding the critical thickness can therefore be demonstrated without forming any lattice misfit dislocations.

In long-haul telecommunication systems, long-wavelength semiconductor lasers are of interest because of the minimum fiber dispersion at 1.3 μm and the minimum fiber loss at 1.55 μm [48]. The dispersion-shifted fiber will have both the minimum dispersion and the minimum loss at 1.55 μm [49–51]. The long-wavelength semiconductor lasers are

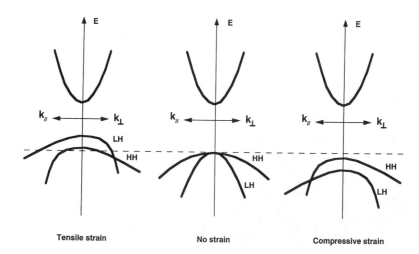

Fig. 2.9   Schematic energy band diagram in k space showing the removal of degeneracy between heavy hole (HH) valence band edge and light hole valence band edge for both compressive- and tensile-strained InGaAs gain medium.

based on $In_xGa_{1-x}As_yP_{1-y}$ active layer lattice-matched to InP cladding layers [52, 53]. By varying the mole fractions x and y, any wavelength ranging between 1.1 and 1.6 μm can be selected. For example (Fig. 2.10), an InGaAsP/InP DH laser at 1.3 μm has an 0.2- to 0.3-μm-thick undoped $In_{0.73}Ga_{0.27}As_{0.63}P_{0.37}$ active layer, 3- or 4-μm n-type cladding layer doped by Sn at 2×10^{18} cm^{-3}, 2-μm p-type cladding layer doped by Zn at 1×10^{18} cm^{-3}, and 0.5-μm-thick p-type InGaAsP contact layer doped by Zn at 1×10^{19} cm^{-3} [54]. The contact layer bandgap corresponds to a wavelength of 1.1 μm. An etched-mesa buried-heterostructure (EMBH) laser [54] made with this epitaxial structure has a threshold current of approximately 15 mA at room temperature and a single-mode output power of approximately 10 mW per facet, as shown in Fig. 2.11. One problem with the long-wavelength semiconductor lasers is the threshold current sensitivity to temperature, at room temperature (small T_0), due to poor carrier confinement and large Auger nonradiative recombination [55]. Improved thermal characteristics [56] and higher modulation speed [39] have been demonstrated when a MQW active layer is used for the long-wavelength semiconductor lasers.

Red visible semiconductor lasers operating in the range of 635–700 nm can find applications in bar-code scanners, laser printers, and laser pointers. They can also be used for plastic fiber data links because of the minimum loss at 650 nm in the plastic fiber [57]. With the emergence of digital video disk (DVD) technology for data storage [58], the market demand for both 635- and 650-nm semiconductor lasers is expected to soon catch

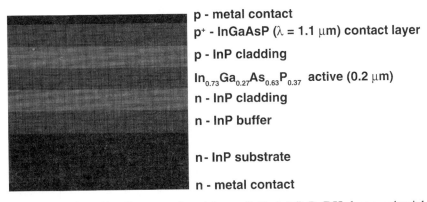

p - metal contact
p$^+$ - InGaAsP (λ = 1.1 μm) contact layer
p - InP cladding
$In_{0.73}Ga_{0.27}As_{0.63}P_{0.37}$ active (0.2 μm)
n - InP cladding
n - InP buffer

n - InP substrate

n - metal contact

Fig. 2.10 Schematic diagram of a 1.3-μm InGaAsP/InP DH laser epitaxial structure.

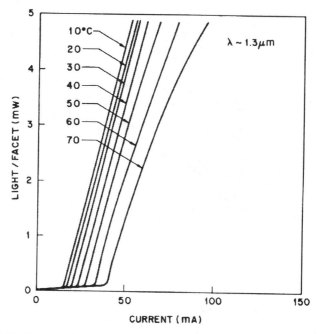

Fig. 2.11   Laser output power vs current of a 1.3-μm InGaAsP/InP DH.

up with the demand for the 780-nm CD lasers. Several material systems, such as AlGaAs [59], InGaAsP [60, 61], and InAlGaP [62–64], have been demonstrated to work in this wavelength region, but InAlGaP is regarded as the most appropriate material because it has a large direct energy bandgap while completely lattice-matched to a GaAs substrate. For example (Fig. 2.12), an InAlGaP DH laser at 650 nm consists of an $In_{0.5}Ga_{0.5}P$ active layer sandwiched between the $In_{0.5}(Al_xGa_{1-x})_{0.5}$ p-cladding layers grown on a GaAs substrate with $x = 0.7$. The active layer thickness is approximately 100 nm, and the cladding layer thickness is between 0.5 and 1 μm. Both the n-cladding and the p-cladding layers are doped at a concentration of 5×10^{17} cm^{-3}. Between the p-cladding layer and the p-GaAs contact layer is a p-type InGaP layer for improving the device I–V characteristics. This p-type InGaP intermediate layer is doped at 5×10^{17} cm^{-3}, and the p-type GaAs contact layer is doped as high as possible. The typical light output vs current characteristics is shown in Fig. 2.13.

High-temperature performance and high-power operation have been the concerns for InAlGaP visible semiconductor lasers due to carrier leakage

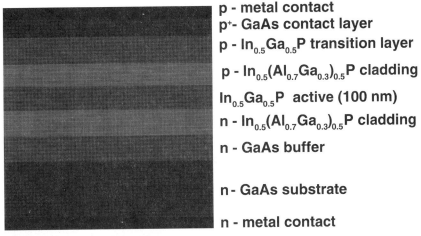

p - metal contact
p⁺- GaAs contact layer
p - $In_{0.5}Ga_{0.5}P$ transition layer

p - $In_{0.5}(Al_{0.7}Ga_{0.3})_{0.5}P$ cladding

$In_{0.5}Ga_{0.5}P$ active (100 nm)

n - $In_{0.5}(Al_{0.7}Ga_{0.3})_{0.5}P$ cladding

n - GaAs buffer

n - GaAs substrate

n - metal contact

Fig. 2.12 Schematic diagram of a visible InGaP laser epitaxial structure.

into the *p*-cladding layer. Methods utilizing components, such as strained active layer [65, 66], off-angle substrate [67, 68], MQW active structure [69], and multiquantum barrier structure [70], have been developed to improve the laser performance.

The new DVD standard has increased the data storage capacity from 650 Mb to 4.7 Gb on a single-sided disk 12 cm in diameter. This storage capacity increase is attributed more to the tightening of system margin

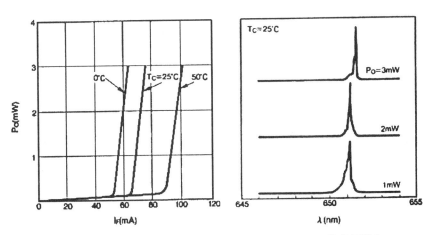

Fig. 2.13 Laser output power vs current of a 650-nm InGaP DH laser.

than to the shortening of laser wavelength from 780 to 635 or 650 nm. A 135-min high-definition motion picture, however, needs a storage capacity of 15 Gb. Because the current DVD standard has squeezed the system margin to the minimum, the future generation of DVD technology will rely to a great extent on laser wavelength shortening to expand the storage capacity in order to maintain the same disk size. Several groups have been investigating green/blue lasers using wide bandgap II–VI compound materials such as ZnCdSe/ZnSSe/ZnMgSSe grown on GaAs substrate [71–73], and good performance lasers have been demonstrated. The device, however, suffers serious reliability problem because of stacking fault-like defects that occur at and near the heterointerface between the GaAs substrate and the II–VI materials. Most II–VI semiconductor lasers degrade rapidly within minutes when running at CW. By reducing the grown-in defects, the device CW operation lifetime has been extended to 100 h [74] but is not yet long enough for any commercial applications.

Recent advancement in blue LED devices based on III–nitride materials [75] has prompted research in the blue/violet semiconductor lasers using InGaN/AlGaN MQW [76, 77] or InGaN/GaN DH structures [78]. The III–nitride epitaxial structures are grown on c-plane or a-plane sapphire substrates with a thick GaN buffer layer in between because there are no lattice-matched substrates available. Lasers on a spinel ($MgAl_2O_4$) substrate have also been demonstrated [79]. Crystal quality, p-contact resistivity, carrier and current confinement, and facet mirror reflectivity have been the four major problems in the III–nitride semiconductor laser development [80]. Continuous wave operation at room temperature has been achieved at a wavelength at approximately 400 nm by improving the p-contact resistance and thus reducing the device operating voltage [81]. The device has a threshold current of approximately 3 or 4 kA/cm^2 and a lifetime of approximately 20 h at 1.5-mW constant power when running at room temperature. The improvement in reliability relies on further reducing the contact resistance and reducing the grown-in crystal defects. The search for a lattice-matched substrate will help accelerate the device development cycles.

2.4.2. *VERTICAL CAVITY SURFACE-EMITTING LASERS*

The majority of the VCSELs being developed today are in the near-infrared wavelength range based on either GaAs/AlGaAs or strained InGaAs active materials. GaAs VCSELs at 850 nm are preferred as the light sources for short-distance optical communications because either Si or GaAs positive–intrinsic–negative (PIN) detectors can be used in the receiver end to reduce the total system cost. A typical GaAs VCSEL epitaxial layer structure is shown in Fig. 2.14, and the correspondent etched-mesa-type device structure is shown in Fig. 2.15. It includes three major portions: bottom DBR, active region, and top DBR.

Generally, the epitaxial material is grown by either MOCVD or MBE techniques. The bottom DBR is n-type doped and grown on an n-type doped GaAs substrate. Typically, there is a GaAs buffer layer grown between the n-DBR and the substrate. Silicon (Si) and selenium (Se) are two commonly used n-type dopants. The n-DBR is composed of 30.5 pair [82, 83] of $Al_{0.16}Ga_{0.84}As/AlAs$, which starts and stops with the AlAs layer alternated by the $Al_{0.16}Ga_{0.84}As$ layer. Each DBR layer has an optical thickness equivalent to a one-fourth of the designed lasing wavelength. The intrinsic cavity region is composed of two $Al_{0.6}Ga_{0.4}As$ spacer layers and three or four GaAs quantum wells with each quantum well sandwiched between two $Al_{0.3}Ga_{0.7}As$ barriers. With the quantum well width of 100 Å and the quantum barrier width of 70 Å, the lasing wavelength is approximately 850 nm. The $Al_{0.6}Ga_{0.4}As$ spacer can be replaced by $Al_xGa_{1-x}As$ spacer, with x graded from 0.6 to 0.3 to form a graded-index separate confinement heterostructure. The total thickness of the spacers is such that the laser cavity length between the bottom and the top DBRs is exactly one wavelength or its multiple integer. The p-type doped DBR is grown on top of the active region. It consists of 22 pairs of $Al_{0.16}Ga_{0.84}As/$AlAs alternating layers that start with the AlAs layer and stop with the $Al_{0.16}Ga_{0.84}As$. Like the n-DBR, each layer has an optical thickness of one-fourth the wavelength. The most common p-type dopants utilized are carbon (C), zinc (Zn), and Beryllium (Be). Typically, C is used for the p doping in the top DBR when using MOCVD growth techniques with the top several layers doped by Zn for better metalization contact [84]. Finally, a GaAs cap is used to prevent the top AlGaAs layer of the p-DBR from oxidization. The cap is highly p doped with Zn as the dopant and is kept to 100 Å thick because the material will be highly absorptive to the optical mode if it is too thick.

The function of a DBR in a VCSEL is equivalent to a cleaved facet in an edge-emitting laser: to reflect part of the laser emission back into the laser cavity and to transmit part of the laser emission as the output. It is in essence similar to a dielectric mirror. A pair of low-refractive index and high-refractive index layers with the optical thickness of each layer to be

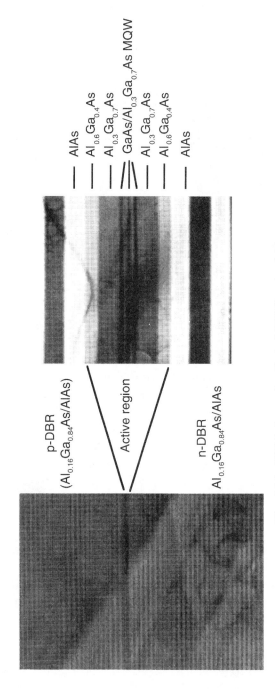

Fig. 2.14 TEM photo of a GaAs VCSEL epitaxial layer structure.

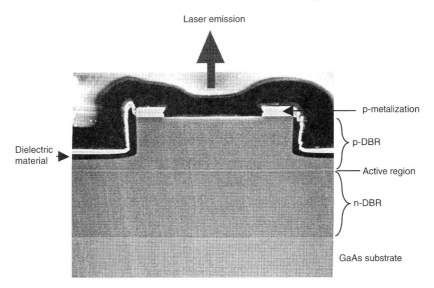

Fig. 2.15 Cross-section SEM photo of an etched-mesa GaAs VCSEL structure.

one-fourth of a specific wavelength will enhance the reflectivity to this wavelength. Many pairs of such alternating layers stacked together will form a mirror with its reflectivity reaching above 99% centered at the designed wavelength with a certain bandwidth. The mirror can be designed to achieve higher reflectivity and wider reflection bandwidth with a larger refractive index difference. $Al_xGa_{1-x}As$ is an ideal semiconductor material that can be monolithically grown on a GaAs substrate to provide a similar function as a dielectric mirror because the refractive index of this material can vary continuously from 3.6 to 2.9, with x varying from 0 to 1. For example, if one-fourth wavelength-thick layer is made of GaAs and the other one-fourth wavelength-thick layer is made of AlAs, 20 pairs of such alternating layers will provide a reflectivity of 99% at the desired wavelength with a bandwidth of larger than 70 nm as long as the desired wavelength is longer than the GaAs absorption band edge, which is approximately 875 nm at room temperature.

Although reflectivity of a natural cleaved facet in an edge-emitting laser is only approximately 30%, it is enough to ensure the lasing action due to the large net gain provided by the long active length, which is usually several hundred micrometers. The reflectivity of a VCSEL DBR has to be more than 99% to reduce the cavity loss due to the extremely short gain length in the active region. For the example shown in Fig. 2.14, the optical gain is provided by the three GaAs quantum wells, with each quantum well thickness of only 100 Å. The net gain length of the VCSEL is thus only one-tenth of 1000 of that of an edge-emitting laser. Due to the reasonable refractive index difference between $Al_{0.16}Ga_{0.84}As$ and AlAs, the 22-pair DBR in this VCSEL has a reflectivity of 99.9% at 850 nm with a bandwidth of about 70 nm (Fig. 2.16). The large bandwidth provides some tolerance to any variation in the lasing wavelength due to growth variation in quantum well thickness and laser cavity length, the center wavelength mismatching between the top and the bottom DBRs, and the growth reactor/tooling variance. A typical GaAs VCSEL output power vs input current is shown in Fig. 2.17 for a mesa diameter of 10 μm and a laser emission aperture of 7 μm.

Although the structure shown in Fig. 2.14 is for GaAs VCSELs, many others have been working with strained InGaAs VCSELs operating at approximately 980 nm. The strain in the active region provides higher gain, which allows lower lasing threshold, and higher differential gain, which allows larger intrinsic modulation bandwidth. In addition, the InGaAs VCSEL has a wavelength transparent to the GaAs substrate, allowing the light emission toward the substrate side and thus the epitaxial side down packaging scheme. In this way, heat generated in the p-DBR mirror and

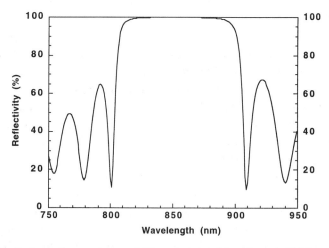

Fig. 2.16 Reflectivity spectrum of 22 periods of $Al_{0.16}Ga_{0.84}As/AlAs$ one-fourth wavelength DBR mirror stacks centered at a wavelength of 850 nm.

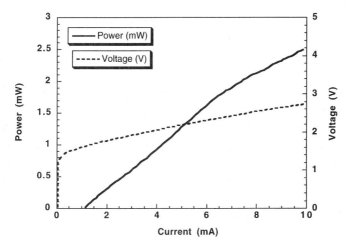

Fig. 2.17 A GaAs VCSEL output power vs input current for an etched-mesa structure with a mesa diameter of 10 μm and an emission aperture of 7 μm. The laser wavelength is approximately 850 nm.

the active junction region can be dissipated more efficiently, resulting in lower junction temperature and higher output power. A disadvantage of this type of VCSELs is that the low-cost Si or GaAs PIN detectors cannot be used when the VCSELs are used as the light sources for data communications. A typical InGaAs VCSEL structure consists of two DBR mirrors with reflection bands centered at approximately 970 nm [17, 85]. Each mirror is composed of one-quarter wavelength-thick layers alternating between AlAs and GaAs. The top p-doped DBR mirror contains 15 periods, whereas the bottom n-doped DBR mirror contains 18.5 periods. The cavity between the mirrors is filled by spacer layers of $Al_{0.3}Ga_{0.7}As$ that are used to center three 8-nm $In_{0.2}Ga_{0.8}As$ quantum wells (almost the thickness limit for coherently strained materials) separated by 10-nm GaAs barriers to form a one-wavelength-long cavity. The $Al_{0.3}Ga_{0.7}As$ spacer on the p-DBR mirror side is p-type doped, and the $Al_{0.3}Ga_{0.7}As$ spacer on the n-DBR mirror side is n-type doped. Above the p-DBR mirror stack, a heavily p-doped (3×10^{19} cm^{-3}) GaAs phase-matching layer is deposited to provide a nonalloyed Ohmic contact to the hybrid Au mirror, which also acts as a p contact. The DBR mirrors are uniformly doped to 1×10^{18} cm^{-3} except for the digital grading region, which is uniformly doped to 5×10^{18} cm^{-3}. The n dopant is Si and the p dopant is carbon. The whole epitaxial structure is grown by either MBE or MOCVD technique on a n-doped GaAs substrate. The device is designed to emit laser toward the substrate side.

Clearly, highly reflective semiconductor DBR mirrors are necessary to ensure low cavity loss, thus allowing the VCSEL to reach lasing threshold at a reasonable threshold carrier density level in the gain medium. Although the refractive index difference between the two constituents of the DBR structures is responsible for high optical reflectivity, the accompanied energy bandgap difference that scales approximately linearly with the index difference results in electrical potential barriers in the heterointerfaces. These potential barriers impede the carrier flowing in the DBR structures and result in a large series resistance, especially in the p-type doping case. The large series resistance gives rise to thermal heating and thus deteriorates the laser performance.

The series resistance due to the heterojunctions in the DBR mirror can be minimized by grading and selectively doping the interfaces. In practice, the simplest approach to grade the interface is to introduce an extra intermediate-composition layer between the two alternating DBR constituents [86]. Instead of transiting directly from GaAs to AlAs, for example, a thin layer of $Al_{0.5}Ga_{0.5}As$ layer can be grown between the two to help smooth the interfaces. With the introduction of more $Al_xGa_{1-x}As$ layers of intermediate Al composition at each interface, greater performance over a single intermediate transition layer can be achieved. The advancement of MOCVD technology has allowed the continuous grading of an arbitrary composition profile. Very low-resistance DBRs have been achieved using this technique [87]. In the structure shown in Fig. 2.14, the heterointerfaces are linearly graded from 15% Al composition to 100% over a distance of 12 nm. The total optical thickness of a pair of alternating layers is maintained to be one-half wavelength. The effect of this interface linear grading on the DBR mirror reflectivity is minimal. Likewise, heterointerface parabolic grading [88–90] and sinusoidal grading [91] have also been used in some cases to flatten the valence band and therefore reduce the p-DBR series resistance.

The graded interfaces can sometimes be heavily doped (5×10^{18} cm^{-3}), whereas the remainder of the mirror is lightly doped (1×10^{18} cm^{-3}) to reduce scattering and free-carrier absorption loss in the mirror, and there is also a reduction in series resistance [92]. A simplified delta doping scheme [93, 94] has worked successfully to reduce the series resistance in the p-DBR mirror. In this scheme, p doping is carried out at interfaces where the nodes of the optical intensity are located, at levels as high as the crystal can incorporate. This heavy doping at the heterointerfaces causes the valence band edge to shift upward and the thermionic emission current to increase. The excess resistance at the higher bandgap side (AlAs) of the heterointerfaces is also reduced together with the relaxation of the carrier depletion in this region. Furthermore, the delta doping introduces a thinner potential barrier that allows for an increased tunneling current. Because the carrier density is increased only locally, the excess free-carrier absorption in the DBR mirror is minimized.

The intracavity metal contact technique [95–97] is an alternative approach to achieve low series resistance. This technique allows the electrical contact to bypass the resistive p-DBR mirror stack layers. Furthermore, the mirror stack above the metal contact does not require any doping, thereby reducing the intracavity free-carrier absorption loss and the optical scattering loss. Good performance VCSELs using this technique have been demonstrated in the research environment.

The intracavity metal contact structure starts with a conventional VCSEL design similar to that shown in Fig. 2.14 [95]. p-Type and n-type bulk layers that are multiples of half wavelengths are inserted on either side of the active region to provide electrical paths for the current to reach the active region from the ring contacts that are deposited on top of the inserted layers. A current blocking region must be formed to force the current into the optical mode. This current blocking region can be formed by ion implantation or undercutting using wet etching between the p-type insertion layer and the active region. A resistive layer is further introduced between the conductive current distribution layer and the active region to overcome any residual current crowding effects near the contact periphery so that the injected current can be more uniformly distributed in the optical mode area. The current blocking layer in the p-type region can also be a thin n-type GaAs or AlGaAs layer that is inserted into the top p-doped cladding layer [96]. A second growth is needed to complete the epitaxial structure after a current flow path is opened by either wet or dry etching. The final VCSEL device will have a reverse-biased p–n junction inside the cavity in the p-doped region for current blocking (Fig. 2.18).

One of the simple approaches to make intracavity metal contact is to start with a VCSEL with an only partially grown p-DBR mirror stack (Fig. 2.19). After the metal contact has been deposited onto the p-DBR, a dielectric mirror stack consisting of quarter-wavelength-thick alternating

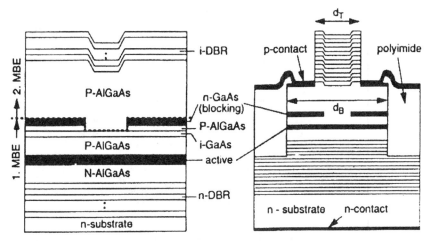

Fig. 2.18 Schematic diagram of an intracavity contact VCSEL structure (after Ref. [96]).

layers TiO_2/SiO_2 is used to complete the device [97]. In this instance, resistance of 50 Ω has been achieved with a laser emission aperture of 5 μm and a proton implantation aperture of 20 μm.

Within the microcavity structure of a VCSEL, only one Fabry–Perot cavity mode exists in the designed DBR reflective bandwidth. A laser can only be sustained at the wavelength of the cavity mode. Clearly, a temperature-insensitive VCSEL can be demonstrated by taking advantage of the microcavity mode characteristics [16]. Typically, the peak of the active gain profile shifts with the temperature at a rate of 3–5 Å/°C, and the cavity resonant mode shifts at a rate of 0.5–1 Å/°C [98]. If the resonant cavity mode is designed to initially sit at the longer wavelength side of the gain profile, the gain peak will gradually walk into the cavity mode with the rise of temperature (Fig. 2.20). Conversely, the gain peak usually decreases with the temperature. Together, the actual gain for the VCSEL cavity mode will vary little with temperature, and the VCSEL threshold current will stay almost constant within a certain temperature range. This temperature insensitivity allows the VCSEL to be designed to operate optimally at the system temperature midpoint region. For example, with modest speed optical data links, the VCSELs can be designed to operate with minimum threshold current at approximately 40°C in a required working range of 0–70°C (Fig. 2.21) [17, 99]. The system can be implemented without any auto power control (APC) circuitry, thereby simplifying the

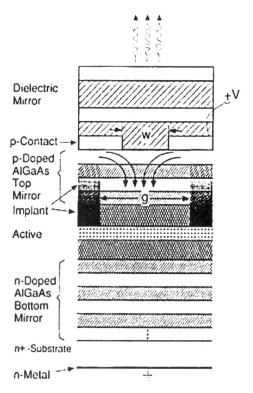

Fig. 2.19 Schematic diagram of a VCSEL structure with partial monolithic semi-conductor DBR and partial dielectric DBR on the *p*-doped contact side. (Reprinted with permission from Ref. [97]. Copyright 1995 American Institute of Physics.)

packaging and reducing the system cost [14]. The application of this method has also allowed the demonstration of VCSELs operating at a record high temperature of 200°C [83].

Apart from VCSELs at 830–870 nm based on GaAs MQWs and VCSELs at 940–980 nm based on strained InGaAs MQWs, VCSELs operating at other wavelengths, such as 780 nm based on AlGaAs MQWs, 650–690 nm based on InAlGaP MQWs, and 1.3–1.5 nm VCSELs based on InGaAsP MQWs, have received attention in the research community.

The vast majority of the semiconductor laser market is at 780 nm, which is predominantly used for CD data storage and laser printing. As a result, the development of VCSELs at 780 nm is of strategic importance from a commercial standpoint. A typical VCSEL at 780 nm has an epitaxial layer structure similar to that of a VCSEL at 850 nm [100–102]. The larger

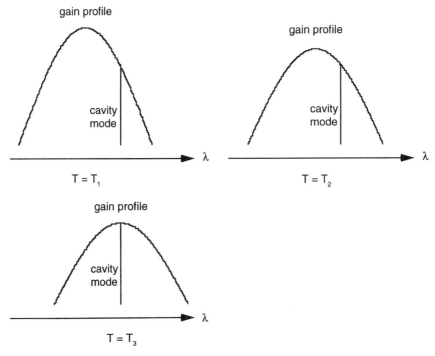

Fig. 2.20   VCSEL cavity mode vs gain profile for temperatures T_1, T_2, and T_3, with $T_3 > T_2 > T_1$.

bandgap requirement for 780 nm drives the MQW active region to the AlGaAs ternary system. The active region usually consists of three or four periods of $Al_{0.12}Ga_{0.88}As$ quantum wells sandwiched between the $Al_{0.3}Ga_{0.7}As$ barriers. The DBR mirror stack consists of 27 pairs of p-type doped $Al_{0.25}Ga_{0.75}As/AlAs$ and 40 pairs of n-type doped $Al_{0.25}Ga_{0.75}As/AlAs$, with the bandwidth centered at 780 nm. The laser performance of a 780-nm VCSEL is similar to that of an 850-nm GaAs VCSEL (Fig. 2.22). The increased aluminum concentration in both the active region and the DBR mirror stack over that used in the 850-nm VCSEL raises a concern with the 780-nm VCSEL device reliability because of the poor edge-emitting semiconductor laser performance at 780 nm. There has been no reliability data published so far for the 780-nm VCSELs, and study is ongoing to address the issue.

   Red visible VCSELs are of interest because of their potential applications in plastic fiber, bar-code scanner, pointer, and most recently the DVD format optical data storage. The epitaxial structure of a red visible VCSEL

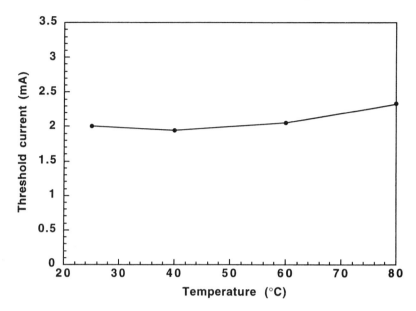

Fig. 2.21   Threshold current of a typical GaAs VCSEL varying with ambient temperature with minimum threshold current at 40°C.

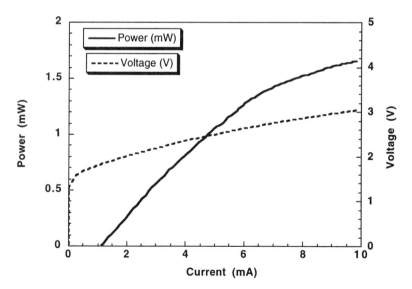

Fig. 2.22   Etched-mesa structure VCSEL output power vs input current at a wavelength of 780 nm.

is grown on a GaAs substrate misoriented 6° off (100) plane toward the nearest |111 > A or on a (311) GaAs substrate [103–106]. It consists of three or four periods of $In_{0.56}Ga_{0.44}P$ QWs with InAlGaP or InAlP as barriers, InAlP as both p-type and n-type cladding layers, and two DBR mirrors (Fig. 2.23). The active QW layer is either tensile or compressive strained to enhance the optical gain. Typically, the QW thickness is 60–80 Å and the barrier thickness is 60–100 Å. The total optical cavity length including the active region and the cladding layers ranges from one wavelength or its multiple integer up to eight wavelengths. The DBR mirrors are composed of either InAlGaP/InAlP or $Al_{0.5}Ga_{0.5}As$/AlAs. The $Al_{0.5}Ga_{0.5}As$/AlAs DBR mirror performs better because of a relatively larger index difference between the two DBR constituents—thus a higher reflectivity and a wider bandwidth. In general, because the index difference between $Al_{0.5}Ga_{0.5}As$ and AlAs is much smaller than that used for the 850-nm VCSELs, more mirror pairs are needed to achieve the required DBR reflectivity. Typically, 55 pairs are needed for the n-DBR and 40 pairs are needed for the p-DBR to ensure a reasonable VCSEL performance. As a rule of thumb, the more pairs in the DBR mirror, the higher series resistance and thus more heat generated in the active region. This implies that the active junction temperature will be higher. Currently, sub-mA threshold red VCSELs have been demonstrated. More than 5-mW output power from a red VCSEL has also been reported. Unfortunately, the carrier confinement of the red visible VCSELs is poor because of the smaller

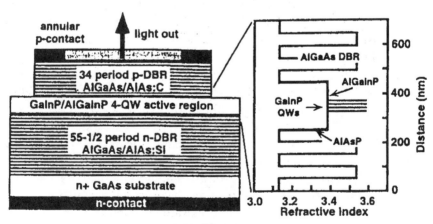

Fig. 2.23   A visible VCSEL structure. (Reprinted with permission from Ref. [105]. Copyright 1995 American Institute of Physics.)

bandgap offset between the quantum well and the barrier and between the active and the cladding. Therefore, the red visible VCSELs are extremely temperature sensitive, and more studies are needed to improve the red visible VCSEL high-temperature performances. VCSELs with wavelengths shorter than 650 nm pose more problems because of even worse carrier confinements. Designing a VCSEL that can effectively confine the carriers in the active region is a challenging topic in today's research community.

Long-wavelength VCSELs at 1.3 and 1.55 μm have drawn attention because of their potential applications in telecommunications and medium- to long-distance data links, such as local area networks and wide area networks, where single-mode characteristics are required. The long-wavelength VCSELs are based on an InP substrate, with InGaAsP MQWs used as the active region. However, the lattice-matched monolithic InGaAsP/InP DBR mirrors do not have sufficient reflectivity for the long-wavelength VCSELs because of the small index difference between the two DBR mirror pair constituents, InGaAsP and InP. In addition, the Auger recombination-induced loss becomes evident due to smaller energy bandgap for the long-wavelength VCSELs. To overcome the difficulty, dielectric mirrors with 8.5 pairs of MgO/Si multilayers and Au/Ni/Au on the p side and 6 pairs of SiO$_2$/Si on the n side have been used instead of the semiconductor DBR. A continuous-wave 1.3-μm VCSEL has therefore been demonstrated at 14°C [107]. To further improve the device perfor- mance, wafer fusing techniques have been adopted to bond GaAs/AlAs DBR mirrors onto a structure with an InGaAsP MQW active layer sand- wiched between the InP cladding layers that are epitaxially grown on the InP substrate [108, 109]. The InP substrate is removed to allow the GaAs/ AlAs DBRs to be bonded onto one or both sides of the InGaAsP active region (Fig. 2.24). Because the DBR mirrors are either n-type or p-type doped, the completed fused wafer can be processed like a regular GaAs VCSEL wafer. In this way, a 1.5-μm VCSEL has been successfully fabri- cated that operates CW up to 64°C [28, 29]. Manufacturing yield and reliabil- ity are still currently unknown with the VCSEL wafer fusion technique. For commercial interest, the CW operation must be driven to at least the 100°C range for the junction, in addition to a number of other issues such as wall-plug efficiency, reliability, consistency, etc.

APC is one of the important features that is easily accomplished with edge-emitting lasers because of the backward emission that can be moni- tored from the cleaved facet. With VCSELs of wavelength shorter than 870 nm, the laser beam emits only toward the top epitaxy side. The backward

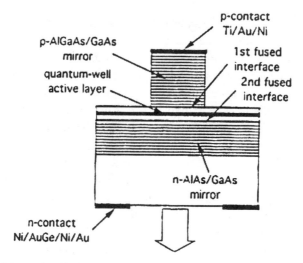

Fig. 2.24 Schematic diagram of a wafer-fused long-wavelength VCSEL (after Ref. [28]).

emission is absorbed by the GaAs substrate, unless the substrate is removed. However, due to the unique vertical stacking feature of VCSELs, a detector can be integrated underneath or above the VCSEL structure during the epitaxial growth [102, 110–113] (Fig. 2.25). For example, a VCSEL can start with a p-type GaAs substrate, with a PIN detector structure grown first on top of the substrate. The PIN detector has a GaAs intrinsic layer of approximately 1 μm and p-doped AlGaAs cladding of approximately 2000 Å between the substrate and the intrinsic absorption layer. The de-

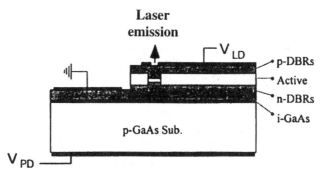

Fig. 2.25 Schematic diagram of a VCSEL with integrated detector (after Ref. [102]).

tector structure stops at a *n*-type doped cladding layer of approximately 2000 Å. A regular GaAs VCSEL epitaxial structure follows the PIN detector, with layers of *n*-DBR, *n* cladding, active, *p* cladding, and *p*-DBR grown in order. The detector cathode in this structure shares a common contact with the VCSEL cathode, with two independent anodes for both the PIN detector and the VCSEL. In practical applications, the anode of the detector can be either reverse biased or without any bias if detector speed is not a major concern. The VCSEL backward emission transmitted through the *n*-DBR is normally in proportion to the VCSEL forward emission. It will be received by the integrated PIN detector and generate a current. The VCSEL output power and the integrated PIN detector response are shown in Fig. 2.26. There is a one-to-one relationship between the PIN detector current and the VCSEL output power up to certain point when the VCSEL output power saturates, but the detector current keeps rising due to the effect of spontaneous emission. Consequently, VCSEL operation with APC can be accomplished by monitoring the current variation generated in this detector when the VCSEL operates below the saturation [102, 110].

Super-low-threshold microcavity-type VCSELs have been proposed that utilize the spontaneous emission enhancement due to more spontaneous emission being coupled into the lasing mode [114, 115]. Although a thresholdless laser is theoretically possible when the spontaneous emission coupling effciency β is made approaching unity, the proposed structures are difficult to make in practice. One of the successful examples in research today is the use of oxidized lateral carrier confinement blocks by oxidizing an AlAs layer in the DBR or the cladding regions [23, 24, 116] (Fig. 2.27).

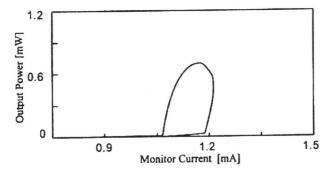

Fig. 2.26 VCSEL output power in relationship with current response of an integrated detector (after Ref. [102]).

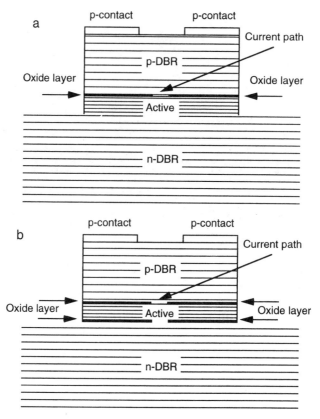

Fig. 2.27 VCSEL with native aluminum oxide for lateral current confinement. (a) Current confinement on *p* side, and (b) current confinement on both *p* side and *n* side.

This technology will be discussed in more detail in Chapter 15. Typically, sub-100-μA threshold can be achieved with this technique. A VCSEL with an extremely low threshold of 8.7 μA has been reported with an active area of 3 μm^2 [24]. It should be noted that there is still a debate on the exact mechanism that has generated this result. VCSELs with oxidized mirrors have been demonstrated with extremely simple epitaxy layers [117, 118]. In this structure, only four to six pairs of GaAs/AlAs DBR stacks are grown on one or both sides of an active region that is made of strained InGaAs MQWs at 970 nm (Fig. 2.28). The AlAs layers in the DBR mirrors

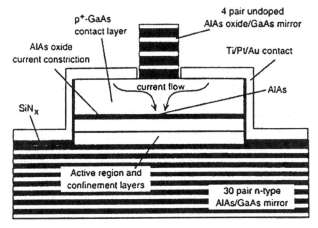

Fig. 2.28 VCSEL with an AlAs oxide–GaAs DBR mirror (after Ref. [117]).

are oxidized during the fabrication procedure. The extremely large index difference between GaAs and the oxidized AlAs layer makes it possible that only four pairs of GaAs/AlAs stacks will provide sufficiently high reflectivity with very large bandwidth for proper device operation. The VCSEL electrical contacts in this case will have to be made laterally inside the cavity as opposed to at the top of the DBR mirror stacks because the electrical conduction through the DBR mirror is prohibited once the AlAs constituent of the mirror is oxidized.

High-speed data transmission requires that a VCSEL be modulated at multi-GHz. The cavity volume of a VCSEL is significantly smaller than that of an edge-emitting laser, resulting in a higher photon density in the VCSEL cavity. The resonance frequency of a semiconductor laser typically scales as the square root of the photon density, thus indicating that a VCSEL has a potential advantage in high-speed operation. However, the parasitic series resistance caused by the semiconductor DBR and the device heating limit the maximum achievable VCSEL modulation bandwidth. Currently, a modulation speed of larger than 16 GHz has been reported with an oxide-confined VCSEL at a current of 4.5 mA [119]. Modeling results indicate that a gain compression limited-oxide VCSEL with a diameter of 3 μm has an intrinsic 3-dB bandwidth of 45 GHz [120] and a measured 3-dB bandwidth of 15 GHz at 2.1 mA due to the parasitic resistance and the device heating (Fig. 2.29).

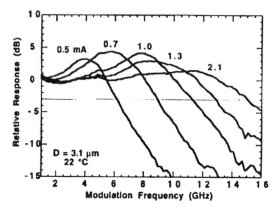

Fig. 2.29 Small signal modulation response of a 3-μm VCSEL at various bias current. The maximum 3-dB bandwidth is approximately 15 GHz (after Ref. [120]).

References

1. Gowar, J. 1984. *Optical communication systems.* Englewood Cliffs, NJ: Prentice Hall.
2. Miller, S. E., and A. G. Chynoweth, eds. 1979. *Optical fiber telecommunications.* New York: Academic Press.
3. Lasky, R., U. Osterberg, and D. Stigliani, eds. 1995. *Optoelectronics for data communication.* New York: Academic Press.
4. Hall, R. N., G. E. Fenner, J. D. Kingsley, T. J. Soltys, and R. O. Carlson. 1962. Coherent light emission from GaAs junctions. *Phys. Rev. Lett.* 9:366.
5. Nathan, M. I., W. P. Dumke, G. Burns, F. H. Dill, Jr., and G. Lasher. 1962. Stimulated emission of radiation from GaAs p–n junctions. *Appl. Phys. Lett.* 1:62.
6. Holonyak, N., Jr., and S. F. Bevacqua. 1962. Coherent (visible) light emission from $Ga(Al_{1-x}P_x)As$ junctions. *Appl. Phys. Lett.* 1:82.
7. Quist, T. M., R. H. Rediker, R. J. Keyes, W. E. Krag, B. Lax, A. L. McWhorter, and J. J. Zeiger. 1962. Semiconductor maser of GaAs. *Appl. Phys. Lett.* 1:91.
8. Kroemer, H. 1963. A proposed class of heterojunction injection lasers. *Proc. IEEE* 51:1782.
9. Hayashi, I., M. B. Panish, P. W. Foy, and S. Sumuski. 1970. Junction lasers which operate continuously at room temperature. *Appl. Phys. Lett.* 17:109.
10. Alferov, Zh. I., V. M. Andreev, D. Z. Garbuzov, Yu. V. Zhilyaev, E. P. Morozov, E. L. Portnoi, and V. G. Triofim. 1971. Investigation of the influence of the AlAs-GaAs heterostructure parameters on the laser threshold current and the realization of continuous emission at room temperature. *Sov. Phys. Semiconductor* 4:1573.

11. Anderson, S. G. 1996. Annual review of laser markets. *Laser Focus World* 32:50.

12. Soda, H., K. Iga, C. Kitahara, and Y. Suematsu. 1979. GaInAsP/InP surface emitting injection lasers. *Jpn. J. Appl. Phys.* 18:2329.

13. Iga, K., F. Koyama, and S. Kinoshita. 1988. Surface emitting semiconductor lasers. *IEEE J. Quantum Electron.* QE-24:1845.

14. Lebby, M., C. A. Gaw, W. B. Jiang, P. A. Kiely, C. L. Shieh, P. R. Claisse, J. Ramdani, D. H. Hartman, D. B. Schwartz, and J. Grula. 1996. Use of VCSEL arrays for parallel optical interconnects. *Proc. SPIE* 2683:81.

15. Orenstein, M., A. C. Von Lehmen, C. Chang-Hasnain, N. G. Stoffel, J. P. Harbison, and L. T. Florez. 1991. Matrix addressable vertical cavity surface emitting laser array. *Electron. Lett.* 27:437.

16. Shieh, C. L., D. E. Ackley, and H. C. Lee. 1993. Temperature insensitive vertical cavity surface emitting laser. U.S. Patent No. 5,274,655.

17. Young, D. B., J. W. Scott., F. H. Peters, B. J. Thibeault, S. W. Corzine, M. G. Peters, S. L. Lee, and L. A. Coldren. 1993. High-power temperature-insensitive gain-offset InGaAs/GaAs vertical-cavity surface-emitting lasers. *IEEE Photon. Tech. Lett.* 5:129.

18. Tai, K., G. Hasnain, J. D. Wynn, R. J. Fischer, Y. H. Wang, B. Weir, J. Gamelin, and A. Y. Cho. 1990. 90% coupling of top surface emitting GaAs/AlGaAs quantum well laser output into 8 μm diameter core silica fibre. *Electron. Lett.* 26:1628.

19. Iga, K., S. Ishikawa, S. Ohkouchi, and T. Nishimura. 1984. Room temperature pulsed oscillation of GaAlAs/GaAs surface emitting laser. *Appl. Phys. Lett.* 45:348.

20. Koyama, F., S. Kinoshita, and K. Iga. 1988. Room-temperature CW operation of GaAs vertical cavity surface emitting laser. *Trans. Inst. Electron. Commun. Eng. Jpn.* E71:1089.

21. Peters, F. H., M. G. Peters, D. B. Young, J. W. Scott, B. J. Thibeault, S. W. Corzine, and L. A. Coldren. 1993. High power vertical cavity surface emitting lasers. *Electron. Lett.* 29:200.

22. Grabherr, M., B. Weigl, G. Reiner, R. Michalzik, M. Miller, and K. J. Ebeling. 1996. High power top-surface emitting oxide confined vertical-cavity laser diodes. *Electron. Lett.* 32:1723.

23. Huffaker, D. L., J. Shin, and D. G. Deppe. 1994. Low threshold halfwave vertical-cavity lasers. *Electron. Lett.* 30:1946.

24. Yang, G. M., M. H. MacDougal, and P. D. Dapkus. 1995. Ultralow threshold current vertical-cavity surface-emitting lasers obtained with selective oxidation. *Electron. Lett.* 31:886.

25. Lear, K. L., K. D. Choquette, R. P. Schneider, S. P. Kilcoyne, and K. M. Geib. 1995. Selectively oxidised vertical cavity surface emitting lasers with 50% power conversion efficiency. *Electron. Lett.* 31:208.

26. Jäger, R., M. Grabherr, C. Jung, R. Michalzik, R. Reiner, B. Weigl, and K. J. Ebeling. 1997. 57% wallplug efficiency oxide-confined 850 nm wavelength GaAs VCSELs. To be published.

27. Iga, K. 1992. Surface emitting lasers. *Opt. Quantum Electron.* 24: S97.

28. Babic, D. I., K. Streubel, R. P. Mirin, N. M. Margalit, J. E. Bowers, E. L. Hu, D. E. Mars, L. Yang, and K. Carey. 1995. Room-temperature continuous-wave operation of 1.54-μm vertical-cavity lasers. *IEEE Photon. Tech. Lett.* 7:1225.

29. Margalit, N. M., D. I. Babic, K. Streubel, R. P. Mirin, R. L. Naone, J. E. Bowers, and E. L. Hu. 1996. Submilliamp long wavelength vertical cavity lasers. *Electron. Lett.* 32:1675.

30. Vakhshoori, D., J. D. Wynn, G. J. Zydik, and R. E. Leibenguth. 1993. 8 × 18 top emitting independently addressable surface emitting laser arrays with uniform threshold current and low threshold voltage. *Appl. Phys. Lett.* 62:1718.

31. Uchiyama, S., and K. Iga. 1985. Two-dimensional array of GaInAsP/InP surface-emitting lasers. *Electron. Lett.* 21:162.

32. Deppe, D. G., J. P. van der Ziel, N. Chand, G. J. Zydzik, and S. N. G. Chu. 1990. Phase-coupled two-dimensional $Al_xGa_{1-x}As$-GaAs vertical-cavity surface-emitting laser array. *Appl. Phys. Lett.* 56:2089.

33. Orenstein, M., E. Kapon, N. G. Stoffel, J. P. Harbison, L. T. Florez, and J. Wullert. 1991. Two-dimensional phase-locked arrays of vertical-cavity semiconductor lasers by mirror reflectivity modulation. *Appl. Phys. Lett.* 58:804.

34. Iizuka, K., K. Matsumaru, T. Suzuki, H. Hirose, K. Suzuki, and H. Okamoto. 1995. Arsenic-free GaAs substrate preparation and direct growth of GaAs/ AlGaAs multiple quantum well without buffer layers. *J. Cryst. Growth* 150:13.

35. Cheng, W. H., and J. H. Bechtel. 1993. High-speed fibre optic links using 780 nm compact disc lasers. *Electron. Lett.* 29:2055.

36. Soderstrom, R. L., S. J. Baumgartner, B. L. Beukema, T. R. Block, and D. L. Karst. 1993. CD lasers optical data links for workstations and midrange computers. *ECTC'93,* 505, June, Orlando, FL.

37. Nakata, N. 1987. Laser diodes have low noise and low astigmatism. *JEE,* August, 49.

38. Wang, S. 1989. *Fundamentals of semiconductor theory and device physics,* 51. Englewood Cliffs, NJ: Prentice Hall.

39. Morton, P. A., R. A. Logan, T. Tanbunek, P. F. Sciortino, A. M. Sergent, R. K. Montgomery, and B. T. Lee. 1993. 25 GHz bandwidth 1.55-μm GaInAsP p-doped strained multiquantum-well lasers. *Electron. Lett.* 29:136.

40. Ralston, J. D., E. C. Larkins, K. Eisele, S. Weisser, S. Buerkner, A. Schoenfelder, J. Daleiden, K. Czotscher, I. Esquivias, J. Fleissner, R. E. Sah, M. Maier, W. Benz, and J. Rosenzweig. 1996. Advanced epitaxial growth and device processing techniques for ultrahigh-speed (>40 GHz) directly modulated semiconductor lasers. *Proc. SPIE* 2683:30.

41. Hersee, S. D., B. de Cremoux, and J. P. Duchemin. 1984. Some characteristics of the GaAs/GaAlAs graded-index separate-confinement heterostructure quantum well laser structure. *Appl. Phys. Lett.* 44:476.

42. Coleman, J. J. 1995. Quantum-well heterostructure lasers. In *Semiconductor lasers: Past, present, and future,* G. P. Agrawal, Chapter 1. Woodbury, NY: AIP Press.

43. Nagarajan, R., T. Fukushima, J. E. Bowers, R. S. Geels, and L. A. Colden. 1991. High-speed InGaAs/GaAs strained multiple quantum well lasers with low damping. *Appl. Phys. Lett.* 58:2326.

44. Miller, B. I., U. Koren, M. G. Young, and M. D. Chien. 1991. Strain-compensated strained-layer superlattices for 1.5 μm wavelength lasers. *Appl. Phys. Lett.* 58:1952.

45. Zhang, G., and A. Ovtchinnikov. 1993. Strain-compensated InGaAs/GaAsP/GaInAsP/GaInP quantum well lasers (1 ~ 0.98 μm) grown by gas-source molecular beam epitaxy. *Appl. Phys. Lett.* 62:1644.

46. Tsuchiya, T., M. Komori, R. Tsuneta, and H. Kakibayashi. 1994. Investigation of effect of strain-compensated structure and compensation limit in strained-layer multiple quantum wells. *J. Cryst. Growth* 145:371.

47. Bessho, Y., T. Uetani, R. Hiroyama, K. Komeda, M. Shono, A. Ibaraki, K. Yodoshi, and T. Niina. 1996. Self-pulsating 630 nm band strain-compensated MQW AlGaInP laser diodes. *Electron. Lett.* 32:667.

48. Nagel, S. R., J. B. MacChesney, and K. L. Walker. 1985. Modified chemical vapor deposition. In *Optical fiber communications,* ed. T. Y. Li, Vol. 1, Chapter 1. Orlando: Academic Press.

49. Cohen, L. G., C. Lin, and W. G. French. 1979. Tailoring zero chromatic dispersion into the 1.5–1.6 μm low-loss spectral region of single-mode fibres. *Electron. Lett.* 15:334.

50. Tsuchiya, H., and N. Imoto. 1979. Dispersion-free single-mode fibre in 1.5 μm wavelength region. *Electron. Lett.* 15:476.

51. Okamoto, K., T. Edahiro, A. Kawana, and T. Miya. 1979. Dispersion minimisation in single-mode fibres over a wide spectral range. *Electron. Lett.* 15:729.

52. Hsieh, J. J., J. A. Rossi, and J. P. Donnelly. 1976. Room-temperature cw operation of GaInAsP/InP double-heterostructure diode lasers emitting at 1.1 μm. *Appl. Phys. Lett.* 28:709.

53. Nelson, R. J., P. D. Wright, P. A. Barnes, R. L. Brown, T. Cella, and R. G. Sobers. 1980. High-output power InGaAsP (1 = 1.3 μm) strip-buried hetero-structure lasers. *Appl. Phys. Lett.* 36:358.

54. Hirao, M., S. Tsuji, K. Mizuishi, A. Doi, and M. Nakamura. 1980. Long wavelength InGaAsP/InP lasers for optical fiber communication systems. *J. Opt. Commun.* 1:10.

55. Dutta, N. K., and R. J. Nelson. 1982. The case for Auger recombination in $In_{1-x}Ga_xAs_yP_{1-y}$. *J. Appl. Phys.* 53:74.

56. Dutta, N. K., S. G. Napholtz, R. Yen, T. Wessel, T. M. Shen, and N. A. Olsson. 1985. Long wavelength InGaAsP (1 ~ 1.3 μm) modified multiquantum well laser. *Appl. Phys. Lett.* 46:1036.

57. Bates, R. J. S., and S. D. Walker. 1992. Evaluation of all-plastic optical fibre compute data link dispersion limits. *Electron. Lett.* 28:996.

58. Gwynne, P. 1996. Digital video disk technology offers increased storage features. *R&D Magazine* 38:40.

59. Yamamoto, S., H. Hayashi, T. Hayakawa, N. Miyauchi, S. Yano, and T. Hijikata. 1982. Room-temperature cw operation in the visible spectral range of 680–700 nm by AlGaAs double heterojunction lasers. *Appl. Phys. Lett.* 41:796.

60. Usui, A., T. Matsumoto, M. Inai, I. Mito, K. Kobayashi, and H. Watanabe. 1985. Room temperature cw operation of visible InGaAsP double heterostructure laser at 671 nm grown by hydride VPE. *Jpn. J. Appl. Phys.* 24:L163.

61. Chong, T. H., and K. Kishino. 1990. Room temperature continuous wave operation of 671-nm wavelength GaInAsP/AlGaAs VSIS lasers. *IEEE Photon. Tech. Lett.* 2:91.

62. Kobayashi, K., S. Kawata, A. Gomyo, I. Hino, and T. Suzuki. 1985. Room-temperature cw operation of AlGaInP double-heterostructure visible lasers. *Electron. Lett.* 21:931.

63. Ikeda, M., Y. Mori, H. Sato, K. Kaneko, and N. Watanabe. 1985. Room-temperature continuous-wave operation of an AlGaInP double heterostructure laser grown by atmospheric pressure metalorganic chemical vapor deposition. *Appl. Phys. Lett.* 47:1027.

64. Ishikawa, M., Y. Ohba, H. Sugawara, M. Yamamoto, and T. Nakanisi. 1986. Room-temperature cw operation of InGaP/InGaAlP visible light laser diodes on GaAs substrates grown by metalorganic chemical vapor deposition. *Appl. Phys. Lett.* 48:207.

65. Hatakoshi, G., K. Nitta, Y. Nishikawa, K. Itaya, and M. Okajima. 1993. High-temperature operation of high-power InGaAlP visible laser. *Proc. SPIE* 1850:388.

66. Hashimoto, J., T. Katsuyama, J. Shinkai, I. Yoshida, and H. Hayashi. 1991. Effects of strained-layer structures on the threshold current density of AlGaInP/GaInP visible lasers. *Appl. Phys. Lett.* 58:879.

67. Honda, S., H. Hamada, M. Shono, R. Hiroyama, K. Yodoshi, and T. Yamaguchi. 1992. Transverse-mode stabilised 630 nm-band AlGaInP strained multiquantum-well laser diodes grown on misoriented substrates. *Electron. Lett.* 28:1365.

68. Tanaka, T., H. Yanagisawa, S. Yano, and S. Minagawa. 1993. High-temperature operation of 637 nm AlGaInP MQW laser diodes with quaternary QWS grown on misoriented substrates. *Electron. Lett.* 29:24.

69. Ueno, Y., H. Fujii, H. Sawano, K. Kobayashi, K. Hara, A. Gomyo, and K. Endo. 1993. 30-mW 690-nm high-power strained-quantum-well AlGaInP laser. *IEEE J. Quantum Electron.* QE-29:1851.

70. Arimoto, S., M. Yasuda, A. Shima, K. Kadoiwa, T. Kamizato, H. Watanabe, E. Omura, M. Aiga, K. Ikeda, and S. Mitsui. 1993. 150mW fundamental-transverse-mode operation of 670 nm window laser diode. *IEEE J. Quantum Electron.* QE-29:1874.

71. Nakayama, N., S. Itoh, K. Nakano, H. Okuyama, M. Ozawa, A. Ishibashi, M. Ikeda, and Y. Mori. 1993. Room temperature continuous operation of blue-green laser diodes. *Electron. Lett.* 29:1488.

72. Gaines, J. M., R. R. Drenten, K. W. Haberern, T. Marshall, P. Mensz, and J. Petruzzello. 1993. Blue-green injection lasers contraining pseudomorphic $Zn_{1-x}Mg_xS_ySe_{1-y}$ cladding lasers and operating up to 394 K. *Appl. Phys. Lett.* 62:2462.

73. Haase, M. A., P. F. Baude, M. S. Hagedorn, J. Qiu, J. DePuydt, H. Cheng, S. Guha, G. E. Hofler, and B. J. Wu. 1993. Low-threshold buried-ridge II–VI laser diodes. *Appl. Phys. Lett.* 63:2315.

74. Taniguchi, S., T. Hino, S. Itoh, K. Nakano, N. Nakayama, A. Ishibashi, and M. Ikeda. 1996. 100 h II–VI blue-green laser diode. *Electron. Lett.* 32:552.

75. Nakamura, S., T. Mukai, and M. Senoh. 1991. High-power GaN p–n junction blue-light-emitting diodes. *Jpn. J. Appl. Phys.* 30:L1998.

76. Nakamura, S., M. Senoh, S. Nagahama, N. Iwasa, T. Yamada, T. Matsushita, H. Kiyoku, and Y. Sugimoto. 1996. InGaN MQW structure laser diodes with cleaved mirror facets. *Jpn. J. Appl. Phys.* 35:L217.

77. Itaya, K., M. Onomura, J. Nishio, L. Sugiura, S. Saito, M. Suzuki, J. Rennie, S. Y. Nunoue, M. Yamamoto, H. Fujimoto, Y. Kokubun, Y. Ohba, G. Hatakoshi, and M. Ishikawa. 1996. Room temperature pulsed operation of nitride based multi-quantum-well laser diodes with cleaved facets on conventional c-face sapphire substrates. *Jpn. J. Appl. Phys.* (Part 2) 35:L1315.

78. Akasaki, I., S. Sota, H. Sakai, T. Tanaka, M. Koike, and H. Amano. 1996. Shortest wavelength semiconductor laser diode. *Electron. Lett.* 32:1105.

79. Nakamura, S., M. Senoh, S. Nagahama, N. Iwasa, T. Yamada, T. Matsushita, H. Kiyoku, and Y. Sugimoto. 1996. InGaN multi-quantum-well structure laser diodes grown on $MgAl_2O_4$ substrates. *Appl. Phys. Lett.* 68:2105.

80. Akasaki, I., and H. Amano. 1996. Progress and future prospects of group III nitride semiconductors. *LEOS'96,* Plen2, November, Boston, MA.

81. Nakamura, S., M. Senoh, S. I. Nagahama, N. Iwasa, T. Yamada, T. Matsushita, Y. Sugimoto, and H. Kiyoku. 1996. First room-temperature continuous-wave operation of InGaN multi-quantum-well-structure laser diodes. *LEOS'96,* PD1.1, November, Boston, MA.

82. Hasnain, G., K. Tai, J. D. Wynn, Y. H. Wang, R. J. Fischer, M. Hong, B. E. Weir, G. J. Zydzik, J. P. Mannaerts, J. Gamelin, and A. Y. Cho. 1990. Continuous wave top surface emitting quantum well lasers using hybrid metal/semiconductor reflectors. *Electron. Lett.* 26:1590.

83. Morgan, R. A., M. K. Hibbs-Brenner, T. M. Marta, R. A. Walterson, S. Bounnak, E. L. Kalweit, and J. A. Lehman. 1995. 200 degrees-C, 96-nm wavelength range, continuous-wave lasing from unbonded GaAs MOVPE-grown vertical cavity surface-emitting lasers. *IEEE Photon. Tech. Lett.* 7:441.

84. Zhou, P., J. L. Cheng, C. F. Schaus, S. Z. Sun, K. Zheng, E. Armour, C. Hains, W. Hsin, D. R. Myers, and G. A. Vawter. 1991. Low series resistance

high-efficiency GaAs/AlGaAs vertical-cavity surface-emitting lasers with continuously graded mirrors grown by MOCVD. *IEEE Photon. Tech. Lett.* 3:591.

85. Tan, M. R. T., K. H. Hahn, Y. M. D. Houng, and S. Y. Wang, 1995. Surface emitting laser for multimode data link applications. *HP J.,* February: 67.

86. Tai, K., L. Yang, Y. H. Wang, J. D. Wynn, and A. Y. Cho. 1990. Drastic reduction of series resistance in doped semiconductor distributed Bragg reflectors for surface-emitting lasers. *Appl. Phys. Lett.* 56:2496.

87. Zhou, P., J. Cheng, C. F. Schaus, S. Z. Sun, K. Zheng, E. Armour, C. Hains, W. Hsin, D. R. Myers, and G. A. Vawter. 1991. Low series resistance high-efficiency GaAs AlGaAs vertical-cavity surface-emitting lasers with continuously graded mirrors grown by MOCVD. *IEEE Photon. Tech. Lett.* 3:591.

88. Schubert, E. F., L. W. Tu, G. J. Zydzik, R. F. Kopf, A. Benvenuti, and M. R. Pinto. 1992. Elimination of heterojunction band discontinuities by modulation doping. *Appl. Phys. Lett.* 60:466.

89. Peters, M. G., D. B. Young, F. H. Peters, J. W. Scott, B. J. Thibeault, and L. A. Coldren. 1994. 17.3-percent peak wall plug efficiency vertical-cavity surface-emitting lasers using lower barrier mirrors. *IEEE Photon. Tech. Lett.* 6:31.

90. Peters, M. G., B. J. Thibeault, D. B. Young, J. W. Scott, F. H. Peters, A. C. Gossard, and L. A. Coldren. 1993. Band-gap engineered digital alloy interfaces for lower resistance vertical-cavity surface-emitting lasers. *Appl. Phys. Lett.* 63:3411.

91. Lear, K. L., S. A. Chalmers, and K. P. Killeen. 1993. Low threshold voltage vertical cavity surface-emitting laser. *Electron. Lett.* 29:584.

92. Young, D. B., J. W. Scott, F. H. Peters, M. G. Peters, M. L. Majewski, B. J. Thibeault, S. W. Corzine, and L. A. Coldren. 1993. Enhanced performance of offset-gain high-barrier vertical-cavity surface-emitting lasers. *IEEE J. Quantum Electron.* QE-29:2013.

93. Schubert, E. F., A. Fischer, Y. Horikoshi, and K. Ploog. 1985. GaAs sawtooth superlattice laser emitting at wavelength l > 0.9 μm. *Appl. Phys. Lett.* 47:219.

94. Kojima, K., R. A. Morgan, T. Mullaly, G. D. Guth, M. W. Focht, R. E. Leibenguth, and M. T. Asom. 1993. Reduction of p-doped mirror electrical resistance of GaAs/AlGaAs vertical-cavity surface-emitting lasers by delta doping. *Electron. Lett.* 29:1771.

95. Scott, J. W., B. J. Thibeault, D. B. Young, L. A. Coldren, and F. H. Peters. 1994. High efficiency submilliamp vertical cavity lasers with intracavity contacts. *IEEE Photon. Tech. Lett.* 6:678.

96. Rochus, S., M. Hauser, T. Röhr, H. Kratzer, G. Böhm, W. Klein, G. Tränkle, and G. Weimann. 1995. Submilliamp vertical-cavity surface-emitting lasers with buried lateral-current confinement. *IEEE Photon. Tech. Lett.* 7:968.

97. Morgan, R. A., M. K. Hibbs-Brenner, J. A. Lehman, E. L. Kaiweit, R. A. Walterson, T. M. Marta, and T. Akinwande. 1995. Hybrid dielectric/AlGaAs mirror spatially filtered vertical cavity top-surface emitting laser. *Appl. Phys. Lett.* 66:1157.

98. Dudley, J. J., D. L. Crawford, and J. E. Bowers. 1992. Temperature dependence of the properties of DBR mirrors used in surface normal optoelectronic devices. *IEEE Photon. Tech. Lett.* 4:311.

99. Lebby, M., C. A. Gaw, W. B. Jiang, P. A. Kiely, P. R. Claisse, and J. Ramdani. 1996. Vertical-cavity surface-emitting lasers for communication applications. OSA annual meeting, WR1, October, Rochester, NY.

100. Lee, Y. H., B. Tell, K. F. Brown-Goebeler, R. E. Leibenguth, and V. D. Mattera. 1991. Deep-red CW top surface-emitting vertical-cavity AlGaAs superlattice lasers. *IEEE Photon. Tech. Lett.* 3:108.

101. Shin, H. E., Y. G. Ju, J. H. Shin, J. H. Ser, T. Kim, E. K. Lee, I. Kim, and Y. H. Lee. 1996. 780 nm oxidised vertical-cavity surface-emitting lasers with $Al_{0.1}Ga_{0.89}As$ quantum wells. *Electron. Lett.* 32:1287.

102. Kim, T., T. K. Kim, E. K. Lee, J. Y. Kim, and T. I. Kim. 1995. A single transverse mode operation of top surface emitting laser diode with an integrated photodiode. *Proc. LEOS'95* 2:416.

103. Schneider, R. P., Jr., K. D. Choquette, J. A. Lott, K. L. Lear, J. J. Figiel, and K. J. Malloy. 1994. Efficient room-temperature continuous-wave AlGaInP/AlGaAs visible (670 nm) vertical-cavity surface-emitting laser diodes. *IEEE Photon. Tech. Lett.* 6:313.

104. Choquette, K. D., R. P. Schneider, M. H. Crawford, K. M. Geib, and J. J. Figiel. 1995. Continuous wave operation of 640–660 nm selectively oxidised AlGaInP vertical-cavity lasers. *Electron. Lett.* 31:1145.

105. Schneider, R. P., Jr., M. H. Crawford, K. D. Choquette, K. L. Lear, S. P. Kilcoyne, and J. J. Figiel. 1995. Improved AlGaInP-based red (670–690 nm) surface-emitting lasers with novel C-doped short-cavity epitaxial design. *Appl. Phys. Lett.* 67:329.

106. Crawford, M. H., R. P. Schneider, Jr., K. D. Choquette, and K. L. Lear. 1995. Temperature-dependent characteristics and single-mode performance of AlGaInP-based 670–690-nm vertical-cavity surface-emitting lasers. *IEEE Photon. Tech. Lett.* 7:724.

107. Baba, T., Y. Yogo, K. Suzuki, F. Koyama, and K. Iga. 1993. Near room temperature continuous wave lasing characteristics of GaInAsP/InP surface emitting laser. *Electron. Lett.* 29:913.

108. Dudley, J. J., M. Ishikawa, B. I. Miller, D. I. Babic, R. Mirin, W. B. Jiang, J. E. Bowers, and E. L. Hu. 1992. 144°C operation of 1.3 μm InGaAsP vertical cavity lasers on GaAs substrates. *Appl. Phys. Lett.* 61:3095.

109. Dudley, J. J., D. I. Babic, R. Mirin, L. Yang, B. I. Miller, R. J. Ram, T. Reynolds, E. L. Hu, and J. E. Bowers. 1994. Low threshold, wafer fused long wavelength vertical cavity lasers. *Appl. Phys. Lett.* 64:1463.

110. Shin, H. K., I. Kim, E. J. Kim, J. H. Kim, E. K. Lee, M. K. Lee, J. K. Mun, C. S. Park, and Y. S. Yi. 1996. Vertical-cavity surface-emitting lasers for optical data storage. *Jpn. J. Appl. Phys.* (Part 1), 35:506.

111. Hasnain, G., and K. Tai. 1992. Self-monitoring semiconductor laser device. U.S. Patent No. 5,136,603.

112. Hasnain, G., K. Tai, Y. H. Wang, J. D. Wynn, K. D. Choquette, B. E. Weir, N. K. Dutta, and A. Y. Cho. 1991. Monolithic integration of photodetector with vertical cavity surface emitting laser. *Electron. Lett.* 27:1630.

113. Hibbs-Brenner, M. K. 1995. Integrated laser power monitor. U.S. Patent No. 5,475,701.

114. Bjork, G., and Y. Yamamoto. 1991. Analysis of semiconductor microcavity lasers using rate equations. *IEEE J. Quantum Electron.* QE-27:2386.

115. Ram, R. J., E. Goobar, M. G. Peters, L. A. Coldren, and J. E. Bowers. 1996. Spontaneous emission factor in post microcavity lasers. *IEEE Photon. Tech. Lett.* 8:599.

116. Huffaker, D. L., D. G. Deppe, and K. Kumar. 1994. Native-oxide ring contact for low threshold vertical-cavity lasers. *Appl. Phys. Lett.* 65:97.

117. MacDougal, M. H., P. Daniel Dapkus, V. Pudikov, H. M. Zhao, and G. M. Yang. 1995. Ultralow threshold current vertical-cavity surface-emitting lasers with AlAs oxide-GaAs distributed Bragg reflectors. *IEEE Photon. Tech. Lett.* 7:229.

118. MacDougal, M. H., G. M. Yang, A. E. Bond, C. K. Lin, D. Tishinin, and P. D. Dapkus. 1996. Electrically-pumped vertical-cavity lasers with Al_xO_y-GaAs reflectors. *IEEE Photon. Tech. Lett.* 8:310.

119. Lear, K. L., A. Mar, K. D. Choquette, S. P. Kilcoyne, R. P. Schneider, Jr., and K. M. Geib. 1996. High frequency modulation of oxide-confined vertical cavity surface emitting lasers. *Electron. Lett.* 32:457.

120. Thibeault, B. J., K. Bertilsson, E. R. Hegblom, E. Strzelecka, P. D. Floyd, R. Naone, and L. A. Coldren. 1997. High-speed characteristics of low-optical loss oxide-apertured vertical-cavity lasers. *IEEE Photon. Tech. Lett.* 9:11.

Chapter 3 | Detectors for Fiber Optics

Carolyn J. Sher DeCusatis

*Lighting Research Center, Rensselaer Polytechnic Institute, Troy,
New York 12180*

Ching-Long (John) Jiang

Amp Incorporated, Lytel Division, Somerville, New Jersey 08876

3.1. Introduction

Detectors are an integral part of all fiber optic communication systems.
They demodulate optical signals—that is, convert the optical variations
into electrical variations—that are subsequently amplified and further processed. For such applications the detectors must satisfy stringent requirements such as high sensitivity at operating wavelengths, high response
speed, and minimum noise. In addition, the detector should be compact in
size, use low biasing voltages, and be reliable under operating conditions.
There are many different types of optical detectors available that are sensitive over the entire optical spectra, the infrared and the ultraviolet, including
such devices as photomultiplier tubes, charge coupled devices, and thermal
and photoconductive detectors [1]. However, fiber optic communication
systems most commonly use solid-state semiconductor devices, such as the
positive–intrinsic–negative (PIN) photodiode. A general semiconductor
detector performs three basic processes: (i) carrier (electron–hole pair)
generation by incident light, (ii) carrier transport and/or multiplication by
whatever current-gain mechanism may be present, and (iii) interaction of
current with the external circuit to provide the output signal. In this chapter,
we will begin with an overview of detector terminology. This is followed
by a detailed description of two types of detectors used extensively in
fiber optics, namely the $p–i–n$ photodiode and the avalanche photodiode.
Although in principle one can perform coherent detection of optical signals

87

HANDBOOK OF FIBER OPTIC
DATA COMMUNICATION

by mixing the desired signal with a reference signal at the detector, this technique is not commonly employed in data communications. Rather, we will concentrate on incoherent detection methods, in which the detector responds directly to the amplitude of the incident optical signal [both laser- and light-emitting diode (LED)-based transmitters make use of this approach]. Finally, we will briefly describe some noise sources in datacom receivers and their impact on the detection electronics.

3.2. Detector Terminology and Characteristics

In this section we will define the terminology used in a typical detector specification. Every detector specification should include a picture and/or physical description of the part, including dimensions and construction (i.e., plastic housing). We have tried to be inclusive in the list of terms, which means not all these quantities will apply to every detector specification. Because specifications are not standardized, it is impossible to include all possible terms used; however, most detectors are described by certain standard figures of merit that will be discussed in this section. It is important to consider the manufacturer's context for all values; a detector designed for a specific application may not be appropriate for a different application even though the specification seems appropriate.

There are several figures of merit used to characterize the performance of different detectors. Responsivity, or response, is the sensitivity of the detector to input flux. It is given by

$$R(\lambda) = I/\phi(\lambda), \tag{3.1}$$

where I is the detector output signal (in amps) and ϕ is the incident light signal on the detector (in watts). Thus, the units of responsivity are A/W. Even when the detector is not illuminated, some current will flow; this dark current may be subtracted from the detector output signal when determining detector performance. Dark current is the thermally generated current in a photodiode under a completely dark environment; it depends on the material, doping, and structure of the photodiode. It is the lowest level of thermal noise. Dark current in photodiodes limits the sensitivity (minimum detectable power). The reduction of dark current is important for the improvement of the minimum detectable power. It is usually simply measured and then subtracted from the flux, like background, in most specifications. However, the dark current is temperature dependent so care must be taken to evaluate it over the expected operating conditions. It is not a good idea for the anticipated signal to be a small fraction of the dark current; root mean square (rms) noise in the dark current may mask the signal. Responsivity is defined at a specific wavelength; the term spectral responsivity is used to describe the variation at different wavelengths. Responsivity vs wavelength is often included in a specification as a graph as well as placed in a performance chart at a specified wavelength.

Quantum efficiency (QE) is the ratio of the number of electron–hole pairs collected at the terminals to the number of photons in the incident light. It depends on the material from which the detector is made and is primarily determined by reflectivity, absorption coefficient, and carrier diffusion length. Because the absorption coefficient is dependent on the incident light wavelength, the quantum efficiency has a spectral response. Quantum efficiency is the funadamental efficiency of the diode for convert-ing photons into electron–hole pairs. It affects detector performance through the responsivity, which can be calculated from quantum efficiency:

$$R(\lambda) = QE \; \lambda \; q/hc, \qquad (3.2)$$

where q is the charge of an electron (1.6×10^{-19} coulomb), λ is the wave-length of the incident photon, h is Planck's constant (6.626×10^{-34} W), and c is the velocity of light (3×10^8 m/s). If wavelength is in microns and R is responsivity flux, then the units of quantum efficiency are A/W. Responsivity is the ratio of the diode's output current to input optical power and is given in A/W. A PIN photodiode typically has a responsivity of 0.6–0.8 A/W. A responsivity of 0.8 A/W means that incident light having 50 μW of power results in 40 μA of current; in other words,

$$I = 50 \; \mu W \times 0.8 \; A/W = 40 \; \mu A, \qquad (3.3)$$

where I is the photodiode current. For an avalanche photodiode (APD), a typical responsivity is 80 A/W. The same 50 μW of optical power now produces 4 mA of current:

$$I = 50 \; \mu W \times 80 \; A/W = 4 \; mA. \qquad (3.4)$$

The minimum power detectable by the photodiode determines the lowest level of incident optical power that the photodiode can detect. It is related to the dark current in the diode because the dark current will set the lower limit. Other noise sources are also factors, including those associated with the diode and those associated with the receiver. The noise floor of a photodiode, which tells us the minimum detectable power, is the ratio of noise current to responsivity:

$$\text{Noise floor} = \text{noise/responsivity.} \qquad (3.5)$$

For initial evaluation of a photodiode, we can use the dark current to estimate the noise floor. Consider a photodiode with $R = 0.8$ A/W and a dark current of 2 nA. The minimum detectable power is

$$\text{Noise floor} = (2\ \text{nA})/(0.8\ \text{nA/nW}) = 2.5\ \text{nW.} \qquad (3.6)$$

More precise estimates must include other noise sources, such as thermal and shot noise. As previously discussed, the noise depends on current, load resistance, temperature, and bandwidth.

Response time is the time required for the photodiode to respond to an incoming optical signal and produce an external current. Similarly to a source, response time is usually specified as a rise time and a fall time, measured between the 10 and 90% points of amplitude (other specifications may measure rise and fall times at the 20–80% points or when the signal rises or falls to 1/e of its initial value). The bandwidth of a photodiode can be limited by either its rise time and fall time or its RC time constant, whichever results in the slower speed or bandwidth. The bandwidth of a circuit limited by the RC time constant is

$$B = 1/2\pi RC, \qquad (3.7)$$

where R is the load resistance and C is the diode capacitance. Figure 3.1 shows the equivalent circuit model of a photodiode. It consists of a current source in parallel with a resistance and a capacitance. It appears as a low-pass filter, a resistor–capacitor network that passes low frequencies, and attenuates high frequencies. The cutoff frequency, which is the frequency that is attenuated at 3 dB, marks the 3-dB bandwidth. Photodiodes for high-speed operation must have a very low capacitance. The capacitance

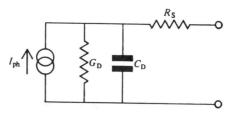

Fig. 3.1   Small-signal equivalent circuit for a reversed biased photodiode.

in a photodiode is mainly the junction capacitance formed at the p–n junction, as well as any capacitance contributed by the packaging.

Bias voltage refers to an external voltage applied to the detector and will be more fully described in the following section. Photodiodes require bias voltages ranging from as low as 0 V for some PIN photodiodes to several hundred volts for APDs. Bias voltage significantly affects operation because dark current, responsivity, and response time all increase with bias voltage. APDs are usually biased near their avalanche breakdown point to ensure fast response.

Active area and effective sensing area are exactly what they sound like: the size of the detecting surface of the detection element. The uniformity of response refers to the percentage change of the sensitivity across the active area. Operating temperature is the temperature range over which a detector is accurate and will not be damaged by being powered. However, there may be changes in sensitivity and dark current that must be taken into account; be sure to consult the manufacturer's specifications for a particular product. Storage temperature will have a considerably larger range; basically, it describes the temperature range under which the detector will not melt, freeze, or otherwise be damaged or lose its operating characteristics.

NEP stands for noise equivalent power. It is the amount of flux that would create a signal of the same strength as the rms detector noise. In other words, it is a measure of the minimum detectable signal; for this reason, it is the most commonly used version of the more generic figure of merit, noise equivalent detector input. More formally, it may be defined as the optical power (of a given wavelength or spectral content) required to produce a detector current equal to the rms noise in a unit bandwidth of 1 Hz:

$$\text{NEP}(\lambda) = i_n(\lambda)/R(\lambda), \tag{3.8}$$

where i_n is the rms noise current and R is the responsivity (defined previously). It can be shown [2] that to a good approximation,

$$\text{NEP} = 2\,h\,c/\text{QE}\,\lambda, \tag{3.9}$$

where this expression gives the NEP of an ideal diode when QE = 1. If the dark current is large, this expression may be approximated by

$$\text{NEP} = h\,c(2\,q\,I)^{1/2}/\text{QE}\,q\,\lambda, \tag{3.10}$$

where I is the detector current. Sometimes it is easier to work with detectivity, which is the reciprocal of NEP. The higher the detectivity, the smaller the signal a detector can measure; this is a convenient way to characterize more sensitive detectors. Detectivity and NEP vary with the inverse of the square of active area of the detector as well as with temperature, wavelength, modulation frequency, signal voltage, and bandwidth. For a photodiode detecting monochromatic light and dominated by dark current, detectivity is given by

$$D = \mathrm{QE}\ q\ \lambda/h\ c(2\ q\ I)^{1/2}. \tag{3.11}$$

The quantity-specific detectivity accounts for the fact that dark current is often proportional to detector area, A; it is defined by

$$D^* = D\ A^{1/2}. \tag{3.12}$$

Normalized detectivity is detectivity multiplied by the square root of the product of active area and bandwidth; this product is usually constant and allows comparison of different detector types independent of size and bandwidth limits. This is because most detector noise is white noise (Gaussian power spectra), and white noise power is proportional to the bandwidth of the detector electronics; thus, the noise signal is proportional to the square root of bandwidth. Also, note that electrical noise power is usually proportional to detector area and the voltage that provides a measure of that noise is proportional to the square root of power. Normalized detectivity is given by

$$D_\mathrm{n} = D(A\ B)^{1/2} = (A\ B)^{1/2}/\mathrm{NEP}, \tag{3.13}$$

where B is the bandwidth. The units are $(\mathrm{cm\ Hz})^{1/2}\ \mathrm{W}^{-1}$. Normalized detectivity is a function of wavelength and spectral responsivity; it is often quoted as normalized spectral responsivity.

Bandwidth, B, is the range of frequencies over which a particular instrument is designed to function within specified limits. Bandwidth is often adjusted to limit noise; in some specifications it is chosen as 1 Hz, so NEP is measured in W/Hz. Wide bandwidth detectors required in optical datacom often operate into a low resistance and require a minimal signal current much larger than the dark current; the load resistance, amplifer, and other noise sources can make the use of NEP, D, D^*, and D_n inappropriate for characterizing these applications.

Linearity range is the range of incident radiant flux over which the signal output is a linear function of the input. The lower limit of linearity is NEP, and the upper limit is saturation. Saturation is when the detector begins to form less signal output for the same increase of input flux. When a detector begins to saturate, it has reached the end of its linear range. Dyanamic range can be used to describe nonlinear detectors such as the human eye. Although datacom systems do not typically use filters on the detector elements, neutral density filters can be used to increase the dynamic range of a detector system by creating islands of linearity whose actual flux is determined by dividing output signals of the detector by the transmission of the filter. Without filtering, the dynamic range would be limited to the linear range of the detector, which would be less because the detector would saturate without the filter to limit the incident flux. The units of linear range are incident radiant flux or power (watts or irradiance). Measuring the response of a detector to flux is known as calibration. Some detectors can be self-calibrated, and others require manufacturer calibration. Calibration certificates are supplied by most manufacturers for fiber optic test instrumentation; they are dated and have certain time limits. The gain, also known as the amplification, is the ratio of electron–hole pairs generated per incident photon. Sometimes detector electronics allows the user to adjust the gain. Wiring and PIN output diagrams tell the user how to operate the equipment by schematically showing how to connect the input and output leads.

3.3.   PN Photodiode

Photodiode detectors used in data communications are solid-state devices; to understand their function, we must first describe semiconductor physics. For the interested reader, other introductory references to solid-state physics, semiconductors, and condensed matter are available [2]. In a solid-state device, the electron potential can be described in terms of conduction bands and valence bands rather than individual potential wells (Fig. 3.2). The highest energy level containing electrons is called the Fermi level. If a material is a conductor, the conduction and valence bands overlap and charge carriers (electrons or holes) flow freely; the material carries an electrical current. An insulator is a material for which there is a large enough gap between the conduction and valence bands to prohibit the flow of carriers; the Fermi level lies in the middle of the forbidden region between bands, called the bandgap. A semiconductor is a material for which the bandgap is small enough that carriers can be excited into the conduction band with some stimulus; the Fermi level lies at the edge of the valence band (if the majority of carriers are holes) or the edge of the conduction

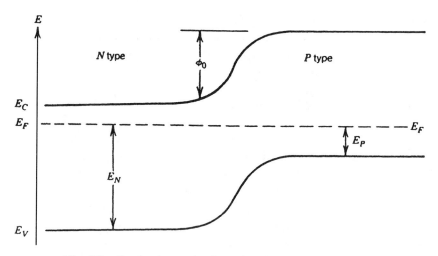

Fig. 3.2   Conduction and valance bands of a semiconductor.

band (if the majority of carriers are electrons). The first case is called a p-type semiconductor, and the second is called n-type. These materials are useful for optical detection because incident light can excite electrons across the bandgap and generate a photocurrent.

The simplest photodiode is the PN photodiode shown in Fig. 3.3. Although this type of detector is not widely used in fiber optics, it serves the purpose of illustrating the basic ideas of semiconductor photodetection. Other devices—the PIN and avalanche photodiodes—are designed to overcome the limitations of the PN diode. When the PN photodiode is reverse biased (negative battery terminal connected to p-type material), very little current flows. The applied electric field creates a depletion region on either side of the p–n junction. Carriers—free electrons and holes—leave the junction area. In other words, electrons migrate toward the negative terminal of the device and holes toward the positive terminal. Because the depletion region has no carriers, its resistance is very high, and most of the voltage drop occurs across the junction. As a result, electrical fields are high in this region and negligible elsewhere.

An incident photon absorbed by the diode gives a bound electron sufficient energy to move from the valence band to the conduction band, creating a free electron and a hole. If this creation of carriers occurs in the depletion region, the carriers quickly separate and drift rapidly toward their respective

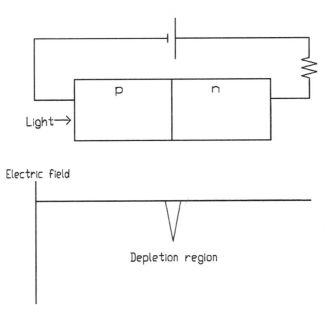

Fig. 3.3 PN diode.

regions. This movement sets an electron flowing as current in the external circuit. When the carriers reach the edge of the depletion region, where electrical fields are small, their movement, however, relies on diffusion mechanism. When electron–hole creation occurs outside of the depletion region, the carriers move slowly toward their respective regions. Many carriers recombine before reaching it. Their contribution to the total current is negligible. Those carriers remaining and reaching the depleted area are swiftly swept across the junction by the large electrical fields in the region to produce an external electrical current. This current, however, is delayed with respect to the absorption of the photon that created the carriers because of the initial slow movement of carriers toward their respective regions. Current, then, will continue to flow after the light is removed. This slow response, due to slow diffusion of carriers, is called slow tail response.

Two characteristics of the PN diode make it unsuitable for most fiber optic applications. First, because the depletion area is a relatively small portion of the diode's total volume, many of the absorbed photons do not contribute to the external current. The created free electrons and holes recombine before they contribute significantly to the external current. Second, the slow tail response from slow diffusion makes the diode too slow for high-speed applications. This slow response limits operations to the kilohertz range.

3.4. PIN Photodiode

The structure of the PIN diode is designed to overcome the deficiencies of its PN counterpart. The PIN diode is a photoconductive device formed from a sandwich of three layers of crystal, each layer with different band structures caused by adding impurities (doping) to the base material, usually indium gallium arsenide, silicon, or germanium. The layers are doped in this arrangement: p-type (or positive) on top, intrinsic (meaning undoped) in a thin middle layer, and n-type (or negative) on the bottom. For a silicon crystal a typical p-type impurity would be boron, and indium would be a p-type impurity for germanium [2–6]. Actually, the intrinsic layer may also be lightly doped, although not enough to make it either p type or n type. The change in potential at the interface has the effect of influencing the direction of current flow, creating a diode; obviously, the name PIN diode comes from the sandwich of p-type, intrinsic, and n-type layers.

The structure of a typical PIN photodiode is shown in Fig. 3.4. The p-type and n-type silicon form a potential at the intrinsic region; this potential gradient depletes the junction region of charge carriers, both electrons and holes, and results in the conduction band bending. The intrinsic region has no free carriers and thus exhibits high resistance. The junction drives holes into the p-type material and electrons into the n-type material. The difference in potential of the two materials determines the energy an electron must have to flow through the junction. When photons fall on the active area of the device, they generate carriers near the junction, resulting in a voltage difference between the p-type and n-type regions. If the diode is connected to external circuitry, a current will flow that is porportional to the illumination. The PIN diode structure addresses the main problem with PN diodes, namely, providing a large depletion region for the absorption of photons. There is a trade-off involved in the design of PIN diodes. Because most of the photons are absorbed in the intrinsic region, a thick intrinsic layer is desirable to improve photon–carrier conversion efficiency (to increase the probability of a photon being absorbed in the intrinsic region). On the other hand, a thin intrinsic region is desirable for high-speed devices because it reduces the transit time of photogenerated carriers. These two conditions must be balanced in the design of PIN diodes.

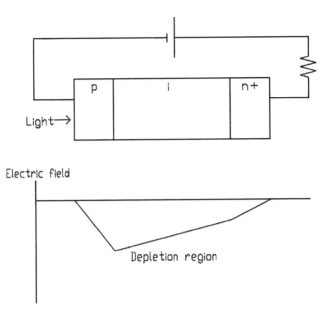

Fig. 3.4 PIN diode.

Photodiodes can be operated either with or without a bias voltage. Unbiased operation is called the photovoltaic mode; certain types of noise, including 1/f noise, are lower and the NEP is better at low frequencies. Signal-to-noise ratio is superior to the biased mode of operation for frequencies below approximately 100 kHz [6]. Biasing (connecting a voltage potential to the two sides of the junction) will sweep carriers out of the junction region faster and change the energy requirement for carrier generation to a limited extent. Biased operation (photoconductive mode) can be either forward or reverse bias. Reverse bias of the junction (positive potential connected to the *n* side and negative connected to the *p* side) reduces junction capacitance and improves response time; for this reason it is the preferred operation mode for pulsed detectors. A PIN diode used for photodetection may also be forward biased (the positive potential connected to the *p* side and the negative to the *n* side of the junction) to make the potential scaled for current to flow less, or in other words to increase the sensitivity of the detector (Fig. 3.5).

An advantage of the PIN structure is that the operating wavelength and voltage, diode capacitance, and frequency response may all be predetermined during the manufacturing process. For a diode whose intrinsic layer

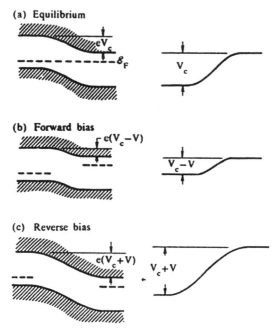

Fig. 3.5 Forward and reverse bias of a diode [2].

thickness is w with an applied bias voltage of V, the self-capacitance of the diode, C, approaches that of a parallel plate capacitor:

$$C = \varepsilon_o \, \varepsilon_1 \, A_o/w, \tag{3.14}$$

where A_o is the junction area, ε_o is the free space permittivity (8.849×10^{-12} F/m), and ε_1 is the relative permittivity. Taking typical values of $\varepsilon_1 = 12$, $w = 50$ μm, and $A_o = 10^{-7}$ m^2, then $C = 0.2$ pF. Quantum efficiencies of 0.8 or higher can be achieved at wavelengths of 0.8 or 0.9 μm, with dark currents less than 1 nA at room temperature. Some typical responsivities for common materials are given in Table 3.1.

The sensitivity of a PIN diode can vary widely by quality of manufacture. A typical PIN diode size ranges from 5 $\times$ 5 mm to 25 $\times$ 25 mm. Ideally, the detection surface will be uniformly sensitive (at the National Institute for Standards and Technology there is a detector profiler that, by using extremely well-focused light sources, can determine the sensitivity of a detector's surface [3]). For most applications, it is required that the detector

Table 3.1   **Common Semiconductor PIN Photodiode Properties**

Material	*Wavelength range (μm)*	*Peak responsivity (A/W)*
Silicon	0.3–1.1	0.5 A/W at 0.8 μm
Germanium	0.5–1.8	0.8 A/W at 0.7 μm
InGaAs	1.0–1.7	1.1 A/W at 1.7 μm

is uniformly illuminated or overfilled. The spectral responsivity of an uncorrected silicon photodiode is shown in Fig. 3.6. The typical QE curve is also shown for comparison. An ideal silicon detector would have zero responsivity and QE for photons whose energies are less than the bandgap or wavelengths much longer than approximately 1.1 μm. Just below the long wavelength limit, this ideal diode would have 100% QE and responsivity close to 1 A/W; responsivity vs wavelength would be expected to follow the intrinsic spectral response of the material. In practice, this does not happen; these detectors are less sensitive in the blue region, which can sometimes be enhanced by clever doping but not more than an order of magnitude. This lack of sensitivity is because there are fewer short wavelength photons per watt, so responsivity in terms of power drops off, and because more energetic blue photons may not be absorbed in the junction region. For color-sensitive applications such as photometry, filters are used so detectors will respond photometrically or to the standardized CIE color coordinates; however, the lack of overall sensitivity in the blue region can potentially create noise problems when measuring a low-intensity blue signal. In the deep ultraviolet, photons are often absorbed before they reach the sensitive region by detector windows or surface coatings on the semiconductor. The departure from 100% QE in real devices is typically due to Fresnel reflections from the detector surface. The long wavelength cutoff is more gradual than expected for an ideal device because the absorption coefficient decreases at long wavelengths so more photons pass though the photosensitive layers and do not contribute to the QE. As a result, QE tends to roll off gradually near the bandgap limit. This response is typical for silicon devices, which make excellent detectors in the wavelength range 0.8 or 0.9 μm. A common material for fiber optic applications is InGaAs, which is most sensitive in the near infrared (0.8–1.7 μm). Other PIN diode materials include HgCdZnTe for wavelengths of 2–12 μm. In the 1940s a popular photoconductive material for infrared solid-state detectors, lead

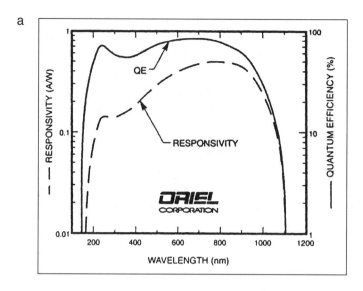

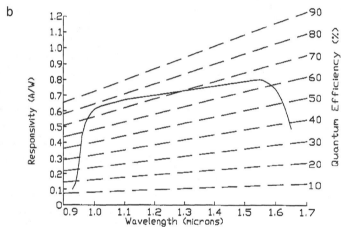

Fig. 3.6 PIN diode spectral response and quantum efficiency: (a) silicon; (b) InGaAs.

sulfide (PbS), was introduced; this material is still commonly used in the region from 1 to 4 μm [6].

PIN diode detectors are not very sensitive to temperature (-25–80°C) or shock and vibration, making them an ideal choice for a data communications transceiver. It is very important to keep the surface of any detector clean.

This becomes an issue with PIN diode detectors because they are sufficiently rugged than they can be brought into applications that expose them to contamination. Both transceivers and optical connectors should be cleaned regularly during use to avoid dust and dirt buildup.

A sample specification for a photodiode is given in Fig. 3.7. Bias voltage (V) is the voltage applied to a silicon photodiode to change the potential

Receiver Section						
Parameter	Symbol	Test Conditions	Min.	Typical	Max.	Units
Data rate (NRZ)	B	—	10	—	156	Mb/s
Sensitivity (avg)	P_{OH}	62.5 μm fiber .275 NA, BER $\leq 10^{-10}$	-32.5	—	-14.0	dBm
Optical wavelength	λ_{IN}	—	1270	—	1380	nm
Duty cycle	—	—	25	50	75	%
Output risetime	t_{TLH}	20 - 80% 50Ω to V_{cc}-2V	.5	—	2.5	ns
Output falltime	t_{THL}	80 - 20% 50Ω to V_{cc}-2V	.5	—	2.5	ns
Output voltage	V_{OL}	—	V_{cc}-1.025	—	V_{cc}-.88	V
	V_{OH}		V_{cc}-1.81	—	V_{cc}-.1.62	V
Signal detect	V_A	$P_{IN} > P_A$	V_{cc}-1.025	—	V_{cc}-.88	V
	V_D	$P_{IN} < P_D$	V_{cc}-1.81	—	V_{cc}-.1.62	V
P_{IN} power levels:						
Deassert	P_B	—	-39.0 or P_B	—	-32.5	dBm
Assert	P_A	—	-38.0	—	-30.0	dBm
Hysteresis	—	—	1.5	2.0	—	dB
Signal detect delay time:						
Deassert	—	—	—	—	50	μS
Assert	—	—	—	—	50	μS
Power supply voltage	V_{cc}-V_{EE}	—	4.75	5.0	5.25	V
Power supply current	I_{cc} or I_{EE}	—	—	—	150	mA
Operating temperature	T_A	—	0	—	70	°C
Absolute Maximum Ratings: Transceiver						
Parameter	**Symbol**	**Test Conditions**	**Min.**	**Typical**	**Max.**	**Units**
Storage temperature	—	—	-40	—	100	°C
Lead soldering limits	—	—	—	—	240/10	°C/s
Supply voltage	V_{cc}-V_{EE}	—	-.2	—	7.00	V

Fig. 3.7 Sample specifications for a PIN diode.

photoelectrons that must be scaled to become part of the signal. Bias voltage is basically a set operating characteristic in a prepackaged detector, in this case −24 V. Shunt resistance is the resistance of a silicon photodiode when not biased. Junction capacitance is the capacitance of a silicon photodiode when not biased. Breakdown voltage is the voltage applied as a bias that is large enough to create signal on its own. When this happens, the contribution of photoelectrons is minimal, so the detector cannot function. However, once the incorrect bias is removed the detector should return to normal.

A variation on the PIN diode structure is shown in Fig. 3.8. This is known as a Schottky barrier diode; the top layer of semiconductor material has been eliminated in favor of a reverse-biased, metal–semiconductor–metal contact. The metal layer must be thin enough to be transparent to incident light (approximately 10 nm); alternate structures using interdigital

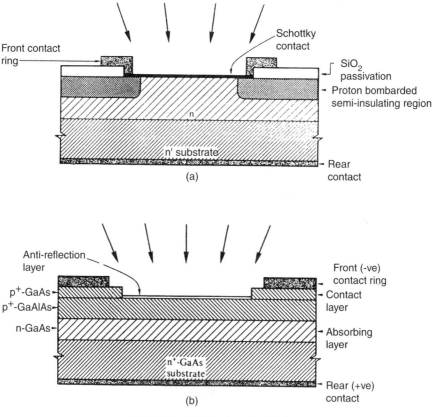

Fig. 3.8 Shottky barrier diode.

metal transducers are also possible. The advantage of this approach is improved quantum efficiency because there is no recombination of carriers in the surface layer before they can diffuse to either the Ohmic contacts or the depletion region. For wavelengths longer than approximately 0.8 μm, a heterojunction diode may be used for this same reason. Heterojunction diodes retain the PIN sandwich structure, but the surface layer is doped to have a wider bandgap and thus reduced absorption. In this case, absorption is strongest in the narrower bandgap region at the heterojunction, where the electric field is maximum; hence, good quantum efficiencies can be obtained. The most common material systems for heterojunction diodes are InGaAsP on an InP substrate or GaAlAsSb on a GaSb substrate.

A typical circuit for biased operation of a photodiode is shown in Fig. 3.9, where C_a and R_a represent the input impedance of a postamplifier. The photodiode impulse response will differ from an ideal square wave for several reasons, including transit time resulting from the drift of carriers across the depletion layer, delay caused by the diffusion of carriers generated outside the depletion layer, and the RC time constant of the diode and its load. If we return to the photodiode equivalent circuit of Fig. 3.1 and insert it into this typical bias circuit, we can make the approximation that R_s is much smaller than R_a to arrive at the equivalent small signal circuit model of Fig. 3.10. In this model the resistance R is approximately equal to R_a, and the capacitance C is the sum of the diode capacitance, amplifier capacitance, and some distributed stray capacitance. In response

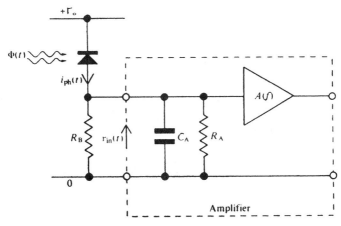

Fig. 3.9 PIN diode and amplifier equivalent circuit.

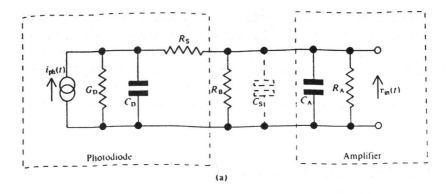

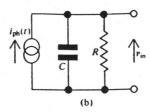

Fig. 3.10 Small-signal equivalent circuit for a normally biased photodiode and amplifier: (a) complete circuit; (b) reduced circuit obtained by neglecting R_s and lumping together the parallel components.

to an optical pulse falling on the detector, the load voltage V_{in} will rise and fall exponentially with a time constant RC. In response to a photocurrent I_p, which varies sinusoidally at the angular frequency,

$$\omega = 2 \pi f, \tag{3.15}$$

the response of the load voltage will be given by

$$V_{in}(f)/I_p(f) = R/(1 + j2 \pi f C R). \tag{3.16}$$

To obtain good high-frequency response, C must be kept as low as possible; as discussed earlier, the photodiode contribution can normally be kept well below 1 pF. There is ongoing debate concerning the best approach to improve high-frequency response; we must either reduce R or provide high-frequency equalization. As a rule of thumb, no equalization is needed if

$$R < 1/2 \pi C \Delta f, \tag{3.17}$$

where Δf is the frequency bandwidth of interest. We can also avoid the need for equalization by using a transimpedance feedback amplifier, which is often employed in commercial optical datacom receivers. These apparently simple receiver circuits can exhibit very complex behavior, and it is not always intuitive how to design the optimal detector circuit for a given application. A detailed analysis of receiver response, including the relative noise contributions and trade-offs between different types of photodiodes, is beyond the scope of this chapter; the interested reader is referred to several good references on this subject [4–12].

3.5. Avalanche Photodiode

PIN diodes can be used to detect light because when photon flux irradiates the junction, the light creates electron–hole pairs with their energy determined by the wavelength of the light. Current will flow if the energy is sufficient to scale the potential created by the PIN junction. This is known as the photovoltaic effect [2, 4–6]. It should be mentioned that the material from which the top layers of a PIN diode is constructed must be transparent (and clean!) to allow the passage of light to the junction. If the bias voltage is increased significantly, the photogenerated carriers have enough energy to start an avalanche process, knocking more electrons free from the lattice that contribute to amplification of the signal. This is known as an APD, it provides higher responsivity, especially in the near infrared, but also produces higher noise due to the electron avalanche process. For a PIN photodiode, each absorbed photon ideally creates one electron–hole pair, which in turn sets one electron flowing in the external circuit. In this sense, we can loosely compare it to an LED. There is basically a one-to-one relationship between photons and carriers and current. Extending this comparison allows us to say that an avalanche photodiode resembles a laser, where the relationship is not one to one. In a laser, a few primary carriers result in many emitted photons. In an APD, a few incident photons result in many carriers and appreciable external current.

The structure of the ADP, shown in Fig. 3.11, creates a very strong electrical field in a portion of the depletion region. Primary carriers—the free electrons and holes created by absorbed photons—within this field are accelerated by the field, thereby gaining several electron volts of kinetic energy. A collision of these fast carriers with neutral atoms causes the accelerated carrier to use some of its energy to raise a bound electron from the valence band to the conduction band. A free electron–hole pair is thus created. Carriers created in this way, through collision with a primary

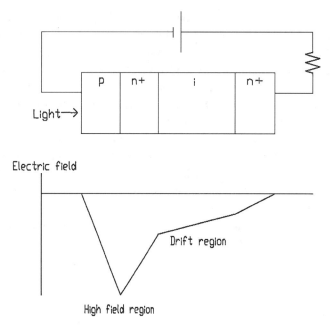

Fig. 3.11   Avalanche photodiode.

carrier, are called secondary carriers. This process of creating secondary carriers is known as collision ionization. A primary carrier can create several new secondary carriers, and secondary carriers themselves can accelerate and create new carriers. The whole process is called photomultiplication, which is a form of gain process. The number of electrons set flowing in the external circuit by each absorbed photon depends on the APD's multiplication factor. Typical multiplication factors range in the tens to hundreds. A multiplication factor of 50 means that, on the average, 50 external electrons flow for each photon. The phrase "on the average" is important. The multiplication factor is an average, a statistical mean. Each primary carrier created by a photon may create more or less secondary carriers and therefore external current. For an APD with a multiplication factor of 50, for example, any given primary carrier may actually create 44 secondary carriers or 53 secondary carriers. This variation is one source of noise that limits the sensitivity of a receiver using an APD.

The multiplication factor varies with the bias voltage. Because the accelerating forces must be strong enough to impart energies to the carriers, high bias voltages (several hundred volts in many cases) are required to create the high-field region. At lower voltages, the APD operates like a PIN diode and exhibits no internal gain. The avalanche breakdown voltage of an APD is the voltage at which collision ionization begins. An APD biased above the breakdown point will produce current in the absence of optical power. The voltage itself is sufficient to create carriers and cause collision ionization. The APD is often biased just below the breakdown point, so any optical power will create a fast response and strong output. The tradeoffs are that dark current (the current resulting from generation of electron–hole pairs even in the absence of absorbed photons) increases with bias voltage, and a high-voltage power supply is needed. Additionally, as one might expect, the avalanche breakdown process is temperature sensitive, and most APDs will require temperature compensation in datacom applications. For these reasons, APDs are not as commonly used in datacom applications as PIN diodes, despite the potentially greater sensitivity of the APD.

3.6. Noise

Any optical detection or communication system is subject to variuos types of noise. There can be noise in the signal, noise created by the detector, and noise in the electronics. A complete discussion of noise sources has already filled several good reference books; because this is a chapter on detectors, we will briefly discuss noise created by detectors. For a more complete discussion, the reader is referred to treatments by Dereniak and Crowe [6], who have categorized the major noise sources. The purpose of the detector is to create an electrical current in response to incident photons. It must accept highly attenuated optical energy and produce an electrical current. This current is usually feeble because of the low levels of optical power involved, often only in the order of nW. Subsequent stages of the receiver amplify and possibly reshape the signal from the detector. Noise is a serious problem that limits the detector's performance. Broadly speaking, noise is any electrical or optical energy apart from the signal itself. Although noise can and does occur in every part of a communication system, it is of greatest concern in the receiver input. This is because the receiver works with very weak signals that have been attenuated during transmission. Although very small compared to the signal levels in most circuits, the noise level is significant in relation to the weak detected signals. The same noise level in a transmitter is usually insignificant because signal levels are very strong in comparison. Indeed, the very limit of the diode's sensitivity is the noise. An optical signal that is too weak cannot be distinguished from the noise. To detect such a signal, we must either reduce the noise level or increase the power level of the signal. In the following sections, we will describe in detail several different noise sources; in practice, it is often assumed that the noise in a detection system has a constant frequency spectrum over the measurement range of interest—this is so-called white noise or Gaussian noise and is often a combination of the effects we will describe here.

3.6.1. NOISE AND AMPLIFICATION

The amplification stages of the receiver amplify both the signal and the noise. Some AC techniques, such as lock-in amplification, can help separate the signal from the noise. To illustrate the principle, we will digress briefly to describe the example of the lock-in amp even though it is not commonly used in datacom systems. Lock-in amplification, or chopping, is a technique used to limit noise, although it can also be used to detect DC signals that are deliberately encoded with a known modulation. A mechanical light chopper (which looks somewhat like a fan) is placed between the signal and the detector. The amplifier is connected to the chopper, and the amplifier "knows" from this reference signal when the signal to be detected is on and when it is off. As you can imagine, this makes for very precise background subtraction. However, that is not the only function the lock-in amplifier performs.

The secret of the lock-in amplifier is that it narrows the bandwidth of the detector, and the narrower that bandwidth, the more precise the measurement. It does this by creating a weak periodic signal and low-pass filtering it (or narrowing the bandwidth). This helps eliminate $1/f$ noise (flicker noise, drifts, etc.). The signal after lock-in amplification is a differential signal, not a linear signal. It is good for determining changes in signal, not signal magnitude. Horowitz and Hill [13] discuss lock-in amplification:

In order to illustrate the power of lock-in detection, we usually set up a small demonstration for our students. We use a lock-in to modulate a small LED of the kind used for panel indicators, with a modulation rate of a kilohertz or so. The current is very low, and you can hardly see the LED glowing in normal room light. Six feet away a phototransistor looks in the general direction of the LED, with its output fed to the lock-in. With the room lights out, there's a tiny signal from the phototransistor at the modulating frequency (mixed with plenty of noise), and the lock-in easily detects it, using a time constant of a few seconds. Then we turn the room lights on (fluorescent), at which point the signal from the phototransistor becomes just a huge messy 120 Hz waveform, jumping in amplitude by 50 dB or more. The situation looks hopeless on the oscilloscope, but the lock-in just sits there, unperturbed, calmly detecting the same LED signal at the same level. You can check that it's really working by sticking your hand in between the LED and detector. It's darned impressive. (p. 631).

Similar tricks to narrow the bandwidth are used in signal averaging, boxcar integration, multichannel scaling, pulse height analysis, and phase-sensitive detection. For example, boxcar averaging takes its name from gating the signal detection time into a repetitive train of N pulse intervals, or "boxes", during which the signal is present. Because noise that would have been accumulated during times when the gating is off is eliminated, this process improves the signal-to-noise ratio by a factor of the square root of N for white, Johnson, or shot noise. (This is because the integrated signal contribution increases as N, whereas the noise contribution increases only as the square root of N) [6]. Narrowing the bandwidth of a fiber optic receiver can also have beneficial effects in controlling relative intensity noise and modal noise (see Chapter 7). The use of differential signaling is also common in datacom receiver circuits, although they typically do not use lock-in amps but rather solid-state electronics such as operational amplifiers (see Chapter 16).

3.6.2 SHOT NOISE

Shot noise occurs in all types of radiation detectors. It is due to the quantum nature of photoelectrons. Because individual photoelectrons are created by absorbed photons at random intervals, the resulting signal has some variation with time. The variation of detector current with time appears as noise; this can be due to either the desired signal photons or by background flux (in the latter case, the detector is said to operate in a background limited in performance mode). To study the shot noise in a photodiode, we will consider the photodetection process. An optical signal and background radiation are absorbed by the photodiode, whereby electron–hole pairs are generated. These electrons and holes are then separated by the electric field and drift toward the opposite sides of the p–n junction. In the process, a displacement current is induced in the external load resister. The photo-current generated by the optical signal is I_p. The current generated by the background radiation is I_b. The current generated by the thermal generation of electron–hole pairs under completely dark environment is I_d. Because of the randomness of the generation of all these currents, they contribute shot noise given by a mean square current variation of

$$I_s^2 = 2q \, (I_p + I_b + I_d) \, B, \tag{3.18}$$

where q is the charge of an electron (1.6×10^{-19} C) and B is the bandwidth. The equation shows that shot noise increases with current and with bandwidth. Shot noise is at its minimum when only dark current exists, and it increases with the current resulting from optical input.

3.6.3. THERMAL NOISE

Also known as Johnson or Nyquist noise, thermal noise is caused by randomness in carrier generation and recombination due to thermal excitation in a conductor; it results in fluctuations in the detector's internal resistance or in any resistance in series with the detector. These resistances consist of the junction resistance (R_j), the series resistance (R_s), the load resistance (R_l), and the input resistance (R_i) of the amplifier. All the resistances contribute additional thermal noise to the system. The series resistance (R_s) is usually much smaller than the other resistance and can be neglected. The thermal noise is given by

$$I_t^2 = 4kTB(\, (1/R_j) + (1/R_l) + (1/R_i) \,), \tag{3.19}$$

where k is Boltzmann's constant (1.38×10^{-23} J/K), T is absolute temperature (Kelvin scale), and B is the bandwidth.

3.6.4. OTHER NOISE SOURCES

Generation recombination and $1/f$ noise are particular to photoconductors. Absorbed photons can produce both positive- and negative-charge carriers, some of which may recombine before being collected. Generation recombination noise is due to the randomness in the creation and cancellation of individual charge carriers. It can be shown [4] that the magnitude of this noise is given by

$$I_{gr} = 2 \, I \, (\, \tau B/N \, (1 +(2 \, \pi \, f\tau)^2) \,)^{1/2} = 2 \, q \, G \, (\varepsilon \, E \, A \, B)^{1/2}, \tag{3.20}$$

where I is the average current due to all sources of carriers (not just photocarriers), τ is the carrier lifetime, N is the total number of free carriers, f is the frequency at which the noise is measured, G is the photoconductive gain (number of electrons generated per photogenerated electron), E is the photon irradiance, and A is the detector active area.

Flicker or so-called $1/f$ noise is particular to biased conductors. Its cause is not well understood, but it is thought to be connected to the imperfect conductive contact at detector electrodes. It can be measured to follow a curve of $1/f^{\beta}$, where β is a constant that varies between 0.8 and 1.2; the rapid falloff with $1/f$ gives rise to the name. Lack of good Ohmic contact increases this noise, but it is not known if any particular type of electrical contact will eliminate this noise. The empirical expression for the noise current is

$$I_f = \alpha \, (i \, B/f^{\beta})^{1/2}, \qquad (3.21)$$

where α is a proprotionality constant, i is the current through the detector, and the exponents are empirically estimated to be $\alpha = 2$ and $\beta = 1$. Note that this is only an empirical expression for a poorly understood phenomena; the noise current does not become infinite as f approaches zero (DC operation).

There may be other noise sources in the detector circuitry as well; these can also be modeled as equivalent currents. The noise sources described here are uncorrelated and thus must be summed as rms values rather than a linear summation (they add in quadrature or use vector addition) so that the total noise is given by

$$I^2_{\text{tot}} = I^2_f + I^2_{\text{gr}} + I^2_s + I^2_t. \qquad (3.22)$$

Usually, one of the components in the previous equation will be the dominant noise source in a given application. When designing a data link, one must keep in mind likely sources of noise, their expected contribution, and how to best reduce them. Choose a detector so that the signal will be significantly larger than the detector's expected noise. In a laboratory environment, cooling some detectors will minimize the dark current; this is not practical in most applications. There are other rules of thumb that can be applied to specific detectors as well. For example, although Johnson noise cannot be eliminated, it can be minimized. In photodiodes, the shot noise is approximately three times greater than Johnson noise if the DC voltage generated through a transimpedance amplifier is more than approximately 500 mV. This results in a higher degree of linearity in the measurements while minimizing thermal noise. The same rule of thumb applies to PbS- and PbSe-based detectors, though care should be taken not to exceed the maximum bias voltage of these devices or catastrophic breakdown will occur. For the measured values of NEP, detectivity, and specific detectivity to be meaningful, the detector should be operating in a high-impedance mode so that the principal source of noise is the shot noise associated with the dark current and the signal current. Although it is possible to use electronic circuits to filter out some types of noise, it is better to have the signal much stronger than the noise by either having a strong signal level or a low noise level. Several types of noise are associated with the photodiode itself and with the receiver; for example, we have already mentioned multiplication noise in an APD, which arises because multiplication varies around a statistical mean.

3.6.5. *SIGNAL-TO-NOISE RATIO*

Signal-to-noise ratio (SNR) is a parameter of describing the quality of signals in a system. SNR is simply the ratio of the average signal power, S, to the average noise power. N, from all noise sources:

$$SNR = S/N. \tag{3.23}$$

SNR can also be written in decibels as

$$SNR = 10 \log_{10} (S/N). \tag{3.24}$$

If the signal power is 20 mW and the noise power is 20 nW, the SNR ratio is 1000, or 30 dB. A large SNR means that the signal is much larger than the noise. The signal power depends on the power of the incoming optical power. The specification for SNR is dependent on the application requirements.

For digital systems, bit error rate (BER) usually replaces SNR as a performance indicator of system quality. BER is the ratio of incorrectly transmitted bits to correctly transmitted bits. A ratio of 10^{-10} means that one wrong bit is received for every 10 billion bits transmitted. Similarly with SNR, the specification for BER is also dependent on the application requirements. SNR and BER are related. In an ideal system, a better SNR should also have a better BER. However, BER also depends on data-encoding formats and receiver designs. There are techniques to detect and correct bit errors. We cannot easily calculate the BER from the SNR because the relationship depends on several factors, including circuit design and bit error correction techniques. For a more complete overview of BER and other sources of error in a fiber optic data link, see Chapter 7.

3.7. Conclusions

In this chapter, we have introduced the fundamentals of fiber optic detectors for data communications. The PN diode was used to illustrate the basic detection mechanisms before proceeding to a description of the more practical PIN diode and avalanche photodiode. We have described the terminology needed to read a basic photodiode specification, including a description of common noise sources and figures of merit for photodiodes. Later chapters will discuss the design of receiver circuits, including such topics as phase-lock loops for clock recovery; we conclude with an example of how the detector and receiver design may dictate other properties of the datacom link.

A fiber optic receiver consists of many other components besides the photodiode, such as an integrated preamplifier and quantizer. The photodetector's signal-to-noise ratio and minimum detectable signal level are factors in the design of other receiver components. For example, most high-speed receivers compare the photodiode output signal with a running average of signal levels to set a threshold that determines whether a 1 or 0 was transmitted. However, this threshold can drift up or down if the incoming data depart from a 50% duty cycle (an equal number of 1's and 0's received). The quantizer may be designed to compensate for duty cycle distortion or pulse-width variation of the incoming signal. In addition, most data links specify an encoding technique for the data that affects the receiver's duty cycle. Manchester codes inherently produce a 50% duty cycle but require a baud rate twice that of the bit rate (a 100 Mbit/s data stream would require a 200-Mbaud transceiver). By contrast, nonreturn to zero (NRZ) encoding requires a lower bandwidth but can result in higher duty cycle variations. Because the long-term duty cycle depends on the encoding method, different applications will specify their own encoding rules. The FDDI standard, for example, uses NRZI and 4B/5B encoding to achieve a duty cycle between 40 and 60%. Fibre Channel Standard uses NRZ and 8B/10B encoding to ensure a duty cycle much closer to 50%, as does the ESCON/SBCON specification. These standards will be presented in more detail in other chapters.

ACKNOWLEDGMENT

The contributions of Bill Reysen, Amp Incorporated, Lytel Division, to the preparation of this chapter are gratefully acknowledged.

References

1. Sher DeCusatis, C. J. 1997. Fundamentals of detectors. In *Handbook of applied photometry,* ed. C. DeCusatis, 101–132. New York: AIP Press.
2. Burns, G. 1985. *Solid state physics,* Chapter 10, 323–324. Orlando, FL: Academic Press.
3. *The NBS photodetector spectral response calibration transfer program,* NBS Spec. Publ. No. SP250-17.
4. Lee, T. P., and T. Li. 1979. Photodetectors. In *Optical fiber communications,* eds. S. E. Miller and A. G. Chynoweth. New York: Academic Press.
5. Gowar, J. 1984. Optical communication systems. Englewood Cliffs, NJ: Prentice Hall.
6. Dereniak, E., and D. Crowe. 1984. *Optical radiation detectors.* New York: Wiley.
7. Graeme, J. 1994 (June 27). Divide and conquer noise in photodiode amplifiers. *Electron. Design,* 10–26.
8. Graeme, J. 1994 (November 7). Filtering cuts noise in photodiode amplifiers. *Electron. Design,* 9–22.
9. Graeme, J. 1987 (October 29). FET op amps convert photodiode outputs to usable signals. *Electron. Design,* 205.
10. Graeme, J. 1982 (May 7). Phase compensation otpimizes photodiode bandwidth. *Electron. Design,* 177.
11. Bell, D. A. 1985. *Noise and the solid state.* New York: Wiley.
12. Burt, R., and R. Stitt. 1988 (September 1). Circuit lowers photodiode amplifier noise. *Electron. Design,* 203.
13. Horowitz, P., and W. Hill, 1980. In *The art of electronics,* chapter 14, 628–631. Cambridge, UK: Cambridge University Press.

Chapter 4 | Logic and Drive Circuitry

Ray D. Sundstrom

Motorola, Incorporated, Chandler, Arizona 85248

Eric Maass

Motorola, Incorporated, Tempe, Arizona 85284

4.1. System Overview

In data communications applications, the data and address information is generally supplied through parallel lines. The parallel information can be directly converted, line for line, into parallel optical signals; alternatively, the parallel information can first be converted into a serial bit stream and then transmitted through a serial optical link.

The parallel optical approach is illustrated through a block diagram in Fig. 4.1a, and the serial optical approach is illustrated in Fig. 4.1b. In the parallel optoelectrical interface, illustrated in Fig. 4.1a, each parallel electrical line has a laser driver and laser associated with it. The light is sent through parallel channels, such as fiber ribbon composed of several optical fibers. The light from each channel is guided to a photo detector, and the signal from each photo detector is amplified through a transimpedance and postamplifier, one for each channel. The array of postamplifiers, including conversion to appropriate digital signal levels, provides a parallel electrical output similar to the parallel electrical lines that were the original inputs.

In the serial optical–electrical interface, illustrated in Fig. 4.1b, the parallel electrical lines are multiplexed to provide a serial signal output. The parallel to serial conversion also requires clock generation. The serialized electrical signal is provided through a laser or light-emitting diode (LED) driver circuit to the laser or LED. The driver circuit may involve feedback circuitry from the laser or LED to keep the drive current at an appropriate level for the laser or LED operation. The resulting serial optical signal is coupled to optical fiber or other light-transmitting medium.

115

HANDBOOK OF FIBER OPTIC
DATA COMMUNICATION

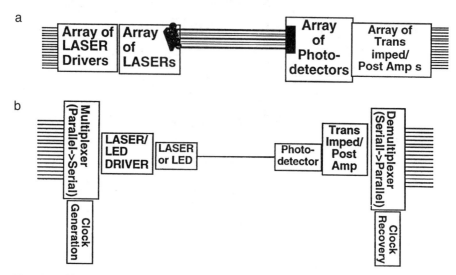

Fig. 4.1 (a) Block diagram for parallel optoelectrical interface, involving arrays of optoelectronic interfaces. (b) Block diagram for serial optoelectrical interface, involving parallel–serial–parallel conversions.

On the receive side, the serial optical signal is coupled to a single photo detector. The signal from the photo detector is generally a very small current, which is converted into a voltage through the transimpedance amplifier, and amplified with a post- or limiting amplifier. The resulting serial electrical signal must then be converted to a parallel signal through a demultiplexer. This serial–parallel conversion requires clock-recovery circuitry. The clock-recovery circuitry involves special requirements, such as recovering clock signals from a variety of bit patterns; these special requirements and the design considerations involved are described later in this chapter and elaborated on further in Chapter 16.

In Fig. 4.2a and 4.2b, the customer requirements associated with the parallel and serial fiber optics systems are listed and mapped to the design and process issues involved in meeting those requirements. The basic customer requirements of distance, frequency or data rate, cost, clean data, low power, and reliability are similar for both approaches. The clean data requirement involves additional jitter requirements associated with the clock-generation and clock-recovery circuitry in the serial approach.

The circuitry associated with these systems can be implemented in various semiconductor technologies, including GaAs, silicon bipolar, compli-

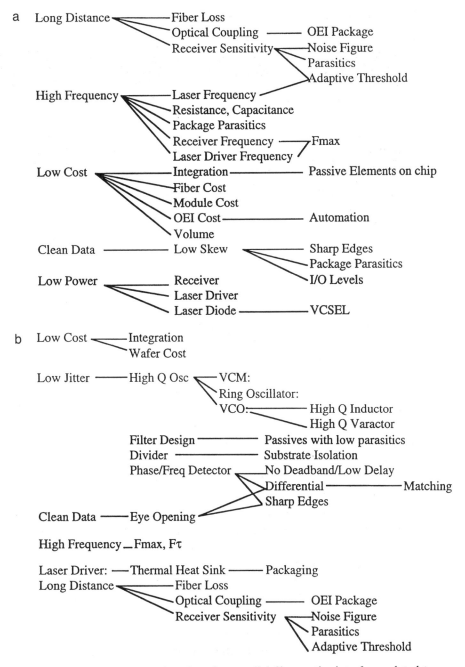

Fig. 4.2 (a) Customer requirements for parallel fiber optics interface related to design and process issues. (b) Customer requirements for serial fiber optics interface related to design and process issues.

mentary metal oxide semiconductor (CMOS), and BiCMOS. GaAs has a particular advantage for the transimpedance function on the receive side in that the photodiode can readily be integrated with this function. The high mobility associated with GaAs facilitates meeting the high-frequency requirements of any circuit in the system. The low transconductance of GaAs field effect transistors (FETs) is a disadvantage for the postamplifier function, generally requiring more amplifier stages. A major disadvantage of GaAs historically has been the high cost relative to silicon solutions.

Bipolar technology has advantages in high bandwidth and high transconductance, which is particularly helpful for the transimpedance function, and high voltage gain, which is helpful for the postamplifier function. The high-frequency performance of bipolar using emitter-coupled logic (ECL) design approaches is useful for the multiplexer and demultiplexer functions. The high-static power consumption associated with high-speed ECL is not an issue if data are expected to be continually transmitted and received. Bipolar integrated circuits tend to cost less than similar GaAs-integrated circuits but more than CMOS circuits.

Small geometry CMOS could be used for several if not all the functions and would be expected to provide the lowest cost solution. The relatively low transconductance of MOS transistors is a disadvantage for both the transimpedance amplifier and the postamplifier function. CMOS would also encounter disadvantages for high-speed, quadrature clock signals for the clock-generation and clock-recovery functions. Using differential techniques with CMOS for the multiplexer and demultiplexer functions has some advantages.

BiCMOS has an advantage at higher levels of integration because it has the flexibility associated with both bipolar and MOS transistors. The main disadvantage of BiCMOS is the relatively high cost and long cycle time due to the process complexity.

The purpose of an optoelectrical interface is to convert electrical information from voltages and currents to a stream of photons in the transmitter and back to voltages and currents in the receiver. For a digital data link, high and low voltage levels are converted to a higher or lower number of photons emitted.

A typical transmitter would include a driver circuit that accepts a standard logic input and converts a high (1) to some known high current and a low (0) to some known low current. This current switch is connected to an LED or Laser diode.

For LEDs, the low current is usually 0 mA, meaning that the diode is completely shut off and emits no light when it is low. The high current is set by the system designer and depends on the system requirements. The higher the current, the higher the light output of the LED. Of course, there is a maximum current limit for every device (beyond which the part's output or reliability will begin to degrade). A normal "on" current for LED's is in the range of 20–100 mA.

For lasers, the low current can also be 0, but it is usually some value large enough to keep the laser active. The advantage of keeping the laser active is that it reduces the turn-on delay of the device. Depending on device size and system requirements, the laser may switch from low = 1 mA to high = 5 mA or as great as low = 100 mA and high = 300 mA.

A typical receiver would include a p-type–intrinsic–n-type (PIN) diode or an avalanche photodiode (APD) to receive the photons and convert them into a proportional current. This current is fed into a transimpedance amplifier (TZA), which converts the current into a voltage.

A PIN diode is similar to a normal diode, in which a layer of *p*-type material forms a junction with *n*-type material, except for the addition of a layer of very lightly doped, or intrinsic material between the *p*-type and *n*-type layers. When the diode is reverse biased, a large depletion region is created. If a photon of sufficient energy is absorbed in the depletion region, an electron will be liberated to assist in current flow.

An APD performs a similar function as a PIN diode, except that the APD also performs the function of current amplification. There are more design issues to account for when designing with APDs, and this discussion will be limited to PIN diodes.

4.2. Electrical Interface

The interface between the electric components and the optical components may be any of the standard logic configurations or a novel interface that solves some particular system requirement such as power consumption, data rate, noise, signal skew, etc.

Transistor–transistor logic (TTL) and CMOS are very common interfaces that are easy to use and do not require termination if the technologies are not pushed to the limit of their bandwidth capabilities. For higher performance, TTL and CMOS can be transmitted over controlled impedances and terminated correctly to reduce problems with noise and reflections. The TTL interface can be used up to approximately 100 MHz, and low-voltage CMOS can be used up to approximately 250 MHz. However, the rail-to-rail voltage swings of these two logic families can consume much power and generate much noise at such high frequencies. Some CMOS designs now include low-voltage differential interfaces to alleviate the noise and power problems.

Differential ECL and CML interfaces are better suited for very high-performance electrical interconnect. These two logic configurations were designed to drive controlled impedance transmission lines and have very good noise immunity due to the differential operation. These two interfaces also have smaller voltage swings, in the range from 0.25 to 1.0 V, that can consume less power at high frequencies. The lower voltage swing also reduces electromagnetic interference. These logic interfaces have been used at data rates as high as 10 Gb/s.

Different parts of a system may be best suited to different logic technologies. In a high-speed serial link, the interface to the parallel data source may be CMOS. The parallel source is often a CMOS microprocessor running less than 200 MHz. A CMOS chip operating at a similar frequency can process the data, encode it, frame it, and prepare it for transmission. For high-speed links, the multiplexer chip is usually bipolar or GaAs to achieve data rates of 1 Gb/s or above. With a parallel 16-bit input at 100 mb/s, the serial output would have to be 1.6 Gb/s. If there is error detection and correction, such as extra parity checking, or framing information, the serial rate must be higher to include this extra information. The small geometries and dense circuit layout of CMOS make it best suited for the lower frequency parallel processing. The high bandwidth of bipolar and GaAs make these processes best suited for the high-frequency serial processing. BiCMOS processes, which combine bipolar and CMOS transistors on the same substrate, make it possible to combine the best technologies on a single chip and provide the designer with greater flexibility—at the expense of greater processing complexity, longer cycle times, and higher wafer cost.

4.3.   Optical Interface

The optical devices used in the transmitter can have a large effect on the system performance. All devices have parasitic properties that the circuit designer must take into account when designing an optical link. Light output of LEDs and lasers varies with temperature and age. Photodiode efficiency is affected by the reverse-bias voltage. Parasitics of the device packages and circuit board can play a major factor in system performance at high frequencies. Higher performance systems usually require that the driver and laser be packaged in a module to reduce lead inductance. The photodiode and TZA also need to be packaged in a module for the same reason.

On the receiver side, noise current at the input of the TZA is crucial in determining how small a signal the amplifier can reliably receive. It is a common practice to specify noise at the input of an amplifier. Input noise is determined by adding the contribution of each noise source in a circuit and calculating an equivalent value as if it were at the input. The noise must be about an order of magnitude less than the minimum input signal expected to be received; this ratio may vary, dependent on the level of errors that can be tolerated. The number of errors observed divided by the number of bits transmitted is commonly know as the bit error rate (BER). A BER of 10E-14 may be a reasonable goal.

There is always a compromise between bandwidth and input noise current when designing a TZA. The parasitic capacitance of the PIN diode is also one of the dominant factors in determining the bandwidth of the receiver. The diode capacitance and input impedence of the TZA often form the dominant pole in the receiver.

On the transmitter side, LEDs are very common and relatively inexpensive. They are usually turned on hard and shut off completely, such that the zero level coincides with no light emission. LEDs generally operate at lower frequencies than lasers and turn on relatively quickly for the frequencies used but shut off more slowly. Unless some method is used to actively turn off the device, shutting off the current results in a long decay in the optical output of the device as the parasitic capacitors discharge through the diode. If the temperature varies, the circuit must vary the current to maintain constant light output. As the device ages, perhaps accelerated by high temperature, the optical output is reduced. If a system has enough margin to allow for temperature and aging effects without compromising system performance, these effects can be left uncontrolled. However, if a higher performance system is required, active feedback circuitry that monitors the temperature or light output may have to be implemented to keep the optical signal at a constant amplitude.

Lasers are usually used for the highest performance systems. Edge-emitting lasers were developed first and have been used very successfully. Edge-emitting lasers cannot be tested in wafer form because the laser does not operate until mirror facets are created when the die is cleaved. The vertical cavity surface-emitting laser (VCSEL) is a promising new development. The mirror facets are developed as layers during wafer processing, and the VCSEL emits light from the top surface of the die, so it can be tested on the wafer. The cleaving operation is no longer a critical process.

Both laser devices have similar design challenges. Lasers have a long turn-on delay if shut off completely. To minimize this turn-on delay, lasers are generally prebiased with a DC bias current and modulated with a switching current. This configuration greatly enhances optical performance but complicates the design of the transmitter and receiver: There are two separate currents that need precise control for optimal performance. The objective is to bias the DC current precisely at the threshold current (where the laser begins to emit light). This allows the maximum difference between a high level and a low level for the least amount of power. Maintaining proper biasing of the laser is not a trivial task.

A. DRIVER CIRCUITRY

The optical driver can vary from a simple switch to a complicated feedback system accounting for many variables. The simple driver may consist of a CMOS or TTL output tied to a resistor and a LED or laser diode, as shown in Fig. 4.3. If system performance is not pushed to the limits, this might be the simplest and therefore perhaps the best solution.

Lasers can be switched on and off, but their performance increases significantly if they are not shut off completely. The performance can be improved by providing both a DC bias current and a switching or modulation current. A differential gate can provide a controlled current source

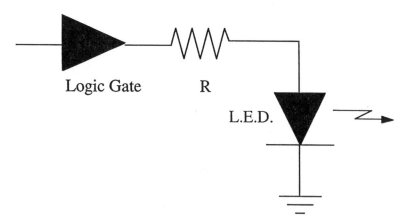

Fig. 4.3 Diagram of simple output.

for modulation and another controlled current source continuously passed through the laser diode for the bias current.

Bias current is a constant current flowing through the laser. The bias current should be set at the threshold current—the point at which the laser just starts to emit light, as illustrated in Fig. 4.4. If the laser is biased below the threshold current, the turn-on delay will increase. If it is biased above the threshold current, the difference between the light emission corresponding to a 0 and a 1 will be decreased. It is important to maximize the difference between the light emission corresponding to a 0 and a 1 because this is one of the factors that determine how far the link can transmit—the larger the difference in the light signals for a 0 and a 1, the larger the signal will be at the receiver.

Figure 4.5 is a transistor-level diagram of an output driver that implements the bias and modulation currents for the laser. A bias control voltage is forced on the base of Q_1, which puts a voltage across R_1. The voltage across R_1 is the bias control voltage minus the base-emitter voltage of Q_1. The current in the emitter of Q_1 is the voltage across R_1 divided by the resistance of R_1. This current minus the base current is the current in the collector of Q_1, which will always flow through the laser diode. This is one method of obtaining the laser bias current. The bias control voltage can be either fixed or controlled by some other monitor circuit.

The modulation current is controlled in a similar fashion by the modulation current control voltage. The modulation current, however, is not always directed through the laser diode. Q_3 and Q_4 make a differential gate that

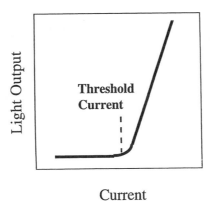

Current

Fig. 4.4 Current vs light curve.

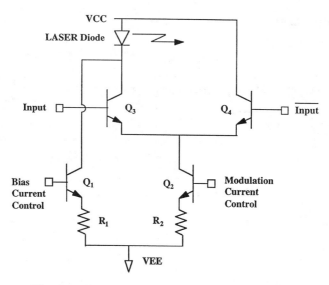

Fig. 4.5 Simplified diagram of laser output driver.

determines whether the current flows through the laser diode. If the voltage on the base of Q_3 is higher than the voltage on the base of Q_4, the current will flow through the laser diode, through Q_3, and into the current source created by Q_2 and R_2. Because the modulation current is flowing through the laser, the optical output is increased corresponding to an optical high level. Alternatively, if the voltage on the base of Q_4 is high and that on Q_3 is low, the current will flow from the power supply through Q_4 and into Q_2. Because the current is not flowing through the laser, the optical output is decreased, corresponding to an optical low state.

As data rates and transmission distances increase, the output of the laser diode needs to be more tightly controlled. Because the laser output power can vary with temperature and lifetime, some method of monitoring the light output and for feedback to the driver is generally incorporated in higher performance systems.

With edge-emitting lasers that emit light from both the front and the back facets, a photodetector or monitor diode is commonly mounted on the back side of the laser (Fig. 4.6). The signal from the monitor diode determines the bias and modulation current adjustments in a closed-loop feedback system.

**Photo Diode
for Monitor**

Edge Emitting Laser

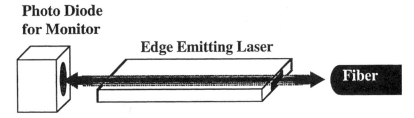

Fig. 4.6 Diagram of laser with monitor diode.

B. RECEIVER CIRCUITRY

The optical reciever design can be virtually an art. The challenge involves trade-offs in improving some parameters at the expense of other parameters. The design is optimized by appropriate compromise so that the receiver has the best combination of high sensitivity, correct bandwidth, low noise, large dynamic range, and low power.

Until direct processing of optical information becomes a technology, optical information must be converted to electrical information for processing. The conversion commonly uses a PIN diode (Fig. 4.7). The intrinsic layer between the p-type and n-type layers allows the formation of a very large depletion region when a reverse DC bias is applied. Photons with sufficient energy are absorbed in the depletion region, liberating electrons. The electric field from the reverse bias on the diode sweeps the liberated electrons across the junction, providing a current proportional to the quantity of photons absorbed. The large depletion region of the PIN diode

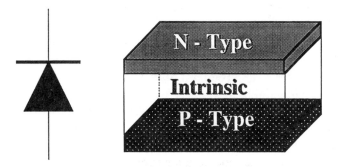

Fig. 4.7 Diagram of PIN diode.

improves the probability of absorbing photons. Intrinsic region thicknesses of approximately 4–15 μm are common.

As alternatives to PIN photodiodes, avalanche photodiodes and photo transistors can also be used for the optical to electrical signal conversion. The mechanism of these devices is somewhat different from that of the PIN diode, although the output remains a current proportional to the light intensity. The main difference is that these two devices also amplify the current.

Optical conversion devices have specifications that describe their performance, limitations, and parasitics. Conversion efficiency is the current produced for a given amount of light. A conversion efficiency of 0.5 A/W means that a PIN diode would produce 0.5 A of current when receiving 1 W of light. Generally, light received is in the mW or μW range, so the output current is in the mA or μA range. Large photodiodes facilitate alignment to the fiber but have more parasitic capacitance. This parasitic capacitance can limit the frequency because it must be charged and discharged by the current produced by the photodiode. As with most high-frequency devices, capacitances should be minimized.

The output from the PIN diode, current proportional to the intensity of light shining on it, is coupled into a TZA. The purpose of the TZA is to convert the current to a voltage. Conceptually, the TZA is an operational amplifier (op amp) with a negative feedback resistor. In the ideal case, the transimpedance is the value of the feedback resistor (measured in ohms).

If the PIN diode in Fig. 4.8 has a conversion efficiency of 0.5 A/W, and the incoming light is 1 mW, it will produce 500 μA of current. The op amp with a 1 kohm transimpedance resistance will convert the 500 μA to produce a voltage of 500 mV. Similarly, if the transimpedance resistance were increased to 10 kohm, the output would be increased to 5 V.

The two most important specifications for a TZA are the bandwidth and input noise level, which involve a trade-off: Increasing bandwidth will increase noise. The key is to first choose the largest transimpedance possible to achieve the bandwidth required because the larger the transimpedance, the lower the input noise of the receiver. Of course, the larger transimpedance increases the input impedance, which reduces bandwidth. The amplifier design should use minimal current because diode shot noise increases with current. Of course, the bandwidth of the amplifier is lowered as the current is reduced.

The range from the smallest detectable signal to the largest allowable signal is called the dynamic range. The largest allowable signal is the maxi-

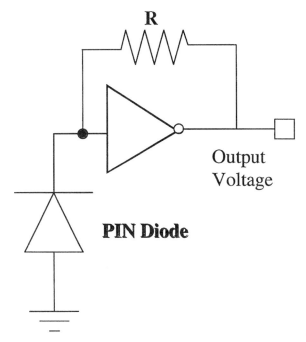

Fig. 4.8   Diagram of PIN diode and TZA.

mum that can be received without saturating the circuitry. The smallest detectable signal is determined by the input noise of the TZA. For example, if the input noise of the receiver integrated over the bandwidth is 250 nA, the smallest detectable signal might be 10 times that current, or 2.5 μA. If the photo detector conversion efficiency is 0.5 A/W, the minimum difference between a low level and a high level would be 2.5 μA/(0.5 A/W) = 5 μW. If this same receiver had a dynamic range spec of 30 dB, the maximum input would be 5 μW(1000) = 5 mW. A 5-mW input corresponds to an input current of 2.5 mA. If the part operates at 5V, the maximum transimpedence gain is 2 kohm (2.5 mA $\times$ 2 kohm = 5 V).

Often, a larger transimpedence is desired to provide more gain at low-input currents and to reduce noise. To achieve this, some method of gain compression can be used at high-input currents. Power supply noise rejection and substrate noise feedback are also extremely important for amplifiers with high gain and high bandwidth. If these are not properly controlled, the part will likely oscillate at low-input signal levels—perhaps being renamed the "transimpedance oscillator."

Figure 4.9 is a diagram of a simple transimpedance amplifier and PIN diode. When light input to the PIN diode increases, base drive is conducted away from Q_1. This starts to shut off Q_1, reducing current I_1 so the voltage across R_1 decreases, which causes the voltage on the base of Q_2 to rise. Transistor Q_2 is an emitter follower that increases current drive capability. When the base of Q_2 rises, the output voltage also rises. When the output voltage rises, the current through R_2 flows into the PIN diode. The voltage change on the output will be approximately equal to the current change in the PIN diode multiplied by the resistance of R_2. This is the voltage change required across R_2 to replace the current being conducted by the PIN diode. Note in this diagram that the PIN diode is biased to a negative voltage. This is necessary in this configuration because the base of Q_1 will be biased to approximately 0.8 V, and the PIN diode generally requires greater than 3 V of reverse bias to operate properly.

If the signal is transmitted over long distances, the optical signal is attenuated and the receiver may receive an extremely small signal. The attenuation is due to fiber loss and coupling losses at the interface between the optical components and fiber. The output voltage from the TZA must be further amplified and compared to a threshold voltage to determine if it is high or low. This is a difficult challenge, particularly if the signal has

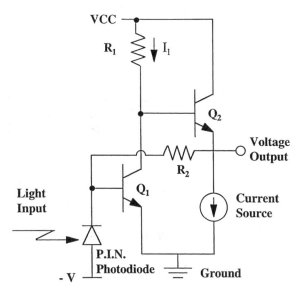

Fig. 4.9   Diagram of TZA and PIN diode.

a substantial difference in the quantity of zeros and ones. If the data are encoded for an average 50% duty cycle, an analog feedback loop similar to that shown in Fig. 4.10 can be used to set the threshold.

If the input data are known to have a 50% duty cycle, output and $\overline{\text{output}}$ (OUT and OUTB) should have an average voltage in the center of the output swing. The resistors and capacitors average the output swings. If the threshold voltage is too high, OUT will tend to be low more than 50% of the time, whereas OUTB will tend to be high more than 50% of the time. The operational amplifier will sense that the noninverting input is lower than the inverting input, and the output will decrease. This will lower the threshold voltage for comparison to the TZA output by the postamplifier. This negative feedback loop will correct the threshold voltage until the outputs have a 50% duty cycle.

If the input data do not have a 50% duty cycle, the feedback circuit will still behave as if it is driving the outputs to a 50% duty cycle, and the threshold will be incorrectly set. Setting the comparison threshold is very challenging if the data are not encoded to maintain a 50% duty cycle.

A low-pass filter is often placed between the TZA and postamp. For a 1-GHz bandwidth, the TZA needs a nominal bandwidth of approximately 1.2 GHz to allow for a 20% process variation. A lot with processing such that the devices are unusually fast may have a bandwidth of 1.4 GHz, and another lot with slower devices may just meet the 1-GHz requirement.

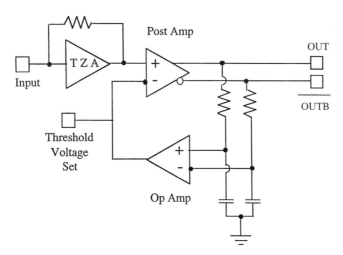

Fig. 4.10 Analog feedback loop.

The "fast" lot will have higher noise when integrated over the 1.4-GHz bandwidth. A filter can be constructed with tight-tolerance discrete components to cut off the bandwidth at 1 GHz, eliminating the extra noise contribution from the excess bandwidth.

Once the threshold is correctly set, the differential data on the outputs of the postamp are the digital equivalent of the data at the input of the laser driver. If this is one channel of a parallel system (as in Fig. 4.1a) the data are ready to use. If it is the output of a serial link (Fig. 4.1b), a clock signal has to be recovered for clocking in the incoming data, and the data need to be demultiplexed (converted from serial to parallel) to recover the parallel data that were present at the transmitter. Because all the data transitions occur at regular intervals, a phase-locked loop (PLL) can be locked to the edges for clock recovery. Once the PLL is locked to the incoming data, a clock is available that is synchronous with the data so that it can be used to clock the data stream into a register. If those data are shifted serially in a shift register and read out at the regular intervals of the clock (every eight clock cycles for an eight-bit system), the data can be converted back to a parallel data stream.

The design of the optoelectrical interface is a challenge of compromises dictated by customer and system requirements. Equations, models, and simulators are absolute necessities. The layout of high-frequency optoelectronics is something of an art, and the designer gains skill through high-frequency circuit design/layout experience. Time spent in the lab troubleshooting circuits and observing the principles first hand gives the designer insight that cannot be learned from books or simulators.

In Chapter 16, some of the design issues and approaches are covered in further detail.

Chapter 5 | Optical Subassemblies

Michael Langenwalter
Georg Jeiter
Siemens Corporation, Semiconductor Division-HL, Berlin, Germany

George DeMario
Siemens Corporation, Semiconductor Division-HL, Wappingers Falls, New York 12590

5.1. Overview of the Optical Subassembly

Key components in a fiber optic link are the electrooptic active elements that transform electrical energy into optical energy and optical energy to electrical energy. These active elements are the light-emitting diodes (LEDS or IREDs in the case of infrared radiation), laser diodes (LDs), and photodiodes (PD). The most serious challenge for proper operation of any fiber optic system is coupling light energy with the highest efficiency possible into a fiber on the transceiver side and to couple from a fiber into PD on the receiver side. Due to fundamental physical laws and design principles for such diodes, a high coupling efficiency is normally achievable with an optical beam transformation system, e.g., lenses (see Sections 5.2 and 5.4). To achieve optimum performance, the optical system must take into account the diode parameters as well as the basic properties of the fiber, such as its core diameter and numerical aperture. Finally, the design of a system with its subelements is needed that ensures precise, reliable alignment, that is cost-effective, and that allows for high manufacturing yield. This system is called the optical subassembly (OSA).

The OSA (also called the coupling unit) contains at least a diode on a header as the electrooptical transformer, a lens or a system of lenses as optical beam transformer, and a precise mechanical port for docking a fiber/ferrule assembly. There are various existing designs for OSAs with different technical solutions for design, alignment mechanism, and mount-

131

HANDBOOK OF FIBER OPTIC
DATA COMMUNICATION

ing. In general, however, all contain at least the previously mentioned three key subelements.

The OSA itself is a complete, separately operating unit and therefore testable at the subassembly level. After successful testing, OSAs may be assembled into fiber optic transceiver modules or they may be used separately in fiber optic receptacles if enough robustness has been designed in. Commonly available fiber optic receptacles mating with different single fiber optic connectors typically contain simply OSAs.

5.2.   Description of the OSA—Principal Considerations and Typical Examples

5.2.1.   COUPLING LIGHT FROM A SURFACE-EMITTING DIODE INTO A FIBER

The purpose of an OSA is to transfer light from a LED, IRED, or laser to a fiber or from a fiber into a light-detecting element such as a photodiode. The overall design of an OSA is dependent on the emitter or detector design, the operation wavelength, and the type of fiber in the link.

Principally all photoemitters generate a more or less divergent optical beam. For example, an IRED as a typical Lambertian source emits its radiation inside the semiconductor chip omnidirectionally. Due to limitations of the refractive index of the chip's bulk material, only a small amount of the total radiated emission leaves the diode surface. For example, in the case of 1300-nm wavelength, refractive index n is equal to 3.22 for indium phosphide. Only approximately 10% of total generated radiation will be emitted in the case of a flat and antireflective coated surface. The divergent beam has a total opening angle of approximately 36°, limited by total reflection.

An optical system is used to collimate this divergent beam and to transform it into a convergent one for effective coupling. A geometrically simple optical system consists of one or more ball lenses made of, for example, glass or sapphire that are highly transparent for the operating wavelength. A typical optical system for coupling light from an IRED into fiber is schematically shown in Fig. 5.1. The first microlens directly in front of the IRED will gather as much light as possible that emerges from IRED's surface and collimate it into a ball lens. This ball lens then focuses the light into the fiber. A graded index (GRIN) lens may be used as an alternative to the ball lens. This is shown schematically in Fig. 5.2. During design of such an optical system, the following parameters should be taken into account:

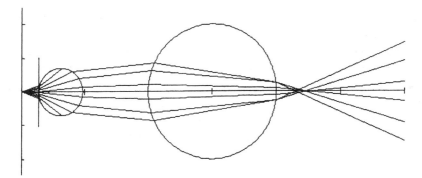

Fig. 5.1 Typical optical system for an IRED with two ball lenses.

Refractive index of the emitter chip material

Dimensions of the chip, especially the position and dimension of the emitting area

Dimensions and distances of the lenses to the chip or fiber, respectively

Refractive indices of the lenses

Focal point positions (GRIN lens)

Core diameter and numerical aperture of the fiber

Boundary conditions

These parameters are basic inputs for calculations by the commonly known technique called ray trace modeling, which typically incorporates the following two modeling simplifications:

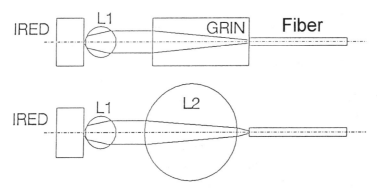

Fig. 5.2 Two lens coupling with GRIN or ball lens for second lens.

1. Linear optical behavior of all elements
2. Axial beams

Such modeling results in geometrical dimensions, aligning tolerances, coupling efficiencies, and other design characteristics depending on the method and the single task. An example of a calculated result from ray trace modeling is shown without values in Fig. 5.3.

The principle of weak focusing behavior of simple spherical lenses is also demonstrated in Fig. 5.3. They are generating somewhat like a beam waist—the so-called "beam caustic"—instead of a narrowly defined focal point. However, for coupling light into multimode fibers this minor weakness is not critical compared to other parameters. Especially in the case of a ball lens, this minor disadvantage is mostly compensated for by its simplicity, availability, and inexpensive price compared to a GRIN lens.

An additional example is the so-called coupling curve shown in Fig. 5.4. The slope of the curve directly indicates tolerances necessary for proper alignment of an OSA. For instance, in the direction perpendicular to the optical axis, a misalignment Δx in the order of 5 μm causes a coupling loss of approximately 20% or 1 dB. In comparison, a misalignment Δz parallel to optical axis is less sensitive by a factor of approximately 10.

It is commonly known that any result of modeling inspires confidence only when the model itself fits the question asked and all the inputs are accurately known. In most cases one should verify the results of beam modeling by experiments, e.g., on an optical bench with actual components.

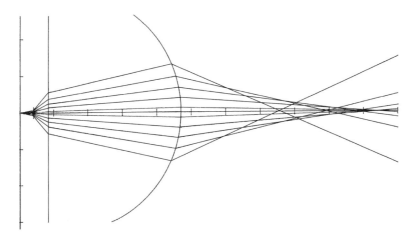

Fig. 5.3 Example of a RAYTRACE modeling result.

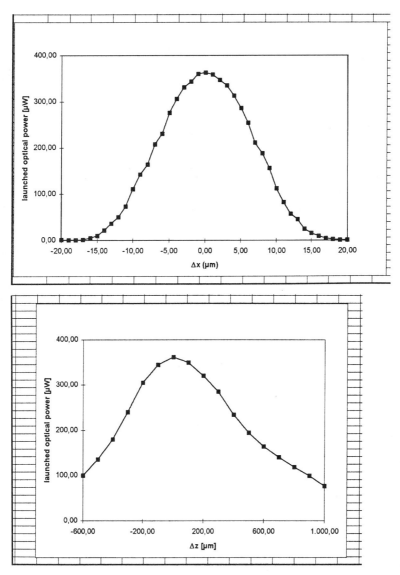

Fig. 5.4 Typical coupling curve for IRED to fiber with two ball lenses: $\Delta x =$ lateral; $\Delta z =$ axial misalignment of IRED with first lens to second lens. Graded index fiber, 62.5/125 μm; IRED, $\lambda = 1300$ nm.

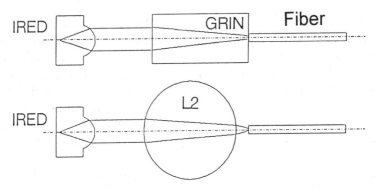

Fig. 5.5 Optical system for an IRED with integrated lens.

The previously mentioned microlens in front of the IRED may be directly replaced by a lens integrated on the chip itself (Fig. 5.5). This lens will be processed on the wafer base along with all other processing steps necessary for IRED wafers. Thousands of lenses are processed simultaneously and precisely aligned to their IREDs. Therefore, the integrated lens is simply saving mechanical parts and critical aligning, assembling, and testing steps that otherwise must be performed on each IRED. The integrated lens also significantly increases the amount of radiation emerging from the chip and allows, in principle, higher coupling efficiency. If a manufacturer masters this tricky processing technology with high and stable yield, the result will be a highly competitive process and product (see Chapter 18).

In contrast to refractive lenses described previously, diffractive lenses are useful for optical beam transformation. Their function is based on the physical effect of interference of light-penetrating slit-like structures. Such structures can be preciously and repeatedly generated by means of photo-lithographical techniques that are well established worldwide in IC found-ries. Performance, availability, and price will also be decisive for success.

5.2.2. COUPLING LIGHT FROM A LASER DIODE INTO FIBER

5.2.2.1. Edge-Emitting Laser Diode

A classical edge-emitting laser diode (EELD) is a dot-like source for a mostly elliptically profiled, highly divergent beam. Depending on the type and design of the EELD, the size of the beam exit window is on the order of $0.5 \times 10 \ \mu m$ and the beam divergence can reach almost 60° (total beam opening angle). For high coupling efficiency the same rules apply as those for an IRED, but the requested precision dramatically increases. For example, the first lens should be aligned within a tolerance perpendicular to the optical axis on the order of 0.1 μm (=100 nm!) for 1 dB maximum loss of coupling efficiency. Consequently, the total design of an LD OSA must consider this tolerance as well as the fact that in almost all cases a single-mode fiber core with a 9-μm diameter is the target for the coupled beam. Also, one should take into account the critical sensitivity of an EELD against back-reflected light (see Chapter 6). Otherwise, even minimal external thermal or mechanical effects may detrimentally influence the OSA's performance.

During the past few years different techniques for coupling light from an EELD into fibers have been investigated and implemented in laser transmitter modules. Especially in the field of telecommunications, in which fiber pigtails are used on modules instead of connectors (as in datacom applications), coupling via a tapered fiber with a microlens on the tip is a commonly used technique that offers high coupling efficiency. Systems with spherical, aspherical, or graded index lenses are used as well as spread-beam systems combined with Faraday insulators preventing back-reflection effects. Each of these has its specific advantages and disadvantages depending on the application.

5.2.2.2.   Vertical Cavity Surface-Emitting Laser Diode

Historically, vertical cavity surface-emitting laser diodes (VCSELs) with approximately 980-nm wavelengths were developed as pumping sources for erbium-doped fiber amplifiers. Compared to an EELD, a VCSEL is a laser with some important characteristics that predestine it to potentially be the universally predominant emitter in the future. Table 5.1 lists some of the important properties of VCSELs.

Regarding OSAs, the small beam divergence in combination with the circular-diameter definable beam is key for cost-effective solutions in the future (see Chapter 6). Worldwide, there are only a few groups working on VCSELs and much more research is required to create inexpensive and reliable VCSELs that can operate in the commonly used systems wavelengths. Especially the second optical window (~1300 nm) and the third optical window (~1550 nm) provide strong development challenges.

5.2.3.   COUPLING LIGHT FROM FIBER INTO A PHOTODETECTOR

The fundamental rules of geometric optics are independent of the direction of the light propagation. Therefore, the considerations for emitters are also valid for detectors. However, practical optical systems for PDs or receivers

Table 5.1   **Comparison of Features, EELD vs VCSEL**

Feature	EELD	VCSEL
Size of active area	0.5–1 × 2–10 μm	Variable, 5–50-μm diameter
Number of modes	Typically 1 or few	1 or even up to many 10s
Beam divergence	Elliptic up to 60 × 20°	Cicular, ca. 5°
Processing of chip	Very specific	Like LEDs
Final processing	Single bar	On wafer
Burn-in and test	Single on heatsink	On wafer
Environmental sensitivity	Extremely high	Moderate
Coupling to fiber	Tricky and sensitive	Easy
Threshold current	Approximately 10 mA	Some mA
Direct modulation bandwidth	High, up to 10 Gb	High
Temperature drift	Fairly high	Tendencially low

are less complicated compared to those for transmitters. The fiber itself operates as a source having limited beam diameter and divergence determined by its core diameter and numerical aperture. Therefore, typically a single focusing lens is sufficient for collimating the beam onto the detector's surface. A typical optical setup for a detector is shown in Fig. 5.6.

The design of the PD chip is an important factor for the receiver's performance. Its upper speed limit is generally determined by the size of its dynamically photosensitive area. The smaller the size of this area, the faster the detector. On the other hand, the precision required for efficient coupling will be higher when the size is smaller. This is because the full dynamic sensitivity of a PD can only be achieved by coupling light into the dynamic sensitive area and nearly no light outside of this area. Even outside the dynamic area, a PD is light sensitive, but it exhibits a low-pass filter characteristic. Depending on the design of the PD, the light coupled to the outside area may noticeably reduce the speed of the PD.

A good compromise between sensitivity and speed as well as alignment precision must be found. This compromise will place the alignment tolerance in the same order or even slightly wider as that for emitters.

In the case of large core fibers and relatively slow links, one can couple light into a PD without any additional lens by so-called butt coupling. The most important geometric boundary condition for high coupling efficiency in this case is aligning and fixing the fiber directly in front of the PD's entrance window.

5.3.   Packaging Considerations

Section 5.2 discussed the need for precision alignment as a basic require-ment for a well-designed OSA. This precision should be maintained and guaranteed during the lifetime and under all mechanical or other environ-mental conditions. All materials for individual parts as well as assembly processes should be selected and tested carefully to achieve this goal. Selec-tion and testing may be performed by so-called "harsh tests" in which the basic stress of operation is accelerated with temperature, humidity, or both. These tests are also applicable for determining absolute maximum stress limits of a material and a manufacturing process.

Additionally, all photoactive semiconductors, such as LEDs, PDs, and in particular LDs, are extremely sensitive to any ionic impurity and/or humidity in their vicinity. Typically, the board assembly process, e.g., wave soldering, is the most stressful for a transceiver module during its total lifetime due to the presence of aggressive chemical agents, high tempera-ture, and sudden temperature changes.

A common way to protect the critical elements is by a hermetic seal—in an environment that will demonstrate a helium leak rate less than 5×10^{-8} hPa cm^3/s. That is usually done by a transistor outline (TO) can with a lens or a window cap such as shown in Fig. 5.6. Of course, no organic material, such as epoxies, may be applied inside this housing.

The completed OSA shown in Fig. 5.7 contains an IRED or PD on a submount that is soldered to the header. Each diode has in front a plano-convex-type etched-silicon lens aligned and fixed in advance by soldering. A window cap is resistance welded hermetically to the header under dry inert atmosphere such as a mixture of 10% helium and 90% nitrogen with a dew point below −50°C. The final assembly of the OSA is performed by adjustment of the diode subunit to the module nose and fixing the optimal alignment position by laser spot welds. The edge of the window cap's front side also functions as a ferrule stop.

An adequate mechanical design of the parts and the application of laser-weldable metallic materials is naturally the premise for such a technique. These materials and the production of mechanical parts are generally costly. However, a closed metallic housing enhances the EMI immunity of an

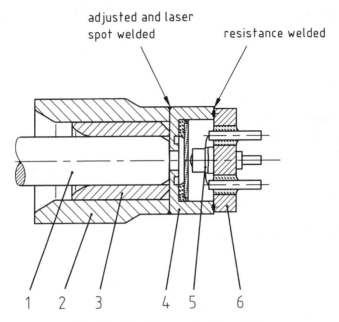

adjusted and laser
spot welded   resistance welded

1   2   3   4   5   6

Fig. 5.6   Classic example of an OSA.

OSA; this is critical, especially in the case of receiver OSA, which is highly sensitive to outside electromagnetic influences.

A remarkable reduction in production costs may be achieved by using injection-molded plastic parts instead of milled metallic parts. The required precision even for multimode fiber applications is achievable, if the mold designer and the injection molder achieve mechanical tolerances on the order of micrometers. Another possibility for cost reduction is the replacement of the ceramic sleeve, especially on the receiver side. However, a significantly different technique must be developed and proven for aligning and fixing an OSA, e.g., by means of light-curing epoxies. The question of long-term stability may be raised and therefore this item should be investigated carefully.

In order to reduce cost drastically in some low-end, large-core fiber applications, all-plastic packaged OSAs may be implemented. The packaging and the lens system are manufactured in one step by means of overmolding a diode mounted and wire-bonded to a leadframe. The molding material typically is a transparent duroplastic or even thermoplastic compound such as PC or PMMA or their derivatives. Some important factors for the applicability of such compounds are their transparency, their transformation or glass transition temperature, their moisture susceptibility, and their shrinkage, which may induce mechanical stress to the diode.

The technique itself is based on a manufacturing process commonly used in the production of millions of optocouplers each year. Careful material and process development tends to also fit the technique for applications with single optosemiconductors and fibers.

5.4. Summary: OSAs Today and Tomorrow

The basic function of a multimode optical subassembly (or coupling unit) is not expected to change in the future. For the transmitter, an electrical signal from a driver circuit is converted into an optical signal and fed into a multimode fiber. For a receiver, the light emerging from a fiber is collected and converted into an electrical signal for a preamplifier circuit. This function is realized with high coupling efficiency for a wide temperature range in a mechanically stable, environment-independent, and reliable manner.

Today, for each subfunction, such as electrical and thermal connection, optoelectrical conversion, sealing, light focusing, ferrule reception, etc., there exists a special part in the OSA. Therefore, the OSA includes a TO header, a ceramic submount soldered to the header, an optoelectronic chip soldered to the submount and bonded to the pins of the header, a glass window TO cap hermetically resistance-welded to the header, and a glass lens press-fitted into the ferrule receptacle laser welded to an adapter between the TO housing and itself (Fig. 5.7).

We see that an OSA consists of up to eight parts that are assembled in different process steps, each process being a single step and single-part process. The result is very costly components that are responsible for almost half the costs of a transceiver.

Cost is the main driver in changing the concepts for coupling units today and the future. The same functions must be realized with less expense (see Chapter 18). The following are several directions to look for cost-effective design improvements:

- Inexpensive materials with inexpensive assembly processes by replacing metal and ceramic parts by plastic parts and replacing laser welding by gluing;

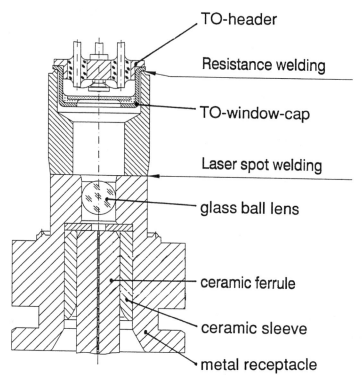

Fig. 5.7   Conventional OSA.

- Reduction in number of assembly parts through integration and with multifunction elements; Reduction in the overall number of processes;
- Increase in yield at the adjustment machine by creating an optical design with broader tolerances;
- Broad-tolerance optical design together with high-precision equipment will enable pick-and-place processes, thus eliminating the need for time-consuming active optical alignment and costly, high-tolerance parts;
- An alternative way to avoid active optical alignment is to use flip-chip bonding of the optoelectronic devices with solder-bump self-alignment;
- Solutions with standard technologies and processes capable of high volume, such as is used in silicon IC production, instead of using special "coupling unit processes";

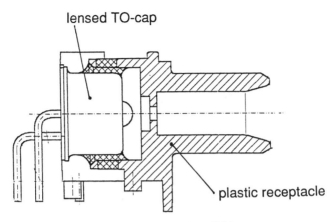

Fig. 5.8 Up-to-date OSA.

• Introduction of batch or even wafer processes for assembly, burn-in, and testing of coupling units/elements based on standard practices of the silicon technology; and

• Fully automated production.

Several of these improvements have already been realized. Most of the LEDs have an integrated etched lens combining the functions of optoelectrical conversion with focusing the light (a second lens is nevertheless necessary). Diodes with lens TO caps glued with a quick, light curing epoxy into plastic receptacles are a common integration of the sealing function of the cap and the optical function of the lens. This is also a step toward using plastic materials (Fig. 5.8).

A further step is the integration of the optical and the receptacle function in a precision-molded plastic component with a 45° internal reflection prism. This results in a 90° beam deflection and integration on the electrical side. Optoelectrical conversion, submount function, and hermeticity are provided by the optoelectrical device mounted together with the ICs and passive components directly onto the PCB and covered with a special transparent glob-top material together with the transceiver module case. Using plastic molding techniques for the manufacturing of the optical lens replaces the spherical lens coupling design. Aspherical lenses with higher coupling efficiency and wider coupling tolerances are possible (Fig. 5.9).

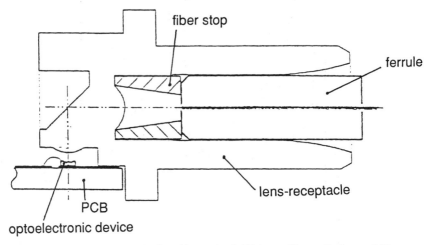

Fig. 5.9 Advanced OSA design (from A. J. Heiney, Jiang, C. L., and Reysen, W. H. 1995. Polymer molded lenses for optoelectronics. *IEEE.* © 1995 IEEE).

The introduction of an optical and mechanical design that permits and requires the desired silicon-based, automated batch processes will be a reality in the future. A first step in this direction might be a subelement consisting of a silicon submount, an optoelectrical device soldered and

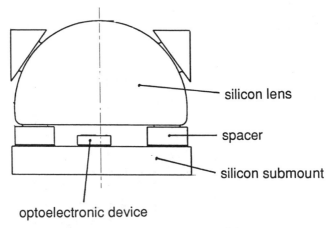

Fig. 5.10 Part of a future OSA.

Table 5.2 Summary of Potential Mechanical Cost-Effective Improvements for OSAs

	Today		Tomorrow	
Function	Part	Material	Part	Material
Electrical and thermal connection to PCB	Pins	Metal	Conductive adhesive or solder	
Base for assembly	Header	Metal	PCB	epoxy
Electrical isolation between diode and header, defines distance diode lens for optics, thermal conductor to heat sink	Submount	Ceramic	PCB or lens	
Optoelectrical conversion	Diode	III–V compound	Diode	III–V compound
Electrical connection	Bonds	Gold	Bonds or solder	Gold/AuZn
Hermetic seal from environment	Cap	Metal	Globe top	
EMI shielding	Cap	Metal	Nickel plating of plastic module case	
Focusing the light	Spherical ball lens	Glass	Lens	Plastic or silicon
Mechanical stable hold and guide for ferrule with fiber, definition of fiber reference plane	Receptacle	Metal with ceramic sleeve	Receptacle or lens receptacle	Plastic

bonded to the submount, and a spacer connecting the submount with a silicon lens. This subelement could be directly mounted to a rigid–flex–rigid PCB. A plastic receptacle may be glued on the rigid part of the PCB with the described subelement (Fig. 5.10).

A summary of potential mechanical cost-effective improvements for OSAs in the future is shown in Table 5.2.

Chapter 6 | Fiber Optic Transceivers

Michael Langenwalter
Kurt Aretz
Carsten Schwantes
Herwig Stange

Siemens Corporation, Semiconductor Division-HL, Berlin, Germany

George DeMario

*Siemens Corporation, Semiconductor Division-HL, Wappingers Falls,
New York 12590*

6.1. Introduction

Fiber optic transceivers have already replaced discrete solutions in most of
the datacom applications with two-fiber, bidirectional transmission. Consisting of a light-emitting diode (LED) or laser transmitter, and in most cases a
positive–intrinsic–negative (PIN) diode receiver, these fiber optic transceivers connect to the transmission media, the multimode or single-mode fiber.

The operation of these fiber optic transceivers in the physical layer, and
some innovative tendencies to integrate more and more functions of the
physical layer into these components, along with some basic applications
are discussed in this chapter.

6.1.1. OVERVIEW OF THE TRANSMISSION SYSTEMS
AND APPLICATIONS

The greatest advantage of fiber and fiber optic transceivers in datacom applications is the substantially higher bandwidth $\times$ distance product compared
with other transmission media. Thus, a wide field of applications has developed in the past 5 years that covers several kinds of networking technologies
with transmission speed from 10 MBd up to 4 GBd, and transceivers are

147

HANDBOOK OF FIBER OPTIC
DATA COMMUNICATION

available for all international and industrial standards. Fiber optic transceivers can be found in systems, supercomputers, mainframes, routers, bridges, switches, etc.; they can be found in any imaginable network system device.

6.1.1.1. Fast Ethernet

Ethernet as the networking standard with the largest installed base worldwide has been increased to 100 MBd during recent years and there is discussion concerning applications of this technology to 1 GBd. Thus, fiber is increasingly used as the transmission media of choice. A group of optical component manufacturers (AMP/Lytel, Siemens, Hewlett-Packard, and Sumitomo Electric) has developed a de facto standard footprint for optical transceiver modules, also known as the multistandard transceiver. These modules are now also used in various applications including fast ethernet to empower the existing Ethernet networks.

6.1.1.2. Fiber Distributed Data Interface

Operating at a nominal data rate of 100 Mb/s, the characteristic of a fiber distributed data interface (FDDI) network is the redundant ring architecture combined with its timed token passing network attach method.

With this architecture (class A concentrator), the FDDI network provides a self-healing structure in single-fault situations. The simpler alternative is the class B concentrator (Fig. 6.1) without failure tolerance, meaning that one physical link failure leads to a disconnected FDDI station. Transmission media is 62.5-μm multimode or 9-μm single-mode fiber with transmission distances from 2 to 40 km at a specified wavelength of 1300 nm connecting a maximum of 500 stations. The standardized fiber optic connector is the FDDI media interface connector (MIC) (Fig. 6.2); however, even for FDDI, the multistandard transceiver form factor is becoming increasingly popular because of its smaller size and pinout (see also Section 6.2.1.1).

6.1.1.3. Enterprise System Connection/Single Byte Command Code Sets Connection

Enterprise system connection (ESCON) was announced by IBM in 1990 as an industry standard for IBM's Enterprise System 390 and has grown into one of the most widely installed fiber optic network standards. ESCON is directly followed by the ANSI single byte command code sets connection

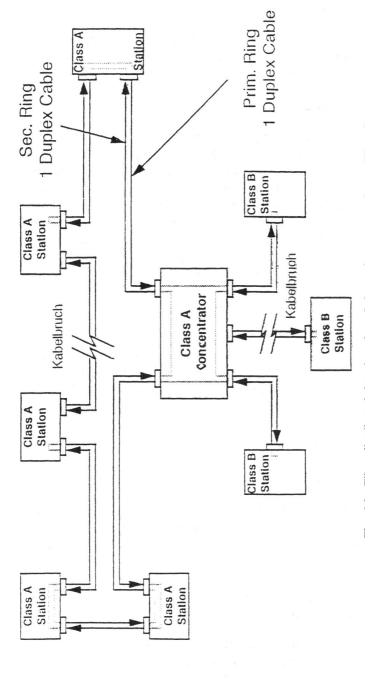

Fig. 6.1 Fiber distributed data interface. Informations-veranstaltung; Fiber Optic Components.

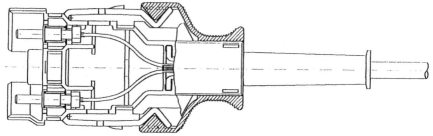

Fig. 6.2 FDDI media interface connector. Fiber optic data transmission; Fiber Optic Components E1/Pred.

architecture (SBCON) standard and therefore ESCON has become an international standard.

The ESCON/SBCON interface is defined as a point-to-point connection. The specified data rate is 200 MBd operating at 1300 nm wavelength with transmission distances up to 3 km for multimode and 10 km for single-mode applications. The standardized connector is the ESCON duplex connector (Fig. 6.3). For single-mode applications the FCS (Duplex SC) connector is also used.

6.1.1.4. Asynchronous Transfer Mode

With asynchronous transfer mode (ATM), a new switching and virtual network topology arrives. ATM is intended to be the "carrier" for most of the future network services worldwide, connecting LANs, MANs, WANs, and connecting between the datacom and telecom worlds. The network

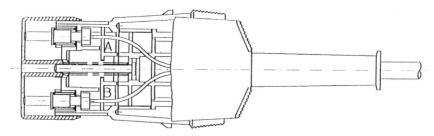

Fig. 6.3 ESCON connector. Fiber optic data transmission; Fiber Optic Components E1/Pred. ESCON is a registered trademark of International Business Machines Corporation.

attach method is asynchronous multiplexing. It supports B__ISDN and SONET activities.

Variable data rates are specified starting from 155 Mb/s up to 2.5 Gb/s. The transmission media are both multimode and single mode, with 62.5- and 9-μm fiber diameter, respectively. Based on the architecture, the transmission distance is unlimited and components for ATM systems are available to support distances between a few hundred meters (LED based) to more than 40 km (laser based) (Fig. 6.4).

6.1.1.5. Fibre Channel

The fibre channel (FC) is logically a bidirectional point-to-point serial data channel structured for high-performance capability. Even though fibre is a general term used to cover all physical media types, such as optic fiber and twisted pair and coaxial cables, the optical fiber solutions have become increasingly dominant. Fibre channel is structured as a set of hierarchical functions called FC-0–FC-4, in which the hierarchies FC-0–FC-2 are defined as the physical layer. The transceivers described in this chapter are defined in FC-0 and FC-1. The FC-0 level covers a variety of media and the associated transceivers are capable of operation at a wide range of speed, from 133 MBd to 4 GBd, and wavelength, 780/1300 nm, thus providing maximum flexibility. Therefore, optical fibers with 62.5, 50 (multimode), and 9 μm (single mode) are specified for various transmission distances from 300 m to 10 km.

6.1.2. PHYSICAL LAYER INTERFACE

Figure 6.5 shows a typical link structure. It shows the original incoming 32-bit-wide transmit word. In the second stage, this transmit data byte gets 8B/10B-coded and thus converted into a 40-bit-wide transmit word. Next, the parallel to serial conversion is performed at the serializer and the serial data stream is fed into the physical media driver, e.g., serial fiber optic transceiver. After passing through the transmission media, e.g., optic fiber, the process runs vice versa.

6.1.2.1. Clock Oscillator/Regenerator

Clock generation/multiplication is mainly realized based on a quartz crystal for very stable clock sources. For example, the ANSI fibre channel requires a frequency accuracy of ±100 ppm. Clock regeneration/synchronization can be realized by either surface-wave filters or phase-locked loops (PLLs), which are more common in today's applications.

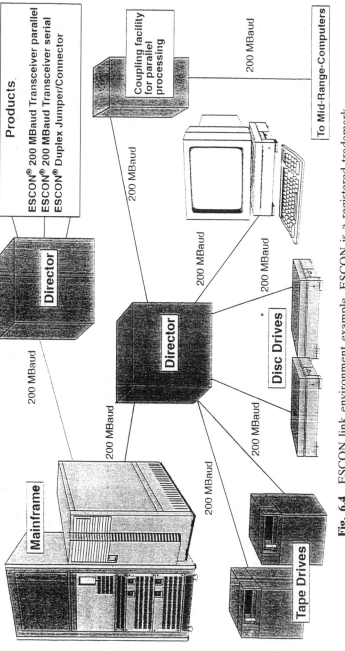

Fig. 6.4 ESCON link environment example. ESCON is a registered trademark of International Business Machines Corporation.

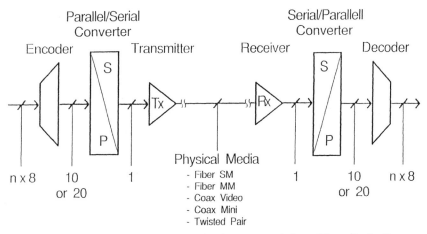

Fig. 6.5 Physical interface. Fiber optic data transmission; Fiber Optic Components.

6.1.2.2.   Serializer/Deserializer (Shifter)

The serializer accepts parallel data from the encoder once each byte time and shifts it into the serial interface output buffers using a PLL multiplied bit clock. The deserializer accepts serial data from the serial receiver input one bit at a time, as clocked by the recovered receiver sync clock, and shifts it back into a parallel data stream.

6.1.2.3.   Encoding/Decoding

The main intention of coding for transmission of high-speed data is to maintain the DC balance by bounding the maximum run length of the code. Typical kinds of coding in data communication include 4B/5B coding (FDDI) or 8B/10B coding (ESCON/SBCON and FC). This means high-speed receiver designs in fiber optic transceivers normally are AC coupled so that each DC component inside the data stream reduces the signal-to-noise ratio in the preamplifier stages.

During clock recovery, the PLLs used in today's application require a certain edge density to ensure that the receiving PLL remains synchronized to the incoming data. In addition, word alignment can be provided by special transmission characters (e.g., K28.5 pattern). In general, two types of characters are defined: data characters and special characters.

6.2.   Technical Description of Fiber Optic Transceivers

When selecting a transceiver, the key parameters are optical output power, operating data rate, saturation, receiver sensitivity, pulse quality (which is defined by parameters such as rise and fall times), jitter, and extinction ratio. Behind serial transceivers, interoperability with the following circuitry is the key point for the application to succeed. Thus, the high-speed characteristics and immunity against external influences have to be carefully evaluated during design. Increasingly, users decide in favor of integrated transceiver solutions covering the high speed and synchronization functions and providing easy-to-use TTL lower-speed parallel interface.

6.2.1.   SERIAL TRANSCEIVERS

Serial fiber optic transceivers are the interface between the serial electrical signal and the transmission media optical fiber. They are composed of the transmitter and receiver functions that operate independently. The implementation of high-speed serial transceivers requires careful design of filter circuitry as well as transmission line considerations for the data input and output tracks to connect to the following circuitry.

Figure 6.6 shows a simple block diagram of a fiber optic transmitter containing a laser-driver circuit and semiconductor laser. Also shown is a

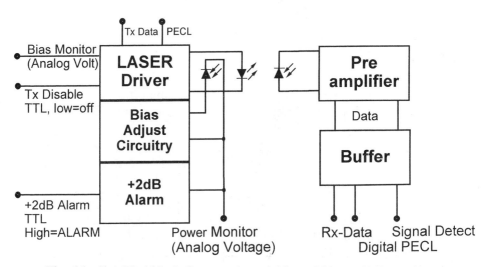

Fig. 6.6   Simplified block diagram of a serial laser driver and PIN receiver.

monitor diode that tracks the laser output, and alarm circuitry that activates if the laser crosses a set threshold.

Figure 6.6 also shows a block diagram of a fiber optic receiver containing the PIN Diode, the transimpedance amplifier that converts extremely small AC currents (nA range) into a larger differential voltage, and the postamplifier for ECL/PECL line driving.

6.2.1.1. Multistandard Serial Transceiver

Figure 6.7 shows a 1 × 9 multistandard serial transceiver along with an ESCON transceiver. This 1 × 9 package is supported by several vendors for fiber optic transceivers, such as Siemens, AMP/Lytel, Hewlett-Packard,

Fig. 6.7 ATM transceiver with duplex SC connector (top) and ESCON transceiver with ESCON connector (bottom).

and Sumitomo Electric, for standards such as Fast Ethernet, FDDI, ATM, FC, B-ISDN, etc.

6.2.1.2.   *ESCON/SBCON Serial Transceiver*

Since the introduction of the IBM industry standard ESCON, the ESCON serial transceiver has reached a huge installed base of more than 2 million transceivers to date; thus, it is one of the best established high-speed fiber optic link applications.

6.2.2.   *PARALLEL TRANSCEIVERS*

During the past 3 years, transceivers with a parallel electrical interface have become increasingly popular in fiber optic link applications and inexpensive and smaller alternatives to existing link card solutions have been developed. Such parallel transceivers include the functionality of a serial fiber optic transceiver plus clock synchronization and serializer/deserializer functions. There is a clear tendency toward higher integration in transceiver solutions, and the integrated 8B/10B encoder/decoder function will be available soon. A block diagram of a parallel transceiver with 8B/10B encoding/decoding is shown in Fig. 6.8.

The parallel transceiver offers many advantages for the system design and application such as higher channel density, low power consumption, easy-to-use TTL lower speed interface, as well as logistical benefits on all levels and the completely integrated physical layer interface from one source with extremely low design risk and simplified development effort.

6.3.   Optical Interface

An interface is the meeting of two objects and, in the case of the optical interface, the fiber comes together with the optical transceiver. The operating link demands that the mechanical and optical properties of these two objects fit together. These attributes are defined in various standards to secure the interoperability of the products of different manufacturers.

A closer look shows that a transceiver has not one but two optical interfaces: the connection of the outgoing fiber at the output of the transmitter and the connection of the incoming fiber at the input of the receiver. The optical subassembly of the transmitter inserts the light into the fiber,

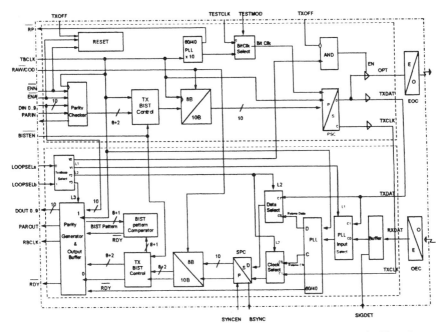

Fig. 6.8 Block diagram of a parallel transceiver with 8B/10B encoder/decoder.

which will be taken up by the optical subassembly of the receiver at the other side of the link.

For a flexible configuration network, it is necessary to have an interface that can be repeatedly connected and disconnected. This function is realized by a connector at the end of the fiber and a port at the transceiver that accepts the fiber connector. There are different designs on the market including the DIN/IEC, FC, SC, SMA, or ST connector systems.

Most systems use a ferrule—a cylindrical tube containing the fiber end—that fits within high tolerance into the flange of the transceiver port. For this system the popular diameter of the ferrule is 2.5 mm. The diameter tolerances within a few microns must be maintained in order to achieve a satisfactory coupling of the light.

Besides the single connector, it is also possible to integrate several fibers into one connector that is used for parallel optical links. In general, the fiber connector has to be fixed, for example, by screwing on a coupling ring or snapping on a latch mechanism. Additionally, some connectors can be provided with a key to distinguish, for example, multimode and single-mode connectors. This design with a connectable optical interface directly at the transceiver module is mostly used for datacom applications.

For bidirectional links it can be useful to integrate the transmitter and the receiver into one device called the transceiver. This is why it is advantageous to combine the two fibers, one for each direction, into one cable with one duplex connector at each side (see Fig. 6.8). This saves space, reduces the effort for assembling the modules and installing the link, and eliminated the possibility of cross-plugging.

Pigtail solutions are preferred for telecom applications. These modules do not have an optical port, but instead have one or more permanently fixed fibers of short length. In this case the optical interface of the transceiver is situated at the connector at the end of the pigtail.

Considering the optical aspects of the interface, it is useful to look at the parameters of the fiber, the medium between transmitter and receiver. An optical fiber can be described through the properties of core diameter, numerical aperture, bandwidth length product, and attenuation. These parameters influence more or less directly the requirements for the output of the transmitter and the input of the receiver.

The core diameter of the fiber defines the maximum waist of the output beam at the transmitter because only light inside this waist will be guided by the fiber. This is why the numerical aperture, which is determined by the index profile of the fiber, limits the maximum divergence of the output beam. Emitted light exceeding this maximum divergence will be lost. An efficient optical coupling between transmitter and fiber can only be achieved if these two pairs of parameters, core diameter and beam waist, as well as numerical aperture and divergence match.

At the receiver side, the optical coupling is again mainly influenced by these two fiber parameters, namely, the waist and divergence of the beam leaving the fiber is determined through the core diameter and the numerical aperture of the fiber. The optical subassembly of the receiver must focus this beam on the sensitive area of the photodiode.

In some cases it is necessary not only to consider the coupling efficiencies at the optical interfaces of the transceiver but also to regard the share of light that is reflected back to the light source. Lasers can be sensitive to backreflections that could disturb the laser emission and cause some noise on the optical signal. To prevent these effects the return loss at the optical interfaces has to be high. The backreflections can be kept low by using physical contacts at the connectors or by inserting an optical isolator in front of the laser. Another method uses a slantwise emitted beam at the optical interface. In this case, the reflected light does not hit the laser.

The fiber bandwidth is caused by the modal dispersion and the spectral dispersion. Modal dispersion occurs in multimode fibers and causes a pulse broadening due to the different speeds of the fiber modes. Therefore, the excitation of many fiber modes leads to a stronger pulse broadening in comparison to the excitation of few modes with low order. This excitation of fiber modes depends strongly on the optical coupling at the transceiver.

The chromatic dispersion limits the spectral bandwidth of the transmitter in order to reduce the pulse broadening. For low bit rates and short distances, LEDs with a typical full width at half maximum of 50–100 nm are sufficient. However, higher bit rates and longer distances require the application of lasers with a typical bandwidth of a few nanometers down to the subnanometer region, which is achieved by distributed-feedback lasers with only one spectral mode.

The transparency of the fiber material at the different wavelengths of light determines the spectral parameters of transmitter and receiver. In the past the three optical windows at 800, 1300, and 1550 nm of fused silica fibers were mostly used. Today, these windows are broader due to the reduced OH concentration in silica fibers and other materials; for example, plastic optical fibers open new windows in the visible region. The wavelength of the transmitter and the spectral sensitivity of the receiver have to fit with these windows of low attenuation. The attenuation of the fiber determines, together with the sensitivity of the receiver, the minimum launched power in the fiber. This defines the minimum optical power of the transmitter at the optical interface if the coupling efficiency of the transmitter is provided.

Especially for laser transmitters, there is also an upper limit for the optical power because lasers can be dangerous to the human eye. This disadvantage must be considered as optical transceivers find more and more applications. There are benefits in the use of lasers in data communication: for example, high data rates and long operating distances. The challenge is how to use these benefits without endangering the human eye.

In principle the answer is simple. The power of accessible laser radiation must not exceed a critical limit that is defined by the mechanisms of eye damages. These mechanisms depend heavily on the wavelength of the laser radiation. Many systems of optical data communication use one of the three optical windows of fused silica glass fibers in the infrared region. The human eye cannot see this infrared light. To be more precise, the eye's receptor cells, the rods and cones of the retina, are not sensitive to this region of the optical spectrum. This infrared light penetrates the cornea and the vitreous humour like visible light, and it is focused by the lens of the eye. To estimate possible hazards, the energy density of this focused light on the retina must be considered. The mechanisms of eye damage are well known because of the experience and use of lasers in medical applications. The standards, which defined the maximum limit of laser light, are based on this knowledge.

Two important standards must be observed; if a laser product or component shall be distributed worldwide, the international standard IEC 825-1 of the International Electrotechnical Commission; and for the U.S. market the regulation 21 CFR Chapter 1, Subchapter J, of the Center for Devices and Radiological Health, a subgroup of the U.S. Food and Drug Administration (FDA). Both standards categorize all laser products into four classes. This classification gives the user an idea about the hazards of the laser and lays down the required precautions to which the user must pay attention. For example, it could be necessary to train the staff to use the laser safely or to wear eye protection.

The safest category of these standards is the laser class 1. Therefore, the laser class 1 products are the least restrictive with respect to the manufacturer and user requirements. The very wide distribution of CD players or laser printers demonstrates that the application of lasers is not an impossible hurdle if the product meets the safe laser class 1.

Laser safety is also an important issue for optical fiber data links. The easiest way to obtain a safe laser class 1 data link is to use components that are inherently safe. In such a case the laser transceiver ensures the laser safety of the complete system. However, it is a difficult task to design a laser transceiver that meets the different class 1 categories of both standards. In general, the eye safety of transceivers with optical connectors can be achieved in the following three ways:

1. Mechanical solutions such as shutters prevent access to laser radiation from the transceiver if no connector is in the port. However, transceivers should remain small and this additional device would increase the size of the port. This can contradict the standards that ensure the compatibility of the transceivers.

2. Optical solutions aim at high coupling efficiency and a divergent beam to reduce the accessible optical power of the unplugged transceiver. This can be achieved by a fiber stub and a split sleeve that enable a physical contact between stub and connector. Even slight pollution that is usual in the operating areas of transceivers can cause high losses or laser-disturbing backreflections. Both effects can disconnect the link.

3. Choose an electronic solution. The advantage is that electronic safety interlocks are easily integrated; however, this type of solution can be much more inexpensive than any mechanical or optical solutions. The laser power of the transceiver is checked electronically by a monitor diode. In any event, this is needed for controlling the laser power to compensate for temperature and aging effects on the laser. In case of failure, the laser is shut down to prevent hazardous laser emission. The transceiver has a redundant safety interlock, which secures eye safety in case of a single-fault condition or any subsequent failures.

Point-to-point optical data links enable the application of another electronic safety interlock. The open fiber control system (OFC) detects link disruptions and forces the transceiver into a repetitive pulsing mode of operation with very low duty cycle. Only after a proper reconnection handshake has checked that the link is closed can the OFC system allow the transceiver to return to normal data traffic. The specification of the OFC system is part of the fibre channel standard. The system can operate with higher optical power that would normally exceed the limits of class 1 lasers. The advantage is a higher power budget for the optical transmission.

If the transceivers are inherently safe, the design of the data communication systems is not restricted by any laser safety requirements. The manufacturer does not have to classify the product, is not obliged to submit reports about laser safety, and need not perform additional tests during production. This allows the safe and simple application of the transceivers.

6.4. Noise Testing of Transceivers

The increasing use of electronics and mobile telecommunication equipment in everyday life has evoked a strong demand for electromagnetic compatibility (EMC) of electronic devices by both manufacturers and customers. With the growing efforts in the field of high-speed data transmission along with ever-increasing computer integration, this demand has also gained major influence on the design and production of today's fiber optic transmission equipment. In close cooperation with computer and telecommunication equipment manufacturers, optical transmission modules of high EMC performance have been designed. This is necessary to meet the standardized EMC requirements for systems that include fiber optic transmission modules and to guarantee reliable operation of these components.

This section describes possible measurement procedures for the EMC performance of fiber optic transmission modules. The measurements presented here deal with noise on supply voltages, emission of electromagnetic noise into the environment, and immunity against electromagnetic noise from the outside.

6.4.1. DESCRIPTION OF THE DEVICE UNDER TEST

For all measurements described, the device under test (DUT) is an electrooptic transceiver or a combination of a transmitter and a receiver. The DUT is plugged onto a PCB (Fig. 6.9) which is populated with power supply filter elements as described in the transmitter/receiver data sheets.

The electrical data output lines of the receiver are connected directly to the electrical data input lines of the nearby transmitter. Top and bottom layers of the PCB are ground. Connections to the measurement setup are fiber optic jumper cables and power supply lines.

6.4.2. NOISE ON VCC

Fiber optic transmission modules are usually configured in an environment with fast-switching circuitry. These switching activities result in high-frequency noise on the DC power supply (Vcc). High-frequency noise on Vcc cannot always be sufficiently suppressed by capacitors or inductors because these components show resonant behavior of their own, with resonant frequencies lying close to the noise frequencies. It is therefore necessary for fiber optic transmission modules to withstand high-frequency noise on Vcc. Noise effects commonly known as ripple are less harmful to fiber optic modules because ripple frequencies are lower than frequencies of noise caused by switching.

To perform a test on a DUT's behavior concerning noise on Vcc, a noise signal must be coupled into the Vcc line using a bias tee as shown in

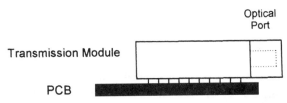

Fig. 6.9 Transmission module (DUT) attached to a PCB for noise testing.

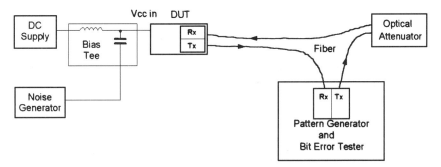

Fig. 6.10   Evaluating DUT for performance with noise on Vcc.

Fig. 6.10. When transceivers are tested, the receiver part should directly act as driver for the transmitter part. If the DUT contains internal decoupling elements, noise of certain frequencies may already be sufficiently suppressed. In these cases, noise voltages at the module Vcc pins are close to zero. At frequencies at which noise suppression is insufficient, noise voltages can be measured at the module pins.

At frequencies that are already sufficiently suppressed, high noise-generator amplitudes would be necessary to keep the noise voltage at the module pins on a constant level. Testing a DUT under such artificial conditions is overly demanding. It is therefore recommended to merely keep the noise-generator amplitude on a constant level for all frequencies.

For the measurement, normal data transmission is established and then optical power is attenuated to a level that causes a certain bit error rate (BER) value. When noise is applied on Vcc the BER increases. This increase in BER is a measure for the DUT's sensitivity to noise. The DUT meets the requirements if the decrease of attenuation necessary to regain the former BER value does not exceed a defined level. The measurement is repeated for different noise frequencies. It is important to always use identical test setups to ensure that measurements are consistent with former measurements.

6.4.3.   ELECTROMAGNETIC COMPATIBILITY

6.4.3.1.   Emission

The emission of electromagnetic radiation may cause inaccurate behavior of components lying near a radiating device. Therefore, maximum allowable electromagnetic noise levels (depending on noise frequency) are defined.

For information technology equipment the emission limits are defined in standard EN 55022 (= CISPR 22 = VCCI = VDE 0878 Teil 3). A common measurement setup for noise emission is shown in Fig. 6.11. The measurements are made in a gigahertz transverse electromagnetic cell for three different orientations of the DUT and the measurement results are correlated to the free field measurement. The data pattern used is a 0101 sequence at nominal data rate. The DUT's emission is measured in a range of 30 MHz to 1 GHz.

According to the previous standard, the "class B" limits of the radio frequency (RF) noise field strength are the following: 30–230 MHz, 30 dB (μV/m) max; and >230 MHz to 1 GHz, 37 dB (μV/m) max. The DUT's emissions are tolerated if noise remains 6 dB below these limits in the 30 MHz to 1 GHz range. An example of measurement results is given in Fig. 6.12.

6.4.3.2.   Immunity

When external electromagnetic fields are applied to the DUT bit errors can occur. The most sensitive part of a DUT is the receiver section around the transimpedance amplifier. Here, very low input currents are converted into low-voltage signals. There are two measurements to determine the

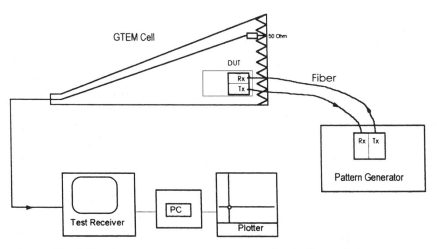

Fig. 6.11   A common measurement setup for testing emission.

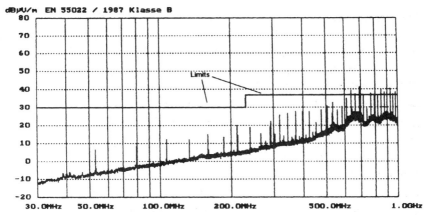

Fig. 6.12 Example of measurement results of emission testing.

DUT's immunity: RF immunity and electrostatic discharge (ESD) immunity measurements.

6.4.3.2.1. Noise Immunity against Radio Frequency Electromagnetic Fields

This immunity test can be performed based on standards ENV 50140 and prENV 50204 (future standards EN 61000-4-3 = IEC 1000-4-3); a possible measurement setup is shown in Fig. 6.13.

In the beginning of the measurement, no noise is applied. Normal data transmission is established (2^7-1 pseudo-random bit sequence at nominal data rate), then the optical attenuator is set to a value at which a bit error rate of 10^{-6} is detected. At this operating point the DUT is more sensitive to external noise than during normal operation. The sine wave noise carrier source is switched on and regulated to generate a field strength of 10 V/m root mean square at the DUT's location. Then amplitude modulation is applied to the carrier signal (modulation factor 80%; sine wave modulation 1 kHz). The noise carrier frequency starts at 80 MHz and is increased in steps of 1% from the last value up to 1 GHz (new value = last value × 1.01). The DUT is tested at its worst-case position (found experimentally) regarding its orientation and polarization of the noise field. BER values are measured for the different noise carrier frequencies.

Test results show that BER values depend on the noise frequency applied. Certain values of BER can be achieved by reducing optical power,

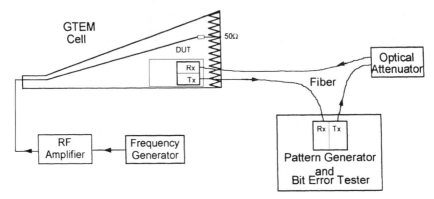

Fig. 6.13   Possible measurement setup for measuring noise immunity to RF fields.

so there is a corresponding optical power level for each BER value. Using this correspondence, test results can finally be presented as the minimum optical power necessary to achieve a certain constant BER value depending on the noise frequency. Figure 6.14 shows a possible test result. The DUT meets the requirements of the previous standard if the sensitivity values given in the DUT's specification are reached for all noise frequencies.

For a more detailed investigation, the difference between the minimum optical input power in the undisturbed case and the worst measurement result with noise applied can be required to be lower than a specified value. This is shown as ΔP_{opt} in Fig. 6.14. (The "baseline" of Fig. 6.14 can be assumed to be similar to the graph in the undisturbed case). This measurement can be used in the development process to indicate improvement or degradation of the DUT.

6.4.3.2.2.   *Immunity against Electrostatic Discharge*

ESD immunity tests deal with noise caused by fast transients from electrostatic discharges. The described tests do not destroy the modules; there are no tests to check the ESD behavior concerning high voltages at the module pins. A transient electromagnetic field based on the human body model is generated and influences the module behavior with respect to bit errors. Two tests are proposed. The first one is based on EN 61000-4-2 = IEC 1000-4-2, and the second one is suitable for a more detailed analysis. In both cases the source of the discharges is a so-called ESD gun. In Fig. 6.15, the DUT is operated with a 2^7-1 pseudo-random bit sequence at nominal data rate. The optical input power is set to the lower specification limit.

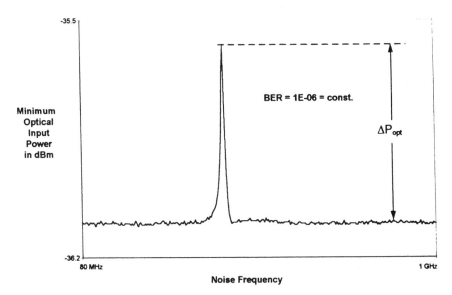

Fig. 6.14 Example of a possible test result of a noise immunity test.

The DUT's receptacle is plugged through a grounded plate simulating a rack panel in a possible application environment. The following are the two steps involved in this test:

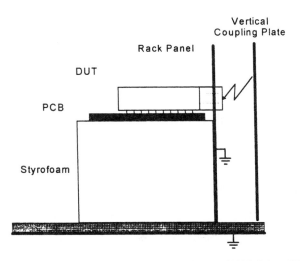

Fig. 6.15 ESD test setup based on the standard, EN 61000-4-2 = IEC 1000-4-2.

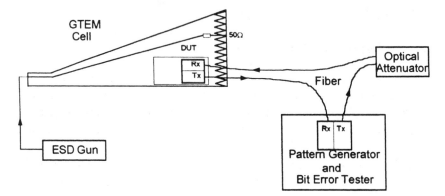

Fig. 6.16   Proposed test setup for more reproducible ESD testing.

1. Without vertical coupling plate: Electrostatic discharges (both po-
   larities) at any point of the rack panel or at the receptacle of the
   DUT are applied to provoke bit errors during a defined time interval of discharge. After the discharge the DUT continues error-free
   operation.
2. Electrostatic discharges are applied at any point of the vertical coupling plate to induce bit errors.

The discharge voltage voltage and the number of bit errors in the time
interval observed describe the ESD immunity of the DUT.

6.4.3.2.3.   *More Reproducible ESD Testing*

To achieve more reproducible results and make thorough analysis possible
the test setup in Fig. 6.16 is proposed. To increase the error probability
the DUT's highest specified data rate is used. The optical input power of
the DUT is attenuated to a level that causes a BER of 10^{-6} and is then
raised again by, e.g., 3 dB. For easy error detection a 0101 pattern is
transmitted. The ESD gun generates noise pulses periodically. The ESD
performance of a DUT is described by the discharge voltage (i.e., the
resulting field strength) and the number of bit errors over a certain period
of time.

6.5. Packaging of Transceivers

6.5.1. BASIC CONSIDERATIONS

As discussed in Chapter 5, the most serious task of any packaging is to maintain functional integrity of a device over its total lifetime under all specified conditions. One of these tasks is to protect the electronic circuitry inside the transceiver (TRX) from electromagnetic influences radiated or conducted into the module as well as to prevent electronics outside the module from being influenced by the module itself. This EMI shielding may be performed by means of a conductive metallic housing, or a conductive plating on a nonconductive housing, or even separate shielding structures inside the housing. Naturally, the electronic circuitry itself should also be decoupled with care via properly selected capacitors and other filter components where needed.

The next task is to avoid possibly aggressive chemical agents and/or humidity from sensitive devices and surfaces during card assembly procedure and during normal operation. Otherwise, for example, corrosion effects in the presence of ionic impurities and humidity will occur. Especially on an electronic circuitry such effects are rapidly enforced by electromigration due to the presence of DC power. Therefore, the appearance of weak shorts or even other catastrophic failures due to complete destruction of joints or components is nearly preconditioned.

Another task of the TRX housing is to avoid influences of external mating or withdrawal forces applied by the optical connector or the cable. This is a question of mechanical stability based on design, materials, and techniques used for assembling the TRX itself and the TRX to a motherboard.

Some typical features may be explained more in detail by the following example. Figure 6.17 shows a TRX module containing in a housing a Tx-OSA and a Rx-OSA as well as a chip on board (COB) electronic circuitry with infrared radiation-emitting diode (IRED) driver IC on the transmitter side and PD preamplifier IC and separated buffer IC on the receiver side. The plastic housing has ground pins that are directly molded in during injection molding. After molding the housing, including the ground pins, is plated with copper and nickel and a flash of gold. Therefore, in combination with the metalized plastic cover a completely closed metallic shielding and an extremely low ohmic-resistance connection to ground pins are

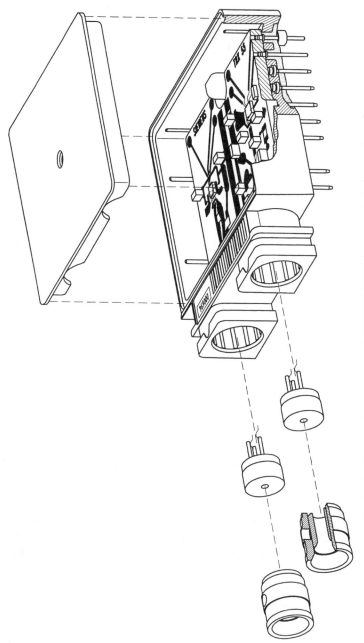

Fig. 6.17 Exploded view of serial ESCON TRX (connector shell not shown).

achieved. Also, the Au-flashed nickel avoids oxidation or corrosion effects on any surface of the housing and on the pins, which guarantees, e.g., proper solderability.

Functional pins are press-fitted into the COB-PCB and penetrate corresponding holes in module housings bottom. Also, the OSAs are press-fitted into noses on the front side of the housing. The PCB is attached to the housing by conductive epoxy and to the cover. For washable tightness, the housing's underside, as well as the nose area, is sealed by a suitable epoxy potting material.

6.5.2. TECHNIQUES FOR ASSEMBLING TRX TO MOTHERBOARD

The TRX shown in Fig. 6.17 is a typical example for a module being assembled by classical through-hole technique to a motherboard. After insertion of the module pins into the corresponding holes in the motherboard, the connector shell is screwed to the board. The TRX is then connected to the board by a common wave soldering or fountain soldering technique followed by water-jet washing, rinsing, and hot-air blower drying. During this procedure the optical ports in the connector receptacle are protected from processing chemicals and water by a process plug.

Especially for TRX with higher pin counts the surface-mount technique has been used. Also, the possibility for assembling both sides of a motherboard is a basic advantage of this technique (Fig. 6.18).

Gull-wing leads principally allow partial soldering techniques like hot ram (Thermode) or hot gas or even laser soldering. The important advantage of these processes is that only the leads are heated up to the melting temperature of the solder and not the TRX itself. Also, no washing procedure after soldering may be needed if less aggressive fluxes are used. Moreover, a visual inspection of the solder joint quality is easily possible.

The main reason why J-leaded TRXs are not well established is due to their basic disadvantage with respect to partial soldering techniques and inspection. This will also be a principal concern for the ball-grid array technique for TRX with high external pin counts caused by very high functional integration.

Finally, in some applications a connector socket is assembled to the motherboard and the counterpart is on the TRX module or TRX card side (see Section 6.5.5). An obvious advantage is the possibility to change a failed TRX quickly and easily without any special tools. However, some

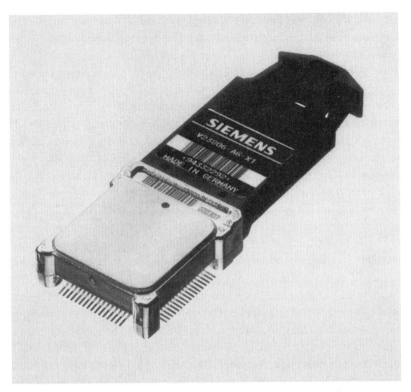

Fig. 6.18   Parallel ESCON TRX with gull-wing leads.

serious challenges in using connectors are introduced in the areas of cross-talk and EMI.

6.5.3. WASHABLE TIGHT MODULE DESIGN

The previously mentioned soldering processes suggest a module that should be designed for hermeticity or washable seal. This means that, on the one hand, no additional care must be taken to protect single electronic devices inside the module, which certainly will reduce cost. On the other hand, the total design and nearly all assembling agents and processes have to be adapted to such a process. For example, the application of a simple soldering process with aggressive fluxes may be strictly forbidden inside a sealed package. Additionally, during manuacturing a suitable and cost-saving procedure for testing the seal must be implemented in a production line. The TRXs shown in Figs. 6.17 and 6.18 are both washable sealed modules.

6.5.4. OPEN MODULE DESIGN

The alternative to the sealed module design is an open design. Such designs are well established worldwide and are becoming increasingly popular. This design is based on the simple consideration that any possible damaging gaseous, liquid, or dust agents going into an open housing may also be washed out. This assumption may be correct in general, however, there are design complications and challenges to open packages.

In open packages, individual protection of single electronic devices such as ICs by separate housing or encapsulation by, e.g., glob top is necessary in the case of a COB technique. Additionally, special care must be taken regarding the circuitry layout and the circuitry assembly and protection, in order to prevent electromigration effects. Finally, the EMI shielding must be managed individually by separate structures such as metallic sheets over sensitive areas of the circuitry.

With open housings, the housing itself may now be a simple injection-molded plastic part and the final assembly of the TRX could be done by a snap-together technique. No additional sealing and corresponding testing procedure will be necessary (Fig. 6.19).

6.5.5. OPEN CARD MODULE DESIGN

As mentioned previously, open module card designs are taking place in fiber optic datacom. The mechanical basis of a card module typically is a PCB combined with a peripheral plastic frame. This frame also carries the OSAs and the connector receptacle on its front side. For the design of such open cards, generally the considerations for the previously mentioned open housings apply.

Figure 6.20 shows as an example the Gigalink Card, which is a laser-based TRX card compliant to ANSI FC-0 100-SM-LL-I standard with 1.0635 GBd at 1300 nm for single-mode fiber application and 20-bit parallel electrical TTL in/out. The module is connected to a motherboard by an 80-pin connector on rear area of the card and secured by a pair of snapping legs. As a remarkable detail, it is reported that total insertion and withdrawal forces for that connector are on the order of nearly 100 N. Therefore, a special withdrawal lever is

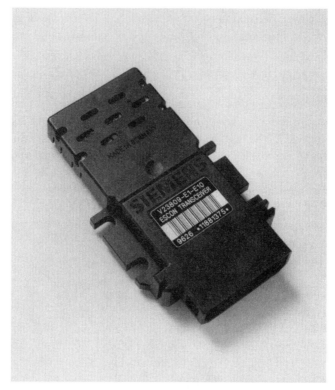

Fig. 6.19   Example of an open module design; the 4th generation of ESCON TRX.

designed as an integral part of the frame in order to prevent damage of module card as well as motherboard during withdrawal.

6.6.   Series Production of Transceivers

6.6.1.   *BASIC CONSIDERATIONS FOR PRODUCTION PROCESSES AND THEIR RELIABILITY*

For every series production a basic precondition for a cost-saving and high-quality output is a continuously running production with qualified equipment as well as qualified and reliable processes. Such equipment and processes are characterized in that within a defined variation of process parameters, the so-called process window, a quite uniform excellent result

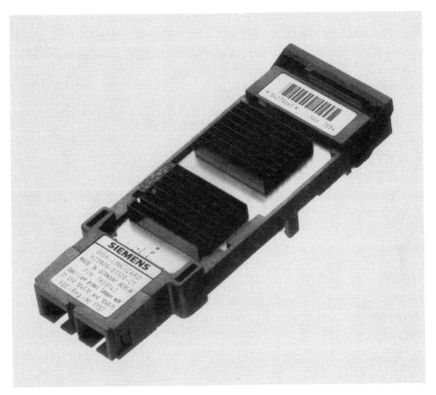

Fig. 6.20   Example for a card module design—the Gigalink Card.

is achieved with a high yield. Naturally, a process is safer as its window is made wider.

Therefore, all equipment and processes must be qualified before the start of series production. This will typically be performed by careful and adequate testing of a statistically significant number of parts, i.e., 40 pieces at the absolute minimum, that had previously been produced on the related equipment by the corresponding process. The test results must be evaluated in a statistical manner in which the number of tested parts defines the confidence level of the result.

6.6.1.1.   Qualification of Processes, the Process Capability

A brief example is provided as an explanation: A mechanical part A shall be attached to another part B by an epoxy adhesive. The key parameter for the process may be the mechanical strength characterized by, e.g., a pull force needed for destruction of the epoxy attachment between parts A and B. During tests the values for the forces are monitored and then their mean value, F_{mean}, and the standard deviation, σ, are calculated. These values are now put into a simple equation for calculating the so-called process capability:

$$C_{pk} = \frac{F_{mean} - F_{min}}{n\,\sigma} \geq S,$$

where S is the capability index, F_{min} is the minimum force specified for the attachment, and $1 \leq n \leq 6$ is the value for the statistical sharpness. For $n = 3$, for instance, S is 1.33 in order to reach a statistical safety of $\pm 3\sigma$ or 99.73%, or even a failure probability of 2700 ppm. In the case of $n = 4$ and $C_{pk} \geq S = 1.66$, the statistical safety will be $\pm 4\sigma$ or 99.9937%, or the failure probability will be 63 ppm. Therefore, an improvement of a process simply is determined by increasing its C_{pk}.

The statistical sharpness indicated by n and the correlated value of S are defined indirectly by the requested product reliability or the permitted maximum number of failures in time, respectively, in the product specification.

6.6.1.2. Correlated Environmental Tests on Subcomponents

Here it makes sense to combine the evaluation of process capability with environmental tests that are also defined typically in a product specification. These tests normally contain long-term temperature and/or humidity stress, temperature cycling, shock stress, mechanical shock, vibration stresses, and also some additional stresses determined by customers' special application. Upper and lower stress limits or values are mostly specified directly or even defined by commonly known MIL or Bellcore generic standards named in the product specification. The major advantage of performing tests as early as possible during development phase and on subcomponent level is clear: Any weakness of a construction detail, a subcomponent, or a manufacturing process will be detected earlier, resulting in time and cost savings, and improvement or general change may be more easily implemented in the design itself and the production flow as well.

Naturally, all tests will also be performed at the end of product development on the completed and final product, and normally no major weakness will occur if during the design phase all pretesting has been performed carefully.

6.6.1.3. Equipment Qualification, the Machine Capability

A similar procedure as for C_{pk} may be used for calculating the so-called machine capability, C_{mk}, which is defined as the statistical safety for processes performed by a tool or a machine. For example, the uniformity of a dispensed epoxy volume will be qualified by evaluation of C_{mk} of the automated dispenser. If C_{mk} exceeds the indicated value of capability index S, then the equipment would be safe and capable of its task. However, a combination with environmental stress here does not make any sense.

6.6.1.4. Frozen Process

After qualification and release of equipment and a corresponding process this production step generally should be frozen. Frozen means here that in case of major changes of equipment, process parameters, agents, or even parts, a partial or total requalification of the production step must be performed. The meaning of major change or even minor change should be defined in close communication with the customer. Also, the release of a changed production step for series production typically should be with customer agreement.

6.6.2. STATISTICAL PROCESS CONTROL AND RANDOM SAMPLING IN A RUNNING SERIES PRODUCTION

During running production some evident quality parameters should be monitored by 100% and some others by statistical random sampling in order to check the uniformity of processes and the quality of their output continuously. Therefore, every deviation from the expected output quality as well as some creeping degradation of process parameters will be detected in time and immediate corrective action could be taken. For key process steps generally a higher percentage of checks should be performed or a higher number of samples should be tested than for proven stable and uniform processes.

A commonly known tool for monitoring the quality level of any production process is the so-called statistical process control card. Herein the actual level of a parameter is compared statistically to previous defined lower and upper limits for this most characteristic parameter that determines the quality of a component or a process. Additionally, the actual level may also be compared to the so-called warning limits that typically are defined as 1σ closer than the lower and upper limit. If during production an actual value is close to a warning limit or even exceeds it, then an early corrective action may happen immediately. Therefore, almost never will components be produced out of specifications.

6.6.3. DOCUMENTATION AND TRACEABILITY

Due to the high requirements for product safety and reliability in the field of telecom as well as datacom, every failed product generally will undergo a failure analysis. This analysis must answer questions about what exactly failed, how, possibly when, and especially why. A more or less detailed failure analysis report is commonly requested by customers. Then, corrective actions must be defined and implemented in order to rule out the occurrence of the same failure in the future. The effectiveness of corrective actions must be proven and documented as well as accepted normally by customers. Especially in answering the question about why a part failed, a detailed documentation of the product's history is very helpful or even basically evident.

This will be managed by an individual series number for each delivered product combined with a lot number or a week code. In association with this number typically other additional information is held in store, for example, about individual subcomponents and their origin and lot number, about date and shift of production, burn-in, and final test, and finally about the individual values of typical electrooptical parameters measured during final outgoing inspection.

The extent of documentation and time for storage are commonly defined in delivery specifications contracted between producer and customer. Also, the generic quality management system specification ISO 9000 describes procedures and contents for documentation and traceability.

6.6.4. ZERO-FAILURE QUALITY, BURN-IN, FINAL OUTGOING INSPECTION, AND SHIP TO STOCK

Any technical system or component behaves with respect to its failure probability according to a well-known time-dependent failure function–the so-called bathtub life curve. This curve describes the fact that a component most probably will fail at a higher rate at the beginning of its lifetime and very late at end of lifetime. In between, the probability of failure is low and constant.

A well-established method to identify early failing parts especially in electronics is to perform a burn-in. During burn-in, the fiber optic TRXs are operated at elevated temperature level over a defined period of time. The time, temperature, and possibly some DC power overload will define the confidence level for effective screening. Before and after burn-in the TRXs will run their normal complete inspection of all relevant electrooptical parameters, and the measured values will be compared. Individual TRXs will be rejected if there is a delta in any parameter exceeding a defined maximum value. Therefore, the early failure rate for field-installed components can be reduced dramatically and the goal of a real zero-failure quality is approached.

In the case of a high level of confidence by the customers, one additional step may be done. If a producer organizes the production according to the previously mentioned considerations and performs burn-in and test of the product by final outgoing inspection, then customers may install the delivered products in their systems without any additional incoming inspection.

In this manner, the producer will be certified to ship the products directly into customers stock with only a simple check. This procedure is an opportunity to reduce cost and time for both supplier and manufacturer.

6.7.   Transceivers Today and Tomorrow

6.7.1.   TRANSCEIVERS TODAY

Today, FO transceivers for applications in the field of datacom are mostly characterized by following a few well-established international standards such as

> SBCON (formerly ESCON Industrial Standard, set in the early 1990s by IBM)
> Fibre channel (with different bitrate levels)
> FDDI
> ATM (with different bitrate levels) with all their associated standards.

These standards define the electrooptical performance for a transceiver as well as its pinout and physical package, including the (duplex) fiber optic connector interface (Fig. 6.21).

Fig. 6.21   Serial ESCON TRX 200 MBd, 1300 nm.

FO transceivers meeting these standards are operating worldwide in mainframes, WANs, and LANs. The volumes for these transceivers have been continuously increasing during the past 5 years and the prices have shown dramatic decreases in the order of 25% per year. Consequently, the goal of all manufacturers is to offer a high level of performance, reliability, and quality while maintaining cost-effective production to meet the market pricing.

6.7.2.   SOME CONCEPTS OF TOMORROW'S TRANSCEIVERS

6.7.2.1.   Geometrical Size of Receptacle/Optical Connector

The bit rates of FO transceivers are increasing. Therefore, a serious problem emerges due to EMI risks caused by electromagnetic radiation in or out of the receptacle opening in a front panel or back panel of a mainframe computer subunit, for example. To minimize such EMI problems the geometrical outline of the FO connector and consequently the size of the receptacle opening will become much smaller than today (by a minimum of a factor of 2 in size, i.e., a factor of 4 in area). In support of this size-reduction trend, customer demand is following the same direction for much smaller transceivers in order to reduce space and volume on their PCBs and to increase the bit rate per square of PCB (see Fig. 6.22).

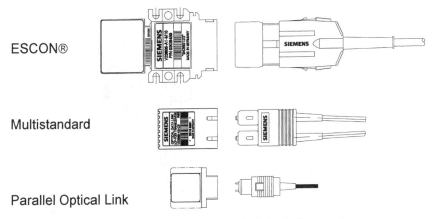

ESCON®

Multistandard

Parallel Optical Link

Fig. 6.22 Comparison of geometrical sizes of transceivers.

6.7.2.2. Functional Integration

A direction for the next generation of transceivers is the inclusion in the module housing of additional electronic functions such as:

Serialization and deserialization of parallel digital bitstreams

Encoding and decoding of bitstreams

Clock regeneration on receiver side

Laser control and laser safety functions in a laser-based transceiver

The main advantage for the customer of such higher levels of integration in transceivers is having no high-speed signals on the system board and all related cost savings. For the developers one of the challenges is to solve the increasingly serious problem of heat dissipation caused by such increased integration of a large number of high-speed digital electronic functions within the package. The direction of the solution is to have IC technologies with a supply voltage of 3.3 V or less, and such ICs are already being introduced (Fig. 6.23).

6.7.2.3. IRED Speed Limits, Lasers, and VCSELs as Optical Sources

If one exceeds the bit rate of approximately 300 MBd in a FO Link, the commonly used IRED on the transmitter side will be too slow and must be replaced by LD. However, the very fast (up to 10 GBd) conventional EELD is more complicated in application than an IRED because of the following:

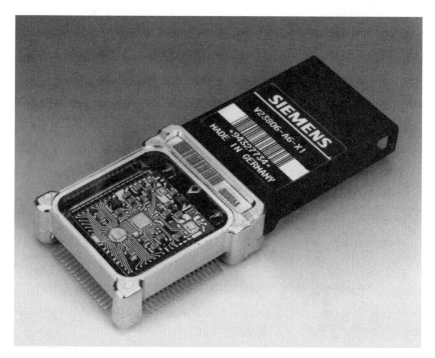

Fig. 6.23   Parallel ESCON TRX 200 MBd, 10-bit parallel electrical in/out.

An EELD needs control circuitry driven by monitoring the optical power due to its significant temperature dependence and electrooptical efficiency characteristic with aging.

Accurate and reliable optical coupling of an EELD to the fiber is much more difficult and therefore more expensive compared to an IRED. The positional accuracy and stability needed for EELD in the range of 0.1 μm is approximately an order of magnitude tighter than that for coupling to an IRED.

Laser safety due to potentially high optical output power has to be taken into account. That can be done by limitation of radiated optical power or by electrical limitation of LD power output (see Section 6.3). In the past some producers also proposed mechanical solutions, such as shutters, that have not been well accepted by customers. Nevertheless, products for most datacom applications cannot be successful in the market without certification of being laser class 1 safe according to IEC 825 or corresponding regulations such as those of the FDA.

Currently, several groups are working on a new type of laser called VCSEL, which originally was developed as a source for pumping energy in erbium-doped fiber amplifiers for long-haul telecom transmission lines. Figure 6.24 shows a cross-sectional diagram of a VCSEL.

One of the key advantages of a VCSEL compared to a edge-emitting laser diode (EELD) is based on the IRED-like technology. This allows one to produce VCSELs with all processing steps, including final testing, completely on a wafer level. Some additional advantages of VCSELs with respect to the EELD are listed in Table 6.1.

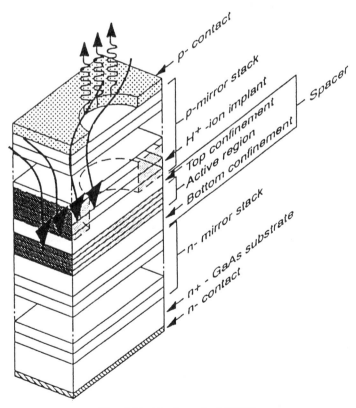

Fig. 6.24 Design of a VCSEL.

Table 6.1   **Comparison of Features, EELD vs VCSEL**

Feature	EELD	VCSEL
Size of active area	0.5–1 × 2–10 μm	Variable, 5–50 μm diameter
Number of modes	Typically 1 or few	1 or even up to many 10s
Beam divergence	Elliptic up to 60 × 20°	Circular ca. 5°
Processing of chip	Very specific	Like LEDs
Final processing	Single bar	On wafer
Burn-in and test	Single on heatsink	On wafer
Environmental sensitivity	Extremely high	Moderate
Coupling to fiber	Tricky and sensitive	Easy
Threshold current	Approximately 10 mA	Some mA
Direct modulation bandwidth	High, up to 10 Gb	High
Temperature drift	Fairly high	Tendencially low

6.7.2.4.   Laser Diodes for Multimode Fibers—Mode Underfill

Worldwide there are many miles of multimode fibers installed in buildings and campuses. However, basically the speed and transmission field length of FO links with MM fibers combined with IREDs is limited due to power budget and bandwidth length limits.

In order to save this investment and use this current cabling even for higher speed transmission over distances of more than 100 m, the concept emerges of using laser diodes as sources for multimode fibers. There are groups studying the idea of extending the limits of multimode fibers by means of the so-called mode underfill launch condition. That means that coupling of optical power from a LD into a multimode fiber with limited focal diameter and numerical aperture will establish only a few low-order propagation modes in the fiber core. The result would be a significant increase of the fiber's bandwidth length limit.

Preliminary experimental investigations are confirming the theoretical assumptions. Therefore, transmission of up to 1 GBd with 1300 nm wavelength over more than 500 m of graded-index MM fibers would work well. This direction is receiving intense discussion in the related standardization groups.

This technique will be established in the market only if the price and performance are drastically improved for LD-based products. One key component would be an inexpensive LD optical subassembly with the previously mentioned performance.

6.7.2.5. Parallel–Optical Links

In order to increase efficiency in ICs and electronic circuitry for encoding, decoding, serializing, and deserializing bit streams, increasingly the direction is to transform the parallel electrical bit streams directly into parallel optical bit streams for transmission via fibers. Even in high-speed short links the aspect of signal propagation time loss due to encoding and serializing supports parallel–optical transmission. A simple block diagram of a parallel optical link is shown in Fig. 6.25.

Due to its special advantages an array of VCSELs will be nearly the ideal light source for parallel–optical transmitters. The processing of VCSELs by means of the well-known IRED technology allows the definition of important design parameters very exactly. Therefore, the pitch of the single diodes in the array should correspond within 1 μm to the pitch of fibers in a fiber ribbon, and the diameter of the active region combined with a small numerical aperture of a VCSEL additionally gives the potential for very robust, simple, and effective optical coupling techniques. This is becoming the basis for cost-saving module designs.

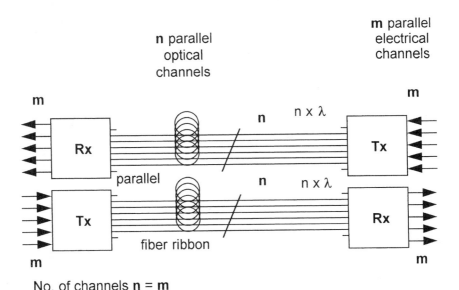

Fig. 6.25 Block diagram of a parallel–optical link.

To reach a real breakthrough for parallel–optical links some effort must be spent on vertical cavity surface-emitting laser (VCSEL) technology for high uniformity of characteristic performances for all VCSELs in an array, or better, in a whole wafer (with thousands of single VCSELs) or a couple of wafers. Only a very high overall yield will drop the unit price as necessary. The batch-type complete processing and testing possibility considerably help reaching this goal.

On the receiver side, an array of, e.g., PIN photodiodes has to be designed with a pitch corresponding to the fiber ribbon. The diameter of the DC photosensitive area of the diodes should be as small as reasonably possible for high detection speed but also must be adapted well to the numerical aperture of the fibers. To maintain the detection speed and sensitivity of the PDs special care must be taken to have a very short and minimum capacitance connection to the first input stages of the preamplifier array. Additionally, some special measures for cross-talk prevention must be implemented into circuitry design.

In the world of datacom a module typically is linked to the fiber transmission line directly via a fiber optic connector as opposed to a fiber pigtail

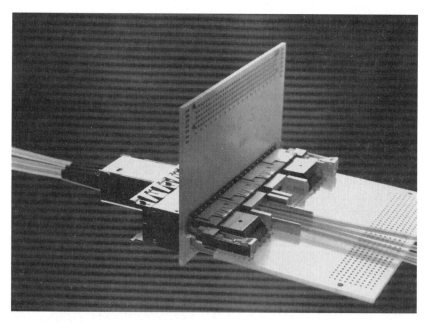

Fig. 6.26   Shielded back panel connector for 4 × 12 fibers.

as is mostly the case in the telecom world. Consequently, for parallel-optical links one needs a multifiber connector fitted to fiber ribbon cables. A serious challenge for the design is to get a low level of insertion loss combined with high fiber-to-fiber uniformity as well as good mating repeatability and reliability. All this must be achieved for a price of under $10 per channel.

During discussions in standardization groups such as HIPPI 6400, the well-known MT connector has been singled out to be a good basis for future multifiber connectors. An example of an electrically shielded multifiber connector for 4 $\times$ 12 multimode fibers 62.5 μm thick is given in Fig. 6.26. This connector is fully compatible with the well-established back-panel system SIPAC.

Given the fundamental advantages offered by fiber optic technology and the advancements presented here as well, as those yet to come, this technology will be critical to advancements in both data communications and telecommunications in the future.

Part 2 | The Links

Chapter 7 | Fiber Optic Link Requirements

Casimer DeCusatis

IBM Corporation, Poughkeepsie, New York 12590

7.1. Introduction

In previous chapters, an overview of the basic technologies used in optoelectronic communications has been given. With these building blocks, we can
begin to assemble practical fiber optic data communication links. In this chapter, I will describe the performance requirements of a fiber link and consider
some of the design trade-offs involved in link planning and implementation.
For the purposes of this chapter, I will consider the performance of point-to-
point data links as illustrated in Fig. 6.4; this is the most common building
block of data communication networks, which is used to build up more complex logical topologies such as switched stars, rings, and so forth. Network
management is a much more complex topic beyond the scope of this book;
in any case, the performance of a network system depends on the interaction
of many elements, including both hardware and software, and is thus highly
application dependent. Most data links are full duplex, permitting two-way
communication between attached devices; a separate transmitter/receiver
pair is typically used for each direction. Datacom systems do not make efficient use of the fiber bandwidth; one user per link is typical. Data link design
is not driven by optimization of the fiber bandwidth but rather by the cost of
optoelectronic transceivers and adapters. By contrast, the telecommunications world makes extensive use of multiplexing to support multiple users on
a single fiber. This minimizes cabling costs and is most clearly illustrated by
transoceanic fiber links that carry hundreds or thousands of voice signals on
a single optical fiber. In other words, the number of electronic transceivers
per kilometer of fiber is much higher in datacom applications than in telecom
systems. Multiplexing techniques are only just beginning to be introduced in

HANDBOOK OF FIBER OPTIC
DATA COMMUNICATION

the datacom market and currently are only practical for large users (see the IBM 9729 Wavelength Division Multiplexer [1] for an example).

There are other important differences between the telecom and datacom applications, even though at first glance they appear to use the same technologies. First, datacom links must maintain a much lower bit error rate (BER), defined as the number of errors per second. This is because the consequences of an error in a data link may be very serious; bad data in the wrong place could potentially crash the computer operating system or cause small but significant errors such as moving the decimal point in a bank account balance. Although transmission rates vary, typical low-end fiber optic links operate around 10–15 Mb/s. By contrast, telecom systems transmit voice signals with a bandwidth of only about 4 kHz and the ultimate receiver is the human ear; excess line noise can often be ignored, such as the background static present on a long-distance call overseas. For this reason, datacom systems must support higher fidelity transmission and a BER of 10e-12 or better (approximately one error every hour for a 200 Mb/s link), as compared with about 10e-9 for telecom systems (approximately one error every 5 s). Some datacom links will support a BER as low as 10e-15; even testing such a link becomes challenging when an error occurs only once every 57 days! I will discuss optical link budgets and BER further in the following sections.

Another unique requirement of datacom systems is distance. Before the introduction of fiber optics in the data center, copper coax or parallel wires were limited to a maximum transmission distance of perhaps a few hundred meters (this is one reason why computer data centers tended to keep equipment clustered together in a single large room). Fiber optic links are capable of error-free transmission over tens of km at data rates significantly higher than those of copper systems (Table 7.1).

Table 7.1 **Comparison of Datacom and Telecom Requirements**

	Datacom	*Telecom*
BER	10e-12 to 10e-15	10e-9e
Distance	20–50 km	Varies with repeaters
No. transceivers/km	Large	Small
Signal bandwidth	100 Mb–1 Gb	3–5 Kb
Field service	Untrained users	Trained staff
No. fiber replugs	250–500	<100 over lifetime

This increase in the data rate-distance product has opened up the data center and is partly responsible for recent distributed computing strategies; it has also enabled new applications, such as real-time data backup at a remote location for emergency disaster recovery. Finally, telecom systems are serviced only by trained personnel and are completely controlled by the local telephone company. In contrast, datacom systems are often services by the users themselves (who would never think of crawling into a manhole and reconnecting wires if their telephone service failed to work). Thus, datacom equipment must be rugged enough to withstand a large number of connections and disconnections, high pulling forces on fiber cables, and other user-inflicted stress. The equipment should also be easy to use and maintain; for example, connectors are keyed to prevent misplugging, and receptacles can be designed to minimize the need for frequent cleaning. Industry standards are an important part of the multivendor datacom environment; however, they do not always guarantee that equipment will work together as intended or promised. Whereas most telecommunications carriers purchase their equipment from only a handful of suppliers, a large data center may be composed of a wide variety of equipment from different vendors. There are standardized procedures for testing and specifying the physical interface of fiber optic equipment; a partial list of some useful fiber optic test procedures (FOTP) standards is given in table Table 7.2. However, this does not guarantee interoperability of equipment from different sources; it is not possible to test every combination of equipment from different sources. A system composed of parts from multiple sources will often fail at its weakest point; thus, different reliability requirements in fiber optic components should be considered (remember that even if a product is intended for only a 40,000-hour lifetime, it must still ensure at

Table 7.2 **Comparison of Serial Optical Fiber vs Parallel Copper Datacom**

	Fiber	*Copper*
Distance, unrepeated	20–50 km	10–100 m
Data rate	100–1000 Mb/s	<100 Mb/s
Weight	20 g/m	1.5 kg/m
Size (outer jacketed diameter)	<5 mm	10 mm–5 cm
Flexibility (bend radius)	10–20 mm	>2 cm (limited by cable thickness)
Noise immunity	Excellent	Good to poor

least 100,000-hours mean time between failures). This problem is somewhat more acute in data communication systems than in telecom systems because data communication networks tend to have more active devices per kilometer of fiber than telecommunications links. The requirements of the international market further complicate interoperability. In most parts of the world, it is possible to make an international phone call without any special effort because the local phone companies have agreed to work together passing calls along their respective networks. There are strict standards that must be met by datacom equipment in order to operate over the public network; it is impossible to even export equipment to some countries, regardless of the intended application, if it fails to meet certain standardized tests. This process is generally called type approval or homologation, and the requirements vary greatly among different countries; Table 7.1 illustrates some requirements for shipping datacom products outside the United States.

7.1.1. TESTING STANDARDS

The standards for specification testing of fiber optic cable and components are well defined by the FOTP series of standards. There are more than 170 such standard documents describing how to measure the specifications of fiber optic link components. I have provided a partial list of some useful FOTP standards (Table 7.3); contact ANSI for a complete listing. In addition to testing of the link components, fiber optic equipment must demonstrate compliance with national and international standards intended to guarantee performance in typical applications (Table 7.4). In particular, the standards required in order to export fiber optic datacom products from the United States are especially detailed and can be difficult to meet. Table 7.5 lists some of the technical standards required in order to obtain homologation and type approval to import or export datacom equipment from the United States to various international countries. Not all nations require approval to all these standards, and the country by country requirements change frequently; consult the local telecom authority (PTT) for the latest information by country. However, the European Commission has recently produced a set of directives intended to harmonize legal product standards in Europe and create a single market by ensuring free sale of compliant products anywhere in Europe. The manufacturer must demonstrate compliance with these standards for all devices containing datacom equipment or adapters; failure to comply may result in denial of import–export permission, legal action, fines, or other actions by individual coun-

Table 7.3   **Selected FOTP Standards for Fiber Optic Testing**

FOTP Standard Number	Purpose
27	Methods for measuring outside diameter of optical waveguide fibers
30	Frequency domain measurements of multimode optical fiber information transmission capacity
45	Microscopic method for measuring fiber geometry of optical waveguide fibers
47	Output far-field radiation pattern measurement
51	Pulse distortion measurement of multimode glass optical fiber information transmission capacity
54	Mode scrambler requirements for overfilled launching conditions to multimode fibers
80	Cutoff wavelength of uncabled single-mode fiber by transmitted power
107	Return loss for fiber optic components
127	Spectral characteristics for multimode lasers
168	Chromatic dispersion measurement of multimode graded-index and single-mode optical fibers by spectral group delay measurement in the time domain
170	Cable cutoff wavelength of single-mode fiber by transmitted power
176	Measurement method of optical fiber geometry by automated gray-scale analysis

Note. The table provides a partial list of industry standard FOTP published by the EIA. Documents can be ordered directly from EIA. Equivalent test procedures as specified in CCITT G.651 or G.652; all FOTP XXX are equivalent to EIA/TIA-455-XXX.

tries. There is also a European standard governing the shipment of electronic products to the European Community (EC) after January 1996—the Communaute Europeenne (CE) mark is applied to products that conform to the EC directives 89/336/EEC [radio frequency interference (RFI) and electrical safety for electronic equipment] or 91/263/EEC (RFI and electrical safety for telecommunications equipment). Products that fail to earn the CE mark by demonstrating compliance at a certified laboratory will not be allowed to enter the EC, and products that claim compliance and are found not to meet these standards are subject to both criminal and civil penalties. Note that this is a telecom CE mark, an additional requirement to

Table 7.4   Recent Homologation Requirements for Leased Dark Fiber by Country

Country	Dark fiber
Argentina	x
Austria	c
Belgium	c
Brazil	—
Canada	x
China, PR	x
Denmark	—
Finland	—
France	—
Germany	c
Ireland	—
Israel	c
Italy	x
Japan	x
Korea	x
Mexico	x
Norway	—
Poland	c
Portugal	—
Spain	x
Switzerland	c
United Kingdom	x

Note. All information currrent as of January 1996; subject to change by local telecom authorities without notice. c, available on case-by-case basis; −, no homologation; x, service not available.

Table 7.5   General Technical Specifications Required for Datacom Product Homologation

Safety compliance: IEC 950/EN-60950 (2nd ed.); LVD (low voltage directive) 73/23/EEC; csa-c22.2, no. 950-m89

ISDN only: ETS 300-046/047

Optical fiber only (optical safety): EN-60825/IEC 825

EMC compliance: EN-55022/CISPR-22 for information technology equipment
   Related standards: EN-60555; EN-61010-1 (IEC 1010-1); EN-50082-1;
   FCC part 15

Analog only: VDE 0878

the more common CE mark applied to products that only meet radiated noise requirements; it also requires products to have immunity to any signals likely in their normal environment, which has not been a legal requirement in the United States. These standards are strict, and datacom products not designed for EMC compliance are unlikely to meet them. Not only must the product itself comply but also it must comply when configured in a typical system, including cables and external interfaces. For example, this can impact fiber optic transceivers that may radiate switching noise outside the protective covers of a host box or fiber optic connectors that contain metal that can reradiate electrical noise when plugged into the transceiver. Manufacturers must sign a declaration of conformity specifying the product, the directives it meets, and the standards applied. If a product meets standards in full, self-certification may be acceptable; otherwise, a technical construction file explaining the compliance rationale must be approved by a designated competent body. Products must be supported by a design file containing test certificates/reports, design information, users guide, schematics, and related information; this documentation must be retained for at least 10 years after the manufacture of the last unit to which it applies. Manufacturers outside Europe must appoint an authorized representative within Europe to take responsibility for all CE mark issues. There may also be additional export standards for datacom and telecom products to meet, such as ISO 9000 series certification on development or manufacturing, the related QS-9000 standards, or the European Environmental Conformance Standard ISO 14000, which requires that all products entering the EC will not pollute the environment or endanger local wildlife habitats. Many other nations are also considering adopting similar standards, so these product requirements should not be considered specific to the European market. Datacom products may also be affected by other standards in the United States, such as Federal Communications Commission (FCC) compliance standards and Environmental Impact Assessments (EIA) on datacom products (because lasers and photodetectors commonly contain gallium arsenide compounds, which qualify as toxic substances for disposal). Because these standards change often, consult the standards bodies or approval labs for the latest information.

7.1.2. BIT ERROR RATE

All fiber optic channels consist of an optical source or transmitter, the cable plant, and an optical receiver. These components have been described in previous chapters; I will concentrate on their interaction when designing a fiber link. The transmitter is capable of launching a limited amount of optical power into the fiber, and there is a limit to the weakness of a signal that can be detected by the receiver in the presence of noise. Thus, a fundamental consideration is the optical link power budget, or the difference between the transmitted and received optical power levels. Some power will be lost due to connections, splices, and bulk attenuation in the fiber. There may also be optical power penalties due to dispersion, model noise, or other effects in the fiber and electronics. The optical power levels define the signal-to-noise ratio at the receiver, Q; this is related to the bit error rate by the well-known Gaussian integral

$$\text{BER} = \frac{1}{\sqrt{2\pi}} \int_{Q}^{\infty} e^{-\frac{Q^2}{2}} \, dQ \cong \frac{1}{Q\sqrt{2\pi}} e^{-\frac{Q^2}{2}}. \tag{7.1}$$

From Eq. (7.1), we see that a plot of BER vs received optical power yields a straight line on semilog scale, as illustrated in Fig. 7.1. Nominally, the slope is approximately 1.8 dB/decade; deviations from a straight line may indicate the presence of nonlinear or non-Gaussian noise sources. Some effects, such as fiber attenuation, are linear noise sources; they can be overcome by increasing the received optical power, as seen from Fig. 7.1, subject to constraints on maximum optical power (laser safety) and the limits of receiver sensitivity. There are other types of noise sources, such as mode partition noise or relative intensity noise (RIN), that are independent of signal strength. When such noise is present, no amount of increase in transmitted signal strength will affect the BER; a noise floor is produced, as shown by curve B in Fig. 7.1. This type of noise can be a serious limitation on link performance.

As can be seen from Fig. 7.1 that receiver sensitivity is specified at a given BER, which is often too low to measure directly in a reasonable amount of time (for example, a 200 Mb/s link operating at a BER of 10e-15 will take only one error every 57 days on average, and several hundred errors are recommended for a reasonable BER measurement!). For practical reasons, the BER is typically measured at much higher error rates, where the data can be collected more quickly (such as 10e-4 to 10e-8) and then extrapolated to find the sensitivity at low BER. This assumes the absence of nonlinear noise floors, as cautioned previously. The relationship between optical input power, in watts, and the BER, is the complimentary Gaussian error function.

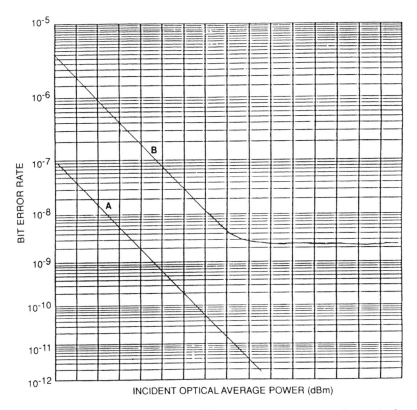

Fig. 7.1 Bit error rate as a function of received optical power. Curve A shows typical performance, whereas curve B shows a BER floor [5].

$$BER = 1/2 \text{ error function } (P_{out} - P_{signal}/RMS \text{ noise}), \qquad (7.2)$$

where the error function is an open integral that cannot be solved directly. Several approximations have been developed for this integral, which can be developed into transformation functions that yield a linear least-squares fit to the data. The resulting fitted equation can then be used to extrapolate the receiver sensitivity. Examples of two transformation functions that have been developed are the following:

$$T(BER) = 10^{(0.526966 \log (-1.837794 \log (3.7678 \, BER)))} \qquad (7.3)$$

$$T(BER) = -C_2 + \sqrt{C_2^2 + 4C_1(C_3 + \ln(2BER))} / 2 \, C_1, \qquad (7.4)$$

where $C_1 = 0.4926119$, $C_2 = 0.2498322$, and $C_3 = 0.7912445$. The collected raw data are used to calculate the least-squares curve coefficients, which are used to assemble a curve fit according to

$$m = \Sigma\ XY - (\Sigma X\ \Sigma Y/N)/\Sigma\ X^2 - ((\Sigma\ X)^2/N) \qquad (7.5)$$

$$c = (\Sigma\ Y - m\ \Sigma\ X)/N, \qquad (7.6)$$

where N is the number of data points, X is optical power in watts (not dBm), $Y = T(\text{BER})$ is the transformed BER data, and

$$T(\text{BER}) = m\ (\text{optical power}) + c. \qquad (7.7)$$

Rearranging this equation, the receiver sensitivity cannot be extrapolated to the desired BER:

$$\text{Optical sensitivity} = 1/m\ (T(\text{Spec. BER}) - c). \qquad (7.8)$$

The same curve-fitting equations can also be used to characterize the eye window performance of optical receivers. Clock position/phase vs BER data are collected for each edge of the eye window; these data sets are then curve fitted with the previous equations to determine the clock position at the desired BER. The difference in the two resulting clock position on either side of the window gives the clear eye opening. Further information on curve fitting of low BER can be obtained in the literature [2–4].

After consulting Fig. 7.1, we might be tempted to conclude that in the absence of nonlinear noise effects, higher optical power is always more desirable at the receiver. If we plot BER vs receiver sensitivity for increasing optical power, we obtain a curve similar to that in Fig. 7.2, which shows that for very high power levels, the receiver will go into saturation. This can happen if a laser transmitter designed for long-distance applications is inadvertently connected over a short link; the situation also arises when using optical loopbacks or wrap plugs for transceiver diagnostics. We see from the characteristic "bathtub"-shaped curve that there is a window of operation with both upper and lower limits on the received power, using a powerful laser does not provide a brute-force solution to link distance problems.

In describing Figs. 7.1 and 7.2, I have also made some assumptions about the receiver circuit. Most data links are asynchronous and do not transmit a clock pulse along with the data; instead, a clock is extracted from the incoming data and used to retime the received data stream. I have made the assumption that the BER is measured with the clock at the center of

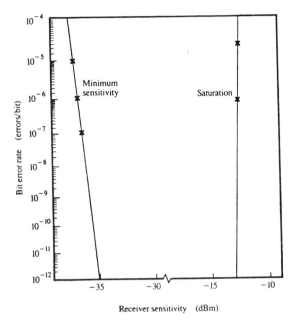

Fig. 7.2 Bit error rate as a function of received optical power illustrating range of operation from minimum sensitivity to saturation. [5].

the received data bit; ideally, this is when we compare the signal with a preset threshold to determine if a logical "1" or "0" was sent. When the clock is recovered from a receiver circuit such as a phase-lock loop, there is always some uncertainty about the clock position; even if it is centered on the data bit, the relative clock position may drift over time. The region of the bit interval in the time domain where the BER is acceptable is called the eye width; if the clock timing is swept over the data bit using a delay generator, the BER will degrade near the edges of the eye window. Eye width measurements are an important parameter in link design, and transceiver vendors will often show a sample eye pattern on their product data sheets. I will describe timing jitter as part of the discussion on optical link budget modeling. In order to design a proper optical data link, the contribution of different types of noise sources should be assessed when developing a link budget. There are two basic approaches to link budget modeling. One method is to design the link to operate at the desired BER when all the individual link components assume their worst-case performance. This conservative approach is desirable when very high performance is required or when it is difficult or inconvenient to replace failing components near the end of their useful lifetimes. The resulting design has a high safety margin; in some cases, it may be overdesigned for the required level of performance. Because it is very unlikely that all the elements of the link will assume their worst-case performance at the same time, an alternative is to model the link budget statistically. For this method, distributions of transmitter power output, receiver sensitivity, and other parameters are either obtained from the vendors or estimated by measuring a small sample of parts. They are then combined statistically using an approach such as the Monte Carlo method, in which many possible link combinations are simulated to generate an overall distribution of the available link optical power. The industry standard is a 3-σ design, in which the combined variations of all link components are not allowed to extend more than 3 standard deviations from the average performance target in either direction. The statistical approach results in greater design flexibility and generally increased distance compared with a worst-case model at the same BER. The statistical approach is only as valid as the input data, however, and these can be difficult to obtain; many vendors consider such data proprietary because it gives an indication of their overall manufacturing process performance. A good desgin compromise is to combine the two methods, using worst-case for some parameters and statistical for others. I will describe the link budget planning process in greater detail in the following sections.

7.2. Link Budget Planning

I now discuss important factors in the modeling and design of fiber optic data links, with emphasis on the link budget requirements, following the treatment of ref. [5]. It is convenient to break down the link budget into two areas: installation loss and available power. Installation or DC loss refers to optical losses associated with the fiber cable plant such as connector loss, splice loss, and bandwidth considerations. Available optical power is the difference between the transmitter output and receiver input powers, minus additional losses due to optical noise sources on the link (also known as AC losses). With this approach, one may choose to treat the installation loss budget as statistical and the available power budget as worst case.

7.2.1. *INSTALLATION LOSS BUDGET*

First, I consider the installation loss budget. Some of the important factors to be considered include the following:

Fiber cable loss

Fiber attenuation as a function of wavelength

Connector loss

Splice loss

Link measurement uncertainty

A typical point-to-point fiber optic link connecting two devices is shown in Fig. 7.3, consisting of both trunk and jumper cables. Trunk cable is used for long installations (hundreds of meters to several kilometers, usually between buildings). A single trunk cable may contain many individual fibers. Jumper cables are used for short connections between devices that are close together or between a device and a distribution panel; they typically contain only one or two fibers. Jumper cables are usually limited to less than 100 m in length and are therefore allowed to have a slightly higher attenuation loss than trunk cable. Typical trunk cable will have a loss of approximately 0.5 dB/km near 1300 or 1550 nm, whereas jumper cable in this range may average 0.8 dB/km or more. Fibers are available with loss as low as 0.2 dB/km for some applications.

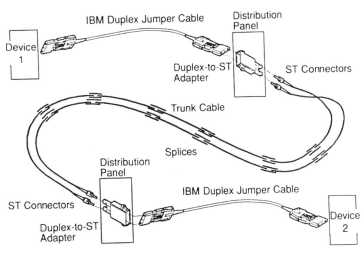

Fig. 7.3 Example of a point-to-point data link, ESCON environment [5].

Transmission loss is perhaps the most important property of an optical fiber or cable; it affects the link budget and maximum unrepeated distance. The number and separation between optical repeaters and regenerators are largely determined by this loss. The mechanisms responsible for this loss (material absorption and linear and nonlinear scattering) have been discussed in Chapter 1. It was also noted that fiber loss varies with operating wavelength, providing transmission windows for optical communication. In most optical transmitters, the central wavelength is subject to variation due to the manufacturing process as well as environment factors such as temperature. Even for laser sources with a spectral width of 6 nm or less, the center wavelength may vary by 80 nm or more. This causes additional loss that must be accounted for in the link budget. Although wavelength-dependent attenuation data are available from most fiber manufacturers, it is often not practical to perform measurements of the fiber at each wavelength of interest, especially during a repair activity. An accurate model for fiber loss as a function of wavelength has been developed by Walker [6]; this model accounts for the effects of linear scattering, macro-bending, material absorption due to ultraviolet and infrared band edges, hydroxide (OH) absorption, and absorption from common impurities such as phosphorous. Using this model, it is possible to calculate the fiber loss as a function of wavelength for different impurity levels; an example plot is shown in Fig. 7.4. Using this method, the fiber properties can be specified

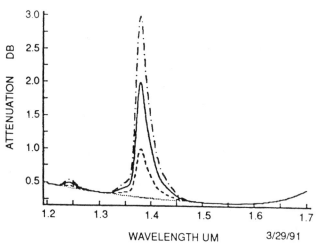

Fig. 7.4 Spectral loss characteristics of single-mode fiber for different impurity levels [5].

along with the acceptable wavelength limits of the source to limit the fiber loss over the entire operating wavelength range; design trade-offs are possible between center wavelength and fiber composition to achieve the desired result. Typical loss due to wavelength-dependent attenuation for laser sources on single-mode fiber can be held below 0.1 dB/km.

In Chapter 1, optical losses due to microbends and macrobends in optical fiber were discussed; clearly, these are to be avoided during the fiber installation. Another important environmental factor is exposure of the fiber to ionizing radiation damage, which is important in some environments such as research labs, nuclear power plants, medical facilities, military applications, and space satellites. Although optical fiber is by definition immune to electromagnetic interference, there is a large body of literature concerning the effects of ionizing radiation on fiber links [7–11]. There are many factors that can affect the radiation susceptibility of optical fiber, including the type of fiber, type of radiation (gamma radiation is usually assumed to be representative), total dose, dose rate (important only for higher exposure levels), prior irradiation history of the fiber, temperature, wavelength, and data rate. Optical fiber with a pure silica core is least susceptible to radiation damage; however, almost all commercial fiber is intentionally doped to control the refractive index of the core and cladding as well as dispersion properties. Trace impurities are also introduced that become important only under irradiation; among the most important are Ge dopants in the core of graded-index fibers, in addition to F, Cl, P, B, OH content, and the alkali metals [7]. In general, radiation sensitivity is worst at lower temperatures and is also made worse by hydrogen diffusion from materials in the fiber cladding.

Because of the many factors involved, there does not exist a comprehensive theory to model radiation damage in optical fibers. The basic physics of the interaction has been described [8, 9]; there are two dominant mechanisms—radiation-induced darkening and scintillation. First, high-energy radiation can interact with dopants, impurities, or defects in the glass structure to produce color centers that absorb strongly at the operating wavelength. Carriers can also be freed by radiolytic or photochemical processes; some of these become trapped at defect sites, thus modifying the band structure of the fiber and causing strong absorption at infrared wavelengths. This radiation-induced darkening increases the fiber attenuation; in some cases, it is partially reversible when the radiation is removed, although high levels or prolonged exposure will permanently damage the fiber. A second effect is caused if the radiation interacts with impurities to produce stray light, or scintillation. This light is generally broadband but will tend to degrade the BER at the receiver; scintillation is a weaker effect than radiation-induced darkening. These effects will degrade the BER of a link; they can be prevented by shielding the fiber or partially overcome by a third mechanism, photobleaching [10, 11]. The presence of intense light at the proper wavelength can partially reverse the effects of darkening in a fiber. It is also possible to treat silica core fibers by briefly exposing them to controlled levels of radiation at controlled temperatures; this increases the fiber loss, but makes the fiber less susceptible to future irradiation. These so-called radiation-hardened fibers are often used in environments where radiation is anticipated to play an important role.

Models for the fiber's response to high radiation levels have been proposed [11]. There have also been studies of the long-term reliability and lifetime effects of background radiation levels, such as those encountered in buried or undersea fibers [12]. Recently, several models have been advanced [13, 14] for the performance of fiber under moderate radiation levels; the effect on BER is a power law model of the form

$$BER = BER_0 + A \, (\text{dose})^b, \tag{7.9}$$

where BER_0 is the link BER prior to irradiation, the dose is given in rads, and the constants A and b are empirically fitted. I will not discuss in detail modeling work in this area, which is available in several references [13, 14]. In most applications, the link can be sufficiently shielded from exposure to mitigate these effects, or specially fabricated radiation-hardened fiber can be used to reduce the effects. The loss due to normal background radiation exposure over a typical link lifetime can be held below approximately 0.5 dB.

There are also installation losses associated with fiber optic connectors and splices; both of these are inherently statistical in nature. There are many different kinds of standardized optical connectors, some of which are shown in Fig. 7.5. Generally, higher variability has been observed for screw-type connectors compared with push–pull or bayonette-type designs. By "making and breaking" the connection many times, a connector loss distribution can be measured; for large number of connections, the distribution is often approximately Gaussian and can be characterized by a mean and σ value as shown. Typically, published manufacturer's connection loss values assume similar fiber parameters on both sides of the connection. This may not always be the case; for example, due to different tolerances on jumper and trunk cables, there may be a mode field diameter mismatch

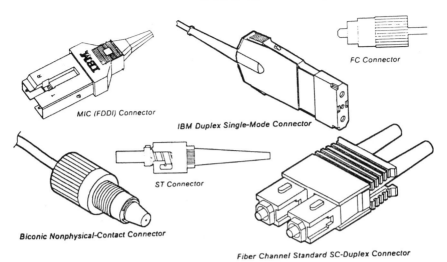

Fig. 7.5 Common fiber optic connectors: MIC (FDDI), ST, FC, ESCON, SC duplex, and biconic (nonphysical contact) [5].

between a jumper-to-trunk connection at a distribution panel. Usually, the variation due to mechanical tolerances on the connectors will dominate any excess connector loss due to mode field diameter mismatch; for some applications, this represents an addition statistical penalty that must be accounted for in the installation loss budget. Usually, the net effect is limited to a few tenths of a dB. There are many different models that have been published for estimating connection loss due to fiber misalignment [15–20]; most of these treat loss due to misalignment of fiber cores, offset of fibers on either side of the connector, and angular misalignment of fibers. The loss due to these effects is then combined into an overall estimate of the connector performance. Although some of these models have experimental verification, there is still no general model available to treat all types of connectors that then allows the modeling information to be used in improving the connector design. The development of optical connectors remains something of an art; even so-called "standardized" connectors exhibit widely different properties depending on the manufacturer. The standard merely ensures that connectors will be mechanically compatible and interoperate; it does not guarantee system performance or other desirable properties such as ease of plugging and durability.

Optical splices are required for longer links, because fiber is usually available in spools of 1–5 km, or to repair broken fibers. There are two basic types, mechanical splices (which involve placing the two fiber ends in a receptacle that holds them close together, usually with epoxy) and fusion splices (in which the fibers are aligned and then heated sufficiently to fuse the two ends together). Fusion splices have become much more common because of their low loss and the advent of new portable tools for efficiently performing fusion splices in the field. Typical splice loss values are given in Table 7.6.

An additional effect of lossy connectors and splices is modal noise. Because high-capacity optical links tend to use highly coherent laser transmitters, different modes propagating in the fiber may interfere with one another. Random coupling between fiber modes causes fluctuations in the optical power coupled through splices and connectors; this phenomena is known as modal noise [18]. As one might expect, modal noise is worst when using laser sources in conjunction with multimode fiber; recent industry standards [19] have allowed the use of shortwave lasers on 50-μm fiber because of potential cost savings on this type of link hardware. Shortwave lasers are currently readily available because of their wide use in compact disk systems for optical storage, and the relaxed tolerances of multimode fiber makes connectorization problems simpler and less expensive to solve. The main disadvantage to such systems is modal noise, along with the limitation of using only 50-μm fiber (the bandwidth of 62.5-μm fiber seriously limits link distance). Modal noise is usually considered to be nonexistent in single-mode systems. However, modal noise in single-mode fibers can arise when higher order modes are generated at imperfect connections or splices. If the lossy mode is not completely attenuated before it reaches the next connection, interference with the dominant mode may occur. This is especially important in local area networks (LANS) in which large numbers of connectors and/or splices are used over relatively short distances. The effect of modal noise have been modeled previously [18], assuming that the only significant interaction occurs between the LP01 and LP11 modes for a sufficiently coherent laser. For N sections of fiber, each of length L in a single-mode link, the worst-case σ for a modal noise can be given by

$$\sigma_m = \sqrt{2} \, N\eta \, (1 - \eta) \, e^{-aL}, \qquad (7.10)$$

where a is the attenuation coefficient of the LP11 mode, and η is the splice transmission efficiency, given by

$$\eta = 10^{-(\eta_0/10)}, \qquad (7.11)$$

Table 7.6 **Typical ESCON Cable Plant Optical Losses [5]**

Component	Description	Size (μm)	Mean loss	Variance (dB²)
Connector[a]	Physical contact	62.5–62.5	0.40 dB	0.02
		50.0–50.0	0.40 dB	0.02
		9.0–9.0[b]	0.35 dB	0.06
		62.5–50.0	2.10 dB	0.12
		50.0–62.5	0.00 dB	0.01
Connector[a]	Nonphysical contact (multimode only)	62.5–62.5	0.70 dB	0.04
		50.0–50.0	0.70 dB	0.04
		62.5–50.0	2.40 dB	0.12
		50.0–62.5	0.30 dB	0.01
Splice	Mechanical	62.5–62.5	0.15 dB	0.01
		50.0–50.0	0.15 dB	0.01
		9.0–9.0[b]	0.15 dB	0.01
Splice	Fusion	62.5–62.5	0.40 dB	0.01
		50.0–50.0	0.40 dB	0.01
		9.0–9.0[b]	0.40 dB	0.01
Cable	IBM multimode jumper	62.5	1.75 dB/km	NA
	IBM multimode jumper	50.0	3.00 dB/km at 850 nm	NA
	IBM single-mode jumper	9.0	0.8 dB/km	NA
	Trunk	62.5	1.00 dB/km	NA
	Trunk	50.0	0.90 dB/km	NA
	Trunk	9.0	0.50 dB/km	NA

[a] The connector loss value is typical when attaching identical connectors. The loss can vary significantly if attaching different connector types.
[b] Single-mode connectors and splices must meet a minimum return loss specification of 28 dB.

where η_o is the mean splice loss (typically, splice transmission efficiency will exceed 90%). The corresponding optical power penalty due to modal noise is given by

$$P = -5 \log (1 - Q^2 \sigma_m^2), \tag{7.12}$$

where Q corresponds to the desired BER.

One final factor that deserves attention as part of the installation loss budget is the uncertainty in performing link loss measurements. There are many different standardized techniques to measuring link loss [21]. One approach involves using a known source and detector to measure the loss of a sample cable and then replacing the cable with the link under consideration and measuring the link loss. By taking the difference between these measurements, random factors such as the connection loss can be reduced. However, these effects cannot be totally eliminated, and the replacement loss technique will itself produce a distribution of measured loss values centered about the mean or average link loss. If this variability is small, as for the large tolerances associated with multimode fiber, then the statistics of the replacement loss technique can be measured and incorporated into the installation loss budget. However, for some applications (particularly single-mode fiber) this approach is not accurate enough. Instead of using typical jumper cables, special reference jumper cables must be used that are manufactured to very tight tolerances. Although reference jumpers are often expensive, they are required to provide the level of accuracy needed in most single-mode measurements. The requirements of reference jumpers must be determined on an individual basis for different applications; even a good reference jumper results in some link loss variability, which must be accounted for in the installation loss budget. Note that due to the nature of the measurement process, it is possible to measure link attenuation that is below the true value; enough link budget margin must be provided to allow for such an error. The design of reference jumper cables and dependence of the assembly loss on the type of cable have been the subject of much research [22], and ongoing work continues in this area.

7.2.2. *OPTICAL POWER PENALTIES*

Next, we will consider the assembly loss budget, which is the difference between the transmitter output and receiver input powers, allowing for optical power penalties due to noise sources in the link. The most important of these effects, and the most important fiber characteristic after transmission loss, is dispersion, which refers to the broadening of optical pulses as they propagate along the fiber. As pulses broaden, they tend to interfere with adjacent pulses; this limits the data rate and bandwidth of the optical link. In multimode fibers, there are two dominant kinds of dispersion, modal and chromatic. Modal dispersion refers to the fact that light can propagate along many different paths, or modes, in a multimode fiber; because not all these paths are the same length, different modes will travel at different velocities and cause pulse broadening. The fiber manufacturer will measure and specify modal bandwidth of the fiber in units of MHz/km. This has been characterized according to

$$BW_{modal} = BW_1/L^\gamma, \tag{7.13}$$

where BW_{modal} is the modal bandwidth for a length L of fiber, BW_1 is the manufacturer-specified modal bandwidth of a 1-km section of fiber, and γ is a constant known as the modal bandwidth concatenation length scaling factor. The term γ usually assumes a value between 0.5 and 1, depending on details of the fiber manufacturing and design as well as the operating wavelength; it is conservative to take $\gamma = 1.0$. Modal bandwidth can be increased by mode mixing, which promotes the interchange of energy between modes to average out the effects of modal dispersion. Fiber splices tend to increase the modal bandwidth, although it is conservative to discard this effect when designing a link. There have been many attempts to fabricate fibers with enhanced modal bandwidth, most of which have not yet produced commercially viable products [23, 24].

The other major contribution is chromatic dispersion, BW_{chrom}, which occurs because different wavelengths of light propagate at different velocities in the fiber. For multimode fiber, this is given by an empirical model of the form

$$BW_{chrom} = \frac{L^{\gamma c}}{\sqrt{\lambda_w}\,(a_0 + a_1\,|\lambda_c - \lambda_{eff}|)}, \tag{7.14}$$

where L is the fiber length in km, λ_c is the center wavelength of the source in nm; λ_w is the source full width at half maximum spectral width in nm; γ_c is the chromatic bandwidth length scaling coefficient, a constant; λ_{eff} is the effective wavelength, which combines the effects of the fiber zero dispersion wavelength and spectral loss signature; and the constants a_1 and a_0 are determined by a regression fit of measured data. From Ref. [25], the chromatic bandwidth for 62.5/125 μm fiber is empirically given by

$$BW_{chrom} = \frac{10^4\,L^{-0.69}}{\sqrt{\lambda_w}\,(1.1 + 0.0189\,|\lambda_c - 1370|)}. \tag{7.15}$$

For this expression, the center wavelength was 1335 nm and λ_{eff} was chosen midway between λ_c and the water absorption peak at 1390 nm; although λ_{eff} was estimated in this case, the expression still provides a good fit to the data. For 50/125 μm fiber, the expression becomes

$$BW_{chrom} = \frac{10^4 \, L^{-0.65}}{\sqrt{\lambda_w} \, (1.01 + 0.0177 \, |\lambda_c - 1330|)}. \tag{7.16}$$

For this case, λ_c was 1313 nm and the chromatic bandwidth peaked at $\lambda_{eff} = 1330$ nm. Recall that this is only one possible model for fiber bandwidth; refer to Refs [26] and [27] for further details. The total bandwidth capacity of multimode fiber BW_t is obtained by combining the modal and chromatic dispersion contributions, according to

$$\frac{1}{BW_t^2} = \frac{1}{BW_{chrom}^2} + \frac{1}{BW_{modal}^2}. \tag{7.17}$$

Once the total bandwidth is known, the dispersion penalty can be calculated for a given data rate. This is also an empirical process; the type of receiver used can also compensate to some extent for the fiber's bandwidth. One possible expression for the dispersion penalty in dB is

$$P_d = 1.22 \left[\frac{Bit \; Rate \; (Mb/s)}{BW_t \; (MHz)} \right]^2. \tag{7.18}$$

For typical telecommunication-grade fiber, the dispersion penalty for a 20-km link is approximately 0.5 dB. Figure 7.6 shows the dispersion penalty as a function of the fiber bandwidth and length.

By definition, single-mode fiber does not suffer modal dispersion. However, chromatic dispersion is an important effect, even given the relatively narrow spectral width of most laser diodes. The dispersion of single-mode fiber corresponds to the first derivative of group velocity τ_g with respect to wavelength, and is given by

$$D = \frac{d\tau_g}{d\lambda} = \frac{S_0}{4} \left(\lambda_c - \frac{\lambda_0^4}{\lambda_c^3} \right), \tag{7.19}$$

where D is the dispersion in ps/km-nm and λ_c is the laser center wavelength. The fiber is characterized by its zero dispersion wavelength, λ_0, and zero dispersion slope, S_0. Usually, both center wavelength and zero dispersion wavelength are specified over a range of values; it is necessary to consider both upper and lower bounds in order to determine the worst-case dispersion penalty. This can be seen from Fig. 7.7, in which I plot D vs wavelength for some typical values of λ_0 and λ_c; the largest absolute value of D occurs at the extremes of this region. Once the dispersion is determined, the intersymbol interference penalty as a function of link length, L, can be determined to a good approximation from a model proposed by Agrawal *et al.* [28]:

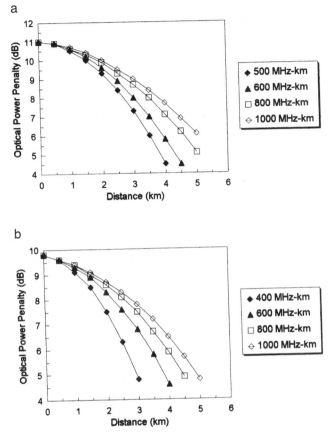

Fig. 7.6 Dispersion penalty vs distance for different fiber bandwidths: (a) 50-μm fiber; (b) 62.5-μm fiber.

$$P_d = 5 \log(1 + 2\pi(BD \; \Delta\lambda)^2 L^2), \tag{7.20}$$

where B is the bit rate and $\Delta\lambda$ is the RMS spectral width of the source. More detailed models of dispersion penalty have been proposed [29] taking into account the design details of the receiver and equalizer circuits. By maintaining a close match between the operating and zero dispersion wavelengths, this penalty can be kept to a tolerable 0.5–1.0 dB in most cases.

Group velocity dispersion contributes to another optical penalty that remains the subject of continuing research—mode partition noise and mode hopping. This penalty is related to the properties of a Fabry–Perot-type

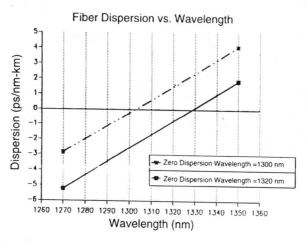

Fig. 7.7 Single-mode fiber dispersion as a function of wavelength [5].

laser diode cavity; although the total optical power output from the laser may remain constant, the optical power distribution among the laser's longitudinal modes will fluctuate. This is illustrated by the model depicted in Fig. 7.8: When a laser diode is directly modulated with injection current, the total output power stays constant from pulse to pulse; however, the power distribution among several longitudinal modes will vary between pulses. We must be careful to distinguish the behavior of the instantaneous laser spectrum, which varies with time, from that of the time-averaged spectrum, which is normally observed experimentally. The light propagates through a fiber with wavelength-dependent dispersion or attenuation, which deforms the pulse shape. Each mode is delayed by a different amount due to group velocity dispersion in the fiber; this leads to additional signal degradation at the receiver in addition to the intersymbol interference caused by chromatic dispersion alone. This is known as mode partition noise; it is capable of generating bit error rate floors, such that additional optical power into the receiver will not improve the link BER. This is because mode partition noise is a function of the laser spectral fluctuations and wavelength-dependent dispersion of the fiber, so the signal-to-noise ratio due to this effect is independent of the signal power. For this reason, mode partition noise has been closely studied by many authors; I will not attempt a complete treatment here, but rather summarize some of the

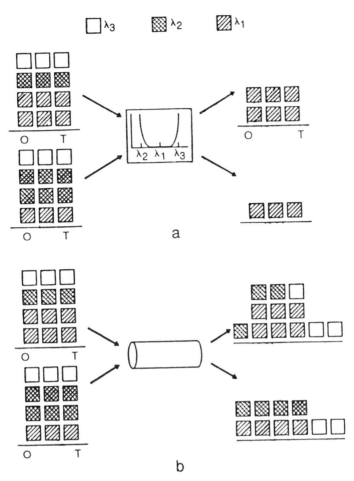

Fig. 7.8 Model for mode partition noise; an optical source emits a combination of wavelengths, illustrated by different color blocks: (a) wavelength-dependent loss; (b) chromatic dispersion [5].

recent models of this effect. The power penalty due to mode partition noise was first calculated by Ogawa [30] as

$$P_{mp} = 5 \log \left(1 - Q^2 \, \sigma_{mp}^2 \right), \qquad (7.21)$$

where

$$\sigma^2_{mp} = \frac{1}{2} k^2 (\pi B)^4 [A_1^4 \Delta\lambda^4 + 42 A_1^2 A_2^2 \Delta\lambda^6 + 48 A_2^4 \Delta\lambda^8] \quad (7.22a)$$

$$A_1 = DL \quad (7.22b)$$

$$A_2 = \frac{A_1}{2(\lambda_c - \lambda_0)}. \quad (7.22c)$$

The mode partition coefficient, k, is a number between 0 and 1 that describes how much of the optical power is randomly shared between modes; it summarizes the statistical nature of mode partition noise. According to Ogawa [30], k depends on the number of interacting modes and root mean square (rms) spectral width of the source, the exact dependence being complex. However, subsequent work has shown that Ogawa's model tends to underestimate the power penalty due to mode partition noise because it does not consider the variation of longitudinal mode power between successive baud periods and because it assumes a linear model of chromatic dispersion rather than the nonlinear model given in the previous equation. A more detailed model has been proposed by Campbell [31], which is general enough to include effects of the laser diode spectrum, pulse shaping, transmitter extinction ratio, and statistics of the data stream. Whereas Ogawa's model assumed an equiprobable distribution of zeros and ones in the data stream, Campbell showed that mode partition noise is data dependent as well. Recent work based on this model [25] has rederived the signal variance:

$$\sigma^2_{mp} = E_{av} \left(\sigma_0^2 + \sigma_{+1}^2 + \sigma_{-1}^2 \right), \quad (7.23)$$

where the mode partition noise contributed by adjacent baud periods is defined by

$$\sigma_{+1}^2 + \sigma_{-1}^2 = \frac{1}{2} k^2 (\pi B)^4 [1.25 A_1^4 \Delta\lambda^4 + 40.95 A_1^2 A_2^2 \Delta\lambda^6$$
$$+ 50.25 A_2^4 \Delta\lambda^8], \quad (7.24)$$

and the time-average extinction ratio $E_{av} = 10 \log (P_1/P_0)$, where P_1 and P_0 represent the optical power by a "1" and "0," respectively. If the operating wavelength is far away from the zero dispersion wavelength, the noise variance simplifies to

$$\sigma^2_{mp} = 2.25 \frac{k^2}{2} E_{av} (1 - e^{-\beta L^2})^2, \quad (7.25)$$

which is valid provided that,

$$\beta = (\pi BD \, \Delta\lambda)^2 \ll 1. \tag{7.26}$$

Mode partition effects deserve careful consideration because they can often limit the performance of a link or generate BER floors. However, many diode lasers have been observed to exhibit mode hopping or mode splitting in which the spectrum appears to split optical power between two or three modes for brief periods of time. The exact mechanism is not fully understood, but stable Gaussian spectra are generally only observed for CW operation and temperature-stabilized lasers [32]. During these mode hops the previous theory does not apply because the spectrum is non-Gaussian, and the model will overpredict the power penalty; hence, it is not possible to model mode hops as mode partitioning with $k = 1$. There is no currently published model describing a treatment of mode hopping noise, although several recent papers [33, 34] suggest approximate calculations based on the statistical properties of the laser cavity. In a practical link, some amount of mode hopping is probably unavoidable as a contributor to burst noise; empirical testing of link hardware remains the only way to reduce this effect.

In the previous discussion, I mentioned the effect of receiver extinction ratio on mode partition noise. There is a much more direct effect of this parameter as well. The receiver BER is a function of the modulated AC signal power; if the laser transmitter has a small extinction ratio, the DC component of total optical power is significant. Gain or loss can be introduced in the link budget if the extinction ratio at which the receiver sensitivity is measured differs from the worst-case transmitter extinction ratio. If the extinction ratio, E_1, at the transmitter is defined as the ratio of optical power when a one is transmitted vs when a zero is transmitted,

$$E_t = \frac{\text{Power (1)}}{\text{Power (0)}}. \tag{7.27}$$

Then we can define a modulation index at the transmitter M_1 according to

$$M_t = \frac{E_t - 1}{E_t + 1}. \tag{7.28}$$

Similarly, we can measure the linear extinction ratio at the optical receiver input and define a modulation index M_r. The extinction ratio penalty is given by

$$P_{er} = - 10 \log \left(\frac{M_t}{M_r}\right), \tag{7.29}$$

where the subscripts t and r refer to specifications for the transmitter and receiver, respectively. By careful choice of the extinction ratios, an effective gain of up to 1 dB or more can be realized in the link budget.

Another important property of the optical link is the amount of reflected light from the fiber end faces that returns up the link back into the transmitter. Whenever there is a connection or splice in the link, some fraction of the light is reflected back; each connection is thus a potential noise generator because the reflected fields can interfere with one another to create noise in the detected optical signal. The phenomenon is analogous to the noise caused by multiple atmospheric reflections of radio waves and is known as multipath interference noise. To limit this noise, connectors and splices are specified with a minimum return loss. If there are a total of N reflection points in a link and the geometric mean of the connector reflections is α, then based on the model of Ref. [35] the power penalty due to multipath interference (adjusted for bit error rate and bandwidth) is closely approximated by

$$P_{mpi} = 10 \log (1 - 0.7N\alpha). \tag{7.30}$$

Multipath noise can usually be reduced well below 0.5 dB with available connectors, whose return loss is often better than 25 dB. A far more serious effect occurs when stray light is reflected back into a Fabry–Perot-type laser diode, giving rise to intensity fluctuations in the laser output. This is a complicated phenomena, strongly dependent on the type of laser; it is called either reflection-induced intensity noise or RIN. This effect is very important because it can also generate BER floors. The power penalty due to RIN is the subject of ongoing research; because the reflected light is measured at a specified signal level, RIN is data dependent, although it is independent of link length. Because many laser diodes are packaged in windowed containers, it is difficult to correlate the RIN measurements on an unpackaged laser with those of a commercial product. There have been several detailed attempts to characterize RIN [36–38]; typically, the RIN noise is assumed Gaussian in amplitude and uniform in frequency over the receiver bandwidth of interest. The RIN value is specified for a given laser by measuring changes in the optical power when a controlled amount of light is fed back into the laser; it is signal dependent and is also influenced by temperature, bias voltage, laser structure, and other factors that typically influence laser output power [38]. If we assume that the effect of RIN is to produce an equivalent noise current at the receiver, then the additional receiver noise, σ_r, may be modeled as

$$\sigma_r = \gamma^2 S^{2g} B, \qquad (7.31)$$

where S is the signal level during a bit period, B is the bit rate, and g is a noise exponent that defines the amount of signal-dependent noise. If $g = 0$, noise power is independent of the signal, whereas for $g = 1$ noise power is proportional to the square of the signal strength. The coefficient γ is given by

$$\gamma^2 = S_i^{2(1 - g)} 10^{(RIN_i/10)}, \qquad (7.32)$$

where RIN_i is the measured RIN value at the average signal level S_i, meaured for a given application including worst-case backreflection conditions and operating temperaures. Clearly, the RIN model contains many parameters that are application dependent; if these can be determined, the Gaussian BER probability due to the additional RIN noise current is given by

$$P_{error} = \frac{1}{2}\left[P_e^1\left(\frac{S_1 - S_0}{2\,\sigma_1}\right) + P_e^0\left(\frac{S_2 - S_0}{2\,\sigma_0}\right)\right], \qquad (7.33)$$

where σ_1 and σ_0 represent the total noise current during transmission of a digital 1 and 0, respectively, and P_e^1 and P_e^0 are the probabilities of error during transmission of a 1 or 0, respectively. The power penalty due to RIN may then be calculated by determining the additional signal power required to achieve the same BER with RIN noise present as without the RIN contribution. One approximation from Ref. [38] for the RIN power penalty is given by

$$P_{rin} = - 5 \log\left[1 - Q^2(BW)(1 + M_r)^{2g}(10^{RIN/10})\left(\frac{1}{M_r}\right)^2\right], \qquad (7.34)$$

where the RIN value is specified in dB/Hz, BW is the receiver bandwidth, M_r is the receiver modulation index, and the exponent g is a constant varying between 0 and 1 that relates the magnitude of RIN noise to the optical power level. Because g may vary depending on factors such as the bias point and extinction ratio, this expression provides only an approximation of the true penalty.

Another important area in link design deals with the effects of timing jitter on the optical signal. In a typical optical link, a clock is extracted from the incoming data signal that is used to retime and reshape the received digital pulse; the received pulse is then compared with a threshold to determine if a digital 1 or 0 was transmitted. So far, I have discussed BER testing with the implicit assumption that the measurement was made in the center of the received data bit; to achieve this, a clock transition at the center of the bit is required. When the clock is generated from a receiver timing recovery circuit, it will have some variation in time and the exact location of the clock edge will be uncertain. Even if the clock is positioned at the center of the bit, its position may drift over time. There will be a region of the bit interval, or eye, in the time domain at which the BER is acceptable; this region is defined as the eyewidth [39]. Eyewidth measurements are an important parameter for evaluation of fiber optic links; they are intimately related to the BER, as well as to the acceptable clock drift, pulse width distortion, and optical power. At low optical power levels, the receiver signal-to-noise ratio is reduced; increased noise causes amplitude variations in the received signal. These amplitude variations are translated into time domain variations in the receiver decision circuitry, which narrows the eyewidth. At the other extreme, an optical receiver may become saturated at high optical power, reducing the eyewidth and making the system more sensitive to timing jitter. This behavior results in the typical "bathtub" curve shown in Fig. 7.2; for this measurement, the clock is delayed from one end of the bit cell to the other, with the BER calculated at each position. Near the ends of the cell, a large number of errors occur; toward the center of the cell, the BER decreases to its true value. The eye opening may be defined as the portion of the eye for which the BER remains constant; pulse width distortion occurs near the edges of the eye, which denotes the limits of the valid clock timing. Uncertainty in the data pulse arrival times causes errors to occur by closing the eye window and causing the eye pattern to be sampled away from the center. This is one of the fundamental problems of optical and digital signal processing, and a large body of work has been done in this area [40–42]. For purposes of this chapter, I will provide only a brief outline of the problem; this remains an active area of research, and the reader is encouraged to follow the recent literature on this subject. Some international standards bodies have adopted conventions for the treatment of jitter in data links and often specify the maximum allowed jitter for a given application. For example, the ANSI Fibre Channel Standard contains recommendations on the maximum allowable jitter per bit time. In general, multiple jitter sources will be present in a link; these will tend to be uncorrected. However, jitter on digital signals, especially resulting from a cascade of repeaters, may be coherent. It is often difficult to determine how to assess jitter penalties against the individual components of a link.

The International Telecommunications Standards Body (CCITT) has adopted a standard definition of jitter [43] as short-term variations of the significant instants (rising or falling edges) of a digital signal from their ideal position in time. Longer-term variations are described as wander; in terms of frequency, the distinction between jitter and wander is somewhat unclear. The predominant sources of jitter include the following:

Phase noise in receiver clock recovery circuits, particularly crystal-controlled oscillator circuits; this may be aggravated by filters or other components that do not have a linear phase response. Noise in digital logic resulting from restricted rise and fall times may also contribute to jitter.

Imperfect timing recovery in digital regenerative repeaters, which is usually dependent on the data pattern.

Different data patterns may contribute to jitter when the clock recovery circuit of a repeater attempts to recover the receive clock from inbound data. Data pattern sensitivity can produce as much as 0.5 dB penalty in receiver sensitivity. Higher data rates are more susceptible (>1 Gb/s); data patterns with long run lengths of 1s or 0s, or with abrupt phase transitions between consecutive blocks of 1s and 0s, tend to produce worst-case jitter.

At low optical power levels, the receiver signal-to-noise ratio, Q, is reduced; increased noise causes amplitude variations in the signal, which may be translated into time domain variations by the receiver circuitry.

Low-frequency jitter, also called wander, results from instabilities in clock sources and modulation of transmitters.

Very low-frequency jitter caused by variations in the propagation delay of fibers, connectors, etc., typically results from small temperature variations (this can make it especially difficult to perform long-term jitter measurements).

In general, jitter from each of these sources will be uncorrelated; jitter related to modulation components of the digital signal may be coherent, and cumulative jitter from a series of repeaters or regenerators may also contain some well-correlated components.

There are several parameters of interest in characterizing jitter performance. Jitter may be classified as either random or deterministic, depending on whether it is associated with pattern-dependent effects; these are distinct from the duty cycle distortion that often accompanies imperfect signal timing. Each component of the optical link (data source, serializer, transmitter, encoder, fiber, receiver, retiming/clock recovery/deserialization, and decision circuit) will contribute some fraction of the total system jitter. If we consider the link to be a "black box" (but not necessarily a linear system) then we can measure the level of output jitter in the absence of input jitter; this is known as the "intrinsic jitter" of the link. The relative importance of jitter from different sources may be evaluated by measuring the spectral density of the jitter. Another approach is the maximum tolerable input jitter for the link. Finally, because jitter is essentially a stochastic process, we may attempt to characterize the jitter transfer function of the link or estimate the probability density function of the jitter. When multiple traces occur at the edges of the eye, this can indicate the presence of data-dependent jitter or duty cycle distortion; a histogram of the edge location will show several distinct peaks. This type of jitter can indicate a design flaw in the transmitter or receiver. By contrast, random jitter typically has a more Gaussian profile and is present to some degree in all data links. All these approaches have their advantages and drawbacks; first, some simple attempts to model jitter in a real link will be considered.

One of the first attempts to model the optical power penalty due to jitter [44] considered the general case of a receiver whose input was a raised cos signal of the form

$$S_t = \frac{1}{2} (1 + \cos(\pi Bt)), \tag{7.35}$$

where B is the bit rate. A decision circuit samples this signal at some interval $t = n/B$. In the presence of random timing jitter, the sampling point fluctuates and the jitter-induced noise depends on the probability density function (PDF) of the random timing fluctuations. Determination of the actual PDF is quite difficult; if we can approximate $Bt \ll 1$, then it has been derived [45] that for a uniform PDF, the jitter-induced noise, σj, is given by

$$\sigma_j^2 = \frac{4}{5} (\pi Bt/4)^4. \tag{7.36}$$

For the less conservative case of a Guassian PDF in the same limit,

$$\sigma_j^2 = 2 (\pi Bt/4)^4. \tag{7.37}$$

The optical power penalty in dB due to jitter noise is then given by

$$P_j = -5 \log \left(1 - 4Q^2 \sigma_j^2 \right), \tag{7.38}$$

where Q is the Gaussian error function. Based on this expression, the penalty for a Guassian system is much larger than that for a uniform PDF; when $Bt = 0.35$, a BER floor appears at 1e-9. However, it was subsequently shown [46] that the approximation on Bt is very restrictive, and the actual PDF is far from Gaussian; indeed, it is given by

$$P_{\text{error}} = \int_{-1/B}^{1/B} (\text{PDF}_j) Q\left(\frac{A \cos (\pi Bt)}{2\sigma_j} \right) dt, \tag{7.39}$$

where the probability density function of the jitter is included under the integral. Numerical integration of this PDF shows that the approximate results given previously tend to underestimate the effects of Gaussian jitter and overestimate the effects of uniform jitter by more than 2 dB! This points out one of the basic problems in attempting to model a power penalty due to jitter effects—namely, the limitations of the assumptions that must be made to reduce the expression to closed form vs the real-world jitter behavior. Using similar approximations, jitter power penalties have been derived for both positive–intrinsic–negative and avalanche photodiode receivers [47], although the limiting assumptions make it difficult to generalize these models. An alternative modeling approach has been to derive a worst-case distribution for the PDF, which will provide an upper bound on the performance of the optical link and can be more easily evaluated for analytical purposes [48].

The problem of jitter accumulation in a chain of repeaters becomes increasingly complex; however, we can state some general rules of thumb. It has been shown [49] that jitter can be generally divided into two components—one due to repetitive patterns and one due to random data. In receivers with phase-lock loop timing recovery circuits, repetitive data patterns will tend to cause jitter accumulation, especially for long run lengths. This effect is commonly modeled as a second-order receiver transfer function [50]; the jitter accumulates according to the relationship

$$\text{Jitter} \propto N + \left(\frac{N}{\zeta} L \right)^2, \tag{7.40}$$

where N is the number of identical repeaters and ζ is the loop damping factor, specific to a given receiver circuit. For large ζ, jitter accumulates almost linearly with the number of repeaters, whereas for small ζ the accumulation is much more rapid. Jitter will also accumulate when the link is transferring random data; jitter due to random data is of two types— systematic and random. The classic model for systematic jitter accumulation in cascaded repeaters was published by Byrne *et al.* [51]. The Byrne model assumes cascaded identical timing recovery circuits; the general expression for jitter power spectrum is given by

$$J_s^N = J_s \left| \prod_{k=2}^{N} H_k(f) + \prod_{k=1}^{N} H_k(f) + \ldots + H_k(f) \right|^2, \quad (7.41)$$

where J_s is the systematic jitter generated by each repeater and $H(f)$ is the jitter transfer function of each timing recovery circuit. The corresponding model for random jitter accumulation is

$$J_r^N = J_r \left[\prod_{K=1}^{N} |H_k(f)|^2 + \prod_{K=2}^{N} |H_k(f)|^2 + \ldots + |H_k(f)|^2 \right], \quad (7.42)$$

where J_r is the random jitter generated by each repeater. The systematic and random jitter can be combined as rms quantities;

$$J_t^2 = \left(J_s^N \right)^2 + \left(J_r^N \right)^2, \quad (7.43)$$

so that total jitter due to random jitter may be obtained. This model has been generalized to networks consisting of different components [52] and to nonidentical repeaters [53]. Despite these considerations, for well-designed practical networks the basic results of the Byrne model remain valid for N nominally identical repeaters transmitting random data:

Systematic jitter accumulates in proportion to $N^{1/2}$.

Random jitter accumulates in proportion to $N^{1/4}$.

Details of jitter accumulation modeling are provided in Refs. [43–53].

Except for those penalties that produce BER floors such as mode partitioning and RIN, most penalties can be reduced by increasing the transmitted or received optical power. This brute-force approach is subject to limitations such as maintaining class 1 laser safety; even if it were possible to increase optical power significantly (as I will discuss later using optical amplifiers), a new class of nonlinear optical penalties becomes important at high power levels. I will briefly discuss some of these here; the reader is referred to Ref. [54] for a more complete treatment of nonlinear phenomena. Class 1 laser systems typically do not experience these nonlinear effects, although they may be important for optical fiber amplifiers or systems using open fiber control (OFC), for which there may be significantly higher power levels present in the fiber.

At high optical power levels, nonlinear scattering may limit the link performance. The dominant effects are stimulated Raman and Brillouin scattering. When incident optical power exceeds a threshold value, a significant amount of light scatters from small imperfections in the fiber core. The scattered light is frequency shifted because the scattering process involves the generation of phonons [55]. This is known as stimulated Brillouin scattering; under these conditions, the output light intensity becomes nonlinear as well. When the scattered light experiences frequency shifts outside the acoustic phonon range, due instead to modulation by impurities or molecular vibrations in the fiber core, the effect is known as Raman scattering. Stimulated Brillouin scattering will not occur below the optical power threshold defined by

$$P_c = 21 \ A/G_n L_c \text{ watts,} \qquad (7.44)$$

where L_c is the effective interaction length, A is the cross-sectional area of the guided mode, and G_b is the Brillouin gain coefficient; refer to Ref. [55] for details. Similarly, Raman scattering will not occur unless the optical power exceeds the value given by

$$P_t = 16 \ A/G_r L_c \text{ watts,} \qquad (7.45)$$

where G_r is the Raman gain coefficient. Brillouin scattering has been observed in single-mode fibers at wavlengths greater than cutoff with optical power as low as 5 mW [54]; a good rule of thumb is that the Raman scattering threshold is about three times larger than the Brillouin threshold. As a general guide, the optical power threshold for Brillouin scattering in single-mode fiber is approximately 10 mW and for Raman scattering is approximately 3.5 W. These effects rarely occur in multimode fiber, where the thresholds for Brillouin and Raman scattering are approximately 450 mW and 150 W, respectively.

The final nonlinear effect we will consider is frequency chirping of the optical signal. Chirping refers to a change in frequency with time and takes its name from the sound of an acoustic signal whose frequency increases or decreases linearly with time. There are three ways in which chirping can affect a fiber optic link. First, the laser transmitter can be chirped as a result of physical processes within the laser [56]; the effect has its origin in carrier-induced refractive index changes, making it an inevitable consequence of high-power direct modulation of semiconductor lasers. For lasers with low levels of relaxation oscillation damping, a model has been proposed for the chirped power penalty [56]:

$$P = 10 \log \left(1 + \frac{\pi B^2 \, \lambda^2 L D a}{4c} \right), \tag{7.46}$$

where c is the speed of light, B is the fiber bandwidth, λ is the wavelength of light, L is the length of the fiber, D is the dispersion, and a is the linewidth enhancement factor (a typical value is -4.5); this model is only a first approximation because it neglects the dependence of chirp on extinction ratio and nonlinear effects such as spectral hole burning [57]. Second, a sufficiently intense light pulse will be chirped by the nonlinear process of self-phase modulation in an optical fiber [58]. The effect arises from the interaction of the light and the intensity-dependent portion of the fiber's refractive index; it is thus dependent on the material and structure of the fiber, polarization of the light, and the shape of the incident optical pulse. Based on a model from Ref. [58], the maximum optical power level before the spectral width increases by 2 nm is given by

$$P = \frac{n^2 A}{377 n_2 \, \kappa L_e} \text{ Watts}, \tag{7.47a}$$

where

$$L_e = (1/a_0) \, (1 - \exp{(-a_0 L)}, \tag{7.47b}$$

where n is the fiber's refractive index, κ is the propagation constant (a typical value is 7×10^4 at a wavelength of 1.3 μm) a_0 is the fiber attenuation coefficient, A is the fiber core cross-sectional area, n_2 is the nonlinear coefficient of the fiber's refractive index (a typical value is 6.1×10^{-19}), and L_e is the effective interaction length for the nonlinear interaction, which is related to the actual length of the fiber, L, by Eq. (7.47b). This expression should be multipled by $\frac{5}{8}$ if the fiber is not polarization preserving. Typically, this effect is not significant for optical power levels less than 950 mW in a single-mode fiber at 1.3-μm wavelength. Finally, there is a power penalty arising from the propagation of a chirped optical pulse in a dispersive fiber because the new frequency components propagate at different group velocities. This may be treated as simply a much worse case of the conventional dispersion penalty [59], provided that one of the first two effects exists to chirp the optical signal.

7.2.3. LINK BUDGET EXAMPLE

To illustrate some of the previous principles, the following example discusses the design of a data communication link using single-mode optical fiber operating at 1300 nm. The application is remote backup of data from a hard disk to a tape drive located in a building 10 km away; because of systemwide fault tolerance, the link has a target BER of 10e-12. The link adheres to an industry standard data rate of 531 Mb/s so that the equipment will interoperate. The transmitter is guaranteed to have a worst-case output of −9.0 dBm, accounting for temperature variation and end of life; the worst-case receiver sensitivity is −26.0 dBm. Optical fiber is chosen with parameters shown in Table 7.6, whereas Table 7.7 gives a complete specification for the transmitter and receiver. Note that the link center wavelength is sufficiently close to the zero dispersion wavelength of the fiber that the dispersion penalty is kept to 0.2 dB worst case. The transmitter is class 1, so we can neglect nonlinear effects in the link; connector return loss is held low enough so that multipath effects are not a concern either. Using the

Table 7.7.   **Transmitter, Receiver, and Optical Fiber Specifications used in the Illustrated Example [5]**

Transmitter	
Power output	−3 to − 9 dBm
Extinction ratio	6 dB
RIN	−112 dB/Hz
Center wavelength	1270–1340 nm
Spectral width	6 nm
Receiver	
Sensitivity	−27 dBm
Saturation	−3 dBm
Fiber	
Zero dispersion wavelength	1310 nom. 1295–1322 (range) nm
Zero dispersion slope	0.095 ps/nm-km
Cutoff wavelength	1280 nm
Attenuation	0.5 at 1310 nm dB/km
Max. attenuation delta (1270–1340)	0.06 dB/km

manufacturer's specification for RIN and assuming $g = 0.5$, we can estimate the RIN power penalty of 0.7 dB, which should be tolerable. By clever choice of the extinction ratios, we realize an approximately 0.5-dB gain in the link to recover part of the RIN penalty. The wavelength-dependent attenuation of the fiber was selected so that the penalty would be tolerable at the required 10-km distance. For this distance, we have chosen a single-mode link, so modal noise is not a concern provided we use sufficiently low loss splices and connectors to minimize modal noise. Although partition noise should be verified experimentally, we estimate that the power penalty is less than 1.5 dB (note that $\beta < 1$, so the approximate model is valid). This gives us a total available power of

$$P_{\text{avail}} = Tx(\text{output}) - Rx(\text{sens}) - \text{Penalties} = 14 \text{ dB}. \quad (7.48)$$

Next, we compute the installation loss. The link contains one distribution panel near the source end and uses previously installed connectors as shown in Table 7.6. We will consider a link with four connectors (mean loss = 0.35 dB; variance = 0.06 dB), two fusion splices (mean loss = 0.15 dB; variance = 0.01 dB), and a 20-km link with 240 m of jumper cable (loss = 0.8 dB/km for the jumpers and 0.5 dB/km for the trunk). If we perform a statistical calculation on the installed budget for connectors, splices, and cable transmission loss, the mean installed loss is 11.75 dB with a variance of 0.26 dB. If we use the industry-standard 3 σ design, then the calculated installation loss is 13.5 dB. This allows us to operate the link with 0.5 dB to spare, the difference between the available power and installation loss budgets. A conservative design should always allow some link safety margin, as we have done, to allow for the many assumptions used in deriving the various link performance models we have used. Of course, there are other possible ways to compute the link budget; if we had obtained the statistics for all the link parameters, we could perform a Monte Carlo simulation for the entire link that might yield a different link budget margin. It should be apparent that link design requires not only the latest valid performance models but also a good amount of engineering judgment in how best to apply these tools. Also, real-world problems are often not as clear-cut as the example I have just provided; a link designer must be able to manage the trade-offs involved in frequently working with incomplete information (Table 7.8). Still, this example should illustrate some of the typical steps required to perform link planning and installation.

A summary of the important optical link power penalties is given in Table 7.9; more detailed descriptions of link budget modeling, including

Table 7.8 **Link Budget Loss Sources**

Source	Typical loss
Fiber attenuation	0.5 dB/km
Connector loss	0.25–0.50 dB
Splice loss	0.15–0.40 dB
Transmitter/receiver end-of-life degradation	0.5–1.0 dB

equations for all the link noise parameters and examples of a complete link budget calculation, are available in the literature [5, 25, 26].

7.3. Link Planning Considerations

There are many factors to consider when designing fiber optic data links. The maximum distance for a fiber optic data link is determined by the link budget calculations described in the previous section; it is also affected by established industry standards for the physical layer of common datacom protocols. Generally, there is a trade-off between link distance and data rate; some of the typical link distances are summarized in Table 7.10. In general, multimode fiber is limited to a distance of a few kilometers, whereas single-mode fiber enables distances of 10 km or more. In some cases, the difference in bandwidth between 50- and 62.5-μm fiber will result in a distance limitation; this depends on the application requirements and may not appear for some datacom standards. The SONET physical layer is assumed for all asynchronous transfer mode (ATM) protocols listed in Table 7.10. A complete list of the datacom speeds allowed for each standard is provided in Chapters 11–14.

Table 7.9 **Typical Link Power Penalties**

Penalty	Typical at 20 km, 200 Mb/s (dB)
Chromatic dispersion	1.2
Mode partition	1.5
RIN	0.8
Multipath	0.25

Table 7.10   **Typical Link Distances and Data Rates**

ESCON, multimode	200 Mb/s	2 km, 50-μm fiber, 8 dB 3 km, 62.5-μ fiber, 8 dB
ESCON, single mode		20 km, single-mode fiber, 14 dB
ATM, multimode	155 Mb/s	3 km, 62.5-μm multimode fiber, 7 dB 20 km, single-mode fiber, 15 dB
FDDI (multimode only)	125 Mb/s	2 km, 62.5-μm multimode fiber, 11 dB
FCS, multimode	266 Mb/s	2 km, 50-μm fiber 1 km, 62.5-μm fiber
	1 Gb/s	0.5 km, 50-μm fiber
FCS, single mode	266 Mb/s	10 km
	1 Gb/s	10 km

7.3.1. *LASER SAFETY REQUIREMENTS*

Laser safety is an important criteria for many data communication products because they are accessible to customers who may have little or no laser training. This limits the amount of optical power that can safely be launched into the link and may impact distance and BER performance. This is a very important area; infrared sources of less than 1 mW can cause serious damage to the unprotected eye. Typically, it is required that all datacom fiber optic products be class 1, or inherently safe; there is no access to unsafe light levels during normal operation or maintenance of the product or during accidental viewing of the optical source. This includes viewing of the optical fiber end face while the transmitter is in operation. In the United States, the Department of Health and Human Services of the Occupational Safety and Health Administration and the Food and Drug Administration (FDA) define the laser safety standards; a class 1 product is limited to −2.0 dBm output power. International standards defined by the IEC are somewhat different and limit class 1 exposure to −6.0 dBm. Following recent revisions to the laser safety standards, the FDA requirements currently permit lower optical power levels and high power densities than the IEC standards; however, the IEC requirements include the provision that a product must operate as a class 1 device even under a single point of failure. This requires redundancy in the hardware design; for example, although many ATM-type transceivers use a 1 × 9 pinout, some designs are standardizing around a 2 × 9 pinout to implement clock recovery and redundant laser safety. There is no consensus on how to deal with emerging transceiver technology such as vertical cavity surface-emitting laser (VCSEL) arrays; these lasers produce a beam of circular cross section rather than the elliptical beam profile typical of long-wavelength lasers. As a result, most of the optical power is launched into the lower order modes of an optical fiber and tends to remain there even after fairly long distances. Although this improves modal noise performance, it also makes laser safety compliance more difficult. It has not been determined whether VCSELs should be treated as extended sources for purposes of laser safety classification. A summary of the different laser safety classifications and power levels is given in Fig. 7.9; for details and current revisions of all standards, contact the references provided in this figure. Recent revisions in the standards apply to all optical sources, including light-emitting diodes (LEDs), both surface-emitting diodes and the edge-emitting diodes required for higher speed applications. Currently, there is an increasing amount of laser safety standards activity. The base standard, ANSI Z136.1 ("Standard for the safe use of lasers") was first released in 1988 and most recently reissued in 1993 with some revisions to the safe emission levels that made it easier to qualify as a class 1 laser product (Table 7.11). The related standard ANSI Z136.2 ("Standard for the safe use of optical fiber communication systems utilizing laser diode and LED sources") is being updated to reflect the changes to Z136.1 in 1993; this is expected to be available in late 1996. A related document, ANSI Z136.3 ("Standard for the safe use of lasers in health care facilities"), was reissued in April 1996 with some new applications and modest changes to control procedures. In addition, three new standards are being discussed for release sometime in 1997 or 1998: ANSI Z136.4 will cover the measurement of laser power and energy, Z136.5 will cover safety aspects of lasers used in educational institutions, and Z136.6 will cover outdoor use of lasers for applications such as surveying, laser displays, and military systems.

Class 1 laser certification is granted by testing a product at an independent laboratory. In the United States, products are granted an accession number by the FDA to certify their compliance with class 1 limits. It is a popularly held misconception that the FDA issues "certification" for products such as fiber optic transceivers that contain laser devices. This is not the case. Manufacturers of such equipment provide a report and product description to the FDA and the manufacturer certifies through supporting test results that the product complies with class 1 regulations; the FDA then asigns an accession number in order to track this part and permit

Fig. 7.9 Optical laser safety requirements.

Wavelength λ(nm)	Emission Duration t(s)							
	$<10^{-9}$	10^{-9} to 10^{-7}	10^{-7} to 1.8×10^{-5}	1.8×10^{-5} to 5×10^{-5}	5×10^{-5} to 10	10 to 10^{3}	10^{3} to 10^{4}	10^{4} to 3×10^{4}
200 to 302.5	2.4×10^{-5} J							
302.5 to 315	2.4×10^{4} W	$7.9\times10^{-7}\,C_1\,$J $(t<T_1)$ $\quad$ $7.9\times10^{-7}\,C_2\,$J $(t>T_1)$				$7.9\times10^{-7}\,C_2\,$J		
315 to 400		$7.9\times10^{-7}\,C_1\,$J			7.9×10^{-3} J	7.9×10^{-6} W		
400 to 550	200 W	2×10^{-7} J		$7\times10^{-4}\,t^{0.75}$ J	3.9×10^{-3} J	3.9×10^{-7} W		
and*	10^{11} W.m^{-2} sr^{-1}	$10^{5}\,t^{0.33}$ J.m^{-2} sr^{-1}			2.1×10^{3} J.m^{-2} sr^{-1}	21 W.m^{-2} sr^{-1}		
550 to 700	200 W	2×10^{-7} J		$7\times10^{-4}\,t^{0.75}$ J $(t<T_2)$		$3.9\times10^{-3}\,C_3\,$J $(t>T_2)$	$3.9\times10^{-7}\,C_3\,$W	
and*	10^{11} W.m^{-2} sr^{-1}	$10^{5}\,t^{0.33}$ J.m^{-2} sr^{-1}		$(t<T_2)$ $3.9\times10^{4}\,t^{0.75}$ J.m^{-2} sr^{-1}		$2.1\times10^{5}\,C_3\,$J.m^{-2} sr^{-1} $(t>T_2)$	21 C_3 W.m^{-2} sr^{-1}	
700 to 1050	200 C_4 W	$2\times10^{-7}\,C_4\,$J		$7\times10^{-4}\,t^{0.75}\,C_4\,$J		$1.2\times10^{-4}\,C_4\,$W		
and*	$10^{11}\,C_4\,$W.m^{-2} sr^{-1}	$10^{5}\,t^{0.33}\,C_4\,$J.m^{-2} sr^{-1}			$3.9\times10^{4}\,t^{0.75}\,C_4$ J.m^{-2} sr^{-1}	$6.4\times10^{3}\,C_4\,$W.m^{-2} sr^{-1}		
1050 to 1400	2×10^{3} W	2×10^{-6} J		$3.5\times10^{-3}\times t^{0.75}$ J		6×10^{-4} W		
and*	5×10^{11} W.m^{-2} sr^{-1}	$5\times10^{5}\,t^{0.33}$ J.m^{-2} sr^{-1}			$1.9\times10^{5}\,t^{0.75}$ J.m^{-2} sr^{-1}	3.2×10^{4} W.m^{-2} sr^{-1}		
1400 to 10^{5}	8×10^{4} W	8×10^{-5} J		$4.4\times10^{-3}\,t^{0.25}$ J		8×10^{-4} W		
10^{5} to 10^{6}	10^{7} W	10^{-2} J		$0.56\,t^{0.25}$ J		0.1 W		

* See Item *d*) of Sub-clause 9.3 for Class 1 dual limits requirements.

232

Wavelength λ (nm)	Emission duration t (s)	Class 2 AEL
400 to 700	$t < 0.25$	Same as Class 1 AEL
	$t \geq 0.25$	10^{-3} W

Fig. 7.9 (*Continued*)

233

Wave-length λ (nm) / Emission Duration t (s)	$<10^{-9}$	10^{-9} to 10^{-7}	10^{-7} to 1.8×10^{-5}	1.8×10^{-5} to 5×10^{-5}	5×10^{-5} to 0.25	0.25 to 10	10 to $<10^{3}$	10^{3} to 3×10^{4}
200 to 302.5	1.2×10^{-4} J and 30 J.m^{-2}						$4\times C_2\times10^{-6}$ J and C_2 J.m^{-2}	
302.5 to 315	1.2×10^{5} W and 3×10^{10} W.m^{-2}	$4\times C_1\times10^{-6}$ J and C_1 J.m^{-2}	$(t>T_1)$	$(t<T_1)$		$4\times C_1\times10^{-6}$ J and C_2 J.m^{-2}	$4\times C_2\times10^{-6}$ J and C_2 J.m^{-2}	
315 to 400		$4\times C_1\times10^{-6}$ J and C_1 J.m^{-2}					4×10^{-2} J and 10^4 J.m^{-2}	4×10^{-5} W and 10 W.m^{-2}
400 to 700	1000 W and 5×10^{6} W.m^{-2}	10^{-6} J and 5×10^{-3} J.m^{-2}		$3.5\times10^{-3}\times t^{0.75}$ J and $18\times t^{0.75}$ J.m^{-2}		5×10^{-3} W and 25 W.m^{-2} (Aversion responses protect for emission >0.25 s)		
700 to 1050	1000 W $\times C_4$ W and $5\times C_4\times10^{6}$ W.m^{-2}	$10^{-6}\times C_4$ J and $5\times C_4\times10^{-3}$ J.m^{-2}			$3.5\times10^{-3}\times C_4\, t^{0.75}$ J and $18\times C_4\, t^{0.75}$ J.m^{-2}			$6\times10^{-4}\times C_4$ W and $3.2\times C_4$ W.m^{-2}
1050 to 1400	10^{4} W and 5×10^{7} W.m^{-2}	10^{-5} J and 5×10^{-2} J.m^{-2}				$1.8\times10^{-2}\times t^{0.75}$ J and $90\times t^{0.75}$ J.m^{-2}		3×10^{-3} W and 16 W.m^{-2}
1400 to 10^5	4×10^{5} W and 10^{11} W.m^{-2}	4×10^{-4} J and 100 J.m^{-2}		$2.2\times10^{-2}\times t^{0.25}$ J and $5600\times t^{0.25}$ J.m^{-2}			4×10^{-3} W and 1000 W.m^{-2}	
10^5 to 10^6	5×10^{7} W and 10^{11} W.m^{-2}	5×10^{-2} J and 100 J.m^{-2}		$2.8\times t^{0.25}$ J and $5600\times t^{0.25}$ J.m^{-2}			0.5 W and 1000 W.m^{-2}	

Fig. 7.9 (*Continued*)

234

Wavelength λ(nm) \ Emission duration t(s)	$<10^{-9}$	10^{-9} to 0.25	0.25 to 3×10^4
200 to 302.5	3.8×10^5 W	3.8×10^{-4} J	1.5×10^{-3} W
302.5 to 315	$1.25\times10^4\ C_2$ W	$1.25\times10^{-5}\ C_2$ J	$5\times10^{-5}\ C_2$ W
315 to 400	1.25×10^8 W	0.125 J	0.5 W
400 to 700	3.14×10^{11} W.m^{-2}	$3.14\times10^5\ t^{0.33}$ J.m^{-2} and $<10^5$ J.m^{-2}	0.5 W
700 to 1 050	$3.14\times10^{11}\ C_4$ W.m^{-2}	$3.14\times10^5\ C_4\ t^{0.33}$ J.m^{-2} and $<10^5$ J.m^{-2}	0.5 W
1 050 to 1 400	1.57×10^{12} W.m^{-2}	$1.57\times10^6\ t^{0.33}$ J.m^{-2} and $<10^5$ J.m^{-2}	0.5 W
1 400 to 10^6	10^{14} W.m^{-2}	10^5 J.m^{-2}	0.5 W

Fig. 7.9   (*Continued*)

import–export as a class 1 laser device. International certification is usually performed by a recognized testing laboratory, such as TUV or VDE in Germany. These labs will assign approval numbers to the products in question based on their independent testing of samples provided by the manufacturer and assessment of any supporting manufacturer data. There may be a fee associated with this evaluation, and there is an annual fee that must be paid in order to keep an IEC registration number active; otherwise, it is withdrawn and the product can no longer claim to be IEC class 1 compliant. International safety standards define the requirements for labels that must be affixed to class 1 optical products and the terminology that must be used in product safety literature. In general, class 1 products comply with ANSI 136.X and FDA/CDRH 21 CFR subchapter J in the United States and with IEC/CEI 825 in the rest of the world. There may be additional local or state labor safety regulations as well; for example, the New York State Department of Labor enforces New York Code Rule 50, which requires all manufacturers to track their primary laser components by a state-issued serial number. Check with local and state authorities for applicable regulations.

In order to receive class 1 certification, a product must be single-fault tolerant; that is, the optical power cannot exceed recommended levels if a

Table 7.11   **ANSI Z136.1 (1993) Laser Safety Classifications**

Class 1—Inherently safe; no viewing hazard during normal use or maintenance; 0.4 μW or less; no controls or label requirements

Class 2—Human aversion response sufficient to protect eye; low power visible lasers with power less than 1 mW continuous; in pulsed operation, power levels exceed the class 1 acceptable exposure limit for the entire exposure duration but do not exceed the class 1 limit for a 0.25 second exposure; requires caution label

Class 2a—Low power visible lasers that do not exceed class 1 acceptable exposure limit for 1000 s or less (not intended for viewing the beam); requires caution label

Class 3a—Aversion response sufficient to protect eye unless laser is viewed through collecting optics; 1–5 mW; requires labels and enclosure/interlock; warning sign at room entrance

Class 3b—Intrabeam (direct) viewing is a hazard; specular reflections may be a hazard; 5–500 mW continuous; <10 J per square centimeter pulsed operation (less than 0.25 s); same label and safety requirements as class 3a plus power actuated warning light when laser is in operation

Class 4—Intrabeam (direct) viewing is a hazard; specular and diffuse reflections may be a hazard; skin protection and fire potential may be concerns; >500 mW continuous; >10 J per square centimeter pulsed operation; same label and safety requirements as class 3b plus locked door, door actuated power kill, door actuated filter, door actuated shutter, or equivalent.

Note. Laser training is required in order to work with anything other than class 1 laser products.

single point of failure occurs in the link. There are three common ways in which this can be guaranteed by the manufacturer. First, the optical transceiver can be designed to never emit more than the maximum allowed optical power level; this approach is the simplest and is implemented on most ATM/SONET transceivers. This is only possible due to recent relaxation in the laser safety standards; products manufactured prior to 1994 required alternate solutions. Some products use higher power laser transmitters but never couple more than the maximum power level into the optical fiber; this accounts for coupling loss in the transceiver and does not present a hazard for viewing the fiber end face. The transmitter must have some additional form of protection, such as mechanical shutters that snap into place covering the laser when not in use. This design was used on some single-mode ESCON transceivers prior to 1995. Finally, the ANSI

Fibre Channel Standard defines a laser safety method called OFC. This implements an interlock at the transceiver that detects whether light is being received; both ends of a duplex link must detect light and exchange a handshake sequence before the laser transmitter is activated. If the link is opened for any reason, such as a broken fiber or pulled connector, the lasers turn off on both sides of the link. The link then enters a waiting mode, in which a low-power optical pulse is transmitted every 10 s. When the link is closed once again, this pulse is detected by both ends of the link, which exchange handshakes and activate the laser transmitters at full power again. Using this approach, the link can employ much higher powered lasers without violating class 1 certification limits; this can enable longer distances and improved BER performance. Many FCS products, including the 266 Mb/s and 1 Gb/s optical link cards from IBM, implement this form of laser safety.

7.3.2. OPTICAL CABLE AND CONNECTOR TYPES

Many different types of fiber optic cables are available, as discussed in previous chapters. Selection of the proper fiber type is an important link design requirement. There are three classes of fiber cable, defined by their fire code rating for use within building environments; riser, plenum, and low halogen [there is actually a fourth, data processing (DP) center rated, as defined by the National Electrical Code, which I will discuss shortly]. Riser-rated cable (UL 1666, OFNR rating) has been used in data centers for many years but provides only nominal fire protection. Many insurance companies require the use of more advanced cables in new construction. The most common in the United States is plenum-rated cable, which is certified not to burn until extremely high temperatures are reached (UL 910, OFNP rating). Plenum-rated cable is required for installation in air ducts by the 1993 National Fire Code 770-53; however, there is an exception in Section 3000-22 of the code for raised floor and data processing environments. Other local codes may require plenum cable even in a data processing environment; check with your local fire marshall or insurance carrier to determine the proper cable type for a given installation. There are two basic types of plenum cable, manufactured with either a Teflon-based jacket or PVC-based jacket. Although they are functionally equivalent, the Teflon-based jackets tend to be stiffer and less flexible, which may cause problems during installation. In Europe and some other nations, the emerging standard is low-halogen fiber, which burns at a lower temperature than plenum cable but does not give off toxic fumes. Currently, plenum-rated cable is more popular; there is no cable that meets both the plenum and low-halogen requirements, and demand for low-halogen cable is currently less than 5% of the world market. In fact, there is an exception in the National Electrical Code that allows some cables to be installed under the raised floor in a computer center to be DP rather than plenum rated ["Information Technology Equipment" section, Article 645-5 (d), exception 3]. Because the NEC standards change and are subject to interpretation by local agencies, always consult the code itself before installing any type of fiber optic cable in a building.

There are also hybrid cables available that combine both fiber and copper in a single cable jacket. These include Type 5 cables, which feature a pair of 100/140 μm fibers bound with a pair of copper conductors in the same cable body. Variations with other types of fiber, including 50/125 and 62.5/125 multimode, are available; there may be a large number of fibers or copper conductors in a single cable jacket. These hybrid cables must also meet appropriate safety codes of riser, plenum, or low halogen.

Many different types of commonly used optical connectors have been discussed in Chapter 1. There are many different fiber optic connectors that are used less frequently, including the DIN connector (47256), D-4, Diamond, E2000, HMS-10/HP, and SMA (905 and 906). I will discuss only the more widely used connector types detailed in Chapter 1; planning considerations for optical cable will be addressed in the next chapter. Most are mechanically keyed to prevent multimode fiber from plugging into single-mode receptacles; this is intended for optical safety and is not a guarantee of plugging the proper connector type into the proper receptacle (and there is no keying to prevent plugging single-mode fiber into a multi-mode receptacle). Some popular connector standards, such as the ATM Forum SC duplex, do not specify multimode or single-mode connector keying. For now, I will discuss an important requirement in planning data-com systems—connector color coding and keying. Optical fiber cables are usually color coded with orange jackets for multimode (either 50 or 62.5 μm) and yellow for single mode; these colors were chosen for their visibility in case of fire. Most cables are also labeled with details of the fiber type, diameter, etc. In those cases in which a duplex cable breaks out into a pair of simplex cables, it is conventional for the simplex connectors to be colored black if light is emerging from the connector and white if light is intended to enter the connector; to further assist in identification (and as a guide for the color-blind) the respective connector ends are often labeled "A" and "B" in addition to their color coding. Other standards for connector color are still emerging and subject to change; in some cases, there is not yet a clear consensus for the accepted coloring of optical connectors. In some cases, there may be technical reasons that interfere with obtaining a good color match; some plastic materials with a high glass fill content cannot be easily colored because this alters their thermal properties. Some colored plastics shrink during injection molding or require very tight process tolerances on temperature, pressure, etc. in order to produce a molded part with the required tight tolerances. Fortunately, this is usually an aesthetic issue; when the connectors are viewed from 10 ft or more in indirect lighting, an exact color match is not often required. Table 7.12 lists the current published conventions, as defined by EIA/TIA 658 and the emerging ANSI SBCON standard. Other emerging standards have not yet defined connector specifications as of this writing; these include the ANSI X3T11 workgroup on gigabit interface converters, also known as fiber channel optical converters, and the IEEE 802.3 1.250 Gb/s Ethernet alliance. There is also no general agreement on matching the color of optical transceivers with the corresponding connector color, so it is not always possible to identify cable receptacles or patch panel inserts in this way. The

Table 7.12   **Optical Connector Color Coding**

Application	*Connector type*	*Color*	*Remarks*
ESCON	ESCON duplex	Black	200 Mb/s, 50/62.5 μm, MM, 2 or 3 km
ESCON	ESCON duplex	Gray	200 Mb/s, SM, 20 km
ESCON	SC duplex	Gray	200 Mb/s, SM, 20 km (ANSI SBCON)
IBM ISC	SC duplex	Black	531 Mb/s, 50 μm, MM, 1 km
IBM ISC	SC duplex	Gray or Blue	1 Gb/s, SM, 3 km
ATM	SC duplex	Beige	155 Mb/s or higher, 62.5 μm, MM, 3 km
ATM	SC duplex	Blue	155 Mb/s or higher, SM, 20 km
LC FDDI	SC duplex	Black	125 Mb/s, 62.5 μm, MM, 2 km

Abbreviations used: MM, multimode; SM, single mode; ISC, Intersystem Channel (trade name of IBM Corporation); ATM, asynchronous transfer mode; LC FDDI, low-cost FDDI (color code is an industry convention, not a standard, for LC FDDI); media interface connector.

fiber distributed data interface (FDDI) standard media interface connector (MIC) utilizes three field-installable color-coded keys to identify different types of connections; these are described Table 7.13. The FDDI connectors may also be identified with four colored labels corresponding to the type of keys used. The keys also function as mechanical stops to prevent misplugs of the connectors and avoid installation errors; without a key installed, the MIC connector can be inserted into any type of FDDI receptacle, but with correct keying only certain configurations are allowed. Examples of cabling problems that are difficult to detect without proper keying in an FDDI environment include the following.

Connection of a single attachment station directly to the trunk ring through either an A- or B-type receptacle (this results in a break in the trunk ring)

Connection of the M-type connector directly into a trunk ring by attaching it to either an A- or B-type receptacle (this results in a break in the trunk ring)

Reversal of a dual-attach station within the trunk ring so that the intended positions of the A- and B-type connectors are reversed (this causes the station media access controls to be inserted in the opposite position of the intended trunk ring)

Table 7.13   **FDDI Network Keying**

FDDI connection type	*Key type*
Workstation to wall	S/S
Distribution panel to concentrator M port	M/M
Distribution panel to concentrator A port	A/A
Distribution panel to concentrator B port	B/B
Concentrator A to concentrator M port	A/M
Concentrator B to concentrator M port	B/M
Concentrator 1A to concentrator 2B port	A/B
Concentrator 1B to concentrator 2A port	B/A
Workstation to concentrator M port	S/M

Note. There are four types of keys/labels in the FDDI MIC standard: port A (A), red key/label; port B (B), blue key/label; master (M), green key/label; slave (S), no key/white label.

7.3.3. COST/PERFORMANCE TRADE-OFFS

It is difficult to discuss the cost/performance trade-offs in fiber optic data communication systems, especially in a rapidly changing marketplace. In this section I will attempt to give an overview of this area along with some examples of how much a data link should cost; the reader should consult current product data sheets in this area, and draw his or her own conclusions based on his or her system requirements.

The cost of even basic fiber optic components can vary widely, making direct comparisons with copper links difficult. For example, in some applications fiber optic cable is practically free compared with the cost of attached datacom equipment. Some vendors will supply fiber at no additional charge with the purchase of switches or routers or will provide fiber optic jumper cables that attach to existing patch panels. However, the cost of installing and terminating fiber in new building construction can be prohibitively high; in some parts of the United States, contractors can charge 20–30% more for the installation of optical fiber in buildings compared with "standard" unshielded twisted pair wiring (UTP-3). Even "premium" grade copper wiring (UTP-5) can be 10% more expensive than UTP-3, or more in some markets; it is hoped that this will change with the increased proliferation of high-grade copper and fiber in new buildings. This apparently high cost is not slowing the installation of unterminated optical fiber, however, as part of the building infrastructure, particularly overseas. Japan, for example, has announced a government initiative to wire all new government buildings with optical fiber by the year 2000. Currently available options include hybrid cables with two pairs of optical fiber and two copper coax cables bound in a single jacket. If the cost of purchasing optical fiber is highly variable, so is the cost of renting or leasing it. For some time, telecommunications carriers resisted the idea of providing bandwidth-on-demand services; these are now available in the United States but can be expensive at $150–300 per kilometer per month. The cost of leased or "dark fiber" is even higher overseas and may not be available in some international markets. This has resulted in a demand for bandwidth from alternative carriers; in the United States, for example, many public utility companies are taking advantage of their ownership of the right-of-way into consumer's homes to provide communication services along with electrical power. As the U.S. government considers new deregulation legislation, other alternative carriers such as cable television are expected to begin offering bandwidth on a leased basis.

We should pause to consider the possible competition for optical fiber from other datacom carriers, such as copper wiring and wireless datacom. The Infrared Datacom Association (IrDA) finalized a standard for wireless infrared communications in 1995; more than 100 members, including Microsoft, IBM, Sun, DEC, Motorola, and others have endorsed this new standard and announced plans to build compliant equipment [60–61]. Products that comply with the IrDA standards (either the IrLAP or lr LMP specifications) may be self-certified to carry the IrDA compliance trademark. This is an emerging market and price/availability vary widely; most applications do not compete directly with fiber because they are limited to data rates of a few hundred kb/s and a few meters distance. The primary applications for infrared datacom today are areas where it is too expensive or impractical to retrofit optical fiber, such as large warehouses using computerized data management. In the future this could be an important area. At least one company currently offers 155 Mb/s ATM service on an unguided line-of-sight laser beam connection; a possible application is skyscrapers in large cities where it is prohibitively expensive to rent even a short segment of dark fiber to go across the street. For additional information, contact the IrDA or review the standards and design guidelines.

There has also been some recent research into plastic optical fiber as a low-cost alternative for optical data links; because of the much higher volume of glass fiber being manufactured today, the price of plastic optical fiber remains higher than glass. This may change as demand for low-cost plastic fiber links improves, but today such fibers are limited by their high loss (120–150 dB/km) to fairly short distances (approximately 50 m) rather than tens of kilometers provided by glass fiber. Despite this, there is increasing interest in plastic fiber because these links can be operated using visible light sources that are potentially very low cost; a 680-nm (red) LED transceiver for plastic fiber links can be offered for as low as $5–10. Plastic fiber links have already found some applications in laser printers and the microprocessor control systems for automotive applications. The current areas of interest include the small office/home office environment, where there is a need for a low-cost, high-bandwidth interconnect solution. The electromagnetic noise environment in a typical home or small office is much noisier than a computer machine room because many home appliances (such as microwave ovens) produce a broad radiated noise spectrum that can interfere with sensitive data communication links. Existing home wiring, such as stereo audio cables, may not be sufficient in this environment, especially because the computer equipment itself must not interfere with existing protected services, such as television or ham radio broadcasts. Plastic fiber is also much easier to install than glass because it requires only minimal polishing when terminating a new connection; this could be done by a homeowner with minimal training. With this motivation, the ATM Forum has recently begun investigating new standards for residential broadband networks, sometimes called home area networks, as an alternative to UTP CAT 5 copper cables. Current standards under discussion include a 51.8 Mb/s link over plastic fiber, 50 m long, with 17 dBm link budget; this is considered adequate for even the largest home offices. Several new types of plastic fiber connectors have been proposed, including variations on the F07 connector, SMA 905 and 906, EIAJ Digital Audio, and JIS Industrial; plastic fiber versions of the ST and FC are also available. Some vendors have even proposed that plastic fiber with red VCSEL lasers could operate up to 1 Gb/s over short distances, and this may be the solution for fiber to the desktop, PC/television convergence, and other areas where there is currently a speed bottleneck and cost is the primary driver to enter the market. This area is expected to see increased activity in the coming years.

A direct cost comparison between optical fiber and copper media is difficult, even accounting for the improved performance of optical fiber with respect to distance, noise immunity, and other factors (see Table 7.2). Many different cost estimates have been proposed; a comparison of the application areas where either fiber or copper are currently the dominant media is shown in Table 7.14, along with a typical cost comparison between fiber and copper links in different environments [62]. This table is intended

Table 7.14   **Copper vs Optical Data Link Solutions; Distance and Cost Considerations**

			Distance		
	Kilometers +	**10–100 m**	**1–10 m**	**0.1–1 m**	**<10 cm**
Cost	>$1000/Gb/s	$100/Gb/s	$10/Gb/s	$1/Gb/s	$0.1/Gb/s
Copper	Coax/ repeated	Coax/ twisted pair	Coax TP	Backplane	Chip-on-board
Optics	Single mode	Multimode	Multimode	Waveguide	Free space

Note. From Ref. [1]. Costs are order of magnitude market cost required to compete at a given distance.

as a guide only; application-specific costs may vary. In general, fiber is cost/ performance competitive with copper for links longer than a few tens of meters, shown by the vertical line in this table (approximately $20 for a 50-m simplex data link, including optical transceivers on each end). Transceiver cost has also been falling rapidly, driven by 21% compound annual growth in this industry; a typical FDDI-type transceiver that cost more than $300 in 1988 was available in 1996 for approximately $60. Even at this growth rate, it is difficult to speculate as to whether datacom will ever match the huge volumes available in the telecommunications market; some observers have predicted that the U.S. telecom infrastructure will need to support an aggrate data rate of 1 Tb/s by the Year 2007.

For longer distances, the performance advantages and room for future bandwidth growth of optical fiber make it the preferred alternative. If we try to compare the end-to-end cost of a data link using fiber vs copper, the results again vary widely. For example, consider an application involving attachment of 10 workstations to a large server using FDDI links. If the server is a bipolar-based mainframe, it is possible to obtain converter boxes that attach FDDI links to either parallel copper or ESCON channel ports; one adapter is required per port, and the average cost per FDDI client can be $25,000 or more. However, if the server is a CMOS-based transaction server, adapters are available that allow direct attachment of an FDDI LAN port to the mainframe; the direct attach feature may cost approximately $15,000, but thereafter each additional client can be added for the cost of inserting a standard FDDI adapter card in the workstation— approximately $1,000 per client. The reduction in total cost of computing is significant enough for a large network that it may justify upgrading the bipolar mainframe to a CMOS version! As can be seen from this example, network attachment costs are clearly very application dependent; we have not even considered associated software costs and maintenance fees because they vary too much to permit a clear discussion within the scope of this chapter.

Future emerging technologies are likely to have an important affect on data link costs. One example is the development of parallel optical transmitters and receivers that incorporate between 10 and 32 channels in a compact package. To emphasize the improved performance of parallel optical links, these units are often priced in terms of dollars/Gb/km. Unfortunately, until the development of long-wavelength vertical cavity surface-emitting lasers or similar transceiver technology, these links will likely be limited in distance and bandwidth; using short-wavelength laser arrays, distances of only a few hundred meters are possible. Multiplexing technology has been used effectively by telecommunication carriers to reduce costs by sharing a single data channel among multiple users; a similar effort is under way in the datacom community. One example is the IBM 9672 Optical Wavelength Division Multiplexer; it is perhaps the first device of its kind developed for the datacom industry. This device provides a 20:1 multiplexing of datacom channels onto a single fiber at wavelengths in the 1550-nm range. Because optical fiber is protocol insensitive, many different types of data can be accomodated. This product has only recently been announced, and in small volume the cost remains high (approximately $250,000 per device or $500,000 for a link). Still, this is cost competitive with leased fiber; at a rate of $150/mile/month, assuming a 10-mile link with 10 independent bidirectional channels, the 9729 link pays for itself in less than a year and actually results in a savings of more than $340,000 per year.

There are also emerging standards and technologies that could drastically change the face of datacom over the next decade. Wavelength multiplexing may face competition from various digital subscriber line (DSL) technologies that have recently become available; these promise to increase the data rate for attached modems on PCs, which in turn drives bandwidth requirements of the supporting network infrastructure. Because DSL approaches are fairly new, there are a number of proposed standards as listed in Table 7.15. Other emerging standards include gigabit ethernet;

Table 7.15   **Proposed Versions of Digital Subscriber Line Technology**

Technology	*Standard*	*Data rate*	*Distance (ft)*
DSL/ISDN	ANSI T1.219 (1991)	160 kbps	18,000
384DSL	None	384 kbps	18,000
HDSL	ANSI T1	1.544 Mbps 2.048 Mbps	12,000
SDSL	None	Same as HDSL	10,000
DMT ADSL	ANSI T1.413 (1995)	6.144 Mbps/640 kbps	12,000
RADSL	None	9 Mbps/1.5 Mbps	Varies with data rate
VDSL	Proposed ANSI	12.96, 25.92, or 51.84 Mbps/2–20 Mbps 1,000–4,500	

Note. Speed indicated is full duplex unless otherwise indicated; variable rates are indicated as max. speed downstream/upstream.

this standard is not yet finalized and there are many proposed options at this time, including a proposal for operating 62.5-μm fiber with long-wavelength lasers at gigabit speeds over hundreds of meters. Fiber optics is also adding function and driving down cost for existing standards by providing optical channel extensions for high-performance parallel interace (HIPPI) and serial storage architecture (SSA) links. HIPPI was developed by the ANSI X3T9.3 committee as an 800 MB/s, 25-m parallel copper link for high-performance computing; in 1991 an ad hoc ANSI subcommittee proposed serial HIPPI, a channel extension using long-wave lasers and single-mode fiber. The HIPPI data and control characters are serialized into a 1.1 Gb/s data stream, which is transmitted over the fiber; this extends the link to 10–13 km. Fiber optic channel extenders also exist for other peripheral attachment standards, including the SSA; in the future, SSA links may merge into an ANSI-approved 1 Gb/s version of fiber channel, tentatively called fiber channel enhanced loop. These and other developments are expected to influence the price and availability of fiber optic components in coming years.

Finally, it should be noted that although a performance comparison favors fiber over copper, the cost of fiber data links is volume driven. Fiber cost is more important for telecommunication carriers because the ratio of fiber distance to number of transceivers is large. Conversely, for datacom systems there are many more transceivers per kilometer of fiber, so the transceiver cost dominates the link cost. Datacom transceivers are available in two basic types, an integrated transceiver solution that is packaged by the manufacturer and a "bag of parts" that is available to equipment designers to construct their own data links. The bag of parts solution is cost effective and offers design flexibility by mixing and matching different transmitters, receivers, serializers, deserializers, and other electronics in the application; however, it does not address component interactions such as high frequency noise. For data rates greater than approximately 1 Gb/s, the integrated solution is more widely accepted because it addresses these system problems. Over time, the cost point between these two solutions will probably be driven higher by advanced technology, such that effective bag of parts solutions will be available by the Year 2000 in the 1 or 2 Gb/s range. With more than 700 miles of optical fiber installed in the United States each year and some analysts projecting annual volumes of 300,000–400,000 for ATM transceivers (with a per channel cost of under $300 for ATM adapter cards), fiber optics remains a growth area. The cost of fiber optics is likely to continue decreasing, especially because installed fiber provides the bandwidth requirements to meet the needs of emerging applications well into the next century.

Appendix A: Contact Information for Optoelectronics and Fiber Optics Information

PROFESSIONAL ORGANIZATIONS

American Institute of Physics (AIP) and American Physical Society (APS)
Headquarters, American Center for Physics
1 Physics Ellipse
College Park, MD 20740
http://www.aip.org or http://www.aps.org

Fiber 66
"The first virtual group of fiber optic teachers on the Internet." A worldwide team of professional instructors and educators working through private newsgroups and mailing lists to collaborate on the design of courses in fiber optics, which are then made available on-line as appropriate. To join, send e-mail to fiber66@optoroute.com or contact Marc Duchesne in France at http://www.optoroute.com/fiber66.

Fiber University (Fiber U)
Administered by Fotec Corporation
151 Mystic Ave
Medford, MA 02155
http://www.std.com/fotec
Provides training and professional certification on fiber optic installation and testing.

Institute of Electrical and Electronics Engineers (IEEE)
445 Hoes Lane, P.O. Box 1331
Piscataway, NJ 08855
http://www.ieee.org

Institute for Optical Data Communication (IODC)
Direct inquires to Dr. Casimer DeCusatis
IBM Corporation
522 South Road, MS P343
Poughkeepsie, NY 12601

casimer@vnet.ibm.com
> Or see the Photonics Technical Home link from the Optical Society of America
> (http://www.osa.org/photonics).

National Society of Professional Engineers (NSPE)
1420 King Street
Alexandria, VA 22314
Phone: (203) 684-4811
http://www.nspe.org.
> The NSPE administers the licensed Professional Engineer certification programs across all engineering disciplines; contact the main office listed above or the Web site for information or a list of the state license boards.

New York Academy of Sciences
2 East 63rd Street
New York, NY 10021
http://www.nyas.org

National Institute of Standards & Technology (NIST)
(particularly Boulder, Colorado, laboratory for fiber optic research)
http://www.nist.gov

Optical Society of America (OSA)
2010 Massachusetts Ave NW
Washington, DC
http://www.osa.org

Society of Photooptic Instrumentation Engineers (SPIE)
P.O. Box 10
Bellingham, WA 98227
http://www.spie.org

LASER SAFETY CERTIFICATION INFORMATION

International Electrotechnical Commission (IEC): IEC International Laser Safety & Compliance Laboratories
> VDE (Verband Deutscher Electrotechniker)
> Association of German Electrical Engineers
> Merianstrabe 28 D-6050 Offenbach
> Phone: +49 069 8306-0

> TUV Rheinland of North America, Inc. (U.S. contact for TUV Laboratory)

12 Commerce Road
Newton, CT 06470
Phone: (203) 426-0888

TUV Rheinland
Prufstelle fur Geratesicherheit
Am Grauen Stein
5000 Koln 91
Germany

U.S. Food and Drug Administration (FDA)
Center for Devices and Radiological Health

2098 Gaither Road
Rockville, MD 20850

Issues accession numbers to verify compliance of laser and optical transmitter products with regulations for the administration and enforcement of the Radiation Control for Health and Safety Act of 1968 (Title 21, CFR, Subchapter J); the U.S. laser safety certification is obtained from the standard FDA 21 CFR Part 1040, performance standards for light-emitting products Sections 1040.10 (laser products) and 1040.11 (specific purpose laser products)

Certified Laser Safety Officer Training

Laser Institute of America (LIA)
12424 Research Parkway
Suite 125
Orlando, FL 32826
Phone: (407) 380-1553

(In addition to providing laser safety training, copies of the ANSI Z136.X laser safety standards are available from the LIA).

STANDARDS BODIES

American National Standards Institute (ANSI)
1430 Broadway
New York, NY 10018
or
11 West 42nd Street
New York, NY 10036
http://www.ansi.org

The ANSI Tl Committee can be accessed on-line at http://www.tl.org.

Administers the laser safety standard document: ANSI Standard Z136.X (first issue 1988) for the safe use of optical fiber communication systems utilizing laser diode and LED sources; to purchase and ANSI/TIA/EIA standards, contact the publisher, Global Engineering Documents (see listing below for address).

ATM Forum
303 Vintage Park Drive
Foster City, CA 94404
(Phone: 1-415-578-6860)
http://www.atmforum.com

For discussion of cell-relay based networks, consult the Usenet newsgroup {comp.dcom.cell-relay}; there is also a mailing list, which is a virror of the newsgroup and is identified as {cell-relay}.

The Desktop ATM 25 Alliance can be accessed on-line at http://www.atm25.com.

Bellcore Standards

Bell Communications Laboratories
331 Newman Springs Road
or
Bellcore Customer Service
8 Corporate Place, Room 3A183
Piscataway, NY 08854
http://www.bellcore.com
Phone: 1-800-521-CORE

Bellcore has recently introduced a program to certify that third-party equipment conforms to recognized networking standards; they issue the Bellcore Certification Mark for Certified Excellence to products that meet these standards. They also provide a catalog of technical information, publication number SR-264.

Building Premises Wiring Standards Body (CICSI)
10500 University Center Drive
Suite 100
Tampa, FL 33612

Provides certification to work with premises wiring, LAN equipment, and other fiber optic applications; administers a professional certification program for registered communication distribution designers (RCDD) and LAN specialists; recently began a program to certify fiber optic technicians. Also sponsors regular conference and training sessions and publishes the *Telecommunications Methods Manual* and *LAN Design Manual*

Committee on Optical Science & Engineering (COSE) of the National Research Council (NRC)

http://www.nsa.edu

e-mail cose@nas.edu

Compliance Standards

In addition to contacting the standards bodies or the certification laboratories, copies of the compliance standards for EMC, ESD, Telecom, Product Safety, and others including EMC resource manuals may be ordered from the following:

Compliance Engineering
One Tech Drive
Andover, MA 01801
Phone: (508) 681-6673
   Other information is available on-line at http://world.std.com/ ~techbook (hotlink to EMC regulatory compliance) from the newsgroup sci.engr.electrical.compliance

Electronic Industries Association (EIA)
Engineering Dept.
2500 Wilson Blvd.
Arlington, VA 22201
or see: EIA Headquarters
2001 Pennsylvania Ave. NW
Washington, DC 20006
also see: Electronic Industries Foundation (EIF)
919 18th St. NW Suite 900
Washington, DC 20006
   Publishes FOTP and JEDEC standards, which can be ordered through Global Engineering.

Fiber Channel Association (FCA)
12407 MoPac Expressway, North 100-357
P.O. Box 9700
Austin, TX 78766

http://www.amdahl.com/ext/CARP/FCA/FCA.html
Phone: 1-800-272-4618

Frame Relay Forum
http://frame-relay.indiana.edu/index.html

Gigabit Ethernet Alliance
http://www.gigabit-ethernet.org
Global Engineering Documents
3130 South Harbor Blvd. #330
Santa Ana, CA 92704
or
15 Inverness Way East
Englewood, CO 80112
Phone: 1-800-854-7179 (internationally call 303-397-7956)

Order ANSI, JEDEC, EIA/TIA, and other standards.

HIPPI Network Forum
P.O. Box 10173
Albuquerque, NM 67184
http://www.esscom.com/hnf

IBM Corporation
http://www.ibm.com

International Electrotechnical Commission, Internationale
Elektrotechnische Kommission (IECCE/CB, IECQ [also known as
Commission Electrotechnique Internationale (CEI)
Bureau Central de la Commissions Electrotechnique Internationale
3 rue de Varembe
Geneva, Switzerland
http://www.hike.te.chiba-u.ac.jp/ikeda/iec/home.html

Issues CEI/IEC 825-1 and 825-2, "Safety of Laser Products, Part 1:
Equipment Classification, Requirements & Users Guide," and
"Part 2: Safety of Optical Fiber Communication Systems,"
frequently collaborates/coreferences documents with CENELEC-
Europaisches Kommittee fur Electrotechniscne Normung-CCA,
HAR, CECC.

Infrared Data Association (IrDA)
Walnut Creek, CA
Phone: (510) 943-6546
http://www.irda.org
e-mail irda@netcom.com

Institute for Interconnecting & Packaging Electronic Circuits (IPC)
2215 Sanders Road
Northbrook, IL 60062
http://www.ipc.org
  (Sponsors circuit board designer certification).

International Standards Organization (ISO)
Central Secretariat
1 Rue de Varembe
Case Postale 56
CH-1211 Geneve 20
Switzerland
http://www.iso.ch

International Telecommunications Union (ITU, formerly CCITT)
http://info.itu.ch

Internet standards are available for discussion on a number of on-line
forums:
  ftp://ds.internic.net/ds/dspg1intdoc.htlm
  ftp://ftp.gmd.de/documents
  ftp://ftp.isi.edu
  gopher://gopher.nic.ddn.mil

Japanese-based telecom/datacom standards (JEDEC)
Contact JEDEC Excutive Secretary at EIA headquarters

Military specification standards
For information or to order copies of the military standards (Mil.
Spec.), contact:
Superintendent of Documents
U.S. Government Printing Office
Washington, DC 20036
or
Commanding Officer
U.S. Naval Supply Depot
Att: code DMD
5801 Tabor Ave.
Philadelphia, PA 18201
  the Mil. Spec. standards are controlled by the U.S. Department of
  Defense, Washington, DC.

Network Computer Alliance & NC Reference Prolife:
http://www.nc.ihost.com

National Institute for Science and Technology (NIST)
Headquarters
Gaithersburg, MD 20899
http://www.nist.gov
or
NIST Boulder, Colorado (center for fiber optic research and development)
325 Broadway
Boulder, CO 80303
http://www.boulder.nist.gov
  NIST Boulder is also the headquarters for the Institute for
  Telecommunication Sciences (ITS) and the National
  Telecommunication and Information Administration (NTIA):
  http://its.bldrdoc.gov/its.html.
Optoelectron Industry Development Association (OIDA)
2010 Massachusetts Ave. NW Suite 200

Washington, DC 20036
Phone: (202) 785-9926
http//www.oida.org
Siecor Corporation
http://www.siecor.com

Switched Megabit Data Services (SMDS) interest group
http://www.cerf.net/smds

DEFINITION OF TECHNICAL TERMS

The Information Technology Vocabulary, developed by subcommittee
1, joint technical committee 1, of the International Organization for
Standardization and the International Electrotechnical Commission
(ISO/IEC JTCl/SC1)

The ANSI/EIA Standard 440A: Fiber Optic Terminology (copyright
1989), available from the EIA, 2001 Pennsylvania Ave. NW,
Washington, DC 20006

*The American National Standard Dictionary for Information Systems,
ANSI X3.172* (1990), available from ANSI

REFERENCES ON TECHNOLOGY SUPPLIERS

Electronic Engineers Master Catalog, Vols. A–D, published
annually by

Hearst Business Communications Inc.
645 Stewart Ave.
Garden City, NY 11530

Laser Focus World Buyer's Guide, published annually by
Laser Focus World Magazine
Penwell Publishing Co.
Ten Tara Blvd. 5th Floor
Nashua, NH 03062

Photonics Corporate Guide and Buyer's Guide, published annually by
Photonics Spectra Magazine
Laurin Publishing Co, Inc.
Berkshire Common, PO Box 4949
Pittsfield, MA 01202

Appendix B: Some Accredited Homologation Test Labs

Note. The following list is provided for reference purposes only and does not constitute an endorsement of the services provided by these laboratories. There are many different international standards for optical data communication, ISDN services, cabling products, voice service equipment, etc. Consult the standards bodies or testing labs for the latest information as to what approvals are required to ship a given product outside the United States. Another resource is the home page for Homologation & Type Approval (H&TA) department of IBM Corporation in LaGaude, France, http://w3.lagaude.ibm.com/homologation/wwwhta.htm.

United States
ASL Inc.
Santa Clara, CA
(408) 244-2921

AT&T Bell Laboratories
International Regulatory Division
11900 North Pecos Street
Denver, CO 80234
(303) 538-3906

Certelecom
820 Proctor Ave
Ogdensberg, NY 13669
(800) 348-9433

Global Product Compliance Lab
Room 11C-165, P.O. Box 3030
Holmdel, NJ 07733
(908) 834-1801

Underwriters Laboratories Inc.
333 Pfingsten Road
Northbrook, IL 60062
(847) 272-8000

Canada

Bell Northern Research
Standards & Accredation
P.O. Box 3511
Station C
Ottawa, Ontario, Canada K1Y 4H7
(613) 763-5380

Certelecom
3325 River Road, RR#5
KIV 1H2
Canada
(800) 563-6336

Germany

BZT Labs
Postfach 100443, 66004 Saarbrucken
Germany
49681 598 0

United Kingdom

Assessment Services Limited
Segenworth Road, Tichfield
Fareham
Hampshire PO15 5RH
United Kingdom
44(0) 1329 443322

British Standards Institute (BSI)
P.O. Box 375, Breklands
Linford Woods West
Milton Keynes MK14 6LL
United Kingdom
01908 312636

KTL Corporation
Newlands Science Park
Inglemire Lane
Kingston upon Hull HU6 7TQ
United Kingdom
44 (0)1482 801801

Telecommunications Services
Maylands Avenue
Hernel Hempstead, HERTS HP2 4SQ
United Kingdom
44 1 44 223 0442

Japan

The Japanese telecom specs are available on the Internet at
http://www.mpt.go.jp/telecom

AKZO Kashima Limited
Telecommunications
Yokahama Test Centre
2-30 Hatsunecho
Naka-ku
Yokahama, 231
Japan
81 45 251 8300

Japan Quality Assurance Organization
Telecommunications
21-25, Kinuta 1-Chrome
Setagaya-Ku
Tokyo 157
Japan
81 3 3416 0193

Australia

Telecom specs are available on the Internet at http://www.dca.au
Advantech Electronics Pty. Ltd.

3/14 Salisbury Road
Hornsby NSW 2077
Australia
61 2 477 7757

AUSTEST Laboratories
4/87 Reserve Road

Artamon NSW 2064
Australia
61 2 437 4500

New Zealand

Wakefield Laboratories
P.O. Box 4512
Auckland, New Zealand
64 9 379 5768

References

1. IBM Corporation 9729 Fiber Optic Multiplexer. 1996 (June). *Photon Spectra* (special issue on Photonics Circle of Excellence Awards), 89.
2. Wolke, A. 1994. Improved curve fit algorithm for receiver sensitivity and eye window extrapolation. Amp Lytel Engineering Report ER:94-3634-037R.
3. Wolke, A. 1994. Curve fitting for sensitivity and eye window extrapolation. Amp Lytel Engineering Report ER:94-3634-016R.
4. Chan, D. 1973. The extrapolation of error rate performance of baseband digital communication systems." Bell Northern Research Technical Report TR 1E90-673.
5. DeCusatis, C. 1995. Data processing systems and optoelectronics. In *Optoelectronics for data communication,* eds. R. Lasky, U. Osterberg, and D. Stigliani, Chap. 6. New York: Academic Press.
6. Walker, S. S. 1996. Rapid modeling and estimation of total spectral loss in optical fibers. *IEEE J. Lightwave Tech.* 4:1125–1131.
7. Allard, F. C. 1990. *Fiber optics handbook for engineers and scientists.* New York: McGraw-Hill.
8. Frieble, E. J., *et al.* 1988. Interlaboratory comparison of radiation-induced atteunation in optical fibers, part I. *IEEE J. Lightwave Tech.* 6:165–171. Frieble, E. J., *et al.* 1990. Interlaboratory comparison of radiation-induced attenuation in optical fibers, part II. *IEEE J. Lightwave Tech.* 8:967–988.
9. Frieble, E. J., *et al.* 1984. Effect of low dose rate irradiation on doped silica core optical fibers. *Appl. Opt.* 23:4202–4208.
10. Frieble, E. J., and M. E. Gingrich. 1981. Photobleaching effects in optical fiber waveguides. *Appl. Opt.* 20:3448–3452.
11. Tsai, T. E., and H. B. Lin. 1990. Photoluminscence induced by 6.4 cV photons in high purity silica. *Proc. Mater. Res. Soc. Symp.* 172:125–131.
12. Haber, J. B., *et al.* 1988. Assessment of radiation-induced loss for AT&T fiber optic transmission systems in the terrestrial environment. *IEEE J. Lightwave Tech.* 6:150–154.

13. Amos, N. A., A. D. Bross, and M. C. Lundin. 1990. Optical attenuation length measurements of scintillating fibers. *Nucl. Inst. Methods* A297:396–403.

14. Kyoto, M., *et al.* 1992. Gamma-ray radiation hardened properties of pure silica core single-mode fiber and its data link system in radioactive environments. *IEEE J. Lightwave Tech.* 10:289–294.

15. Young, M. 1973. Geometrical theory of multimode optical fiber to fiber connections. *Opt. Commun.* 7:253–255.

16. Bisbee, D. L. 1971. Measurements of loss due to offsets and end separation of optical fibers. *Bell Sys. Tech. J.* 49:3159–3168.

17. Chu, T. C., and A. R. McCormick. 1978. Measurements of loss due to offset, end separation, and angular misalignment in graded index optical fibers excited by an incoherent source. *Bell Sys. Tech. J.* 57:595–602.

18. Gloge, D. 1975. Propagation effects in optical fibers. *IEEE Trans. Microwave Theor. Tech.* MTT-23:106–120.

19. Fiber channel—Physical layer and signaling specification, ANSI X3T9/91-062, X3T9.3/92-001 rev. 4.2, 1993.

20. Shanker, P. M. 1988. Effect of modal noise on single-mode fiber optic networks. *Opt. Commun.* 64:347–350.

21. Fiber Optic Test Procedure FOTP-171, Attenuation measurement by substitution method (also EIA-TIA-455-171), available from Electronics Industry Association, New York.

22. Kosiorska, H., *et al.* 1991. Statistical method for predicting fiber loss in the field using factory measurements. *Proc OFC'91,* 138.

23. DeCusatis, C., and M. Benedict. 1992 (May). Method for fabrication of high bandwidth optical fiber. *IBM Tech. Disc. Bull.*

24. Marcuse, D., and H. M. Presby. 1975. Mode coupling in an optical fiber with core distortion. *Bell Sys. Tech. J.* 1:3.

25. Refi, J. J. 1986. LED bandwidth of multimode fiber as a function of source bandwidth and LED spectral characteristics. *J. Lightwave Tech.* LT-14:265–272.

26. Miller, S. E., and I. P. Kaminow, eds. 1988. *Optical fiber telecommunications II.* New York: Academic Press. Also see Kulikowski, A. 1991. *Fundamentals of geometric design & tolerancing.* New York: Delmar Press.

27. Green, P. 1993. *Fiber optic networks.* New York: Prentice Hall.

28. Agrawal, G. P., *et al.* 1988. Dispersion penalty for 1.3 micron lightwave systems with multimode semiconductor lasers. *IEEE J. Lightwave Tech.* 6:620–625.

29. Midwinter, J. E. 1977. A study of intersymbol interference and transmission media instability for an optical fiber system. *Opt. Quantum Electron.* 9:299–304.

30. Ogawa, K. 1982. Analysis of mode partition noise in laser transmission systems. *IEEE J. Quantum Electron.* QE-18:849–855. Also see Ogawa, K. 1985. Semiconductor laser noise: Mode partition noise. In *Semiconductors and semimetals,* eds. R. K. Willardson and A. C. Beer, Vol. 22C, Chap. 8. Lightwave Comm. Tech. Ser., Tsang, W. T. ed. New York Academic Press.

31. Campbell, J. C. 1988. Calculation of the dispersion penalty of the route design of single-mode systems. *J. Lightwave Tech.* 6:564–573. Also see Jiang, W., R. Feng, and P. Ye. 1990. Improved analysis of dispersion penalty in optical fiber systems. *Opt. Quantum Electron.* 22:23–31.

32. Ohtsu, M., *et al.* 1988. Mode stability analysis of nearly single-mode semiconductor laser. *IEEE J. Quantum. Electron.* 24:716–723.

33. Ohtsu, M., and Y. Teramachi. 1989. Analysis of mode partition and mode hopping in semiconductor lasers. *IEEE. Quantum Electron.* 25:31–38.

34. Olshansky, R. 1976. Mode coupling effects in graded index optical fibers. *Appl. Opt.* 14:935–945.

35. Duff, D., *et al.* 1989. Measurements and simulations of multipath interference for 1.7 Gbit/s lightwave systems utilizing single and multifrequency lasers. *Proc. OFC 1989,* 128.

36. Radcliff, J. 1989. Fiber optic link performance in the presence of internal noise sources. IBM Tech. Report, Glendale Labs, Endicott, NY.

37. Xiao, L. L., C. B. Su, and R. B. Lauer. 1992. Increase in laser RIN due to asymmetric nonlinear gain, fiber dispersion, and modulation. *IEEE Photon. Tech. Lett.* 4:774–777.

38. Hakki, B. W., F. Bosch, and S. Lumish. 1989. Dispersion and noise of 1.3 micron lasers in microwave digital systems. *IEEE J. Lightwave Tech.* 7:804–812.

39. AT&T Technical Report. 1985. Methods of optical data link module testing, 14 pp.

40. Manalino, A. A. 1987. Time-domain eyewidth measurement of an optical data link operating at 200 Mbit/s. *IEEE Trans. Inst. Meas.* IM-36:551–553.

41. Trischitta, P., and P. Sannuti. 1988. The accumulation of pattern dependent jitter for a chain of fiber optic regenerators. *IEEE Trans. Comm.* 36:761–765.

42. Walker, S. D., R. B. P. Carpenter, and P. Cochrane. 1983. Jitter accumulation with SAW timing extraction at 325 Mbaud. *Electron. Lett.* 19:193.

43. CCITT 1984. Recommendations G.824, G.823, O.171, and G.703 on timing jitter in digital systems.

44. Gardner, F. M. 1979. *Phaselock techniques,* 2nd ed. New York: Wiley.

45. Agrawal, G. P., and T. M. Shen. 1986. Power penalty due to decision-time jitter in optical communication systems. *Electron. Lett.* 22:450–451.

46. Schumacher, K., and J. J. O'Reily. 1986. Power penalty due to jitter on optical communication systems. *Electron. Lett.* 23:718–719.

47. Shen, T. M. 1986. Power penalty due to decision-time jitter in receivers using avalanche photodiodes. *Electron. Lett.* 22:1043–1045.

48. Schumacher, K., and J. J. O'Reily. 1989. Distribution-free bound on the performance of optical communication systems in the presence of jitter. *IEEE Proc.* 136:129–136.

49. Bates, R. J. S. 1983. A model for jitter accumulation in digital networks. IEEE Globecon '83, 145–149, San Diego, CA.

50. Trischitta, P. R., and E. L. Varma. 1989. *Jitter in digital transmission systems.* Boston: Artech House.

51. Byrne, C. J., B. J. Karafin, and D. B. Robinson, Jr. 1963. Systematic jitter in a chain of digital regenerators. *Bell Sys. Tech. J.* 43:2679–2714.

52. Bates, R. J. S., and L. A. Sauer. 1985. Jitter accumulation in token-passing ring LANs. *IBM J. Res. Dev.* 29:580–587.

53. Chamzas, C. 1985. Accumulation of jitter: A stochastic model. *AT&T Tech. J.* 64.

54. Stolen, R., 1979. Nonlinear properties of optical fiber. In *Optical fiber communications,* eds. S. E. Miller and A. G. Chynoweth, Chap. 5. New York: Academic Press.

55. Cotter, D. 1982. Observation of stimulated Brillouin scattering in low loss silica fibers at 1.3 microns. *Electron. Lett.* 18:105–106.

56. da Silva, H. J. A., and J. J. O'Reily. 1989. System preformance implications of laser chirp for long haul high bit rate direct detection optical fiber systems, Proc. IEEE Globecom, 19.5.1–19.5.5.

57. Kurashima, T., T. Horiguchi, and M. Tateda. 1990. Thermal effects of Brillouin gain spectra in single-mode fibers. *IEEE Photon. Tech. Lett.* 2:718–720.

58. Stolen, R. H., and C. Lin. 1978. Self-phase modulation in silica optical fiber. *Phys. Rev. A* 17:1448.

59. Fisher, R. A., and W. K. Bischel. 1975. Numerical studies of the interplay between self-phase modulation and dispersion for intense plane-wave laser pulses. *J. Appl. Phys.* 46:4921.

60. IrDA Serial Interface Test Suite, Link Access Protocol (Ir LAP), Secondary Station Implementation Test Specification Ver. 3.0; and IrDA Serial Interface Test Suite, Link Management Protocol (Ir LMP), Secondary Station Implementation Test Specification Ver. 3.0; both available from Genoa Technology, Moorpark, CA (July 1995) or by anonymous FTP at ftp.gentech.com in the /pub directory.

61. *IrDA data link design guide.* 1995. Palo Alto, CA: Hewlett–Packard.

62. Hibbs-Brenner, M. K., Y. Liu, A. Cox, D. Greenlaw, L. Galarneau, J. Bristow, J. A. Lehman, T. Werner, and R. A. Morgan. 1996. Components and packaging for low-cost optoelectronic modules, Paper WB3, 132. Proceedings of the OSA Annual Meeting, Rochester, NY.

Chapter 8 | Planning and Building the Optical Link

R. T. Hudson
D. R. King
T. R. Rhyne
T. A. Torchia

Siecor Corporation, Hickory, North Carolina 28601

8.1. Introduction

With the rapidly advancing information age, the importance of building a reliable and long-lasting communications infrastructure is becoming critical. Organizations that best develop and understand information technology will have a distinct advantage over their competitors.

Significant planning and foresight is required to build a stable communications platform. Predicting both today's and the future requirements for a building or campus infrastructure is very difficult due to the tremendous rate of progress in information technology. Planning is essential in order to implement a practical solution meeting today's needs while providing a migration path to tomorrow's requirements.

Employing a structured cabling system utilizing optical fiber can meet these needs. Optical fiber is the only medium with the bandwidth to meet current and future needs. Fiber is a very attractive medium due to its ability to support the widest range of applications at the fastest speeds and for the longest distance. Other advantages that fiber offers include electromagnetic interference and radio frequency interference immunity. In addition, it is extremely secure because it is almost impossible to tap. Fiber optic cables are also smaller and lighter in weight than their copper counterparts.

This chapter contains a brief introduction to structured fiber optic cabling systems used in voice, video, and data communication applications. It is divided into three sections: private network (premises) applications, standards for fiber optic building wiring, and installation/handling issues.

HANDBOOK OF FIBER OPTIC
DATA COMMUNICATION

8.2.   Private Networks

A.   INTRABUILDING

This section deals with cabling systems within a common building. It includes backbone and horizontal applications as well as the specific types of cables used in each.

1.   Intrabuilding Backbone

The intrabuilding backbone connects the main cross-connect in the building to each of the other cross-connects. An optical fiber intrabuilding backbone has emerged as the medium of choice due to its ability to support multiple high-speed networks in a smaller cable without cross-talk concerns. Also, more users are utilizing fiber to support voice and private branch exchange (PBX) applications by placing small remote PBX shelves within cross-connects supported by the intrabuilding backbone.

2.   Intrabuilding Topologies

The intrabuilding backbone design between the building main cross-connect (MC) or intermediate cross-connect (IC) and the horizontal cross-connect (HC) is usually straightforward; however, various options exist. A single hierarchical star design between the building cross-connect and the HC is strongly recommended (Fig. 8.1). The only possible exception occurs in an extremely large building, such as a high-rise, where a two-level hierarchical star may be considered (Fig. 8.2).

The same options and decision processes apply here. Sometimes, building pathways link HC to HC in addition to IC to HC, especially in buildings with multiple HCs per floor. Only in specific applications should a user design an HC-to-HC link. For example, this pathway might be of value in providing a redundant path between HC and IC, although direct connection between HCs should always be avoided (Fig. 8.3).

3.   Horizontal Cabling

The benefits of fiber optics (high bandwidth, low attenuation, and system operating margin) allow greater flexibility in cabling design. Because of the distance limitations of copper and small operating margin, passing tests over unshielded twisted pair (UTP) requires compliance with stringent

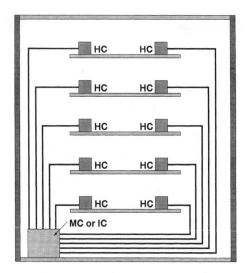

Fig. 8.1   Intrabuilding backbone (one-level hierarchy).

design rules. In addition to the traditional horizontal cabling network, optical fibers support designs specifically addressing the use of open-systems furniture and/or centralized management.

The traditional network consists of an individual outlet for each user within 300 ft of the telecommunications closet. This network typically has data electronics equipment (hub, concentrator, or switch) located in each telecommunications closet within the building. Because electronics are located in each closet, this network is normally implemented with a small fiber count backbone cable (12 or 24 fiber count). This network uses a single cable per user in a physical star from the HC to the telecommunications outlet (Fig. 8.4).

For open-systems furniture, a multiuser outlet is recommended. The multiuser outlet calls for a high-fiber count cable to be placed from the closet to an area in the open office where there is a fairly permanent structure, such as a wiring column or cabinet, in a grid-type wiring scheme. At this structure a multiuser outlet, instead of individual outlets, is used (Fig. 8.5).

Fiber optic patch cords are then installed through the furniture raceways from the multiuser outlet to the office area. This allows the user to rearrange the furniture without disrupting or relocating the horizontal cabling. This type of network is supported in Annex G of TIA/EIA-568-A.

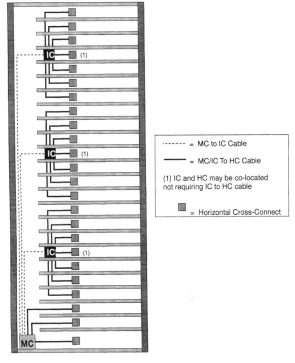

Fig. 8.2   Intrabuilding backbone (two-level hierarchy).

Centralized optical fiber cabling is intended as a cost-effective alternative to the optical HC when deploying 62.5-μm optical fiber cable in the horizontal in support of centralized electronics. Category 5 UTP systems are limited to 100 m total length and require electronics in each closet for high-speed data systems. Fiber optic systems do not require the use of electronics in closets on each floor, and therefore support a centralized cabling network. This greatly simplifies the management of fiber optic networks, provides for more efficient use of ports on electronic hubs, and allows for easy establishment of work group networks. Centralized cabling provides direct connections from the work areas to the centralized cross-connect by allowing the use of pull-through cables, a splice, or an interconnect in the telecommunications closet instead of a HC. Although each of these options has its benefits, splicing a low-fiber count horizontal cable to a higher fiber count intrabuilding cable in the telecommunications closet is often the best choice.

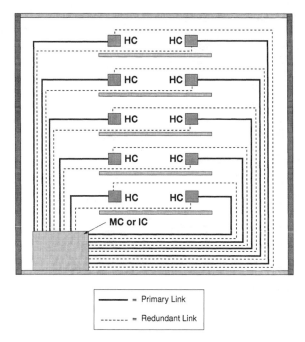

Fig. 8.3   Intrabuilding backbone (redundant routing).

This type of network is currently under study and being considered by the committee responsible for TIA/EIA-568-A (Fig. 8.6).

4.   Applications of Intrabuilding Cables

Premises cables are generally deployed in one of three intrabuilding areas: backbone, horizontal, or interconnect. Higher fiber count tight buffered cables can be used as intrabuilding backbones that connect a MC or data center to IC and telecommunications closets. Likewise, lower fiber count cables can link an IC to a HC to multiple workstations.

Interconnect cables patch optical signals from the optical end equipment to the hardware that contains the main distribution lines. These cables can also provide optical service to workstations or transfer an optical signal from one patch panel to a second patch panel.

▼ = Outlet

☒ = Patch Panel

| = Cables

┊ = Patch Cords/Equipment Cables

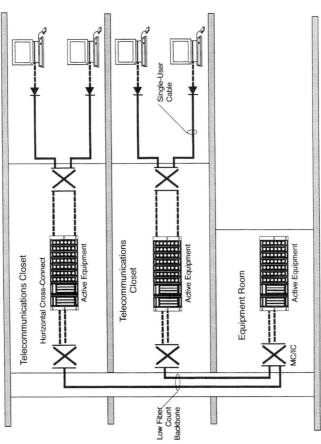

Telecommunications Closet

Horizontal Cross-Connect

Active Equipment

Telecommunications Closet

Active Equipment

Equipment Room

Active Equipment

MC/IC

Single-User Cable

Low Fiber Count Backbone

Fig. 8.4 Traditional horizontal cabling.

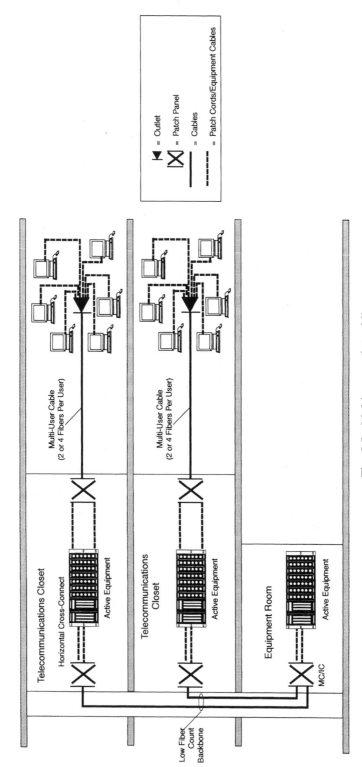

Fig. 8.5 Multiuser outlet cabling.

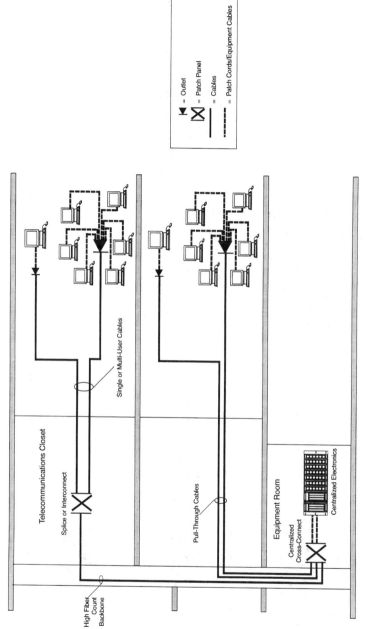

Fig. 8.6 Centralized optical fiber cabling.

5. Specific Cables for Intrabuilding

a. *Tight-Buffered Cable Construction*

Tight-buffered cables have several beneficial characteristics that make them well suited for premises applications. Tight-buffered cables are generally small, lightweight, and flexible in comparison to outside plant cables. As a result, they are relatively easy to install and easy to terminate with optical fiber connectors. Tight-buffered cables are capable of passing the most stringent flame and smoke generation tests and are normally listed as type OFNP (*o*ptical *f*iber *n*onconductive *p*lenum listing) or type OFNR (*o*ptical *f*iber *n*onconductive *r*iser listing). Tight-buffered cables were the first generation of premises cables. The name tight buffered is derived from the layer of thermoplastic or elastomeric material that is tightly applied over the fiber coating. This method contrasts sharply with the loose tube design cable, in which 250-μm fibers are loosely contained in an oversized buffer tube.

A 250- or 500-μm-coated optical fiber is overjacketed with a thermoplastic or elastomeric material, such as polyvinyl chloride (PVC), polyamide (nylon), or polyester, to an outer nominal diameter of approximately 900 μm. All tight-buffered cables will contain these upjacketed fibers. The 900-μm coating on the fiber makes it easier to terminate with an optical fiber connector. The additional material also makes the fiber easier to handle due its larger size.

The 900-μm tight-buffered fiber is the fundamental building block of the tight-buffered cable. Several different tight-buffered cable designs can be manufactured and the design depends on the specific application and desires of the user.

In one cable design, a number of 900-μm optical fibers are stranded around a central member, a serving of strength member (such as aramid yarn) is applied to the core, and a jacket is extruded around the core to form the cable (see Fig. 8.7). As an alternative, the fibers in the cable can be allocated into smaller groups, called subunits. Each subunit contains the same number of 900-μm tight-buffered fibers, a strength member, and a subunit jacket. Each of these individually jacketed groups of fibers (subunits) are stranded around another central member and the composite cable is enclosed in a common outer jacket (Fig. 8.8).

Lower fiber count cables are necessary to connect the optical end equipment to the optical fiber distribution system. They can also be used to bring fiber to a workstation. Tight-buffered interconnect cables with 900-μm

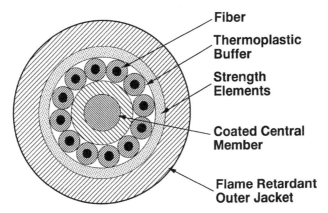

Fig. 8.7   Twelve-fiber tight-buffered cable.

fibers are again well suited for this application. One or two 900-μm tight-buffered fibers are independently jacketed with a serving of strength member to construct a tight-buffered interconnect cable commonly called a jumper (Figs. 8.9 and 8.10). For applications requiring higher fiber counts, a number of these individually jacketed fibers can be stranded around a central member to construct a fan-out cable (Fig. 8.11).

Materials are chosen for premises cables based on flame resistance, mechanical performance, chemical content, chemical resistance, cost, and

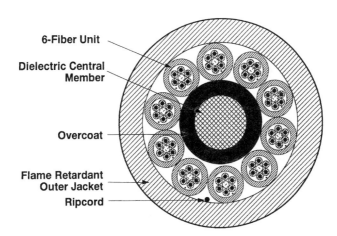

Fig. 8.8   Sixty-fiber unitized tight-buffered cable.

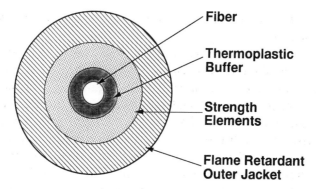

Fig. 8.9   Single-fiber Interconnect cable.

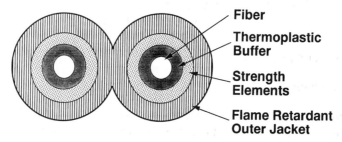

Fig. 8.10   Two-fiber zipcord cable.

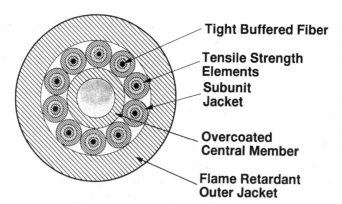

Fig. 8.11   Tight-buffered fan-out cable.

so on. The selection of material is based on the application for which the cable is designed.

The cable jacket can be constructed from PVC, fluoropolymers, polyurethane, flame-retardant polyethylene (FRPE), or other polymers. The specific jacket material used for a cable will depend on the application for which the cable is designed. Standard, indoor cables for riser or general-purpose applications may use PVC for its rugged performance and cost-effectiveness. Indoor cables designed for plenum applications may use fluoropolymers or filled PVCs due to the stringent flame resistance and low smoke requirements. Filled PVCs contain chemical flame and smoke inhibitors and provide a more flexible cable that is generally more cost-effective than fluoropolymer-jacketed cables. Polyurethane is used on cables that will be subjected to extensive and unusually harsh handling but do not require superior flame resistance. FRPE provides a flame-resistant cable that is devoid of halogens. Halogen-free cables and cable jackets do not produce corrosive, halogen acid gases when burned.

B. INTERBUILDING

1. Interbuilding Backbone

In campus environments, such as universities and industrial parks, optical fiber is used extensively in the interbuilding backbone that provides communications between a number of buildings. This backbone can be used for voice, data, and video applications.

2. Interbuilding Topologies

The interbuilding backbone cabling is the segment of the network that typically presents the designer and user with the most options. It is also the most constrained by physical considerations such as duct availability, right-of-way, and physical barriers. For this reason, this section will present a number of options for consideration.

In a smaller network (both in number of buildings and in geographical area), the best design involves linking all the buildings requiring optical fibers to the MC. The cross-connect in each building then becomes the IC, linking the HC in each building to the MC. The location of the MC should be in close proximity to (if not colocated with) the predominant equipment, for example, the data center or PBX. Ideally, the MC is centrally located among the buildings being served, has adequate space for the cross-connect hardware and equipment, and has suitable pathways linking it with the other buildings. This network design would be compliant with the TIA/EIA-568-A standard. Some of the advantages of a single hierarchical star for the interbuilding backbone include:

- Provides a single point of control for system administration
- Allows testing and reconfiguration of the system's topology and applications from the MC
- Allows easy maintenance for security against unauthorized access
- Allows for the easy addition of future interbuilding backbones

Larger interbuilding networks (both in number of buildings and in geographical area) often found at universities, industrial parks, and military bases may require a two-level hierarchical star. This design provides an interbuilding backbone that does not link all the buildings to the MC. Instead, it uses selected ICs to serve a number of buildings. The ICs are then linked to the MC. This option may be considered when the available pathways do not allow for all cables to be routed to an MC or when it is desirable to segment the network because of geographical or functional communication requirement groupings. In large networks, this often translates to a more effective use of electronics, such as multiplexers, routers, and switches, to better utilize the bandwidth capabilities of the fiber or to segment the network (Fig. 8.12).

It is recommended that no more than five ICs be used unless unusual circumstances exist. If the number of interbuilding ICs is kept to a minimum, the user can experience the benefits of segmenting the network without significantly sacrificing control, flexibility, or manageability.

It is strongly recommended that when such a hierarchical star for the interbuilding backbone is used, it be implemented by a physical star in all segments. This ensures that flexibility, versatility, and manageability are maintained. However, the following are two main situations in which the user may consider a physical ring for linking the interbuilding ICs and MC:

- Existing conduit supports it
- Primary (almost sole) purpose of the network is fiber distributed data interface (FDDI) or token ring

It is seldom recommended that the outlying buildings use a second physical ring (Fig. 8.13).

The ideal design for a conduit system that provides a physical ring routing would dedicate X number of fibers of the cable to a ring and Y number of fibers to a star by expressing (not terminating) fibers through the ICs

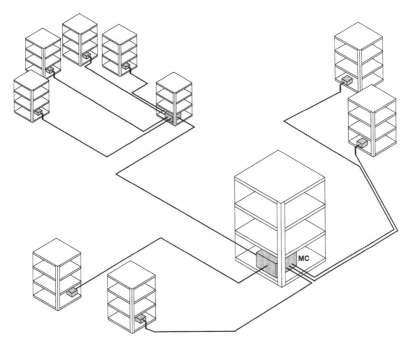

Fig. 8.12 Interbuilding backbone (two-level hierarchy).

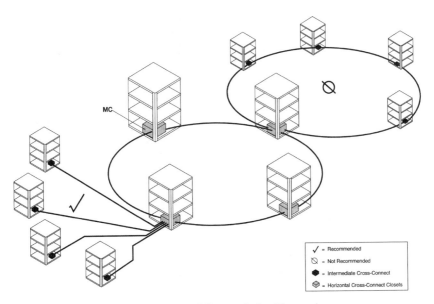

Fig. 8.13 Interbuilding main backbone ring.

directly back to the MC. This design requires the end user to have a more exact knowledge of present and future communication requirements (Fig. 8.14).

3. Applications of Interbuilding Cables

Interbuilding cables must be capable of withstanding a variety of environmental and mechanical extremes. The cable must offer excellent attenuation performance over a wide range of temperatures. The cable must also be sufficiently strong to endure the rigors of installation and provide protection against ultraviolet (UV) radiation, gnawing rodents, and other mechanical disturbances. Furthermore, the cable should have a high packing density to maximize the use of available installation space.

4. Specific Cables for Interbuilding

a. Loose-Tube Cable Construction

Loose-tube cables are designed primarily for outside plant environments and interbuilding applications. Fibers are placed in gel-filled buffer tubes to isolate them from external forces. The loose-tube design provides stable and highly reliable optical transmission characteristics over a wide temperature range. Additionally, the loose tube design ensures long cable life by isolating the fibers from mechanical stresses.

After optical fibers have been placed in loose buffer tubes, the buffer tubes are then stranded around an antibuckling central member. The central member may be either steel or glass-reinforced plastic (GRP). GRP central members are used when the customer wants an all-dielectric cable (no metallic components).

For those designs that do not rely solely on the central member for its tensile strength, high tensile strength yarns, such as fiberglass and aramid yarns, are typically helically wrapped around the cable core before a sheath is placed on the cable. Other designs include embedding the strength elements in the cable jacket. These elements could be various sizes of steel wire or GRP rods. Embedding the strength elements in the sheath is typically employed in a central tube cable.

A water-blocking material protects the interstitial areas of the cable. This compound or tape blocks the ingress and migration of water in the cable. This allows the cable to be installed in up to 10 ft of standing water without special precautions.

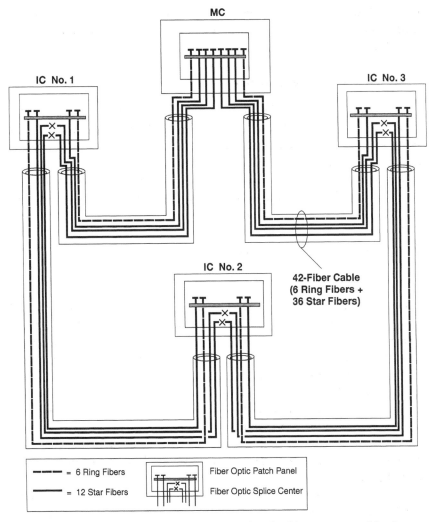

Fig. 8.14 Main backbone ring and redundant backbone star combined.

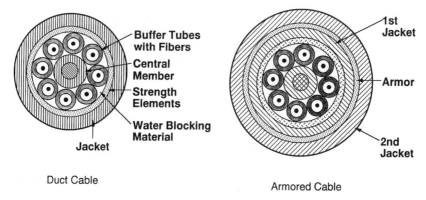

Duct Cable

Armored Cable

Fig. 8.15   Stranded loose-tube cable.

The entire cable core is jacketed with an outer sheath. The jacket is the first line of defense for the fibers against any mechanical and chemical influences on the cable. Several jacket material options could be selected. The best material depends on the application. The most commonly used cable jacket material for outdoor cables is polyethylene. The polyethylene sheat contains carbon black to provide excellent UV resistance (Figs. 8.15 and 8.16).

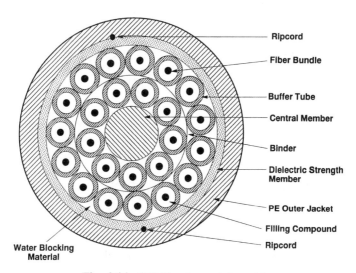

Fig. 8.16   288-fiber loose-tube cable.

5.   Loose-Tube Cable Options

Several cable options are available for applications that require additional protection or special installation conditions.

a.   Armoring

Steel armoring provides additional mechanical strength and resistance to rodents (see Fig. 8.15).

b.   Self-Supporting

If no messenger is available, a self-supporting loose-tube cable can be installed for aerial applications (Fig. 8.17).

c.   Flame Retardant

Some loose-tube cables are designed for use in both outside plant and premises applications. Variations of the loose-tube cables that are halogen free, flame retardant, or that possess water-blocking characteristics are also available.

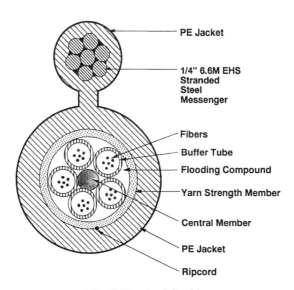

Fig. 8.17   Aerial cable.

C. DATA CENTERS

A specific application of data communications that utilizes both interbuilding and intrabuilding cabling systems is the data center. Communication systems are a vital part of today's corporate structure. Most companies require data communications supported by an efficiently structured and managed cabling system to support future growth.

A data center installation can have multimode or single-mode fibers going to different places within the site, including the following:

- Within a computer raised floor
- Local area network wiring closets
- Between floors in a building (intrabuilding)
- Between buildings in a campus (interbuilding)
- Between campuses

Data center equipment can be connected with either jumper cables, which directly connect two pieces of equipment, or through trunk cables and distribution panels. Jumper cables appear to be the least expensive approach until a reconfiguration occurs. Each jumper must be rerouted under a raised floor in this case. Usually, the most cost-effective connectivity solution is to use trunk cables and distribution panels. Individual equipment is connected to the distribution panel via short jumpers. In the event of a reconfiguration, one simply has to move the equipment and reconnect the jumpers while leaving the trunk cabling system in place. An example of a data center configuration is shown in Fig. 8.18.

8.3. Standards

Building and fire codes must be considered not only with intrabuilding telecommunications cabling, but also, more important, when interbuilding telecommunications cabling enters the building. One of the most significant documents, although advisory in nature, is the "National Electrical Code" (NEC).

A. NATIONAL ELECTRICAL CODE

The "National Electrical Code" is a document issued every 3 years by the National Fire Protection Association. The document specifies the requirements for premises wire and cable installations to minimize the risk and

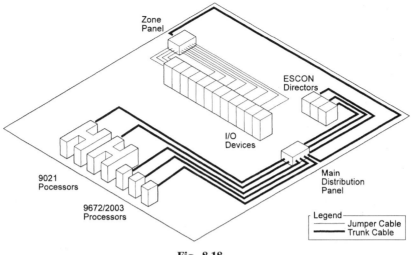

Fig. 8.18

spread of fire in commercial buildings and individual dwellings. Section 770 of the NEC defines the requirements for optical fiber cables. Although the NEC is advisory in nature, its content is generally adopted by local building authorities.

The NEC categorizes indoor spaces into three general areas — plenums, risers, and general building. A plenum is an indoor space that handles air as a primary function. Examples of plenum areas are fan rooms and air ducts. A riser is an area that passes from one floor of a building to another floor. Elevator shafts and conduits that pass from one floor to another floor are examples of risers. Indoor areas that are not classified as plenums or risers are also governed by the NEC but are not specifically named.

These different installation locations — plenums, risers, and general buildings — require different degrees of minimum flame resistance. Because plenum areas provide a renewable source of oxygen and distribute environmental air, the flame-resistance requirements for cables installed in plenum areas are the most stringent. Likewise, riser cables must demonstrate the ability to retard the vertical spread of fire from floor to floor. Other indoor cables (not plenum or riser) must meet the least demanding fire-resistant standards.

Optical fiber cables that can be installed in plenums and contain no conductive elements are listed as type OFNP. These cables are generally referred to as plenum cables. Likewise, optical fiber cables that can be installed in riser applications and contain no conductive elements are listed as type OFNR. These cables are generally referred to as riser cables.

Optical fiber cables that can be installed indoors in nonplenum and nonriser applications and contain no conductive elements are listed as type OFN. These cables may be referred to as general-purpose cables and meet the least demanding of the fire-resistance standards.

Premises cables must be subjected to and pass standardized flame tests to acquire these listings. The tests are progressively more demanding as the application becomes more demanding. Therefore, plenum cables must pass the most stringent requirements, whereas general-purpose cables must meet less demanding criteria.

To obtain a type OFN listing, the cable must pass a standardized flame test, such as the UL 1581 flame test. The UL 1581 test specifies the maximum burn length the cable can experience to pass the test. Type OFN cables can be used in general-purpose indoor applications.

Similar to the type OFN listing is the type OFN-LS listing. To obtain this listing, the cable must pass the UL 1685 test. This test is similar to the UL 1581 test, but the UL 1685 test includes a smoke-generation test. The cable must produce limited smoke as defined in the standardized test to achieve the listing. A cable with a type OFN-LS listing is restricted to general-purpose indoor applications only.

To obtain a type OFNR listing, the cable must pass a standardized flame test, such as the UL 1666 flame test. The UL 1666 test contains specifications for allowable flame propagation and heat generation. The test is more stringent than the UL 1581 test. Type OFNR cables can be used in riser and general-purpose applications.

To obtain a type OFNP listing, the cable must pass a standardized flame test, such as the NFPA 262-1990 (UL 910) flame test. The NFPA 262 test contains a specification for allowable flame propagation and a specification for the average and peak smoke that can be produced during the test. The plenum test is the most severe of all flame tests. Type OFNP cables can be used in plenum, riser, and general-purpose applications.

Section 770 of the NEC contains a variety of other requirements on the placement and use of optical fiber cables in premises environments. The section also contains several exceptions that permit the use of unlisted cables under specific conditions. The NEC and the local building code should always be consulted during the design of the premises cable plant.

B. *TIA/EIA-568-A, "COMMERCIAL BUILDING TELECOMMUNICATIONS CABLING STANDARD"*

1. Brief Description of Cabling Topologies and Applications

One of the most challenging series of decisions a telecommunications manager makes is the proper design of an optical fiber cabling plant. Optical fiber cable, which has an extremely high bandwidth, is a powerful telecommunications media that supports voice, data, video, and telemetry/sensor applications. However, the effectiveness of the media is greatly diminished if proper connectivity, which allows for flexibility, manageability, and versatility of the cable plant, is not designed into the system.

The traditional practice of installing cables or cabling systems dedicated to each new application is all but obsolete. Instead, designers and users alike are learning that proper planning of a structured cabling system can save time and money. For example, a cable that is installed for point-to-point links should meet specifications, or later upgrades, as it becomes part of a much larger network. As a result, duplications of cable, connecting hardware, and installation labor can be avoided by anticipating future applications and providing additional fibers for unforeseen applications as well as interfacing with local service providers.

A structured cabling design philosophy that uses a physical hierarchical star topology is recommended. The hierarchical star offers the user a communications transport system that can be installed to efficiently support all logical topologies (star, bus, point-to-point, and ring). The guidelines presented here follow the general recommendations set forth by the TIA/EIA-568-A "Commercial Building Telecommunications Cabling Standard." This section will discuss the physical implementation of these logical topologies in the interbuilding backbones and intrabuilding backbones within the premises environments.

2. Network Topologies

Network topologies are shown in Fig. 8.19.

3. Logical Topology

To be universal for all possible applications, a structured cabling plant must support all logical topologies. These topologies define the electronic connection of the system's nodes. Fiber applications can support the following logical topologies:

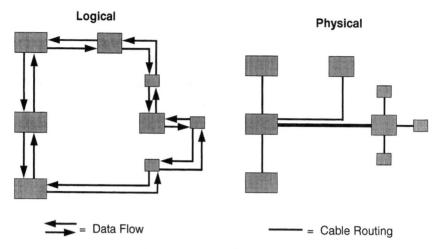

Fig. 8.19   Networking topologies.

- Point-to-point
- Star
- Ring
- Bus

a.   Point-to-Point

Point-to-point logical topologies are still common in today's customer premises installations (Fig. 8.20). Two nodes requiring communication are directly linked by the fibers—normally a fiber pair (one to transmit and one to receive). Common point-to-point applications include

- Fiber channel
- Terminal multiplexing
- Satellite up/down links
- ESCON
- Telemetry/sensor

b.   Star

An extension of the point-to-point topology is the logical star topology (Fig. 8.21). This is a collection of point-to-point topology links, all of which have a common node that is in control of the communications system. Common star applications include

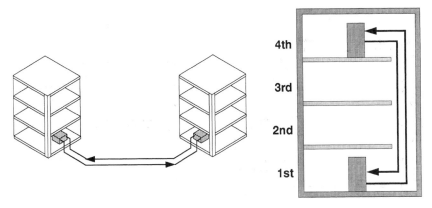

Fig. 8.20   Logical point-to-point topology.

• A switch such as a PBX, asynchronous transfer mode, or data switch
• A security video system with a central monitoring station
• An interactive video conference system serving more than two locations

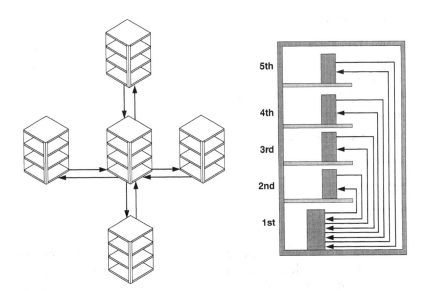

Fig. 8.21   Logical star topology.

c. Ring

In this topology, each node is connected to its adjacent nodes in a ring (Fig. 8.22). The logical ring topology, especially prevalent in the data communications area, is supported by two primary standards:

- Provides Token Ring (IEEE 802.5)
- FDDI (ANSI X3T9)

d. Logical Bus

The logical bus topology is also utilized by data communications and is supported by the IEEE 802.3 standards. All nodes share a common line rather than in one direction, as on a ring. When one node transmits, all the other nodes receive the transmission at approximately the same time. The most popular systems requiring a bus topology include Ethernet and Manufacturing Automation Protocol.

e. Token Bus

A token bus topology is shown in Fig. 8.23.

4. Physical Topologies and Logical Topology Implementation

All the logical topologies noted earlier are easily implemented with a physical star cabling scheme, as recommended by TIA/EIA-568-A "Commercial Building Telecommunications Cabling Standard." Imple-

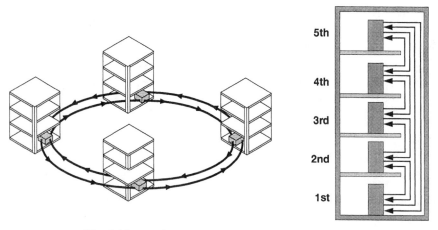

Fig. 8.22 Logical counter-rotating ring topology.

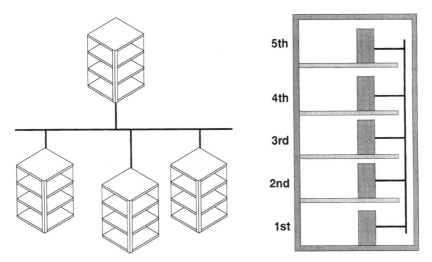

Fig. 8.23   Logical bus topology.

menting point-to-point and star topologies on a physical star is straight-forward.

The use of data networks that utilize bus or ring topologies are very prominent in the market, i.e., Ethernet, Token Ring, and FDDI. With the benefits of physical star cabling, electronic vendors and standards have developed electronic solutions designed to interface with a star network. These applications are typically implemented with an "intelligent hub" or concentrator. This device, in simple terms, establishes the bus or ring in the back plane of the device and the connections are made from a central location(s). Therefore, from a physical connection, these networks appear to be a star topology and are quite naturally best supported by a physical star cabling system.

5.   Physical Stars versus Physical Rings

There are some situations in which a physical ring topology would seem appropriate. As seen in the following comparison of an optical fiber physical star topology to a physical ring topology, a physical star topology is recommended for supporting the varied requirements of a structured cabling system.

Physical star
  Advantages
      Flexible — supports all applications and topologies
      Acceptable connector loss with topology compatibility
      Centralized fiber cross-connects make administration and re-arrangement easy
      Existing outdoor ducts are often configured for the physical star, resulting in easy implementation
      Easily facilitates the insertion of new buildings or stations into the network
      Supported by TIA/EIA-568-A, "Commercial Building Telecommunications Cabling Standard"
  Disadvantages
      Cut cable will result in node failure unless redundant routing is used
      More fiber length is used in a logical ring topology
Physical ring
  Disadvantages
      Less flexible
      Unacceptable connector loss if star or bus logical topologies are required
      Connectors are located throughout system, making administration and rearrangement difficult
      Physical implementation can be difficult, often requiring new construction
      Insertion of new buildings or stations into a physical ring can cause disruptions to current applications or inefficient cable use
      Even with ring topology applications, the majority of nodes, if not all, must be present and active
      Node survivability in the event of cable cut, if cable truly utilizes redundant paths
      Less fiber lengths are used between nodes

6. Physical Star Implementation

The more important recommendations of the TIA/EIA-568-A standard as they relate to optical fiber are summarized as follows: The standard is based on a hierarchical star for the backbone and a single star for horizontal distribution (Fig. 8.24).

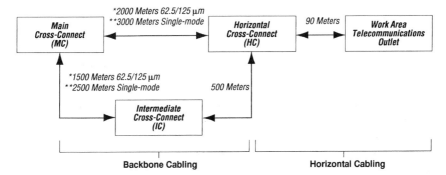

Fig. 8.24 "Commercial Building Telecommunications Cabling Standard" (TIA/EIA-568-A). *Backbone distances based on optical fiber cable. When IC-to-HC distance is less than 500 m, then the MC to IC can be increased accordingly but the total MC to HC shall not exceed 2000 m for 62.5/125 μm or 3000 m for single mode. **Single-mode fiber can support distances up to 60 km (37 miles); however, this is outside the scope of TIA/EIA-568A.

The rules for backbone cabling include a 2000-m (3000 m for single mode) maximum distance between the MC and the HC for 62.5/125μm and a maximum of one IC between any HC and MC. The MC is allowed to provide connectivity to any number of HCs.

The standard does not distinguish between interbuilding (outside) or intrabuilding (inside) backbones because these are determined by the facility size and campus layout. However, in most applications, one can envision the MC to IC being an interbuilding backbone link. The exceptions to this would be major high-rise buildings in which the backbone may be entirely inside the building or a campus containing small buildings with only one HC per building, thereby eliminating the requirement for an IC (Fig. 8.25).

The horizontal cabling is specified to be a single star linking horizontal cross-connect closet to the work-area telecommunications outlet with a distance limitation of 90 m (Fig. 8.26). This distance is not based on fiber capabilities but rather it is based on copper distance limitations to support data requirements. Currently, TIA is studying two alternate horizontal cabling topologies when using fiber in the horizontal: the multiuser outlet and centralized optical fiber cabling.

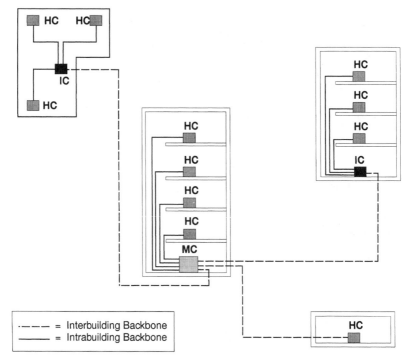

Fig. 8.25   Backbone cabling.

C.   TIA/EIA-598-A, "OPTICAL FIBER CABLE COLOR CODING"

A common denominator for all cable designs is identification/color coding of fibers and fiber units. TIA/EIA-598-A defines identification schemes for fibers, buffered fibers, fiber units, and groups of fiber units within outside-plant and premise optical fiber cables. This standard has been adopted by the Rural Utility Service within 7 CFR 1755.900 and Insulated Cable Engineers Association, Incorporated S-87-640-1992, "Standard for Fiber Optic Outside Plant Communications Cable."

This standard allows for fiber units to be identified by means of a printed legend. This method can be used for identification of fiber ribbons and fiber subunits. The legend will contain a corresponding printed numerical position number and/or color for use in identification (Table 8.1).

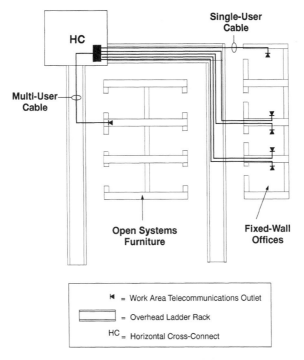

Fig. 8.26   Horizontal cabling.

8.4.   Handling and Installing Fiber Optics

A.   *CABLE HANDLING CONSIDERATIONS*

1.   Minimum Bend Radius

Optical fiber cable installation is as simple as, and in many cases much easier than, installing coaxial or UTP cable in the horizontal. The most important factor in optical fiber cable installations is maintaining the cable's minimum bend radius. Bending the cable tighter than the minimum bend radius may result in increased attenuation and broken fibers. If the elements of the cable are not damaged, when the bend is relaxed, the attenuation should return to normal. Cable manufacturers specify minimum bend radii for cables under tension and for long-term installation (Table 8.2).

Table 8.1   **Color Code**

Position Number	Base Color and Tracer	Abbreviation
1	Blue	BL
2	Orange	OR
3	Green	GR
4	Brown	BR
5	Slate	SL
6	White	WH
7	Red	RD
8	Black	BK
9	Yellow	YL
10	Violet	VI
11	Rose	RS
12	Aqua	AQ
13–24	Colors 1–12 repeated with black tracer	BL/BK, OR/ BK etc

Table 8.2   **Minimum Bend Radii**

		Minimum Bend Radius			
		Loaded		Unloaded	
Application	Fiber Count	cm	in.	cm	in.
Interbuilding backbone	2–84	22.5	8.9	15.0	5.9
	86–216	25.0	9.9	20.0	7.9
Intrabuilding backbone	2–12	10.5	4.1	7.0	2.8
	14–24	15.9	6.3	10.6	4.2
	26–48	26.7	10.5	17.8	7.0
	48–72	30.4	12.0	20.3	8.0
	74–216	29.4	11.6	19.6	7.7
Horizontal cabling	2	6.6	2.6	4.4	1.7
	4	7.2	2.8	4.8	1.9

Note. Specifications are based on representative cables for applications. Consult manufacturer for specifications of specific cables. For open-systems furniture applications where cables are placed, Siecor's 2- and 4-fiber horizontal cables have a 1-in. bend radius. Please contact your Siecor representative for further information.

2. Maximum Tensile Rating

The cable's maximum tensile rating must not be exceeded during installation. This value is specified by the cable manufacturer. Tension on the cable should be monitored when a mechanical pulling device is used. Hand pulls do not require monitoring. Circuitous pulls can be accomplished through the use of backfeeding or centerpull techniques. For indoor installations, pull boxes can be used to allow cable access for backfeeding at every third 90° bend (Table 8.3).

3. Maximum Vertical Rise

All optical fiber cables have a maximum vertical rise that is a function of the cable's weight and tensile strength. This represents the maximum vertical distance the cable can be installed without intermediate support points (Fig. 8.27). Some guidelines for vertical installations include the following:

- All vertical cable must be secured at the top of the run. A split mesh grip is recommended to secure the cable.
- The attachment point should be carefully chosen to comply with the cable's minimum bend radius while holding the cable securely.

Table 8.3 **Maximum Tensile Load**

| | | Maximum Tensile Load | | | |
| | | Short Term | | Long Term | |
Application	*Fiber Count*	N	lbs	N	lbs
Interbuilding backbone	2–84	2700	608	600	135
	86–216	2700	608	600	135
Intrabuilding backbone	2–12	1800	404	600	135
	14–24	2700	608	1000	225
	26–48	5000	1124	2500	562
	48–72	5500	1236	3000	674
	74–216	2700	600	600	135
Horizontal cabling	2	750	169	200	45
	4	1100	247	440	99

Note. Specifications are based on representative cables for applications. Consult manufacturer for specific information.

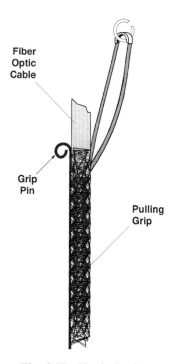

Fiber
Optic
Cable

Grip
Pin

Pulling
Grip

Fig. 8.27   Vertical cable.

- Long vertical cables should be secured when the maximum rise has been reached.

4. Cable Protection

If future cable pulls in the same duct or conduit are a possibility, the use of innerduct to sectionalize the available duct space is recommended. Without this sectionalization, additional cable pulls can entangle an operating cable and could cause an interruption in service. Care should be exercised to ensure the innerduct is installed as straight as possible, without twists that could increase the cable pulling tension.

When the cable is installed in raceways, cable trays, or secured to other cables, consideration should be given to movement of the existing cables. Although optical fiber cable can be moved while in service without affecting fiber performance, it may warrant protection with conduit in places exposed to physical damage (Fig. 8.28).

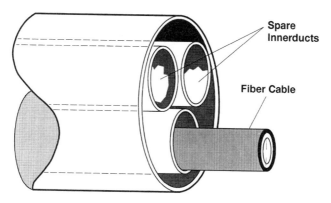

Fig. 8.28 Spare innerducts.

5. Duct Utilization

When pulling long lengths of cable through duct or conduit, less than a 50% fill ratio by cross-sectional area is recommended. For example, one cable equates to a 0.71-in. outside-diameter cable in a 1-in. inside-diameter duct. Multiple cables can be pulled at once as long as the tensile load is applied equally to all cables. Fill ratios may dictate higher fiber counts in anticipation of future needs. One sheath can be more densely packed with fiber than multiple cable sheaths. In short, for customer premises applications, the cost of extra fibers is usually small when these extra fibers are not terminated until needed. For a difficult cable pull, extra fibers installed now but not terminated may be the most cost-effective provision for the future (Fig. 8.29).

6. Preconnectorized Assemblies

Of special consideration is the use of preconnectorized cables. Although the use of factory-terminated cross-connect and interconnect jumper assemblies is acceptable, the use of preconnectorized backbone and distribution cable presents special installation techniques. These connectors must be protected when installing the connectorized end of these cables. Protective pulling grips are available to protect connectors, but the grips' outside diameter may prevent installation in small innerducts or conduits. The size of the preconnectorized assembly and pulling grip should be considered before ordering factory connectorized cables. There may also be additional installation requirements imposed on the grip by the manufacturer, in terms

Innerduct/Conduit

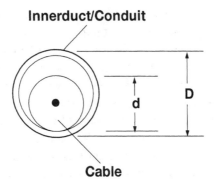

Cable

Fig. 8.29   Duct utilization: $\dfrac{d^2}{D^2} < 50\%$, where d is the cable diameter and D is the innerduct diameter.

of minimum bend radius and tension, that would be the limiting parameters in an installation.

7.   Slack

A small amount of slack cable (20–30 ft) can be useful in the event that cable repair or relocation is needed. If a cable is cut, the slack can be shifted to the damaged point, necessitating only one splice point in the permanent repair rather than two splices if an additional length of cable is added. This results in reduced labor and hardware costs and link loss budget savings.

Additional cable slack (approximately 30 ft) stored at planned future cable drop points will result in savings in labor and materials when the drop is finally needed. Relocation of terminals or cable plant can also take place without splicing if sufficient cable slack is available.

B.   CABLE SPLICING METHODS

In the commercial building and campus environment, the designer/installer can often avoid the requirement of fiber-to-fiber splicing by installing a continuous length of cable. This is normally the most economical and convenient solution. However, because of the cable plant layout, length, raceway congestion, or requirements to transition between nonlisted and UL-listed cable types at the building entrance point, splices cannot always be avoided. This requires splicing cables together, of which there are two basic techniques. Field splicing methods for optical fibers can be grouped into two major categories: fusion and mechanical.

Fusion splicing consists of aligning the cores of two clean (stripped of coating), cleaved fibers and fusing the ends together with an electric arc. The fiber ends are positioned and aligned using various methods. Alignment can be performed using fixed V-groove, profile alignment, or light injection. It can be manual or automatic and is normally accomplished with the aid of a viewing scope, video camera, or a specialized type of optical power meter for local injection and detection of lights. High-voltage electrodes, contained in the splicer, arc across the fiber ends as the fibers are moved together, thus fusing the fibers together (Fig. 8.30).

A mechanical splice, by comparison, is an optical junction, where two or more optical fibers are aligned and held in place by a self-contained assembly approximately 2 in. in length. Single-fiber mechanical splices rely on the alignment of the outer diameter of the fibers, making the accuracy of core/cladding concentricity critical to achieving low splice losses.

This method aligns the two fiber ends to a common centerline, thereby aligning the cores. The cleaned (stripped of coating) fiber ends are cleaved, inserted into an alignment tube, and butted together. The tube has factory-installed index-matching gel to reduce reflections and loss at the splice

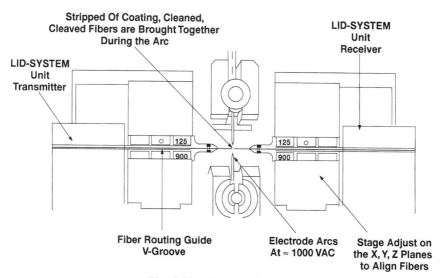

Fig. 8.30 Fusion splicing.

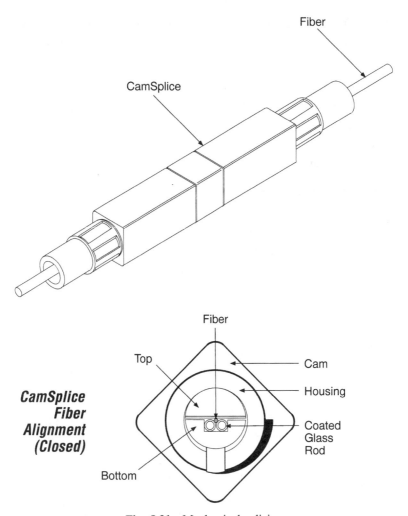

Fig. 8.31   Mechanical splicing.

point. Usually, the fibers are held together by compression or friction, although some older methods rely on epoxy to permanently secure the fibers (Fig. 8.31).

C. HARDWARE

The selection of proper hardware for a commercial building or campus environment is essential to provide a flexible cabling system that will conform to the TIA/EIA-598-A "Commercial Building Telecommunications

Cabling Standard." The hardware selection process can be narrowed down to a few products based on the following four fundamental categories surrounding the connecting hardware application:

- Indoor or outdoor environment
- Field connectorize or splicing pigtails
- Wall or rack mounted
- Fiber count

D. INDOOR HARDWARE

Indoor hardware selection is complex because of the variety of applications and topologies encountered. Selecting the proper indoor hardware will ensure a flexible fiber optic cable plant. Indoor hardware can be divided into the application area's MC. The MC is the hub of the cable plant. Consequently, it is the main administrative point and is characterized by a high concentration of fibers as well as multiple indoor and outdoor cables. Adds, moves, and charges are also frequent. Hardware must have a high termination density and capacity. The capacity should be modular, however, to provide solutions that can grow as additional capacity is required. Different termination methods may be employed. Typically, indoor cable is terminated by field connectorization, outdoor cable is terminated by pigtail splicing, and interconnection cables (jumpers) are terminated by factory termination. The hardware must be readily compatible with the different methods, but must be flexible to provide only necessary components for the method employed. The MC generally requires rack-mounted hardware and equipment. Hardware should be wall-mounted if rack space is limited or not available. As the main administrative point, this area will contain a high number of jumpers for both cross-connection of backbone fibers and interconnection to the electronics. Hardware must provide ample room for jumper storage as well as neat, secure routing throughout the rack.

1. Intermediate Cross-Connect

This is the second hierarchical level of cross-connects in the backbone wiring. The IC is generally where the interbuilding backbone and intrabuilding backbone meet. In general, the IC has lower fiber counts than the MC. The required number of jumpers for cross-connection of backbone fibers and interconnection to the electronics is reduced. Racks with fiber routing provisions may be unnecessary.

2.   Telecommunications Closet

This is the location where the backbone wiring and horizontal wiring meet. Typically, this is the location in the network where the fiber backbone transitions to the horizontal distribution to the work areas. The backbone wiring is typically low-fiber count indoor cable. Thus, low-capacity termination housings are usually required. The standard termination method for indoor cables is field connectorization. Thus, only path panels are necessary — the purchase of a splice housing is rarely, if ever, required. The selection of wall- or rack-mountable hardware depends almost entirely on the architectural spaces provided. Jumpers will be required only to interconnect at the electronics. Jumper routing and storage capacities are not as critical.

3.   Work Area Telecommunications Outlet

This is the endpoint of the horizontal wiring. Again, the cables may be fiber only or a combination of fiber and copper. Although the work area usually requires only one outlet, the outlet must usually accommodate two or three different types of media. Mounting is usually in fixed-wall offices or in open-systems furniture; however, floor outlets are sometimes used. Hardware must be compatible with the most common mounting applications and allow a variety of cable entry locations. The work area is a high traffic zone and therefore hazardous to the cable, connectors, and jumpers. Units must be low profile to minimize the risk of damage. The hardware must provide adequate routing and storage space so that the minimum bend radius of the fiber is maintained.

Outdoor hardware consists of a line of splice closures, wall-mountable distribution centers, and pedestal-mountable cross-connects. These units provide environmental protection for splices, connectors, and jumpers in the outside plant environment, often required in industrial and other special applications.

In some indoor circumstances, space is limited for mounting hardware. Specially designed furcation (or fan-out) kits provide protection and pull-out strength for bare fibers, and they are direct connectorized. These are most useful when the fiber counts are low and all of the fibers will be patched into other hardware or electronics in the same area.

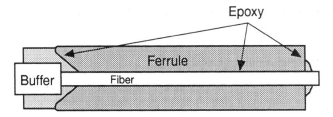

Fig. 8.32 Heat-cured connector.

E. CONNECTORS

1. Field and Factory Connectorization

With the advent of easy-to-install field connectors and new methods to fan-out loose-tube cables, field connectorization has become the common method for terminating fiber optic cables in the premises market. There are numerous types of connectors on the market, each with slightly different installation procedures.

With regard to all connectors, the fiber must be epoxied into the connector. The epoxy keeps fiber movement over temperature at a minimum, allows polishing without fracturing the fiber, and seals the fiber from the effects of the environment. In addition, it allows the fiber to be aggressively cleaned on the end face.

In addition, the connector end face must be polished. A physical contact (PC) finish is recommended and is specified by TIA/EIA-598-A. This means that the fibers will be physically touching inside the connector adapter as they are held under compression. Lack of a PC finish results in an air gap between the fibers and increased attenuation.

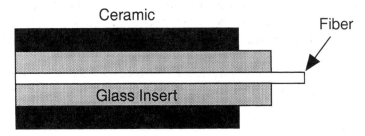

Fig. 8.33 Glass-insert connector.

There are several polishing methods recommended, which are typically dependent on the ferrule material used. If a ferrule material is very hard, such as ceramic, it is common for the ferrule to be preradiused. Softer ferrule materials, such as composite thermoplastic or glass in ceramic, may be polished flat. These materials wear away at approximately the same rate as the fiber and can be polished aggressively and still maintain a PC finish.

Heat-cured connectors have the advantage of being a cost-effective way to make cable assemblies or to install in a location where a large number of fibers are terminated at one time. Heat-cured connectors typically require more time for the epoxy to harden and skill to polish, and they need epoxies that have limited pot lives. Therefore, they are best used in a controlled environment or where large numbers of connectors are done to take advantage of the low connector costs and special ovens. Connectorized assemblies are almost exclusively heat-cured connectors (Fig. 8.32).

UV-cured connectors can be termed glass-insert connectors. The ferrule features a glass insert surrounded by ceramic. The glass insert allows the installer to do two things. First, the glass insert propagates light. Therefore, a UV-curable adhesive can be used to bond the fiber into the ferrule in a mere 45 s. Second, the glass insert protrudes beyond the ceramic outer sleeve. This means the glass insert is polished along with the fiber. The glass is approximately the same hardness as the fiber and polishes at the same rate. This results in a flat PC polish (Fig. 8.33).

No-cure, no-polish connectors have the advantage of no epoxy, no polish in the field, no consumables, few tools needed, and minimal setup required. These types of connectors are actually a mini-pigtail housed in a connector body. There is a fiber stub already bonded into the ferrule in the factory, where the end face of the ferrule is polished to a PC finish. The other end

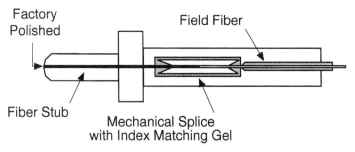

Fig. 8.34 No-cure, no-polish connector.

of the fiber is cleaved and resides inside the connector. The field fiber is cleaved and inserted into the connector until it "butts up" against the fiber stub. A simple cam actuation process completes the connector with no epoxy or polishing required (Fig. 8.34).

References

1. Englebert, J., S. Hassett, and T. Summers. Optical fiber and cable for telecommunications. In *The electroncis handbook*, ed., J. Whitaker, Chap. 10. Boca Raton, FL: CRC Press.
2. IBM. 1996. *FTS direct attach physical and configuration planning,* 2nd ed. New York: IBM.
3. Siecor Corporation. 1995. *Universal transport system design guide, release III.* Hickory, NC: Siecor.
4. Siecor Corporation. 1996. *Premises fiber optic products catalog,* 6th ed. Hickory, NC: Siecor.

Chapter 9 | Testing Fiber Optic Local Area Networks

Jim Hayes

Fotec, Incorporated, Medford, Massachusetts 02155

Greg LeCheminant

Hewlett-Packard, Santa Rosa Systems Division, Santa Rosa, California 95403

9.1. Introduction

With the widespread usage of fiber optics in data communications, the test and troubleshooting of fiber optic networks has become quite important. Although fiber optic networks have some major physical differences from copper-based networks, testing and troubleshooting them can be easily mastered by technicians with some basic understanding of the technology and training in the procedures.

9.2. Fiber Optic Testing Defined

Fiber optic testing concerns primarily four areas: tracing/continuity, optical power, optical loss, and fault location. All are used in various stages of design, installation, and maintenance to ensure the fiber optic link will operate properly.

Tracing and continuity checking simply require using a visible light source to determine if the fibers in the cable are continuous and/or to identify the fibers at a connection. Because the biggest problem in installation is connecting fibers correctly, tracers are extremely valuable tools for installation and upgrades. Tracers may simply be a flashlight that fixtures a fiber close enough to couple light into the fiber, a visible light-emitting diode (LED), or a visible laser diode. The sensor is the tester's eyes, looking for visible proof of the continuity or finding the proper fiber.

304

HANDBOOK OF FIBER OPTIC
DATA COMMUNICATION

Optical power is the most basic parameter of fiber optic measurements and is equivalent to voltage in electronics. Optical power is measured at the transmitter to determine if the transmitter power output is within specification and at the receiver to determine if adequate power is available for reliable transmission. The difference between the typical output power of the transmitter coupled into the fiber and sensitivity of the receiver determines the "loss budget" of the link.

Shown in Fig. 9.1 is the performance of a typical receiver over variations in input power. As the power increases, the bit error rate (BER) decreases, indicating better performance. At higher power levels, the receiver may saturate, with subsequent higher BER. Proper performance means the receiver power must fall within the range shown on the graph in Fig. 9.1. The difference between the transmitted and received power is the amount of loss tolerable in the cable plant connecting them. An optical power meter adapted for connection to fiber optic cables is used to measure power.

The optical loss budget is determined during the design of the link by the typical power output of the transmitter and sensitivity of the receiver. Because the cable plant is installed before the network equipment is installed, it is tested with a test source similar to the transmitter to simulate loss in the cable plant that approximates what the link will actually see in operation.

Fault location uses high-powered visible diode laser fault locators or optical time domain reflectometers (OTDRs). Visible fault locators use the eye to detect large light losses in cables that are of short length, visible, and readily accessible. OTDRs work like radar to find problems in the cables at longer distances and in location where the cable is not visible, such as underground runs in conduit.

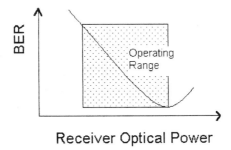

Fig. 9.1   Bit error rate performance of optical link.

9.3. Practical Considerations

Most data communications applications of fiber optics use LED sources and multimode fibers. These components allow links to operate at distances up to 3 or 4 km and at data rates up to several hundred megabits per second. Beyond these limits, links using laser diode sources and single-mode fiber are required. The distance limitation for most multimode links is bandwidth. The LED/multimode fiber combination is usually distance limited not by the loss of the cable plant but rather by the bandwidth of the LED/multimode fiber combination. Datalinks are classified into several basic categories as summarized in Table 9.1.

Most datacom applications are based on multimode fiber, not the single-mode fiber used in telecommunications or CATV networks. The multimode fiber in use is almost always what is called "62.5/125," with the dimensions of the core and cladding in μm. This fiber design is almost 20 years old and was designed for the last generation of multimode telecom systems of the early 1980s. As a result, it was optimized for both high bandwidth and long distance, with the dispersion characteristics optimized at 1300 nm, the lowest loss window feasible at the time.

Datacom links first used 850-nm LEDs and multimode fiber. When datacom links reached higher speeds, 100 Mb/s or more with fiber distributed data interface (FDDI) and enterprise system connection (ESCON), it was necessary to use 1300-nm sources to properly utilize the bandwidth characteristics of the fiber. Lower speed networks, such as Ethernet and Token Ring, continued to use 850-nm LEDs because the lower bandwidth and higher attenuation of the fiber worked in concert to limit transmission

Table 9.1 **Categories of Data Links**

Description	Example	Specifications	Distance (km)
Low speed, short distance	10baseF, FOIRL	850-nm LED, multimode	2
High speed, short distance	FDDI, ESCON on campus	1300-nm LED, multimode	2–4
High speed, longer distance	FDDI, ESCON-based WAN	1300-nm laser, single mode	2–20

distances. Laser sources were not considered for datacom because their higher price and lower reliability was considered undesirable for low-cost networks.

Bandwidth limitations are primarily caused by chromatic dispersion in the fiber, a complicated interaction of the wavelength and spectral width of the LED source, combined with the dispersion characteristics of the fiber. Fiber bandwidth is specified in two ways. Modal dispersion refers to the dispersion caused by the different modes of light propagating in the fiber, whereas chromatic dispersion refers to the dispersion caused by the different velocities of various wavelengths of light from the source.

LEDs are made by two different semiconductor processes. Edge-emitting LEDs (E-LEDs) are fabricated like semiconductor lasers, emitting light from the edge of the LED chip, but lack the reflective cavity required to stimulate lasing. Surface-emitting LEDS (usually just called LEDs) are made to emit light directly from the surface of the chip. The ELED has a narrower spectral width, which allows it to have lower chromatic dispersion and thereby higher bandwidth than the surface emitter. Coupled with the fact that it has a higher typical power output, ELEDs are preferable for systems that need to go longer distances.

Figure 9.2 shows the theoretical bandwidth of fiber optic cable as a function of distance for several source types, illustrating the variations in

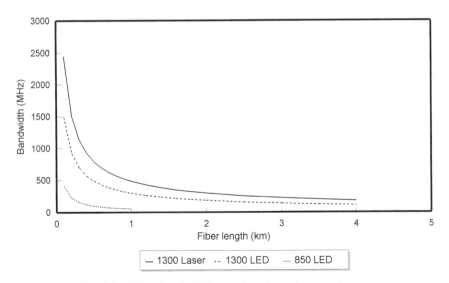

Fig. 9.2 Fiber bandwidth as a function of source type.

bandwidth capability with source characteristics. These kinds of analyses led to maximum distances being specified for high-speed networks such as FDDI and ESCON. These maximum distances are quite "safe," ensuring the networks would be "plug and play" without requiring users to consider any parameters other than cable plant loss. Knowledgeable users often push these limits by 1 km or more with no problem.

For these fixed limits, the typical cable plant will have much less optical loss than the transmitters and receivers will tolerate. The loss budget is adequate to allow the full fiber length and six or more connections between network equipment with considerable margin for safety.

Although multimode links are bandwidth- and not loss-limited, it is only feasible to test attenuation on installed networks. The variables needed to test bandwidth are so complex that test equipment becomes too expensive and cumbersome for applications outside the engineering or manufacturing environment. However, it is possible to simulate bandwidth performance of fibers and sources with computer programs such as that used to generate the data in Fig. 9.2.

9.4. Testing

9.4.1. REAL-WORLD TESTING

There are mandatory tests for every stage of the design, manufacture, and installation of the fiber optic components, link, cable plant, and network.

9.4.2. COMPONENT TESTING

Fiber optic data links are composed of three components: a transmitter, a receiver, and the interconnecting cable plant. These components must be compatible with the parameters of the intended application. The loss budget must be adequate for the expected loss in the cable plant and the dynamic characteristics must meet the bandwidth requirements set by the network data transmission rate. These two factors are interrelated.

The loss budget is set by the output power of the transmitter and the sensitivity of the receiver. On a static basis, this is determined by the difference in optical power levels, but in reality the issue is a dynamic one, determined by the performance of the components at the data transmission rate of interest. Therefore, one uses a power meter to measure the power levels and a BER tester to verify the quality of the data transmission.

Between these two measurements, a curve similar to that in Fig. 9.1 will be generated for the data link.

9.4.3. LINK TESTING

The total link performance specifications will include the loss budget, minimum receiver power, and maximum length of fiber optic cable that can be used for the link. When generating bit error rate curves, one generally uses a variable attenuator to simulate the loss in the cable plant. These data do not include the effects of dispersion in the cable plant. One should also perform the tests with lengths of fiber of nominal characteristics to verify the performance under more realistic conditions, up to the maximum length specified.

9.4.4. CABLE PLANT TESTING

The cable plant is tested after installation to ensure the proper installation of all the components: cables, connectors, splices, patch panels, etc. Even before installation, tracers are used to test the continuity of most fibers in a cable before pulling, to ensure that the cable has not been damaged during transport.

As the cables are pulled and termined, each is tested with a test source and power meter to ensure the cable was installed without damage and the terminations have been done properly. After the entire cable plant is installed, end-to-end runs that will connect network equipment are patched in and tested for total loss. This test will determine if the networking equipment will operate properly upon installation.

9.5. Standardization of Testing Procedures

Most fiber optic test procedures have been thoroughly tested and codified as industry standards. These standards are part of EIA/TIA RS455 and are being adapted into IEC standards. Most of these standards refer to tests of individual components under a variety of environmental conditions; therefore, only a few related to testing the performance of the installed cable plant. With datacom networks, we are primarily concerned with three fiber optic test procedures (FOTPs): FOTP-95 for measurements of optical power, FOTP-171 for testing patchcords, and OFSTP-14 for testing the loss of the installed cable plant. These are described in the following section.

9.6.   Fiber Optic Test Equipment Needed for Testing

The fiber optic power meter is a special light meter that measures how much light is coming out of the end of the fiber optic cable. The power meter needs to be able to measure the light from the fiber optic cable at the proper wavelength and over the appropriate power range. Most power meters used in datacom networks are designed to work at 850 and 1300 nm, the wavelengths used in datacom networks. Power levels are modest, in the range of −15 to −35 dBm for multimode links and −10 to −40 dBm for single-mode links. Power meters generally can be adapted to a number of different connector styles for convenience in testing.

Optical power is usually measured in dBm, or decibels referenced to one milliWatt. This log scale is used because of the large dynamic range of fiber optic links, a range of 1000 or more. Some power levels may be given in μW, which many meters measure directly. Power meters measure average optical power, not peak power; therefore, they are sensitive to the duty cycle of the data being transmitted. It is important to specify the test conditions for measuring the optical power of a transmitter or at a receiver in terms of the data transmitted. Most networks have a diagnostic test signal for just this purpose.

Fiber optic power meters, such as DMMs, come in a variety of types. The measurement uncertainty of practically all fiber optic power meters is the same because it is limited by the physical constraints of transferring standards with optical connections. Most meters have an uncertainty of ±5%, or approximately 0.2 dB, no matter what the resolution of the display may be. Lower cost meters usually have a resolution of 0.1 dB, mid- to high-range meters display 0.01 dB, and a resolution of 0.001 dB is available on a few specialized meters.

The appropriate resolution for a measurement should be chosen according to the test. Laboratory measurements of low-loss patchcords, connectors, and splices can be made to 0.01 dB resolution and an uncertainty of 0.05 dB or less if great care is used in controlling the test conditions. Field measurements of absolute power are no better than the absolute calibration uncertainty, but relative power measurements can be made to 0.1 dB. Cable plant loss is affected by not only the meter uncertainty but also the characteristics of the test source and may have an uncertainty of 0.5 dB or more.

The test source is a portable version of the source that is used in the communication network attached to the fiber. It simulates the signal in the fiber for loss testing with a power meter. It should be a LED for multimode fiber or single-mode fibers less than 5 km long. A laser should be used for long single-mode fibers.

One must always match the source wavelength to the network requirements. Although one refers to 850, 1300, or 1550 nm routinely, actual source wavelength may vary. On long lengths of fiber typical of wide-area networks (WANs) or long-distance networks, the difference becomes important. The fiber attenuation coefficent is a function of wavelength, so source spectral characteristics can be important. When testing long lengths of fiber, one may want to make corrections to nominal source wavelength losses.

In multimode fiber, one may also need to be concerned by "mode power distribution," or how the light is transmitted in the various modes transmitted in multimode fibers. If one uses a test source with a LED typical of the types of LEDs used in actual link transmitters, the differences in mode power distribution will be minimized.

It is possible to obtain a source and meter together in a "test kit." One can also obtain an optical loss test set (OLTS), which is a meter and source in one package. However, because the source and meter must be at opposite ends of the fiber being tested in the actual installation, an OLTS requires two "sets" of instruments at a higher cost to perform the same measurement.

Reference test cables (sometimes called test jumpers) are needed to test unknown cables. A "launch" cable attaches to your source to set reference conditions for all loss tests. In some tests, you will also need to add a receive cable on the meter after you set your launch reference. These cables must match the fiber type and size in the cable and mate to the connectors on the cables to test. These cables must be tested often to make certain they are good themselves, otherwise all measurements will be in error. Most multimode fibers are the same size (62.5-μm core) and mate to ST connectors. Most single-mode fibers are the same, simplifying test cable inventory.

Splice bushings are also called bulkhead splices and some other names! They allow two cables with connectors to mate. They are also very important to good loss testing because they contribute to the mated connector loss. Bushings with plastic alignment inserts are not good for testing because they wear out quickly. Use only those with metal or ceramic alignment sleeves. One can get hybrids too, to mate ST to FC or SC connectors, to simplify the testing of different connector types.

9.7. Measuring Optical Power

Measuring power at the transmitter or receiver requires only a fiber optic power meter, an adapter for the fiber optic connector on the cables used, and the ability to turn on the network electronics. Remember that when measuring power, the meter must be set to the proper range (usually dBm, but sometimes μW) and the proper wavelength.

When you are ready to measure power, attach the meter to the cable at the receiver to measure receiver power or to a short test cable (tested and known to be good) that is attached to the source (Fig. 9.3) to measure transmitter power. Turn on the source and note the power that the meter measures. Compare it to the specified power for the system and make sure it is enough power but not too much.

9.8. Testing Loss

Loss testing requires measuring the optical power lost in a cable with a source and power meter. In order to measure loss, it is necessary to create reproducible test conditions for testing fibers and connectors that simulate actual operating conditions. This simulation is created by choosing an appropriate source and mating a known good test cable, called a launch reference cable, to the source. The reference cable creates controlled test conditions that allow testing other cables mated to a known good connector

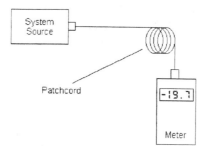

Fig. 9.3 Measuring optical power.

with a matching fiber type and known source characteristics. (If you have read about fiber optics, you have probably noted references to a tester called an OTDR. This instrument can be used in some situations to measure loss indirectly by analyzing backscattered light in the cable. Because the measurement is indirect and does not simulate actual operating conditions, it should not be used to measure actual link loss. It is also limited by resolution, making it useful only on long-distance networks and primarily for finding and diagnosing faults.)

Correct setting of the reference launch power is critical to making accurate loss measurements. The output power of the source and launch cable should be measured by the power meter and recorded for loss calculations (Fig. 9.4). It is also advisable to periodically recheck this source output to ensure the source has not varied, and it must be rechecked if the launch cable is changed. Some meters will automatically store loss references and calculate loss directly. Even with these meters, it is advisable to record the power level from the launch cable in case the reference value in the meter is lost.

Another important issue in fiber optic testing is cleanliness. Optical fiber is only the size of a human hair and typical airborne particles are almost as big as the light-carrying core of the fiber. Dirt particles on the end of the fiber will create an attenuator when measuring power or when mating connectors to measure loss. It is absolutely mandatory to clean the mating surfaces of the connectors before making measurements (and a good idea to do it again before making cable plant connections!). All connectors and mating splice bushings are supplied with dust caps that should not be removed except when testing or connecting the components.

No loss tests can be done reliably unless the launch cable is of high quality. Always test reference cables by the single-ended method described

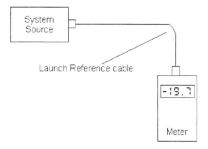

Fig. 9.4 Setting the reference power for loss measurements.

in Section 9.9 to make sure they are good before testing other cables. "Reference-quality" launch cables are not necessary; only cables that exhibit low loss are necessary (less than 0.5 dB maximum; 0.3 dB or better is preferred).

The method described previously for setting the reference for loss measurements is the only method that will give reliable results. Some old reference books and manuals show setting the reference power for loss using both a launch and receive cable mated with a splice bushing or even with a launch and receive cable and a "reference system cable."

The reference power for loss measurements is supposed to represent the amount of power launched by the source and should be measured directly by a meter from the output of the launch cable. The other methods introduce uncertainties in the measurements that will adversely affect the measurements.

When using the method of two-cable reference, if either the launch or receive cable is bad, setting the reference with both cables hides the fact. Even if the connectors were so bad that the loss of the mated connectors was 3 dB, the reference value would not reflect this. Thus, you could begin testing with bad launch cables, making all your loss measurements inaccurate. Even in the best case, this technique will underestimate connector loss by one connector compared to how the link loss was specified.

The three-cable reference has the same problems as the two-cable reference, and it also has the problem that if the system reference cable is bad, it will cause "loss" measurements to show a "gain" with respect to the reference. Fortunately, this method has rarely been referenced.

9.9. Testing Cable Loss

There are two methods that are used to measure loss, called "single-ended loss" and "double-ended loss." Single-ended loss uses only the launch cable, whereas double-ended loss also uses a receive cable attached to the meter.

Single-ended loss (Fig. 9.5) is measured by mating the cable you want to test to the reference launch cable and measuring the power out of the far end with the meter. When you do this, you measure the loss of the connector mated to the launch cable and the loss of any fiber, splices, or other connectors in the cable you are testing. This method is described in

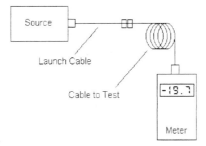

Fig. 9.5 Single-ended loss.

FOTP-171. Because you can reverse the cable to test the connector on the other end, it becomes the best possible method of testing patchcords because it tests each connector individually.

In a double-ended loss test (Fig. 9.6), you attach the cable to test between two reference cables—one attached to the source and one to the meter. This way, you measure two connectors' loses (one on each end) plus the loss of all the cable or cables in between. This is the method specified in OFSTP-14, the test for loss in an installed cable plant. Remember that the reference is set the same way for both tests—by measuring the output of the launch cable with the power meter.

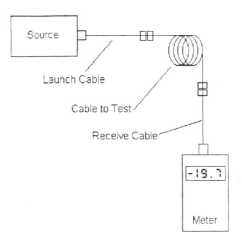

Fig. 9.6 Double-ended loss.

9.10.   OTDR Testing

The OTDR is both the best known and least understood fiber optic instrument. The OTDR does not measure loss, but instead implies it by looking at the backscatter signature of the fiber. It does not measure cable plant loss that can be correlated to power budgets.

The OTDR works like radar or sonar, sending out a pulse of light from a very powerful laser, that is scattered by the glass in the core of the fiber. Approximately one-millionth of the light is scattered back up the fiber to the OTDR, where it is captured, averaged to improve the signal-to-noise ratio, and analyzed (Fig. 9.7).

Because the group velocity of light in the fiber can be determined, the round-trip time for the OTDR pulse can be expressed in distance. The pulse is attenuated on the outward path and again on return after backscattering; therefore, one can calculate the loss of the fiber and any intermediate connections from the backscatter signature of the fiber, assuming the backscatter coefficient is constant.

The OTDR suffers from several serious uncertainties in measurements and physical limitations. The measurement uncertainties come primarily from the variations in backscatter of the fiber. The backscatter coefficient is a function of the material properties of the glass in the core and the diameter of the core. Variations of the fiber materials or geometry can

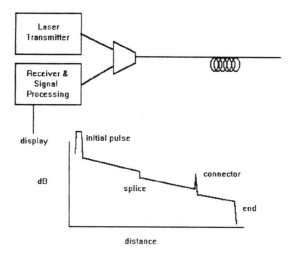

Fig. 9.7   OTDR block diagram.

cause major changes in the backscattered light, making splice or connector measurements uncertain by as much as ± 0.4 dB. This has often led to confusion by showing a virtual gain at a connector, where the fibers involved have different backscatter coefficients. Connector or splice loss must be measured from both directions and averaged to remove this source of error.

In multimode cable, the OTDR laser source does not fill the higher order modes in the fiber, which are the highest loss. Therefore, the loss measured underestimates the loss an actual LED source will have in actual use. Also, the OTDR does not see the connectors on each end of the cables, further underestimating the actual cable plant loss. Although it has proven impossible to correlate OTDR and source/power meter loss techniques, expect the OTDR to underestimate the loss of a link by as much as 3 dB.

Even in troubleshooting, OTDRs can be confusing. Although they can tell where connectors, splices, or breaks are in cables, the measurement is not simple to interpret. The uncertainty of the distance is due to not only instrument limitations but also the fact that optical cable has approximately 1% more fiber than the length of the cable to prevent stress on the fiber during pulling. Thus, the actual distance to an event must be corrected for the excess fiber in the cable.

The laser pulse width of the OTDR must be wide in order to include enough light to process. The pulse width may be as little as 5 m or as much as 250 m on long-distance OTDRs. This means that the OTDR cannot distinguish most patchcords in a local area network (LAN) cable plant or any connectors, splices, or breaks close to the OTDR itself.

Within these limitations, OTDRs are used primarily to find faults in long ($\sim$km) fiber runs, especially when they are remove, buried, or otherwise inaccessible. They should never be used to measure the cable plant loss for power budgeting.

9.11.   Troubleshooting Hints for the Cable Plant

If you have high loss in a cable, the most likely problem is bad connectors. To diagnose the problem cable, reverse it and test in both directions using the single-ended method. Because the single-ended test tests only the connector on one end, you can isolate a bad connector—it is the one at the launch cable end on the test when you measure high loss.

High loss in the double-ended test should be isolated by retesting using the single-ended method, reversing the direction of test to see if the end connector is bad. If the loss is the same, you need to either test each segment separately to isolate the bad segment or, if it is long enough, use an OTDR.

If you see no light through the cable (very high loss—only darkness when tested with your visual tracer), it is probably the result of a bad connector, and you have few options. The best option is to isolate the problem fiber, cut the connector of one end (flip a coin to choose), and hope it was the bad one. Alternatively, if the cable is long enough (>10 m), it can be tested with an OTDR with long (>100 m) cables on either side to find the bad connector. Of course, the OTDR will find breaks or cuts in the cable if it is long enough.

9.12. Dynamic Link Testing

9.12.1. BIT ERROR RATIO MEASUREMENTS

In any digital communications transmission system, the fundamental measure of the quality of the system is found in the probability that transmitted bits will be correctly received as logic ones or zeroes. The parameter that describes this measure of quality is the BER. Simply stated, BER is the ratio of the number of bits received incorrectly compared to the number of bits transmitted (for a specified time interval or quantity of bits). Typical acceptable BER levels range from 1e-9 (one error per 1 billion transmitted bits) to 1e-12.

The test equipment required to perform a BER test include a pattern generator and an error detector. The pattern generator will produce a known data sequence to the system or device under test. Test patterns can be designed such that they intentionally stress some aspect of the system. For exmple, a data sequence may be generated that is difficult for a clock recovery system to synchronize. Also, data patterns that attempt to mimic real traffic can be produced. A common pattern used is the pseudo-random binary sequence (PRBS). A PRBS has the characteristic of producing all possible combinations of ones and zeroes for a given pattern length. For example, a 2^7-1 PRBS pattern will produce all possible combinations of 7 bit binary numbers from 0000001 to 1111111. (The 0000000 sequence is not produced due to limitations of the logic hardware used to generate the pattern.) Pattern generators also typically produce longer PRBS patterns, some as long as $2^{31}-1$, that will generate all combinations of 31 bit binary numbers except the case of all zeroes.

The error detector is used to determine if the data received match the transmitted pattern. The error detector will receive the output of the receiver of the system under test. The system receiver under test will have attempted to determine the logic level of each bit transmitted by the system under test. The output of the receiver decision circuit is then routed to the error detector. In its most basic building blocks, the error detector consists of an internal pattern generator and an exclusive or logic gate. The error detector's internal pattern generator will produce a reference pattern identical to the pattern fed from the pattern generator feeding the test drive. The reference pattern is synchronized to the test device output and then compared bit for bit with the exclusive or gate. Determining the BER is then a simple matter of counting the total bits received and the number of bits that were incorrectly received (Fig. 9.8).

In typical digital communications systems, the power from the transmitter is large enough that if it were to arrive unattenuated at the system receiver, there would be error-free communication. System performance in terms of BER is then often characterized in terms of the amount of attenuation between the transmitter or receiver. Similarly, the BER can be characterized in terms of power level at the receiver. Figure 9.9 shows a typical BER characterization of a high-speed system. As the received power is decreased, the signal-to-noise ratio is reduced and the probability of a bit being received in error increases.

Because BER measurements are used to characterize the effects of system noise, they require analysis from a statistical perspective. Specifically, one must consider how many errors must be detected to render a statistically valid measurement. Ideally, a minimum of 1000 errors should be measured to yield a high confidence level in the accuracy of the BER measurement, although good measurements with as few as 100 errors can be achieved. This then leads to one of the primary difficulties encountered when performing a BER measurement: If a BER of 1e-10 is required, and at least 1000 errors are to be detected, then at least 10 terabits must be transmitted during the measurement. If the transmission rate is 2488 Mb/s, the time required for the measurement will be more than 1 h. If the data rate is 155 Mb/s, the test time would be almost 18 h!

As BER levels are made more stringent (some systems require BERs of 1e-13), test times become proportionally longer, measured in days instead of hours. Referring to Fig. 9.9, a common technique for characterizing systems is to make measurements at signal levels that result in high BERS. Because many errors are generated, the time required to make the measure-

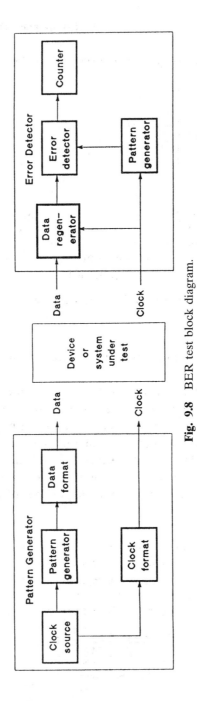

Fig. 9.8 BER test block diagram.

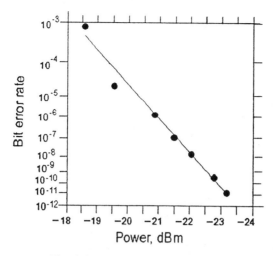

Fig. 9.9   BER vs received power.

ments is comparatively low. From these results, conditions for very low BER can be estimated.

BER measurements may also be used to characterize system performance in terms of parameters other than attenuation or signal power. Any parameter that can impair system performance might be considered. Examples include BER vs jitter, sampling instant, and dispersion. Figure 9.10 shows a jitter tolerance measurement. In this measurement, jitter is sequentially increased until the BER is degraded, indicating the system/device under test's ability to perform in the presence of high levels of jitter.

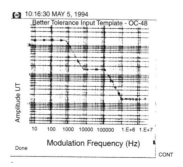

Fig. 9.10   Jitter tolerance measurement.

9.13. Characterizing Digital Communications Waveforms

Complete systems and system receivers are characterized using BER tests. Characterization of transmitters is to a large extent achieved through analysis of the waveforms they produce. The instrument used to view a high-speed digital communications waveform is typically a high-speed digital sampling oscilloscope (DSO). Waveforms can be viewed in one of two ways depending on how the oscilloscope is triggered. A pattern generator, similar to that used in a BER system, is used to produce a data signal for the transmitter being tested. Pattern generators may produce two types of signals used for timing references (triggers) for the oscilloscope. A pattern trigger is a timing edge that occurs once for every repetition of the data pattern. A clock trigger produces a timing edge for every transmitted bit.

Use of a pattern trigger allows one to view a specific portion of the data pattern. Figure 9.11 shows a data waveform measured with a DSO. The measurement shows the waveform for a sequence of several ones and zeroes. From this measurement we can see the relative separation between the two logic levels. Rise and fall times, indicating the speed of the transmitter, can also be quantified.

By using a pattern trigger, a good snapshot view of a small sequence of bits in a transmitted waveform is provided. However, it is important to understand how the transmitter behaves under all operating conditions and not just a select pattern. This is achieved with the eye diagram display.

The eye diagram, shown in Fig. 9.12, is an overlay of many transmitted waveforms. By using a clock signal as the trigger input to the DSO, the transmitted waveform can be sampled over virtually the entire data pattern generated by the transmitter. Thus, all the various bit sequences that might be encountered can be sampled to build-up the eye diagram.

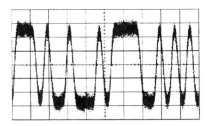

Fig. 9.11 Pulse train from a high-speed laser transmitter.

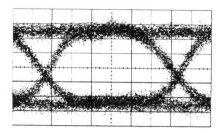

Fig. 9.12   Eye diagram displayed by a DSO.

For eye diagram analysis the sample rate of a DSO is typically orders of magnitude slower than the data rate of the signal being characterized. For example, a DSO used to measure a 2488-Mb/s signal may sample the waveform at a 40,000 sample/s rate. When a point is sampled, the next adjacent point in time sampled will occur thousands of bits later in the transmitted waveform. It should be noted that the intent of the eye diagram is not to display individual transmitted bits. Instead, it is constructed from a multitude of waveform samples. It then allows a good visual representation of the overall quality of the waveform.

The eye diagram includes a significant amount of information about the transmitter output. The relative separation between the two logic levels is easily seen in the vertical "opening" of the eye. Obviously, the more open the eye is, the easier it will be for the receiver to determine the signal logic level. Rise and fall times can be measured on an eye diagram to determine the transmitter speed. Modern DSOs have the ability to automatically construct histograms to statistically analyze the eye diagram. Jitter describes the relative instability of the bit period or data rate of a signal. This can be quantified by measuring the variance in the relative time in which the rising and falling edges occur. By constructing a histogram at the eye diagram crossing point, jitter is determined (Fig. 9.13).

Other histogram-based measurements include eye height and width, which describe the vertical and horizontal openings of the eye, respectively. A critical measurement for high-speed laser transmitters is extinction ratio. In simple terms, extinction ratio is the ratio of the power level of the logic "1" signal to the power of the logic "0" signal. Extinction ratio is an indicator of how well available laser power is being converted to modulation or signal power. When the extinction ratio is high, virtually all the transmitted power is used as signal power. Typical extinction ratio values range

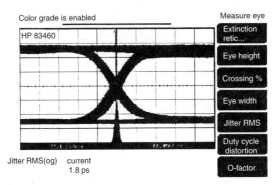

Fig. 9.13 Using histograms to measure jitter.

from 7 (8.5 dB) to 30 (14.8 dB). Extinction ratio is easily characterized using DSO histogram capabilities (Fig. 9.14).

As extinction ratios approach high levels, measurement error can increase. Fundamentally, a high extinction ratio implies either a very high logic 1 signal power or a very low logic 0 signal power. The latter case is typical. An accurate extinction ratio measurement then requires an accurate measurement of a very small signal. Measurement errors can occur from offset signals or waveform distortion in the instrumentation. Today's modern DSOs are designed to minimize these potential errors.

Another technique used to characterize transmitted waveforms is to define and measure the shape of the eye diagram. This is achieved with masks and pulse templates. To define the allowable shape of the eye diagram, polygons are constructed by the DSO to define regions where the

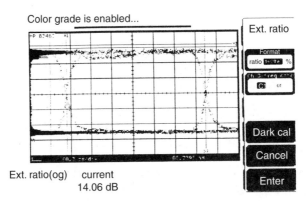

Fig. 9.14 DSO measurement of extinction ratio.

waveform may not exist. Figure 9.15 shows a mask test for a high-speed laser transmitter. Note that the mask consists of three polygons. The polygon in the middle of the eye determines the minimum allowable opening of the eye. The polygons above and below the eye determine allowable overshoot, undershoot, and ringing.

Many communication systems using electrical waveforms will character-ize waveshapes by examining individual pulses rather than an eye diagram. Instead of a mask test, pulses are required to fit within a template as shown in Fig. 9.16.

Today's DSOs have built-in automatic capabilities to perform mask and template tests. Masks or templates are selected from menus and automati-cally aligned, and test results are logged.

It is important when characterizing the shape of an eye diagram that the measurement systems have a consistent and well-behaved frequency response. If the measurement channel does not have a flat frequency re-sponse, it may introduce waveform peaking or overshoot. Also, a waveform may look much different when measured on a very wide bandwidth system in comparison to a similar measurement made on a lower bandwidth system. To achieve measurement consistency, standards committees often define the frequency response of the measurement system. For example, the opti-cal receiver used with a DSO for laser transmitter measurements may be followed by a low-pass filter and the 3-dB point of the test system is set to be 75% of the data rate. All those using such a measurement setup should have comparable results.

Masks and pulse templates are set up and defined by standards bodies to specify equipment to be used in systems such as SONET, SDH or

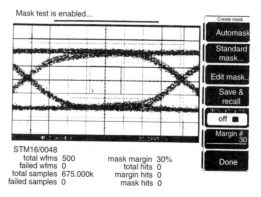

Fig. 9.15 Eye diagram mask testing.

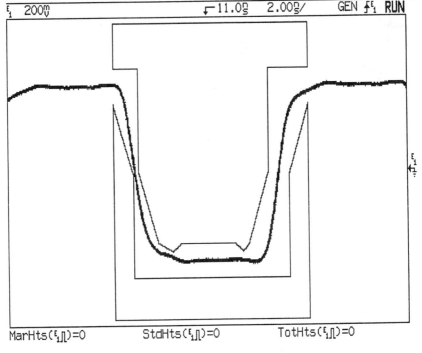

Fig. 9.16 Pulse template testing.

Fibre Channel. The intent of these tests, coupled with a variety of other specifications and tests, is to guarantee a minimum level of performance in transmitters to allow multiple vendors to supply equipment for a communications system.

9.14. Testing and Troubleshooting Networks

The installed network can be tested quickly and easily with a fiber optic power meter, trace, and attenuators. If no power meter is available, a fiber tracer or even a flashlight can be used to determine if the cables are transmitting light to isolate the problem to either the cable plant or the electronics.

When using a power meter for diagnostics, the network transmitter should be set to transmit a bit stream of unknown duty cycle. Set the power meter calibration on the proper wavelength and measurement range. To test the received power, the most critical element in the network, merely disconnect the fiber optic cable connector at the receiver, attach the power meter, and measure the power (Fig. 9.17).

If the receiver power is low, the transmitter power should be measured by disconnecting the source jumper cable at the first available connector and measuring the power with the fiber at that point. Alternatively, one can disconnect the cable at the transmitter and use a test jumper of known loss to measure the coupled power. If the output is measured through a short network jumper cable (less than 10 m), no compensation for fiber loss in the jumper cable is necessary. For longer jumpers, some compensation may be necessary.

If receiver power is low but transmitter power is high, there is something wrong with the cable plant. The cable plant must be tested at every connection to isolate the bad cable(s) and or connectors. This can be done from either end but is more convenient starting from the transmitter end. Starting from the transmitter or receiver end, follow the network cables to every patch panel. Disconnect the connector and measure the power at each point. By making measurements in dB, one can easily calculate the loss of the cable network to each point by subtracting successive readings.

When a suspect cable is found, by noting a larger than expected loss in the cable link, the cable needs testing by the appropriate method described previously. If a cable has attenuation that is higher than specifications, but still transmits light, check connectors on a microscope to determine if they have been damaged and should be replaced. If the connectors look good, the best solution may be to replace the cable or switch to a spare. If a visible laser fault locator is available, it can be used to visually locate breaks in the fiber and find broken connectors. In some circumstances, such as high loss in long jumper or trunk cables, an OTDR can be used to diagnose cable faults.

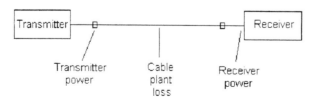

Fig. 9.17 Network troubleshooting.

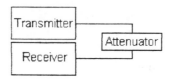

Fig. 9.18   Loopback testing.

9.15.   Transceiver Loopback Testing

Loopback testing is a powerful tool for data link troubleshooting. It involves looping a signal back across a section of the network to see if it works properly. In a fiber optic link, one wants to test network loopback first from electronics on each end of the data link because it includes both transceivers (transmitter/receiver pair) and the cable plant.

The datacom capabilities of the transceiver can also be tested with a loopback test (Fig. 9.18). This test uses an attenuator between the transmitter and the receiver and transmits a signal looped back on itself. Thus, the transceiver can be tested to see if it can transmit data over the optical loss specified in the loss budget. If transceivers on both ends of a data link pass a unit loopback test but fail a network loopback test, the problem is in the cables, which then need testing using the methods described previously.

9.16.   Conclusion

Testing fiber optic communications require some specialized knowledge of test and handling procedures but is otherwise straightforward. If networks are properly designed for testability, fiber optic networks are no more difficult to test and troubleshoot than any copper-based network.

Part 3 | The Applications

Chapter 10 | Intramachine Communications

John D. Crow

IBM Corporation, Yorktown Heights, New York 10598

Alan F. Benner

IBM Corporation, Poughkeepsie, New York 12601

10.1. Introduction

Despite many years of predictions of the imminent arrival of optical computers and optically interconnected computers, there has been relatively little
penetration of optoelectronics in the computer interconnect field. There
are incidents, but the impressive and continuous growth in the capability
of integrated circuit (IC) chip and packaging technology has combined to
shrink both the size of computer system building blocks to well within the
capabilities of copper wire and electrical connections and the price of
computing power to the point that it is difficult for any new technology to
be cost-effective. Nevertheless, there has been a recent paradigm shift
in the interconnect architecture of computing systems that may provide
opportunities for optical interconnect technology. This chapter describes
these opportunities and their requirements.

Historically, the data processing subsystem of a computer has been
confined to a single equipment frame with highly integrated ICs and dense
packaging. In this configuration, wiring density is essential, and the aggregate bandwidth of electronic interconnects per unit area has been greater
than that which optical interconnects could achieve for distances within a
machine. High aggregate bandwidth for electronic interconnects is achieved
by tightly packing wires in two-dimensional formats [e.g., hundreds of pins
per cable connector and >700 pin ball grid array (BGA) chip carriers].
Optical interconnects have until recently been almost exclusively single-
fiber connections. Because optical links are ultimately driven by electronic

331

HANDBOOK OF FIBER OPTIC
DATA COMMUNICATION

circuits, and electronic circuits cannot be made to operate hundreds of times faster, regardless of only having to drive within a module, there has been no reason to use optics.

Over the past few years, it has become increasingly difficult to define "intramachine" connection due to the advent of distributed computing systems. When, for example, an ocean/atmosphere weather simulation is run, with the ocean model being executed on an Intel Touchstone Delta and the atmosphere model executed on a Cray YMP—with interface data being exchanged constantly during the simulation [1]—it is somewhat difficult to say whether one machine or two are being used. Even more difficult is describing the operation of a distributed memory parallel processing machine such as the IBM RS/6000 SP-2, in which each node has its own separate operating system, memory, disk, and input/output (I/O) adapters, but specialized high-performance hardware and software lets the machine be programmed as a single computing resource. On such a machine, parallel computing jobs are started, debugged, checkpointed, run, and stopped on one, several, or all nodes simultaneously. Here, the question of whether one machine or many are being used at any particular time is more a statement of system management and system software rather than cabling and addressability.

There is an increasing trend in the high-performance computing world toward aggregating computing resources and performing computing over physically distributed processing engines. This trend naturally leads toward increasing requirements on the computer interconnects to allow close coupling between computing facilities and allow software to view them as a single distributed computing resource.

In this chapter, we describe the general characteristics of computer networks that couple distinct classic computers [central processing unit (CPU) + memory] into a single closely coupled distributed computing resource. Because the requirements on such a network are fairly well established and unique, we introduce the notion of a system area network (SAN) and enumerate characteristics that distinguish it from a local area network (LAN) and from the internal wiring of a single computer. As the performance requirements for LANs increase to the level currently required for SANs, much of the physical layer of cables, and signaling, will be extended from SANs to LANs.

We also enumerate the characteristics of optical components and optical data links that would be required to satisfy the requirements of a system area network and show that the requirements (particularly aggregate bandwidth) can be most cost-effectively addressed using parallel optical interconnects (POIs). Parallel optical interconnects are a relatively new technology, consisting fundamentally of an array of lasers in a single package connected by a single connector to a ribbon of optical fibers and to an array of receivers. We describe the current state of the art in parallel optical interconnects and directions for the technology.

10.2.   Current Intramachine Optics Applications

Before discussing system area networks and parallel optical interconnect hardware, we will discuss various intramachine computer interconnect applications in which optical components have already been applied. We concentrate here on the applicability of these technologies to computer systems and on the reasons why, despite their higher cost [2], they have been able to replace electronic components in those applications.

10.2.1.   OPTICAL CLOCK DISTRIBUTION

A large computing resource such as a parallel supercomputer will generally consist of a number of distinct processors (CPU + memory) that may be physically separated into different racks. All of the digital logic will require clocking signals, and the clock frequencies at all processors will generally be at least approximately equal.

There are two common strategies for generating such clock signals. Either each processing element may have its own oscillator and clock distribution system or the whole system may be clocked from a single oscillator with a large clock distribution system. With the first method, the clock distribution problem is easier (a clock signal may not have to propagate outside a single rack), but the machine will require asynchronous buffers between clock domains and idle cycles over the communication links between clock domains to compensate for slightly different clock frequencies in different domains. A single distributed clock signal over the entire machine can significantly increase performance for low-overhead operations compared to multiclock designs. For machines constructed using a single oscillator, a clock distribution system must be constructed that takes a single high-frequency oscillator signal and distributes it over the full complex of digital circuitry with low jitter and without radiating electromagnetic interference (EMI) from the computing complex. For such a system, the high fan-out capabilities and low EMI of an optical clock distribution system can be attractive, particularly at high (>500 MHz) clock frequencies. Several commercial Cray machines have been built with optical clock distribution. In such a system, a single oscillator drives several lasers (for redundancy), and the optical clock signal is fanned out and distributed to each portion of the parallel computer, where a receiver converts the signal to electronic format and an electronic network further distributes the clock signal over the digital logic.

10.2.2. I/O BUS EXTENSION

The other main area in which optical components have been successful has been in the extension of I/O buses, such as a personal computer interface (PCI) bus, out to longer distances than can be supported using electronic cabling. There are two main applications for optical bus extenders: LAN communications and long-distance disk I/O. Optical LAN cables have become increasingly prevalent as multimode fiber cabling inside a building has become competitive in price with copper cabling [3].

A significant fact in the usage of optics for disk I/O has been the decreasing size of disk drives. Disk drives platters have recently become prevalent with platter diameters approximately half the size of a parallel small computer system interface (SCSI) connector. This has driven the advent of serial disk interfaces [e.g., Fibre Channel or storage system architecture (SSA)], which use a single signal line and can be either optical or electronic. The simplicity of converting between electronic and optical cabling has encouraged the flexible placement of disk drives at relatively long distances from the processors to which they are attached. The latency of the physical cabling medium (~5 ns/m) is relatively insignificant relative to the access latency of the disks, which is measured in milliseconds.

The primary reason for optical bus extenders has generally been the desire to separate processing components to different parts of a single machine room or to different rooms or buildings. Some examples are optical channel extenders for high-performance parallel interface (HIPPI) optical implementations of Fibre Channel and serial HIPPI, and, at the high end, optical extenders for the Cray GigaRing.

A second primary reason for optical bus extenders has been the need to simplify cabling for multichannel installations. Mainframe installations, such as the IBM S/390, often need to interconnect multiple separate processing engines with multiple sets of disk drives and multiple external connections. Wiring of such installations can be made much simpler and more reliable if intermachine connections are converted to serial optical format and ganged together in multifiber ribbons. As many as 144 1-Gb serial optical lines can be ganged together in a single connector housing, significantly simplifying wiring, configuration, and reconfiguration of large systems [4].

Parallel optical interconnects are currently used in I/O extenders to serve a line concentration function when many I/O channel source and sink in the same physical location (see Chapter 13). There is little or no synchronism between the fiber lines, and typical direct access storage device (DASD) I/O channel data rates are used. A similar use for POI is found in large telecommunications multiframe switching systems.

10.3. System Area Networks

We introduce the concept of a system area network here to distinguish the types of systems in which parallel fiber interconnects are most likely to be used first.

The 1990s have seen the widespread adoption of paralleling processing modalities for large systems. In these systems, multiple processors are coupled together to solve parts of a large-scale problem. Because each processor can work semi-independently, and because the group of processors can be programmed together, a group of processors can solve problems faster than a single processor. In the best cases, adding processors to a multiprocessor system can linearly speed up the problem-solving speed. (There are even some cases in which there is better than linear speed-up because each processor's limited caching space is better utilized when operating over a smaller portion of the full job.)

The evolution of parallel processing machinery has seen a steady evolution toward a system model that might be termed "clustered multiprocessors." In this architecture, a cluster of shared memory processors (SMPs) operate as a single computing resource. Each SMP contains between 2 and 16 processors, sharing (logically) a single system address bus, and having a single multithreaded operating system. Building SMPs larger than 16-way becomes increasingly difficult and less rewarding due to the difficulty of scaling the operating system and increased competition for the system bus.

To further increase scalability, the SMPs are clustered together over a high-speed network. The high-speed network is termed a system area network. The characteristics of this network are intermediate in performance and operation between those of a system bus and a LAN. The performance of the entire parallel processing machine is dependent, aside from the CPU performance and parallelizability of the algorithm being executed, on the performance of this network.

This kind of network is given the name system area network to emphasize that this network operates within the bounds of a set of processors that manage as a single system but shares architectural similarities with a classical network rather than a bus.

Figure 10.1 illustrates how a SAN might be implemented functionally and how it relates to data storage and data communications networks. In this figure, building blocks of microprocessor + local memory are interconnected to form a "single-image system." The single-image concept arises because a user (application) connected to any one node of the SAN does not know which or how many of the nodes are processing his or her job. The system architecture and software determines how to distribute his or

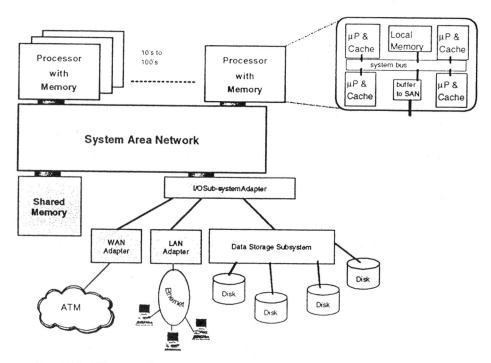

Fig. 10.1 Diagram of a system area network, illustrating how it interfaces to other data communications networks.

her job to optimize resource. To maintain this transparency of the system nodes and interconnects to the user, the performance demands on the interconnect hardware can be large. Conversely, the connections with the storage system and the global data communications network, shown in Fig. 10.1, are isolated or buffered from the computing operation by interface adapters, that contain both hardware and software. The user/application definitely knows when data are sent over these interconnects because of the communication protocols that must be added to the data and also because of (usually) a serious degradation in system performance.

10.3.1. SAN DISTINGUISHING CHARACTERISTICS

The following are the characteristics that distinguish a system area network. Performance values are current as of early 1997. Clearly, the absolute numbers will improve, but it appears that the distinction between system bus, SAN, and LAN interconnects as distinct technologies will remain.

10.3.1.1. Bandwidth

The bandwidth of a SAN is higher than that of LAN, ideally as close as possible to the bandwidth of a system bus. Practically, this will mean a bandwidth higher than 100 MBytes/s.

10.3.1.2. Latency

Because communications over the SAN are intended to approach the performance over a system bus, where latency is measured in hundreds of ns, the latency across a SAN (including communication protocol) must be <100 μs, preferably <10 μs.

10.3.1.3. Scalability

A SAN would be expected to support economic scalability from a few nodes up to thousands of nodes. It does not imply direct connectivity to tens of thousands of nodes, given the difficulty of programming applications scalable to this range, although some SIMD machines could be considered to be in this range.

10.3.1.4. Packet-Switched Network Topology

Given the requirement of connecting up to thousands of nodes simultaneously, a SAN cannot depend on a shared medium. Also, the low latency requirement drives toward implementation as a packet-switched network.

Some SANs will incorporate support for circuit switching as well, and advanced implementations will incorporate support for multistage networks for economical scalability over the full range of supported system sizes.

10.3.1.5.   Flow-Controlled Transmission for Congestion Control

At a general level, congestion in a network caused by multiple sources sending to the same destination or over the same link or switch can be handled by either (i) dropping or rerouting low-priority packets for retransmission and later delivery or (ii) inserting link-level flow control so that the sender of data is prevented from sending it until it is assured of being receivable. There will also generally be a mechanism for end-to-end flow control.

For long-distance communications or communications in which it is not vital that every bit be received correctly (e.g., voice or video traffic), flow control is generally used less frequently because link-level flow control requires the transmitter to maintain data buffering of a size equal to at least the link bandwidth times the round-trip time to maintain full transmission bandwidth. For SANs, however, the link round-trip times are short enough that the cost of the buffering is less detrimental than the cost of the protocol processing required for retransmitting dropped packets. A SAN will generally not handle congestion control by dropping packets but will use link-level and end-to-end flow control.

10.3.1.6.   Controlled Topology

Because all nodes on the SAN network are known and managed as a single system, the topology of the network can be determined, fixed, and set in the source routing tables. This is generally the case with a LAN as well. The point is emphasized here to contrast it with the case of internetworked LANs, in which (i) routing information is not made available between LANs and bridges or routers must add source and destination identifiers to routing information, at a cost in latency of many microseconds; and (ii) the source-level protocol must assume that intermediate nodes may or may not be available to route messages to a destination node.

10.3.1.7.   Source Routing

Source routing means that the originator of a message can specify the entire routing through the network at the same time that it is doing the rest of the protocol processing, eliminating the need for any routing look-ups in network switches. This characteristic is in contrast to asynchronous transfer mode (ATM) networks, which require the network switches to convert virtual path identifier and virtual circuit identifier information into routing information and back at each switch stage. Because ATM networks are largely targeted at longer distance networks, where the propagation time may be hundreds of μs, the extra latency of a routing look-up is negligible. In a SAN, the low latency and fast routing requirement dictate that the source be able to specify routing through the entire network. In some instances, it is possible to achieve adequate performance if routing information can be calculated at each switching stage by an extremely fast and simple conversion from the destination identifier.

10.3.1.8. Addressability

Some system area networks may also manage their addressing and routing by maintaining a single physical address space across the network so that a physical address presented on one SMP bus would be responded to with data residing in random access memory chips on another bus. Some may maintain cache coherence over the network in very tightly coupled systems.

10.3.1.9. Optimized User-Space Communication Protocol

The latency and bandwidth capabilities supported by the hardware cannot be utilized with a traditional protocol stack driven by the operating system. For example, the latency for TCP/IP or UDP communications is several hundred microseconds, and the maximum bandwidth supported is below 50 Mb/s for most machines. The protocol-processing performance is improving, of course, but not as quickly as the performance of the network. The ratio between latency and bandwidth performance supported by the network and that supported by the traditional protocol stack is at least a factor of 2 and growing.

To circumvent this problem, a SAN will generally use a highly optimized communication protocol. In a SAN, as opposed to a LAN, some optimizations are possible by (i) utilizing limited addressing, for communicating with a limited and relatively stable set of communication partners, and (ii) minimizing the number of copies for data and control information by collapsing the number of protocol stack layers.

The biggest breakthrough in this area has been the advent of "user-space" communication protocols. These protocols minimize protocol processing overhead and latency by operating entirely in the user's process space through library calls and allowing the user-level applications to directly address network interface resources, eliminating the need for a lengthy processor context switch into kernel-level protocol processing code.

The general picture of a system area network is a multiprocessor computing resource in which similar control mechanisms, bandwidth requirements, and latency requirements to those of a computer's system bus are implemented over a distributed network.

Most important, this requires protocol processing mechanisms that are different (faster and more controlled) than those in a LAN interconnection. The TCP/IP protocol is a sample case in which the processing power of a Cray YMP is only capable of processing the TCP/IP protocol fast enough to send data at 80 Mb/s, and yet the system bus speed for such a system (and for a large-scale SMP of 1996 and beyond) will be more than an order of magnitude faster.

Table 10.1 **Sample of Multiprocessor Computers with Integrated Network**[a]

System	Type	Interconnect performance	Status
DEC 7000 Open VMS	Message passing	STAR coupled 70 MBytes/s	Product 1982 25K+ Installations
IBM ES9000 Sysplex	Message passing	Switched 12 MBytes/s	Product 1994
SUN SPARC cluster 2000	Message passing	Ethernet	Product 1994
Hewlett–Packard Exemplar SPP 1200	Shared memory	Switched SCI, 128 CPUs	Product June 1995
IBM RS/6000 SP-2	Message passing	150 MBytes/s, 512 CPUs	Product March 1996
Tandem Servernet	Message passing	50 MBytes/s	Product 1996
Sequent NUMA-Q	Shared memory	1 GByte/s IQ-Link SCI, 32 CPUs	Product December 1996
Data General NUMALiNE	Shared memory	1 GByte/s SCI, 32 CPUs	Product 1997

[a] The newer machines have networks with SAN-like characteristics.

10.3.2. SAN PROTOTYPE MACHINES

There are a large number of commercially available server machines that have internal connection networks fitting the description of a system area network. Table 10.1 lists typical examples with some of their interconnect features. This is by no means an exhaustive list. The description essentially fits a description of all parallel processing and distributed memory machines. All the systems in Table 10.1 are (or can) be wired for short distances using parallel copper wire technology, but most could also use parallel interconnect technology if it becomes cost-effective.

Clearly, the performance and node count trend is favorable for optical implementation of the interconnect network, but it cannot be overemphasized that without cost-effectiveness and the demonstration of reliable and flexible operation in these systems, the designers will continue to rely on copper and will design assuming copper's limitations.

10.4. SAN Physical Layer Technology Requirements

To date, there has been very little penetration of optical components into the system area network market, but this situation is expected to change quickly with improving optical technology and increasing requirements on computer interconnects. The greatest barrier to entry has been the inability to inexpensively run multiple optical fibers between transceivers. Where copper interconnects cables may have as many as 50 signal lines per direction, fiber optic links have been serial, using discrete device components. Supporting equivalent bandwidth requires supporting much higher modulation bandwidth (which is difficult, because copper and fiber links both use the same generation of circuit technology) and/or multiple independent transceivers (greatly increasing the cost). The first technology that has shown true promise in cost-effectively addressing the problem of optical computer connect at the SAN level is the parallel optical interconnect technology, which is described in Section 10.6.

In this section, we enumerate some of the other characteristics of the physical layer technology that are required to support the higher level capabilities of a SAN-level interconnect. These characteristics serve both as a current requirements list and as a guide for future development of optical link technology.

10.4.1.   LATENCY

The ability to transport memory data blocks in a very few processor clock cycles is a primary requirement for the interconnect network. Although much of latency optimization can only be achieved in the software layers of the protocol stack (i.e., the optimized user-space protocols described previously), it is important that latency be minimized at the hardware layers as well.

Although some hardware delay is caused by speed-of-light propagation over the link cables, which cannot be improved, a greater amount can be caused by media-dependent operations, such as data encoding and decoding, buffering, clock recovery, and link control at the transceivers. Although some amount of data encoding and buffering may be required, depending on link requirements, it is important at the system level to minimize these latencies as much as is practicable.

A further improvement in latency can be gained by incorporating as much of the protocol as possible into hardware (chips) rather than into software (programming). Because chip design is complex, difficult, and expensive, the simpler this can be made, the better. For this reason, a physical layer link module that, for example, supports higher level media access functions as well as media-dependent protocol functions, will likely see more acceptance from system designers than one that supports only electron–photon conversion.

10.4.2.   BANDWIDTH

The bandwidth requirements for optical interconnects for SANs are strongly driven by the capabilities of the system buses and network switches and interface adapters that interface with them. These requirements increase approximately at the rate of the increase in processor performance—approximately a factor of 3 every 2 years. The exact bandwidths required are somewhat dependent on implementation but are currently in the range of 100–1000 MBytes/s.

10.4.3.   DISTANCE

Distance for a SAN is quite a bit shorter than would be required for LAN networks. Two major factors limited the required distance. First, the latency over a long link limits overall system performance—the speed-of-light round-trip delay over a 100-m link can be several hundred processor clock cycles. Second, and more important, because the link-level flow control is generally used for congestion management, longer distances either imply lower link speeds or large flow control buffers, increasing overall system cost.

In practice, most SANs will be implemented within a single machine room, with the large majority of SAN links being less than approximately 20 m long. Some links will have to be longer, to handle widely distributed systems spanning multiple rooms or multiple buildings, in which case links of 100–400 m would be useful.

10.4.4. PACKAGING DENSITY

The density of lines in and out of the modules and in the cables is high to conserve real estate and keep the cables manageable in a machine room or lab environment. This is an increasingly difficult challenge for copper wiring because line size fundamentally must increase with line speed and line distance required. Some signal processing can be applied to data on the copper lines, but this adds to the protocol complexity and the overall latency of the data transfer.

10.4.5. RELIABILITY, AVAILABILITY, AND SERVICEABILITY

The SAN is the "center" of the entire computing system, and failures in any way that stop the system cannot be permitted. This must be true 24 h a day for the life of the system, typically about 5 years. Therefore, the link components must be very reliable compared to typical datacommunications links. Fault tolerance is also designed into the network so that alternate paths can be found around any failed part.

Network failure also refers to the network corrupting data. The system generally recovers from link error by a detect and retransmit operation. This should occur less than a few times in a month, implying that raw bit error ratio on a line be better than 10^{-15}. All systems employ error detection, usually at the packet level (e.g., cyclic redundancy-check), and some systems are considering error-correction schemes.

In general, SANs will be installed, serviced, and maintained by users who are unfamiliar with optical component technology and who are accustomed to the reliability, availability, and service ability characteristics of complementary metal oxide semiconductor (CMOS)-based computer technology. In practice, this means that SAN designers and users are unlikely to use optical interconnect components unless they meet the same order of magnitude performance as CMOS-based components in terms of fault tolerance, hot plugability, servicing and reconfiguration, and component lifetimes.

10.4.6. FLEXIBILITY

As SAN applications expand from the supercomputer arena to commercial applications, and from a few large installations to many smaller installations, other technology drivers also become important. Commercial customers require the capability to grow their system in a modular way, adding processing and/or memory power (i.e., by adding network nodes) as needed or affordable. This approach is also desired for a smooth upgrade path.

This flexibility demand affects the interconnect hardware. Standard PC and workstation cards and boards are more cost-effective (compared to custom high-density multiprocessor cards, which must be paid for at the time of initial installation); this implies that more interconnect distance is needed between the boards and frames. Hot plugability for the interconnect cards is required to add nodes to the network while it is running and upgrade or replace computing nodes. Hot plugable cards also make it easier for the user to install and reconfigure the system. This flexibility in interconnect configuration also implies that there will be more direct customer involvement in installation and reconfiguration.

10.4.7. SAFETY

Designers and users of SAN equipment (and other computer equipment) will strongly resist optical component technology that requires any special handling for safety that is more difficult than is required for CMOS-based components. For optical implementations this translates, in practice, into a requirement for "class 1 laser" eye safety. The specific requirements of class 1 laser safety are quite detailed, but in general they imply that a class 1-compliant component cannot cause eye damage even if someone looks directly into the optical cable.

Class 1 laser safety can be implemented either by ensuring (i) that the laser transmitter power is low enough to be eye safe, or (ii) that there is an open fiber control safety interlock, such as described in the Fibre Channel standard [5], which disables both transmitters of a bidirectional link if the link is opened. Using current open fiber control circuitry places restrictions on the topologies that can be built because it requires that each link be bidirectional, whereas the strategy of lowering transmitter power places fairly strict requirements on the receiver sensitivity.

10.4.8.   COST

The cost of the interconnect is directly borne by the computer system owner. This makes the network customer much more cost sensitive than the typical customer in "shared-media" applications [e.g., metropolitan area network (MAN) or wide-area network (WAN)]. Historically, the user does not ammortize his or her interconnect network over many product cycles and generally does not care about standard solutions (unless standard implies how cost through multiple sources). For the commercial environment, in which low-cost microcomputers are used as the nodes, the interconnect network cost must be even lower, on the order of 10–20% of the cost of the microcomputer (as in the office PC communications card environment).

Today, with copper wire interframe connections, this implies a physical layer module cost in the range of $10–15 per Gigabit/second aggregate bandwidth and cable costs of appropximately $20 per meter (2 bytes, full duplex).

10.5.   Copper vs Optical: Technology Trade-Offs

Computer communications can be viewed as a hierarchy of interconnects:

1. Intrachip connections between transistors, gates, and functional block
2. Interchip connections, ranging from direct chip to chip on a single circuit board to bused connections between multiple chips on multiple circuit boards
3. Rack-to-rack connections, for closely packed racks in large machines within a single machine room
4. Buildingwide or campuswide connections over a LAN
5. Intercampus connections across a metropolitan area network (MAN)
6. WAN machine connections with a large city or between cities

The general characteristics of these types of interconnects are shown in Table 10.2. The actual numbers are approximate and will not of course be equivalent for a personal computer and a supercomputer; however, with

Table 10.2 **Hierarchy of Computer System Interconnect Technologies**

Interconnect level	Technology options	Bit rate per line	Signal lines per connection	Line distance
Intrachip	Metal on CMOS	50–1000 Mb/s	64–256	<25 mm
Chip to chip	Metal on circuit board	30–1000 Mb/s	32–128	1–100 cm
Rack to rack	Twisted pair, coax, multimode fiber	20–200 Mb/s	16–54	1–10 m
LAN	Twisted pair, coax, multimode fiber	10–100 Mb/s	1–10	10–1000 m
MAN	Single-mode fiber, multimode fiber	56 kbps– 1.544 Mbps	1–10	1–10 km
WAN	Single-mode fiber	2488 Mbps	1	>10 km

the increasing convergence between long-end and high-end computer technology, the values are correct over a perhaps surprisingly wide range of computing systems. The basic electronics that drive each of these levels is approximately the same. To within a close approximation, the speed of the transistor that drives a wire to the other side of a very large-scale integration (VLSI) chip is the same as the speed of the transistor that drives a coaxial line or a laser driver.

A number of factors determine whether a given level of interconnect is implemented in electronic or optical transmission technology. These include the following:

1. Optical data transmission technology is, in general, more expensive than electronic data transmission because of the greater costs incurred in electronic/optical/electronic conversion. For the longest distances (multi-km), the cabling and repeater costs of copper outweigh the optical module costs, and optical technology becomes cost-effective.

2. Independent of technology, the cost of an interconnect increases dramatically with distance—for example, a single interchip wire costs less than a penny, whereas a LAN cable costs tens to hun-

dreds of dollars and a WAN line costs thousands of dollars. Copper wire is more cost sensitivie to a data rate · distance figure of merit.

3. Signals are degraded over increasing distance by attenuation, distortion, and cross-talk. These factors are more significant for electronic data transmission than they are for optical.

4. The density of interconnect wires for intrachip and chip-to-chip interconnects is much higher than that for the other levels, driven by the small size of CMOS chips. Electronic interconnect densities can be higher than optical interconnect densities at shorter distances but are lower at longer distances due to cross-talk penalties.

This combination of factors has historically led to optical data transmission technology being used at the higher levels of interconnect (WAN, MAN, and, increasingly, LAN), at which the distance · bandwidth product of the signal wires favored the cost of optical technology. Electronic data transmission has been used at more tightly integrated levels, where the distance · bandwidth requirements have not been quite so high.

With increasing clock speeds in CMOS, and increasingly distributed computing algorithms requiring high communication-to-computation ratios, the bandwidth · distance product for lower levels interconnects is increasing, which increasingly favors optical data transmission technology at lower levels of computer system interconnects.

10.5.1. PACKING DENSITY

Fiber technology in larger distances has been driven by the increase in packing density relative to copper cable interconnects. This has two types: density of fiber connectors vs multiline copper connectors and density of fiber cables vs copper cables.

The HIPPI network, which has been the de facto standard for supercomputer interconnection, uses a 50-pair shielded copper cable for distances up to 25 m. Although adequate for wiring within a machine room, this cable cannot be carried through a building. Fiber extenders have recently been demonstrated that run the same data rate over fiber cabling.

10.5.2. DISPERSION

Pulse dispersion is a basic advantage of optic over electronic data transmission. On copper interconnect, whether on circuit board, twisted pair, or coaxial cable, the propagation speeds of different frequency components of a data pulse are different. A data pulse, which contains a range of frequency components, will then spread out as it passes down a transmission line. In optical transmission, there is still frequency dispersion (in fiber cables, it is referred to as chromatic dispersion), but the amount is lower by several orders of magnitude.

10.5.3. POWER

At extremely short distances (intraboard), the power requirements of a copper line driver and a fiber line driver are approximately equivalent—both drive about 10 mA at 3.3 or 5 V.

With higher distance · bandwidth product interconnect requirements, the effect of the much higher frequency-dependent loss in electronic cables vs optical increases, greatly increasing the difference in drive current required for electronic vs optical data links. Also, for electronic data transmission, distortion-correcting circuitry needs to be added to link interface chips, further increasing electrical driver power consumption. With low-threshold lasers for optical line drivers, the optical drive current can be less than 10 mA.

10.5.4. SKEW

Skew is a factor in multifiber vs multiwire ribbons. Without special expense and selection, the skew in a copper cable can be as much as three 2-ns bit periods over 20 m, implying across-line skew of 6 ns/20 m, or 300 ps/m.

In fiber ribbons, on the other hand, the average skew for even unselected fibers is well under 10 ps/m, and selected fiber ribbons are available with skew of 1 or 2 ps/m. This factor mean the difference between requiring and not requiring skew-compensation circuitry.

Aside from the intrinsic fiber skew, there is also skew in the parallel fiber interconnects due to several other factors. Chromatic dispersion in the fibers, together with lasers in the array operating on a different wavelength, can result in bit skew at the receiver. Skew in the transmitter can be due to variations in laser threshold across the array coupled with a constant bias point from the driver.

10.5.5. DELAY PREDICTABILITY

In copper cables, the delay predictability is relatively low. As evident from the skew variations described previously, in copper the delay variability is up to 6 ns/ (20 m · 4 ns/m) = 7.5%. This means that the speed of propagation of a signal through twisted pair wire can vary unpredictably by 7.5%. In optical fiber, the variation in signal arrival time is between 30 and 200 times better. For some applications, this can make a significant difference.

10.5.6. COST

For the distances and bandwidth requirements discussed in Sections 10.3 and 10.4, it is clear that it is technically possible to build copper cables that will satisfy most current needs.

For interconnect distances within a frame, copper wiring is mature and the costs are very low by current fiber optics standards. It is not expected that optical interconnects will be cost-effective in this environment for approximately 10 years.

For interframe connections, the copper solutions meeting the performance and reliability requirements are not inexpensive—cables may cost several hundred dollars—but they are not beyond the bounds of practicality. There will be some applications in which copper interconnect solutions will be inadequate due to power or space requirements, but these will generally be of limited applicability.

Figure 10.2 represents a "snap-shot" of the relative costs of a copper bus and an optical bus of 500 Mb/s per line and 20 lines wide (nominally 2 bytes wide) as a function of the distance. It illustrates that as the distance

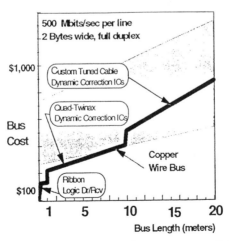

Fig. 10.2 Comparison of copper and fiber bus cost versus distance for 500 Mb/s per line, 20 lines wide, including modules and cables.

increases, the copper version will have to use more expensive cabling, eventually employing discrete passives into the cable to compensate for distortion, and the copper driver and receiver modules will have to employ signal processing to further correct for distortion. The optical modules will have the higher module costs associated with optoelectronic signal conversions, but the cables can be lower in cost with distance extendability.

At some distance to the right of Fig. 10.2 (at a length of <100 m), the copper wire bus cannot support the 500 Mb/s line speed, and a signal regenerator will be required to redrive the copper bus at a significant increase in price.

The reader who is also a user is forewarned that the costs and performance of both copper and optical technology are dynamic, even though copper is a relatively mature technology compared to optical. Historically, the cost of copper components have dropped approximately 20–40% per year. The reader who is also an optical technology developer is also forewarned: Optical interconnects for SANs must compete with copper solution primarily on a cost basis. The historical trail of optical interconnects for computer applications is littered with the bones of exotic, overdesigned, and overpriced optical solutions waiting for the "killer app" to give them life.

10.6.   Parallel Optical Interconnect Hardware

POIs seek to exploit the advantages of optical technology for the interframe and intraframe applications. The fundamental advantages of parallel connections are (i) the potential of higher line density for distances of 1–500 m and lower cable bulk, (ii) higher per line bandwidth, (iii) freedom from susceptibility to electromagnetic noise or generation of electromagnetic noise (e.g., through frame openings for connectors), (iv) DC electrical isolation between frames, and (v) simpler electronic interfaces to the interconnect (due to lower power requirements and less need to correct signal for the distortion, skew, and cross-talk associated with electrical cables). Disadvantages compared to parallel electrical interconnect are (i) extra optoelectronic conversion chips are required (more card space consumed and more card cost); (ii) the technology is new and has not demonstrated the reliability and robustness of electrical interconnect; and (iii) many implementations of the technology are significantly more expensive than electrical options, both in parts and installation costs. The main differences in the optoelectronic technology for POI compared to those for the duplex links covered in other chapters of this handbook are the greater requirement for high-density chips and packages and the lack of a requirement for long-distance (approximately >300 m) lines. Whereas single or duplex line optical links are built from discrete OE devices and low lead count packages, POI technology is fundamentally an integrated technology [6]. It thus represents a new, advanced generation in optical interconnect technology.

POI technology has been pursued in research labs for more than a decade because the challenges of integrating with optoelectronics and optics are formidable. Lately, there have been chip and package technology, breakthroughs leading to a rapid introduction of POI into the marketplace and the potential of becoming cost competitive with high-speed parallel copper interconnects, while offering 10 times smaller and lighter interframe cables and 10 times longer link distances than copper can support. As was emphasized in Section 10.5, cost competitiveness is essential to the use of POI and should be a major consideration in making the performance trade-offs discussed in this section.

This section will review the relevant component design issues and the status of the components. The section will focus on component properties relating to integration, arrays, and uniqueness to parallel applications. The reader can learn general component characteristics from Part I of this handbook.

Figure 10.3 illustrates schematically the elements of a parallel optical interconnect technology. The module may have transmitter or receiver functions or both. There are many parallel electrical data and control lines on and off the module. The lines can be bit synchronized to a clock line; this interface is called synchronous and is usually measured in multiples of bytes (0.5–4 bytes wide have been demonstrated in various prototypes). The SAN application would use synchronous lines. The lines can also be asynchronous (no timing relation between the lines). However, all lines generally operate at the same data rate and with the same protocols. The functions in the module may be simple optical to/from electrical conversions with associated device drivers/receivers, but most users want more functions to condition the interface and make it easier to use. Thus, there can also be functions for logic level receivers and drivers, adding transmission protocols to enhance the reliability of data transmission, data flow control functions, and maintenance, test, and safety functions.

The multiple fibers in the interconnect link are generally captured into one cable and connector structure and hence all lines must terminate at the same physical location. The technology can support associated fan-in

(a)

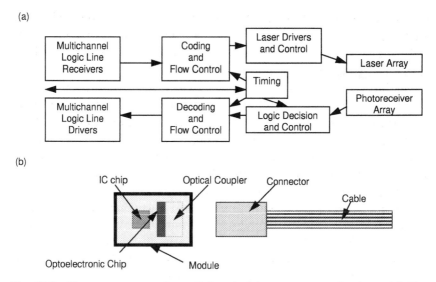

(b)

Fig. 10.3   The components of parallel optical interconnect technology. (a) The logical functions that might be contained in a POI module. (b) The physical components of one end of the interconnect.

and fan-out jumpers to support limited physical spreading of sources and sinks. The cabling technology also supports mid-span connectors, bulkhead connectors, and installation tooling and support.

The reader should be forewarned that the technology of POI is still new and immature. The devices, packaging, and tools continue to evolve at a rapid pace. There are no standards for the components on such fundamental issues as fiber type, connector type, cable pitch, light signaling wavelength, package electrical or mechanical footprint, or power supply. In this sense, this topic is premature for a "handbook." Nevertheless, there are significant resources (by both companies and governments) being put into developing POI for intramachine communications standards, and many system users are integrating optical options into their 1998–2000 product plans. This section will attempt to present an "up-to-the-minute" status (early 1997).

10.6.1.   OPTOELECTRONIC AND IC CHIPS

10.6.1.1.   Light-Emitter Chips

For cost-effective parallel applications, monolithic arrays of light emitters are required. Although arrays of discrete devices have been used in prototype to enhance the yield of arrays [7], the packaging and testing costs of this approach are unattractive. The design issues for the monolithic array are to achieve the required device speed, optical power output, efficiency, and noise characteristics consistent with high water-level functional yield of good die, power supply constraints, driver circuit constraints, and packaging constraints. High yield is achieved through device design (i) with large tolerance for fabrication process error, (ii) for uniformity of functional properties, and (iii) for simplicity in wafer fabrication processes. The laser diode device yield on a wafer must be very high to achieve a practical die yield for the wafer. For example, to achieve greater than an 80% yield for a 2-byte-wide (e.g., 20-device array) die, the individual laser device yield would have to be more than 99% on the wafer.

The emitter device could be either an injection laser diode (LD) or a light-emitting diode (LED). Because the modulation speed of lasers is at least 10 times higher than that of LEDs, and the electrical to optical power conversion efficiency is at least 10 times greater, the LD is the preferred device to fabricate into monolithic arrays, despite the added complexity in LD device processing compared to the LED.

Both edge-emitting and surface-emitting laser arrays have been used in POI prototypes and products. The advantages and disadvantages of each laser type for array applications are listed in Table 10.3.

The lowest cost light-emitter today is the edge-emitting audio disk laser (at less than $1 per packaged device), but unfortunately it has not demonstrated the ability to yield monolithic arrays, multigigabit/s speeds, or high-power conversion efficiency. The volume application for this low-cost emitter is moving to visible (red and eventually blue) wavelengths with only a modest increase in speed. Unfortunately, this is incompatible with low-loss fibers.

The most significant recent development in communications emitters has been the emergence of the vertical cavity surface-emitting laser (VCSEL) from research into production. Although complex in fabrication, the high functional array yields demonstrated, efficient light conversion, wafer-level testability, and symmetric beam have generated considerable enthusiasm in POI suppliers, and products have been introduced using monolithic arrays of this laser.

In synchronous bus operation, the emitters in the array are driven by a bit-synchronized data pattern, which is commonly resynchronized at the receiver using a common clock. Thus, skew between the optical outputs of the array is important. The skew depends on uniformity of lasing threshold, laser turn-on characteristics, and laser bias conditions and is also dependent on the data patterns applied, the temperature, and applied voltage varia-

Table 10.3   **Comparison of Edge- and Surface-Emitting Lasers for POI Applications**

Property	Edge-emitter array	Surface-emitter array
Wavelength (nm)	Made in many material systems, at all communications wavelengths; $\lambda = 650$, 780, 850, 1300, and 1500 nm, but with low to moderate yield	Primarily a GaAs device; 650 nm (low yield), 850 nm (high yield), and 1300 nm (very low yield)
Speed	Tens of GHz	Comparable
Optical beam	Asymmetric and divergent, difficult to couple to fibers; in the plane of the chip, good for low-profile package	Symmetric and low divergence; easy to couple to fibers; normal to chip surface, needs beam redirection for low-profile package
Electrical to optical power conversion efficiency	Moderate (<10%)	High (<40%)
Modal characteristics	Single or multimode	Comparable
Processing	Complex heteroepitaxy growth, regrowths in long wavelength devices; simple or complex postprocessing depending on device; processing on die edge required for passivation	Very complex heteroepitaxy growth, with atomic dimension thickness control required for many layers; simple postprocessing
Temperature sensitivity	Moderate, $\lambda = 780$–850 nm; high, $\lambda = 1300$ nm	Low, $\lambda = 850$ nm
Power consumed	20–100 mW per device	2–10 mW per device
Optical coupling to fiber	Optical lensing component required	Can be butt-coupled
Testing	Must be cleaved into bars to be tested	Automated wafer-level testing demonstrated

Table 10.3 *continued*

Property	Edge-emitter array	Surface-emitter array
Wafer yield	Low to moderate depending on material and structure	Greater than 99.5% in early results at 850 nm
Reliability	High reliability (MTTF > 10^6 h) demonstrated	Insufficient reliability data presented, but early results are encouraging for MTTF > 10^6 h
Cost	Low, $\lambda = 780$ nm; moderate, $\lambda = 850$ nm; high, $\lambda = 1300$ nm	Data not available; should be low

tions. The link skew also depends on the uniformity of laser wavelength across the array because this leads to propagation delay variations. Thus, the laser choice for a specific transmitter is not straightforward. The reader is referred to Ref. [8] for further details about the design trade-offs.

The emitter module for a parallel optical interconnect must satisfy laser safety as defined by the OSHA and International Electrotechnical Community (IEC) class 1 criteria for eye safety, as discussed in Section 10.4. However, the entire optical output of all emitters must be considered. The maximum permissible power depends on the safety standard definition and the emitter wavelength, and the reader is referred to the standards documents themselves for exact definitions [9]. In the case of arrays, the light source is considered "extended" (i.e., not a point source), which leads to special considerations for safe power levels. In general, the total array power must be less than approximately 600 μW for 850-nm wavelengths and 1000 μW for 1300-nm wavelengths.

10.6.1.2. Light-Receiver Chips

As was discussed for light emitters, the photodetectors used in the POI receiver are generally made in monolithic arrays. Two types of photodetectors are used: the positive–intrinsic–negative (PIN) photodiode and the metal–semiconductor–metal (MSM) photodiode.

The MSM photodiode has some unique advantages for monolithic arrays. Because the capacitance per unit area of a transverse structure MSM detector is low, a fast photodiode with large area can be made. This is an important feature for the MSM arrays, in which it is important to have large tolerance to misalignment of the optics from the chip to keep module assembly cost down.

Because the optical and thus photoelectric signals are low (measured in μA of photocurrent for μW of light power), the data rate is high, and channels are on a small pitch, there can be considerable cross-talk between the channels in a parallel array. This makes it advantageous to integrate an amplifier on the chip with the photodetector. This is called an integrated photoreceiver. Other advantages of integrated receivers in the array include the ability to do wafer-level testing and screening and the uniformity of channel-to-channel performance.

A comparison of a typical PIN photodiode array and a typical MSM photoreceiver array is made in Table 10.4. In serial Gb/s links, the integrated

Table 10.4 **Comparison of PIN and MSM Photodetectors in Monolithic Arrays**

	PIN photodiode array	*MSM photoreceiver array*
Bandwidth	Tens of GHz	Same
Photoresponse	Fraction of a μA of current output for each μW of light input (depends on wavelength)	Logic voltage swing output for μW of light input
Wavelength	650–850 nm using Si or GaAs; 1300 nm using InGaAs	780–850 nm using GaAs
Interchannel cross-talk	Dominated by mm-long wirebond cross-talk	Dominated by chip substrate isolation
Device structure	Vertical p–n junction device; mature device for data link products	Horizontal Schottky barrier device; compatible with MESFET IC processes
Testing	Screened at wafer, functional at module	Functional at wafer
Cost	Moderate for InGaAs (1300 nm) arrays; low for Si (780–850 nm) arrays	Low for GaAs-integrated receiver arrays (780–850 nm)

receiver has a lower cost than a 1300-nm photodiode but a more expensive cost than a Si photodiode. However, a significant portion of the cost of a hybrid (photodiode + amplifier) module is in the assembly and test of the two chips.

10.6.1.3. Receiver and Transmitter ICs

The support chips used in transmitter and receiver must supply both digital and analog functions at high speed. The transmitter chip must receiver multichannel logic-level signals from off module—data, control, and clock(s). It provides the bias and analog modulation circuits for each laser channel. It may add coding, to improve transmission integrity, and additional link control functions. The receiver chip amplifies the photocurrent; makes a logic decision using a timing signal that has either been extracted from a data channel or provided on a separate line; and drives logic-level signal off the module.

As mentioned in Section 10.4, it is desireable from both a system latency viewpoint and an ease-of-use viewpoint to provide media-dependent drive and control functions and reliability functions on the chips in the POI modules. Most current prototypes and products do not provide functions beyond optoelectronic signal conversion and some amplification. This is done for the convenience of the POI module supplier, who can then make one module fit many applications. The system integrator then must develop or purchase support modules to provide the link interface functions and develop cards to handle high-speed analog lines and provide special power supplies. The trade-off is design time and cost savings for the POI module supplier versus design time and cost savings for the SAN subsystem user.

As mentioned previously, there are no standards for the functions provided in a POI interconnect, and variations of all the previously described functions are to be found in prototypes and products. Si bipolar and CMOS and GaAs MESFET IC technologies have all been used. General issues under consideration today include the following:

1. Should the links be AC- or DC-coupled optically? DC coupling allows unconstrained codes to be transmitted and requires simpler circuits. AC is more reliable when there are long run-length data patterns, bias control variations, and adverse environmental effects.

2. Can CMOS ICs provide all electronic functions or are there advantages to bipolar ICs?

3. Should the functions be designed as custom or application-specific integrated circuit (ASIC)? To keep the cost of the chip (both in

design and in fabrication) competitive, it is best to choose among ASIC technologies, although many prototype POI modules still use custom designs.

4. An important consideration is the supporting design and test tools for the chip technology chosen, both the accuracy of the device models (especially OE device models) and the ability of the tools to handle both digital and analog simulation. The state-of-the-art in optoelectronic device and circuit simulation tools is far behind the electronics industry expectations at this point in time.

The previous topics are discussed in detail in Ref. [6].

10.6.2. OPTICAL MODULE PACKAGING

10.6.2.1. Optical Components in the Module

The transmitter and receiver modules of a parallel optical interconnect have optical components inside to couple the light emitter and detector chips to the fibers of the cable. The optical component should be multielement to couple all channels with one alignment operation and keep package assembly costs low and module size small. The optical coupling mechanism can be reflective, refractive, or light guiding, and all have been utilized in prototypes. The optical coupler is often used as part of an environmental barrier (e.g., hermetic seal) between the inside and outside of the module.

The fiber optics industry has historically focused on developing single-channel couplers based on miniature optical elements of high precision and high-coupling efficiency and with little regard to cost. Multichannel optics compatible with IC module packaging is only just emerging. Likewise, IC module technology has developed materials and processes without regard for optically transparent materials or optical alignment tolerances. Thus, packaging optical elements into IC packaging will likely be the most expensive part of the POI technology.

The issues that influence cost and performance for the optical coupling elements include the following:

1. *Alignment tolerances:*   As discussed in the other chapters, single-mode fiber optics required submicrometer tolerances for package alignment, multimode fiber optics requires tolerances of a few micrometers, and "large-core" multimode fiber optics requires tolerances of tens of micrometers. When multichannel couplers are required, these tolerances must be maintained over dimensions of

tens of millimeters. This creates a challenge for the fabrication and stability of the monolithic optical coupling element. The lowest cost approach is to use a coupler with greatest tolerance to misalignment consistent with the coupling requirements.

2. *Coupler material and fabrication processes:*   Glass microoptics (lens or reflector) has good optical and mechanical properties but it is difficult to fabricate in multielement arrays and expensive to finish. These couplers usually must be assembled into other structures in order to align them to the OE chip. This adds cost. Glass-filled plastics can be used in a transfer-molded process to make monolithic couplers (also with single-mode precision). The lowest cost approach is to use injection-molded transparent plastic to make an array of coupling elements in a monolithic structure containing the mechanical alignment features. There is no assembly of this coupler. Injection-molded plastic parts are commonly used for consumer and PC applications when low cost is the objective. The challenge is to find transparent materials with correct mechanical and thermal properties for easy molding and durability.

3. *Coupler assembly into module:*   The coupler must be aligned to the emitter and detector chips and to the fiber array (either in the pigtail or in the connector ferrule). The most precise approach is to use active alignment; that is, manipulating the coupler into alignment with an OE device while it is operating and then fixing it into the optimum position. This is generally employed for single-mode systems, but it is time-consuming and expensive. In multimode systems, the coupler can be passively aligned to the chip. Though generally less expensive, this can also be time-consuming due to the number of optical elements to simultaneously align and the number of degrees of freedom in the alignment (both rotational and positional). In some modules, solder bump reflow is used to passively align the chip to the optical component (see Chapter 17). With "relaxed-alignment tolerance" optical components, self-alignment (or snap-together alignment) of components has been demonstrated as the lowest cost approach.

4. *Light bending in small dimensions:*   For multichannel couplers it is desirable to keep channels on a small pitch to conserve chip real estate, but small dimensions and tight pitch are more expensive to implement. To keep the packages profile low in modules in which

the OE chip is mounted on the carrier along with IC chips, a 90°
light turning must be made in a few millimeters, but this implies reflective optics [extra coatings or high numerical aperture wave-
guides], which can adversely affect fiber bandwidth and coupler
loss. If the coupler does not bend the light, i.e., end fire coupling
between the OE device and the fiber, it is generally simpler to fabricate and has lower coupling loss. Unfortunately, for surface-
emitting/detecting devices, this implies that the OE chip must be
mounted perpendicular to the main IC module and the high-speed
electrical traces must be carried around package corners (EMI,
cross-talk, and reliability exposure). This also adds another package layer to the module.

5. *Compatibility with IC chip carriers:*   If the coupler can be
   aligned with the OE chips using standard IC chip carrier features,
   then a special OE submount does not have to be developed, and
   added assembly and testing steps can be avoided.

Most optical couplers are derivatives of the three types described and
illustrated below: type 1, an array of discrete microoptic elements (e.g.,
fibers) affixed into a precision V groove carrier; type 2, a monolithic array
of plastic elements (e.g., lenses or reflectors); and type 3, a multielement
planar lightguided structure.

Type 1:   The V groove with microoptics coupler consists of a piece
of silicon wafer with precision-etched grooves on one surface. Fibers
are laid in the grooves, thus aligning them with each other, and epoxied. The fibers + silicon subassembly is then end-polished at a 45°
angle so that light can be reflected into or out of the fiber from an
OE device above the subassembly. The end face is coated with metal.
Single-mode tolerances can be held over array dimensions up to approximately 5 mm; coupling loss of less than 3 dB can be obtained;
but the assembly costs for this coupler are high. This is illustrated in
Fig. 10.4(b).

Type 2:   The injection-molded coupler is fabricated by injecting plastic into a mold that has the desired optical and mechanical features—an array of microlens dimples, an array of reflectors, or an
array of lightguides. This can be the lowest cost approach to making
optical elements (by 10 times), but for the small dimenions and pitch
of the array considered for POI, there is a challenge in making the
mold with optical precision over distances up to 1 cm and keeping the

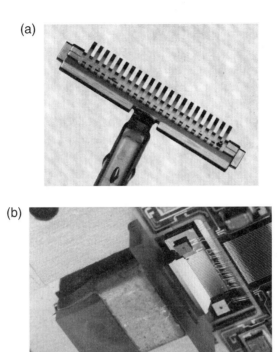

Fig. 10.4 Photographs of two optical couplers for parallel optical interconnect modules: (a) Jitney coupler and (b) OETC coupler.

mold clean. The part cost in this process can be very low (<$1) as long as assembly into a submount is not required.

Type 3: The lightguide coupler can be made by molding but can also be made by laminating or depositing a blanket of the plastic or glass lightguide material on a substance (sometimes containing mechanical alignment features) and photolithographically processing the light-guides.

Figure 10.4 shows two of the optical coupler types—type 1 and type 2. The Jitney coupler is 22 reflector vanes (on a 500-μm pitch) of metallized, transparent plastic. Each vane preserves VCSEL divergence while bending the light path by 90°. Note the "ears" on the end for passive registration into the package.

10.6.2.2. Modules for Array Optoelectronic and IC Chips

The chip carrier modules used for POI have generally been derivatives of IC chip carriers. Lead-frame and pin grid array carriers were used in early prototypes, but the ball gride array is the direction many suppliers are headed. The IC industry is increasingly using plastic packages for low cost, but the mechanical dimensional control and rigidity are concerns for adaptation to optoelectronic array packages. Many optical chip carriers still use more expensive ceramic or metal materials. The technical issues for a good optoelectronic chip carrier are the following:

1. *Module electrical lead bandwidth:* The applications demand data rates per line from hundreds to thousands of megabits/s, leading to low inductance, impedance control, and high electrical isolation requirements for the leads between module and card. Generally, differential signals are utilized, with generous use of grounding and power leads. Surface-mount techniques are best for high-speed electrical signal integrity in leads on and off the module. Through-hole packages (e.g., pin grid arrays) can have cross-talk arising from the lack of close proximity return paths for the signals. Multiple levels of signal and power planes are advantageous for signal integrity in the package but add to module cost.

2. *Optical alignment capability:* If optical components permitting tens of micrometers of misalignment are used in the module, IC packages can have mechanical alignment features (pins and molded surfaces) included in module fabrication. However, for alignments of submicrometer or a few micrometers, a separate level of package—the optical subassembly (OSA)—is needed. In this OSA subpackage the precision alignments are made, and the OSA is electrically attached to the module. OSAs use grooved Si wafer "optical benches" or small leadframes with embedded lightguides. The advantage of this approch is that it partitions the electrical and the optical/mechanical functions. The disadvantage is that it increases package cost by adding many more parts, assembly processes, and testing steps.

3. *I/O count for electrical interface:* The POI module must support enough I/O pins for multibyte data transfers. This excludes discrete device packages (e.g., TO can) or common OE packages [e.g., 1×9 fiber distributed data interface (FDDI) carrier]. Peripheral

leaded packages have 20–150 leads and can support up to approximately 2- to 3-interfaces of approximately 500–700 Mb/s per line. Arial I/O packages (e.g., BGA) have 100–1000 leads and can support wider interfaces or full duplex interfaces, and at speeds greater than 1 Gb/s/line. All these options must be balanced against the higher cost of high lead count packages.

4. *Mechanical strength and rigidity:*   The rigidity of the package for ensuring optical component alignment was discussed previously. With an OSA the rigidity is put into the submount, whereas direct OE chip attach approaches use metal heat slugs or rely on ceramic chip carriers. The trends in low-cost IC packages are to use plastic chip mounts, however. Although some POI modules have been made and installed with optical fiber "pigtails," the approach adds greatly to the expense of card assembly and handling and card robustness (due to the dangling short optical leads). Many parallel link users will not accept such an approach for their applications (especially because all other IC modules on the card employ surface mount for electrical connection and a card-edge connector socket for cables). Thus, the POI module should also include an optical socket for a cable. This adds the mechanical requirement that the module must strain-relieve the cable attached to the module when it is on the card. Some applications require rear-mounted connection of the card to the cable or board (blind connect) and this influences the mechanical module design.

5. *Cooling:*   Virtually all applications are in an air-cooled environment (0–200 linear ft per minute) with forced air of 0–65°C ambient temperature.

6. *Footprint:*   The applications are generally in a dense card part of the system, requiring the optical module to be compatible in size to an IC module or an electrical card-edge connector (i.e., approximately 10 × 40 mm for a 1-byte-wide, full-duplex connection).

10.6.3.   *FIBER OPTIC RIBBON CABLES AND CONNECTORS*

10.6.3.1.   Cables

The cables for POI are generally intraframe harnesses, jumper cables between frames, or installed cables in a raised-floor or horizontal building environment. The installations are dynamic in the sense that there is continual rearrangement (30% per year) of the interconnect system as nodes are added or equipment is upgraded. The downtime tolerated for installation or upgrading of the interconnect network is measured in hours. Thus, a flexible and easy to install cabling solution is a requirement. For raised-floor installations, sufficient durability to withstand the stress of power cables laying on top, impact from floor tiles, etc. must be considered.

The cables are generally connectorized at a factory, although the ability to field-install connectors must be supported. The cost of the cables is especially important because they are generally not amortized over many system upgrades or many users. They must be competitive with the copper cables used in today's systems, even though they may offer more performance.

The parallel optical cables and connectors used in early prototypes were taken from the telecommunications industry's trunking applications. Although the performance and durability was good, the cost has been very high (5–10 times that of equivalent copper solutions). This has caused a number of system developers to reject optical interconnect solutions in favor of developing more advanced copper cabling systems (including signal processing at the driver and receiver to enhance data integrity). Fortunately, there are now cost-effective parallel interconnect cables and connectors emerging from development that focus on the data communications installation environment, limited distance requirements, and pricing structure (both parts and installation costs).

Three fiber types have been used in POI cables: the large-core step and graded-index fiber, the data communications standard multimode fiber, and the telecommunications single-mode fiber. The cable characteristics are compared in Table 10.5. Plastic optical fiber is also a good candidate for low-cost, jumper cable-length connections, but to date there have been no ribbon cables developed.

Note that the bandwidth and loss characteristics in Table 10.5 are for full-mode excitation or equilibrium light propagation conditions. In short-distance links (the definition of "short" length depends on cable type—generally <100 m, but sometimes more than 1 km) under laser excitation (VCSEL or edge-emitting), mode equilibrium is almost never established (even in installed cables) and the effective bandwidth is higher and connector/cable attenuation is less [10].

The large-core fibers have the most relaxed alignment tolerances, permitting plastic connectors and the lowest cost assembly. The step-index fiber bandwidth has been shown to be enhanced with VCSEL laser excitation

Table 10.5   **Comparison of Fibers used in Ribbon Cables**

Property	Large-core multimode	Datacomm multimode	Single mode
Core/cladding dimension (μm)	200/230	62.5/125	9/125
Alignment tolerance (μm) (1 dB added connector loss)	80	5	1
Fiber bandwidth per line (MHz/km at λ)	SI, 50 ($\lambda = 850$) GI; 350 ($\lambda = 850$)	SI, 165 ($\lambda = 850$) GI, 500 ($\lambda = 1300$)	SI, 2000 ($\lambda = 1300$)
Fiber loss (dB/km)	4	3 ($\lambda = 850$) 0.8 ($\lambda = 1300$)	Same
Cable pitch (μm)	500	250	250
Cable dimensions (mm)	1×14 (20 fibers)	2×8 (12 fibers)	2×8 (12 fibers)
Maximum cable distance (m)			
500 Mb/s/line	40	300	No data
1 Gb/s/line	30 (VCSEL excitation)	300 (Skew limited)	
Skew	SI, 1.1 ps/m GI, 4–5 ps/m	GI, 4–5 ps/m GI, 2 pm/m (selected fibers)	
Relative cable cost (20-fiber cable)	$7–10/m	$10–15/m	$10–15/m

Note. SI, step-index core; GI, graded-index, full-mode excitation.

so that 40-m spans are possible at 500 Mb/s. Graded-index versions of this fiber support 500 Mb/s to more than 300 m [11], but their cost is considerably (more than two times) higher. The datacomm standard multimode fiber is used by most POI links and can cover both jumper cable lengths and building wiring spans (up to 300 m). Table 10.5 indicates current specifications, but there are datacomm multimode fibers coming on the market with enhanced short wavelength ($\lambda = 850$ nm) performance of more than 500 MHz/km. The cost of the ribbon cable types of Table 10.5 are approximately comparable (on a per fiber basis) and dominated by the environmental protection required. For interframe connections, round cables are generally preferred over flat cables, but flat ribbon cables are preferred in the frame.

There are no specific ribbon fiber standards for POI, but it is expected that the environmental requirements will be based on duplex cable requirements. There would be separate requirements for building wiring and for computer room installations. For the cable environmental requirements for building wiring and jumper cable requirements, the Fibre Channel Standard may be consulted [12]. For computer machine room installations, the ESCON cabling guide [4] is representative. However, it should be recalled that many of the SAN applications are designed as an entire system so that a standardized interconnect is not necessary (or may not even be desirable for business reasons). For system performance or cost leverage, ad hoc interconnects may make sense. This does not imply that interconnect robustness, integrity, or flexibility will be compromised, but only that the time-consuming process of standardization is avoided.

10.6.3.2. Connectors

The array connectors that have been developed consist of a ferrule, which maintains the relative position of the fibers with respect to registration surfaces, and a body, which provides the latching mechanisms and strain

Table 10.6 **Comparison of Multifiber Array Connector Types**

	Connector type			
	MTP	**mini-MAC**	**Jitney**	**Argus**
Material	Cast plastic	Si or cast plastic	Injection-molded plastic	Ceramic
Alignment tolerance	$<1\ \mu m$	$<1\ \mu m$	$<10\ \mu m$	$<1\ \mu m$
Maximum fiber count demonstrated	12	32	20	18
Release mechanism	Push–pull	Push–pull	Side latch	Push–pull
Dimensions	12.5×7.5 mm	12×8.6 mm	24×5.5 mm	14.4×7.2 mm
Relative cost (including assembly)	$10\times$	$10\times$	$1\times$	$3\times$

relief. Ferrules have been made from V-grooved silicon or ceramic pieces, cast-molded silica-filled plastic, or injected-molded pieces. Their characteristics are compared in Table 10.6. Note that for array connectors, dimensional control must be maintained over larger dimensions than for single-fiber cable connectors, and that there are more ferrule dimensions that must be maintained to tight tolerance (i.e., there is not circular symmetry as in the simplex connector ferrule).

For connector body-latching mechanisms, push–pull-type connectors are preferred, with a positive locking insertion. However, it is required that the connectors release prior to module damage or cable damage should they be subjected to strong pulls in field installation or use.

A major cost factor in ribbon cables is the cost of preparing the cable for connectorizing and the connector assembly procedure. Cables with

(a)

(b)

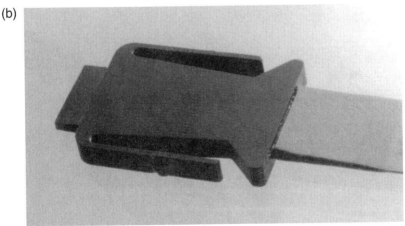

Fig. 10.5 Photographs of two ribbon connectors on their cables; (a) MT connector and (b) Jitney connector.

extruded plastic jackets are generally more expensive to connectorize because the coatings must be stripped off before connectorizing and reapplied to ensure mechanical integrity after connectorizing. Cables with laminated jackets can be made easier to strip, and it has been shown that ferrules can even be inserted during cable manufacture for very low-cost assembly. For large-core MMF, the ferrule dimensions can be controlled adequately to align to the module body. For datacom MMF or SMF, separate alignment pieces (e.g., pins or balls) are usually fitted to the ferrule to increase the precision of registration. This adds to the connector cost and assembly time.

Figure 10.5 shows two typical plastic connectors for ribbon cables; the MTP connector from US Conec, which must be assembled on the end of an extruded coating cable, and the Jitney connector from 3M, which has been fabricated into the cable during manufacture. The figure illustrates how the large core MMF cable can be connectorized using the ferrule alone as an alignment surface, whereas the datacomm MMF requires the use of steel pins for precision registration.

10.6.4. PARALLEL OPTICAL LINK DESIGN AND TESTING TOOLS

The parallel optical link designer uses optical power budgets to determine the transmitters, receivers, link spans, and number of connectors to be used in implimenting the network, just as does the serial link designer. Effective power loss penalties are added for link noise, data pattern distortion, and environmental effects. Specific penalties to consider for parallel interconnects are the effects of interchannel cross-talk statistical variation of devices across the array, and skew. An example of a typical POI power budget is shown in Table 10.7. Note the wide range in dynamic range required for the optical receiver due, in part, to the inaccuracy in accounting for line noise, distortion, and cross-talk. These power budgets are, at best, a rough (and, it is hoped, conservative) estimate of link field performance.

The datacommunications fiber optic industry suffers, in general, from the lack of system modelling tools to aid in design. For POI, with its array of interacting components, determining the cost-effective component specifications using "power budgets" or prototypes adds considerably to the design time for the network, often forcing designers to build excessive margins (and hence component cost) into the network. This situation must change if computer system network designers are to consider optical interconnects as easy to impliment as are copper wire interconnects.

Table 10.7 **Typical POI Optical Power Budget**

Transmitter Budget VCSEL Link	*Best Case*	*Typical*	*Worst Case*
Optical output power [dBm] (10 mA peak drive)	3	3	0
Coupling loss to fiber	−1	−3	−3
Convert to average power	−3	−3	−3
Total transmitted optical signal (power)	−1	−3	−6
Link loss [dB]			
Optical connector loss	−0.3	−0.5	−0.7
Number of connectors	0	2	4
Coupling loss to detector	−1	−2	−3
Optical crosstalk		0	
Fiber attenuation (dB) (at 4 dB/km)		0.2	0.2
Total link losses [dB]	−1	−3.2	−6
Total received optical signal power	−2	−8.4	−12
Receiver penalties [dB]			
Extrapolation BER = 10^{-12} to 10^{-15}	1	1	1
Electrical crosstalk	0.5	1	1
DC offset			
Modal noise penalty	0	0.5	0.5
Jitter penalty	0	1	1
Dispersion penalty			
Skew penalty	0	1	1
ISI penalty	0	0	1
Total penalties	1.5	5	6
Receiver sensitivity × [dBm ave.] BER = 10^{-12}	−20	−19	−18
Link margin [dB]	16.5	5.5	0.5

Though there are no CAD POI simulator products on the market at this time, a number of companies and universities have POI simulation tools [13]. These tools allow much more precise estimations of link performance versus component specification than either power budgets or simple spreadsheets. For example, the IBM or University of Illinois tools include interchannel noise and distortion effects and statistical variations of component parameters (to include manufacturing and environmental effects). Figure 10.6 illustrates how one of these tools can be used to establish the optimum

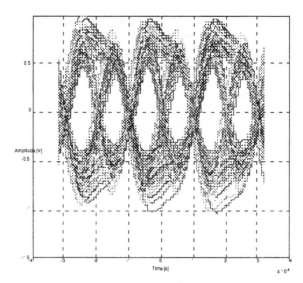

Fig. 10.6 Superimposed eye diagram from 18 separate receiver channels synchronized to a common clock at the transmitter.

decision time for all 18 channels of a synchronous bus, given the statistical variations of all the components and the cross-talk in all channels of the bus.

For POI components and links to be efficiently characterized and qualified for computer interconnects, automated testing tools and testing methodology still need to be developed and standardized. For example, IBM developed prototypes of such automated testers during the ARPA OETC and NIST-ATP Jitney programs, and tens of thousands of channels of receivers and transmitters were characterized at the wafer and module level [14]. Testing times were reduced to 4–7 s per channel, but this still added $5–20 to the cost of the module. Clearly, rapid screening test for manufacturing (including optical, electrical, and mechanical screens) will be required.

10.7.   Examples of Parallel Optical Interconnect Links

10.7.1.   INTERFRAME INTERCONNECTS

There are a number of product and prototype parallel optical interconnects reported, and new prototypes are being introduced regularly. Table 10.8 provides an overview comparison of 10 examples current in the literature.

Table 10.8 Comparison of 10 POIs

	Name	Maker	Status (1997)	Comments	Performance	Relative expected cost [1–5 (highest)]
1.	Optobus	Motorola	Product	2nd-generation product	400 Mb/s/line, duplex link	2 ($100/Gb/s)
2.	MDS2101/ MDR2101	Hitachi	Product prototype	Prototype in field (1993)	250 Mb/s, 100 m, simplex link	5 (>$1K/Gb/s)
3.	POLO-2 [15]	HP, DuPont AMP, SDL, USC,	Technology prototype		1 Gb/s/line, <300 m, duplex	4
4.	OPTM [16]	Oki	Prototype	ATM switch application	2.48 Gbit/s/line	5
5.	Ericsson POI [17]	Ericsson	Prototype	Intraframe applications	4 lines, 400 Mb/s/line	4
6.	OETC	AT&T, IBM, Honeywell, Lockheed–Martin	Technology prototype	Samples in field (1995), IBM and LLNL	1 Gb/s/line, 200 m, simplex	3
7.	NEC POI [18]	NEC	Technology prototype		1 Gb/s/line, >300 m (estimate), simplex	5
8.	Paroli [19]	Siemens	Technology prototype		1.2 Gb/s/line, 300 m, simplex	4
9.	Jitney	IBM,3M,Lexmark	Technology prototype		500 Mb/s/line, 40 m, simplex	1 ($15/Gb/s)
10.	P-VixeLink	Vixel Corp.	Product		1.25 Gb/s/line, 100 m, simplex	3 ($200 Gb/s)

The relative cost column is meant only to compare these examples based on technology complexity and infrastructure (an approximation of the cost/performance figure of merit is given); the current prices cover a much broader range and are marketing strategy-sensitive. Table 10.9 compares their functions, and Table 10.10 compares the technology used. There are currently three general design points for the technology.

The first class of POI subsystem is made from the long-wavelength device, single-mode fiber technologies developed by the telecommunications industry for their long-haul applications. These buses are capable of point-to-point spans of kilometers in asynchronous mode (skew would limit their synchronous mode distance) but the module costs are more than $1000 per gigabyte/second. This technology provides a data rate × distance figure of merit that copper interconnects cannot match. The SAN and most LAN applications do not need, and cannot afford, this technology.

By choosing this design point, the telecom industry suppliers hope to leverage their existing serial link technology, expand its markets, and reduce its high costs. Although very expensive (more than $1000 per gigabyte/second), as a method for efficiently packing multiple high-speed lines into compact cabling and module form factors, they will find a niche market. An inherent advantage of using single-mode technology is its coherence, which makes possible the eventual evolution to coherent optical routing and switching devices to form all-optical network. Examples of this technology class in Tables 10.8–10.10 are Nos. 2,4,5, and 7.

The second class of POI technology uses short-wavelength OE devices and datacom multimode fibers. The specified spans are for up to 300 m, forcing a requirement for graded-index fibers, and thus active alignment in the packaging. The design point data rates are 400–1250 Mb/s per line. This places their performance (as measured by the data rate × distance figure of merit) more than ten times beyond current copper interconnect capabilities. Unfortunately, their cost-effectiveness figure of merit is between $100 and $1000 per gigabyte/second, making them uncompetitive with copper at the distances that copper can reach; thus, it is expected that this approach will be a niche market until after the Year 2000.

Again, both the performance and the cost are beyond copper wire implementations, and the system designer is currently designing for the limits of copper. Suppliers choosing this technology class should try to motivate the system designers to move their system requirements beyond copper capabilities. In any case, it is expected that this second class of design point will eventually be required of POI subsystem for computers. Example Nos. 1,3,6,8, and 10 in Tables 10.8–10.10 are in this catagory.

The third class of POI technology also uses short-wavelength OE devices but drives even further toward low cost by using large-core fibers, plastic optical components, and passive assembly of the package. The performance design points are limited to data rates of 500–1000 Mb/s per line, and distances up to approximately 30–50 m. Estimates indicate that this class of POI is cost competitive with copper interconnects while offering approximately two times more in the (distance × data rate) figure of merit. Thus, this technology class may displace copper from the volume applications in the near term. However, it is limited in growth to higher data rate at longer distance, and it uses a nondatacom cabling, which may limit its acceptance by some users. A possible motivation of the suppliers of this technology class is to enter the applications market as soon as possible with a cost-effective solution, become a technology driver on the existing application road map, and then improve the technology along that road map's time-frame. Only one example (No. 9) is shown in Tables 10.8–10.10.

This section will not discuss all the examples of Tables 10.8–10.10 (the reader can refer to the references) but only mention some of the design motivations and features of a few significant examples.

10.7.1.1. Optobus

The Optobus is the first fully developed product on the market (as opposed to a marketing prototype) backed by a major company in the datacommunications field. It uses common multimode optics technology and common IC packaging technology within the computer industry generally, although there are notable exceptions. Optobus uses a novel, molded-plastic light-guide technology to couple light to and from the fibers, and a separate submount is required to mount the OE chips perpendicular to the IC package. The electrical interfaces are DC-coupled nonlogic interfaces, so the user is required to supply additional interface chips to make the link compatible with a logic standard bus. The modules provide only optical–electrical conversions. Therefore, all link flow control, clocking, and testing functions must be provided in support chips. This gives the Optobus modules the flexibility to be used in many applications, which, it is hoped, will increase the volume of modules needed. However, they are not expected to be particularly user-friendly to integrate into the SAN system. A photo of the Optobus module is shown in Fig. 10.7.

Table 10.9 Functions of 10 POIs

	Data rate per line	No. of lines (simplex)	Electrical interface	Optical interface	Mechanical interface	Function
1. Optobus	400 Mb/s	10	DC coupled Analog	MT connector 62-μm MMF		OE conversion
2. MDS2101/ MDR2101	250 Mb/s	9 data, 1 clock	CMOS	1-m pigtail, MT or MPO connector single-mode fiber 500-ps skew	14-mm (w) by 9 mm (l) by 6 mm (h) footprint 2.6-W power consumption	OE conversion
3. POLO-2	1 Gb/s	10	ECL	MT connect 62-μm MMF	Ceramic BGA 25 mm (w) $\times$ 75 mm (l)	OE conversion
4. Oki OPTM	311 Mb/s	12	ECL, AC-coupled Asynchronous	Pigtail to MPO 50-μm MMF	Ceramic/metal SIP 40 mm (w) $\times$ 100 mm (l) $\times$ 15 mm (h)	OE conversion Hot plug APC
5. Ericsson	400 Mb/s	4	PECL DC coupled	50-μm MMF or polymer lightguides MT connector	DIP	OE conversion Clocking Line coding

#	Name	Speed	Channels	Signaling	Fiber/Connector	Package	Functions
6.	OETC	1 Gbaud	32	ECL LVDS DC coupled	MAC connector 62-μm MMF	Ceramic QFP 33 mm (w) $\times$ 44 mm (l) $\times$ 5 mm (h)	OE conversion Clocking Line coding APC
7.	NEC POI	1 Gb/s	8	DC-coupled Analog	Pigtail module ? connector 62.5-μm MMF	Ceramic QFP 18 mm (w) $\times$ 25 mm (l) $\times$ 5 mm (h)	OE conversion
8.	Paroli	1.26 Gb/s	12	ECL-Tx CMS-Rcv. $-3.2, 5.2$ V	62.5-μm core MMF Pigtail to MT 350-ps skew	5 mm (l) $\times$ 10 mm (w) $\times$ 4 mm (h) 1.32 W	OE conversion
9.	Jitney	500 Mb/s	20	LVDS	Jitney connector 200-μm core MMF	Plastic QFP 30 mm (w) $\times$ 20 mm (l) $\times$ 7 mm (h)	OE conversion Cock Control Line coding Self-test
10.	PVL-4	1.25 Gb/s	4	PECL differential AC coupled	62-mm MMF Pigtail to MTP	PGA	OE conversion

Table 10.10 Optical Chip and Package Technologies Used in the POIs

	Name	Emitter	Receiver	IC	Package	Cable and card mount	Comments
1.	Optobus	$\lambda = 850$ nm VCSEL array	GaAs PIN PD array	Bipolar and CMOS	Plastic PGA Plastic lightguide OE submount	62-μm MMF MT connector	
2.	MCS2101/ MCR2101	$\lambda = 1310$ nm edge laser array	InGaAs PIN PD array	Si bipolar driver and amplifier	Si V groove submount for optics Active alignment Ceramic on metal stud	SMF MT pigtail	NTT low-skew ribbon
3.	POLO-2	$\lambda = 850$ or 980 nm VCSEL array	Si PIN PD array	Bipolar	Ceramic BGA Polymer lightguide couplers	62-μm MMF MT connector	
4.	OPTM	$\lambda = 1300$ nm 12-edge LD array	$\lambda = 1300$ nm 12-InGaAs PD array		Ceramic SIP Optical submount Microlens array Active alignment, welded	Pigtail to MPO 50-μm MMF	
5.	Ericsson	$\lambda = 1300$ nm 4-edge LD array		BiCMOS	Plastic DIP Si V groove chip carrier	Intraframe harness	Quasi-hermetic Synch and asynch types

376

		Source	Detector	Electronics	Packaging	Fiber/Connector	Features
6.	OETC	$\lambda = 850$ nm VCSEL array	MSM photoreceiver	CMOS	Ceramic QFP Si V groove submount optics Active align	62-μm MMF MAC connector 140-μm pitch	No control checks Wide dynamic range design
7.	NEC-POI	$\lambda = 1300$ nm 4-edge laser array (2 required)	InGaAs PIN PD		Flip-chip mount Si V groove carrier QFP, pigtail	62-μm MMF	
8.	Paroli	$\lambda = 850$ nm Edge laser array	InGaAs PIN PD	Si Bipolar for amplifier	Si V groove submount for optics	62-μm fiber MT pigtail Low-skew fiber (1.1 ps/m)	Technology to change to VCSEL
9.	Jitney	$\lambda = 850$ nm VCSEL array	MSM photoreceiver	Si CMOS	Plastic optical coupler Passive alignment QFP	200-μm MMF Low-skew cable (1.1 ps/m) 3M plastic connector	
10.	PVL-4	$\lambda = 850$ nm VCSEL array	MSM photoreceiver		Ceramic OE submount PGA module	62-μm MMF MTP connector	Wafer-testable chips Plastic optics

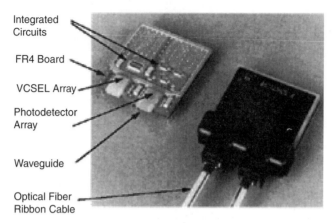

Integrated
Circuits

FR4 Board

VCSEL Array

Photodetector
Array

Waveguide

Optical Fiber
Ribbon Cable

Fig. 10.7   Optobus module.

10.7.1.2.   Jitney

The Jitney prototype is unique in that its design is driven by cost competitiveness with copper [11]. It uses modifications of standard datacom OE chips and computer IC packages. Because Jitney uses the large-core, relaxed-tolerance multimode fiber, it can also use standard plastic optic components (from the consumer/printer industry) and QFP packaging (from the consumer/computer industry). Jitney also develops significant manufacturing infrastructure in CAD and wafer-level testable arrays to drive the cost and design time down. It uses a unique approach to cable deployment, using a step-index, large-core fiber cable to achieve cost below copper wiring but limited to distances below 50 m, together with a more expensive graded-index large-core fiber cable for distances up to 300 m. Both cable types are mateable to the module. This approach may be attractive if the majority of cabling (e.g., >80%) is short jumpers (this is generally the case in machine room wiring, for example). However, the user is required to support two cable types (similar to the current use of copper). It remains to be seen whether SAN users will accept a cost-optimized cabling strategy for their optical interconnects.

The Jitney modules support logic interfaces and a number of self-test and flow-control functions. The electrical interfaces are DC coupled, but the optical lines employ a run length-limited transmission code to increase the robustness of the links to data pattern effects and component and environment variations. This type of coding has also been adopted for

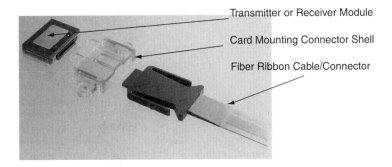

Transmitter or Receiver Module

Card Mounting Connector Shell

Fiber Ribbon Cable/Connector

Fig. 10.8 Jitney module and cable.

virtually all product serial data links. The Jitney modules and cable are shown in Fig. 10.8.

10.7.1.3. PVL-4

The Parallel Vixel Link is a product that falls in the second category of POI—using standard multimode fiber components. Its design approach is to maximize the optical line speed and minimize the number of optical lines; it employs only four channels (to keep cost down) at 1.25 Gb/s line speed. It requires 8B/10B line-coded data patterns (run length-limited and DC-balanced transmission code) to be transmitted to simplify its electronics and to improve optical line robustness. PVL-4 provides only light conversion functions. It is designed to be used with a companion set of multiplexer/demultiplexer chips to re-create a byte-wide, NRZ parallel electrical interface. One drawback of this implimentation is that it requires the SAN designer to support eight 1.25 Gb/s lines on his or her card (full duplex). This is more than twice as fast as that designed for the computer card industry's cost-effective technologies, but perhaps a rapid introduction of Gb/s Ethernet will change this drawback in the near future.

The interesting concept regarding the PVL-4 is the potential follow-up product that integrates the MUX/DEMUX function into the POI module so that the POI module electrical interface is optomized for SAN designer bus width and card speed, whereas the optical interface is optimized to take advantage of the superior optical line driver/receiver speed at a reduced optical I/O count. A PVL-4 is shown in Fig. 10.9.

The conclusion from the sampling shown in Tables 10.8–10.10 is that there is no clear consensus in the specifications from the POI suppliers,

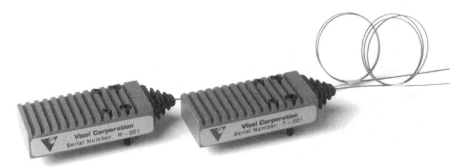

Fig. 10.9   PVL-4 module.

partially reflecting a lack of consensus on what customers want and partially reflecting the suppliers' history and marketing strategy. Currently, most suppliers are positioning their products ten times beyond the current copper wire offerings in the data rate × distance performance figure of merit. The functions in the modules are simple optoelectronic conversions because the suppliers attempt to cover as many users as possible with one module type. This also reflects the lack of user consensus in requirements. The cost of this technology will likely be two or three times that of copper. It remains to be seen, once field reliability has been demonstrated, whether SAN system designers will pay such a premium for optical interconnects.

However, some trends are emerging:

1. LVDS logic interfaces, with 3.3-V power supplies
2. 500 Mb/s, 1-byte wide, with a doubling in aggregate data rate approximately every 2 years
3. 300-m span (with the potential of a cost-reduced shorter span)
4. Surface-mount modules (probably BGA for multibyte-wide interfaces or full duplex from one module)
5. Simple OE conversion functions initially, migrating to complete physical layer function integration as standards develop or a volume user appears—This may be supported by a multiple module set initially, but the demand for smaller footprint will likely lead to the added function being integrated into the optical modules

10.7.2 INTRAFRAME INTERCONNECTS

There have been intraframe versions of POI technology developed (with government sponsorship: United States, ARPA; Europe, RACE). The main additional components are a ribbon wiring harness, which offers smaller connectors and more flexibility than a robbin interframe cable, and an interboard blind-mating card connector.

The OETC project developed a flexible ribbon harness to go with the OETC module (Tables 10.9–10.10, No. 6), called the flexible optical circuit [20]. This harness has fibers embedded in a plastic (Kapton) covering, and the fibers are custom wired to connect a set of transmitters and receivers from the array modules into duplex connections at the card edge. A MAC-type connector was modified to be blind mateable along with a series of electrical card-edge connectors. Figure 10.10 illustrates an OETC transmitter and an OETC receiver module connected into one array edge connector.

Ericsson also developed a blind-mateable ribbon harness for interboard connections [17]. The harness uses either embedded optical fibers or polymer lightguides. The connector for this harness was based on the MT ferrule, and it is molded directly on the harness's ribbon to lower cost (similar to the Jitney approach). The harness attaches to the low-profile modules (described in Tables 10.8–10.10, No. 5).

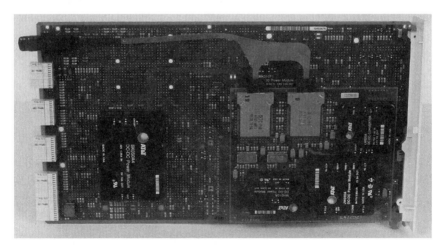

Fig. 10.10 Optical wiring harness.

10.8. Conclusions

This chapter has described some key intramachine opportunities for data-com optoelectronics, notably the system area network. The emergence of the SAN, and its increasing impact on computing system architecture, was not enabled by optical interconnects, and the system designers are currently driven by the capabilities of parallel copper wire technology. However, they are also driven by increasing node performance and node count and increasing flexibility in system layout and growth. These drivers all point to the eventual use of parallel optical interconnects. The questions are when and how.

This chapter has also described the parallel optical interconnect technology in products and protypes today. The technology is immature but moving rapidly, and there are no established standards. With few notable exceptions, the suppliers of this technology are hoping to adapt telecom or LAN datacom module and cable components to the intramachine application and maintain the marketing price and service structure established in their earlier markets, in which many users share the optical network in order to defray its high price. These price models do not fit with the pricing and rapid product migration models required in the computer industry, so the SAN developer is still concerned about POI filling more than a niche in the application market. (A similar situation developed in the LAN FDDI market, and copper wire technology took over as the de facto cost-effective standard.) It is significant that there are versions of parallel optical intercon-nect that can be price competitive due to the arraying of optical devices and component into monolithic structures. The next few years will be critical for both users and suppliers to find the right design point for POI.

References

1. Kleinrock, L., M. Gerla, N. Bambos, J. Cong, E. Gafni, L. Bergman, J. Bannister, S. P. Monacos, T. Buhjewski, P.-C. Hu, B. Kannan, B. Kwan, E. Leonardi, J. Peck, P. Palnati, and S. Walton. 1996 (June). The supercomputer supernet testbed: A WDM-based supercomputer interconnect. *J. Lightwave Tech.* 14(6):1388–1399.
2. Feehrer, J., J. Sauer, and L. Ramfelt. 1995 (October). Design and implementa-tion of a prototype optical deflection network. *Comput. Commun. Rev.* 24(4):191–200.

3. Saunders, S. 1996. *The McGraw-Hill high-speed LANS handbook.* New York: McGraw-Hill.

4. ESCON Physical Layer I/O Interface Specification, SA23-0395.

5. Benner, A. F. 1996. *Fibre Channel: Gigabit communications and I/O for computer networks.* New York: McGraw-Hill.

6. Dagenais M., R. Leheney, and J. Crow, ed. 1995. *Integrated optoelectronics.* New York: Academic Press.

7. Hahn, K. H. 1995. POLO—Parallel optic links for gugabyte data communications. *IEEE Electronic Components and Technology Conference 95*, Las Vegas, NV, May, pp. 368–375.

8. Pepeljugoski, P. K., B. K. Whitlock, D. M. Kuchta, J. D. Crow, and S.-M. Kang. 1995. Modeling and simulation of OETC optical bus. *IEEE Lasers and Electro-Optics Society Annual Meeting,* San Francisco, CA., October 30–31, pp. 185–186.

9. "Performance Standards for Laser Products," 21CFR1040, Center for Devices and Radiological Health (HFZ-312), Rockland, MD. Safety of Laser Products, Parts I (IEC 825-1, equipment classification) and II (IEC 835-2, communications systems)," American National Standards Institute, New York, NY.

10. Schlager, J. 1997 (June). "Round robin results of the multimode fiber restricted launch group—bandwidth," Minneapolis, MN: Telecommunications Industry Association Publication.

11. Crow J. *et al.* 1996 (May). The Jitney parallel optical interconnect. Proceedings of ECTC'96. p. 292, ISSN 0569-5503.

12. ANSI Standard X3T9.3. 1992 (June). "Fibre Channel Physical and Signalling Interface, FC-PH," Rev. 3.0.

13. For example, the University of Illinois tool *iFrost* was developed by the DARPA sponsored OETC Consortium of companies, and operates on various workstation platforms. (PC and UNX). See Whitlock, B., *et al.,* "Computer Modeling and Simulation of the OETC Optical Bus," Proceedings of OFC'96.

14. Staiwasz, K. *et al.* 1997 (May). "Automated Testing Methodologies for Low Cost, Parallel Optical Bus Components," Proceedings of ECTC'97. pp. 217–224.

15. Hahn, K. *et al.* 1996 (June). Gigabyte/s data communication with the POLO parallel optical link. *IEEE Electronic Components and Technology Conference 96*, p. 301.

16. OKI Electric Industry, Inc. 1994 (September). "Low-Power 11 × 320 Mbps Parallel Transmitter/Receiver LSIs for 2.4 Gbit/s Optical Links," *JETRO,* p. 20.

17. Niburg, M. *et al.* 1996 (September). A complete sub-system of parallel optical interconnects for telecom applications. *Proceedings of ECOC'96,* p. 259.

18. Nagahori, T., *et al.* 1996 (June). 1-Gbyte/sec array transmitter and receiver modules for low-cost optical fiber interconnect. *IEEE Electronic Components and Technology Conference 96*, p. 255.

19. Karstenshen, H. *et al.* 1994 (November). PAROLI—High performance optical bus with integrated components. *Proceedings of LEOS'94.*

20. Wong, Y. M., D. J. Muehlner, C. C. Faudskar, B. D. Buchholz, M. Fusgteyn, J. L. Brandner, W. J. Parzygnat, R. A. Morgan, T. Mullally, R. E. Leibenguth, G. D. Guth, M. W. Focht, K. G. Glogovsky, J. L. Zilko, J. V. Gates, P. J. Anthony, B. H. Tyrone, Jr., T. J. Ireland, D. H. Lewis, Jr., D. F. Smith, S. F. Nati, D. K. Lewis, D. L. Rogers, H. A. Aispain, S. M. Gowda, S. G. Walker, Y. H. Kward, R. J. S. Bates, D. M. Kuchta, and J. D. Crow. 1995 (June). Technology development of a high-density 32-channel 16 Gb/s optical data link for optical interconnection applications for the optoelectronic technology consortium (OETC). *IEEE J. Lightwave Tech.* 13(6):995–1016.

21. Grimes, G., *et al.* 1996 (November). Applications of parallel optical interconnects. *Proceedings of the IEEE LEOS'96 Annual Meeting,* Vol. 2, pp. 6–7.

22. Crow, J., and F. Tong. (1997). Data processing and data communications networks—The drive for cost effective photonic technology. *Photonic Networks,* New York: Springer, p. 442.

23. Motegi, Y., and A. Takai. 1994 (April). Optical interconnection modules utilizing fiber-optic parallel transmission to enhance information throughput. *Hitachi Review,* pp. 79–82.

24. Lebby, M., *et al.* 1996 (May). Characteristics of VCSEL arrays for parallel optical interconnect. *Proceedings of ECTC'96,* pp 279–291.

25. Swirhun, S., *et al.* 1996 (May). The P-VixeLink multichannel optical interconnect. *Proceedings of ECTC'96,* pp 316–320.

Chapter 11 | Applications: Asynchronous Transfer Mode and Synchronous Optical Network

Carl Beckmann

*Thayer School of Engineering, Dartmouth College, Hanover,
New Hampshire 03755*

11.1. Introduction

Early communication networks were driven by telephony and telecommunications requirements. The past 20 years have seen rapid growth in computer networking. Integration of these two applications initially started in wide-area networks, where economic considerations dictated that data networks use existing telephony infrastructures. The increasing use of digital media types promises that integration will reach both the corporate office and home/consumer networks increasingly in the future. This chapter starts with a review of the fundamental characteristics of telecommunications and data networks and the resulting requirements on integrated services networks.

11.1.1. TELEPHONY AND CIRCUIT SWITCHING

In the beginning, there were analog networks for supporting telephony. In the 1960s it became apparent that delivering analog telephone calls using analog frequency-division multiplexing techniques was prone to noise and did not make as effective use of the available bandwidth on copper wires as could digital techniques. Then, long-haul digital networks were installed for delivering long-distance telephone service. (Local service remained primarily analog for some time.) The general organization of telephone networks is illustrated in Fig. 11.1.

For telephony, the basic requirements are to establish a point-to-point connection for a "call" that is typically several minutes in duration. The

HANDBOOK OF FIBER OPTIC
DATA COMMUNICATION

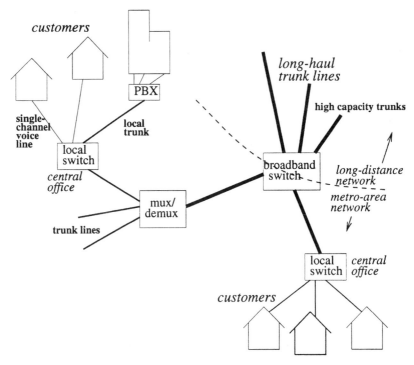

Fig. 11.1 Typical telephone network topology.

delay through the network must be small enough not to interfere with the quality of speech and to avoid perceptible echo effects, on the order of milliseconds or less. In North America, voice is digitized to 8 bits precision at 8000 samples per second (to support an analog bandwidth of 300–3000 Hz), for a data rate of 64 kb/s per call. The capacity of a typical copper wire digital trunk link (a T1 line) is 1.5 Mb/s by comparison.

These basic requirements are served well by using reserved connections along all the links from a source to a destination (circuit switching), with time-devision multiplexing (TDM) used to share the bandwidth among multiple calls on the individual links. TDM keeps the latency on each link very low while dividing the available bandwidth evenly between calls, with only a small amount of framing overhead. American and international standards bodies have adopted a variety of standard data rates and interoperability specifications, which are summarized in Table 11.1.

Table 11.1 **North American and International
Telecommunications Standards**

Name	Organization	Bit Rate	No. Voice Channels
T1	AT&T	1,544	24
T3	AT&T	44,736	672
DS1	ANSI	1,544	24
DS2	ANSI	6,312	96
DS3	ANSI	44,736	672
DS4	ANSI	274,176	4032
E1	CCITT	2,048	32
E2	CCITT	8,448	128
E3	CCITT	34,368	512
E4	CCITT	139,264	2048

11.1.2. DATA COMMUNICATIONS AND PACKET SWITCHING

Although telecommunications networks can be used to carry data as well, local area data networks can be built much more cheaply and efficiently without resorting to switch-based network architectures.

The requirements for data networks come from the ability to transfer files and other packets of information, such as electronic mail messages, and terminal data interfaces, consisting of typed keyboard strokes from a user and displayed textual and graphic information back to the user. Other traffic comes from remote procedure calls for distributed computing and operating system information and distributed file system transfers. Compared with telephony, a much more heterogeneous mix of traffic exists on data networks.

For most data traffic, there are no hard real-time constraints on its delivery. File transfers should happen quickly to allow for smooth operation and rapid response time to interactive users; slower response time is merely annoying but does not render the service useless.

In early systems, network bandwidth and delay were dictated by near-real-time requirements of character terminal input/output: A fast typist can type 60–100 words per minute. If each word is an average of 6 characters long (including the space), and each character is represented in 8-bit ASCII, then this represents a steady throughput of 80 bits per second. Characters come at an average rate of 10 per second. Terminal equipment was usually connected to the computer (often through an intermediate multiplexer) via a 9600-baud serial data line. Thus, the events of interest (defining the maximal acceptable latency) are on the order of hundreds of milliseconds, and the bandwidth required per connection is on the order of tens of kilobits or less per second. One hundred users could be accommodated with approximately 1 Mb/s worth of usable bandwidth. Table 11.2 lists the data rates in use in several widespread local area network (LAN) schemes.

Due to the highly heterogeneous and unpredictable nature of traffic on data networks, the use of connection-oriented communications, in which bandwidth is reallocated only every few minutes per channel, is not efficient. Moreover, higher latencies are tolerable. This makes connectionless communications based on discrete data packets, each containing its own addressing and format information, much more attractive. The bandwidth of the network can, in effect, be reallocated on a packet-by-packet basis on demand as the packets arrive at the network. To make this scheme efficient, a larger number of data bits should be put in each packet that is required for the packet "header" information. This incurs extra latency but is tolerable.

Finally, LANs can be built cheaply and extensibly by connecting all data equipment to a shared medium (such as coaxial cable) and distributing the switch functionality (channel allocation) out to the data nodes. Additional nodes can be added by simply splicing the new equipment into the existing cable. Network interface functionality can be built into each computer

Table 11.2   **Data Rates on Popular Local Area Network (LAN) Technologies**

Name	*Organization*	*Data Rate (Mb/s)*	*Applications*
AppleTalk	Apple Computer	<1	File sharing, printer sharing, electronic mail
Ethernet	ARPA	10	LAN for workstations and mainframes
Fast Ethernet	IETF	100	Buildingwide LAN backbones and servers
FDDI		100	Fast LAN and medium-scale networks (several kilometers)

system node for a fraction of the cost of the CPU, memory, or hard disk subsystems. Internetworks can be constructed by linking LANs together by bridges, routers, or computing nodes connected to multiple LANs or interconnected via telecom lines. Store-and-forward routing of packets is acceptable because of the longer tolerable latencies for typical datacom applications. The organization of typical local area networks is illustrated in Fig. 11.2.

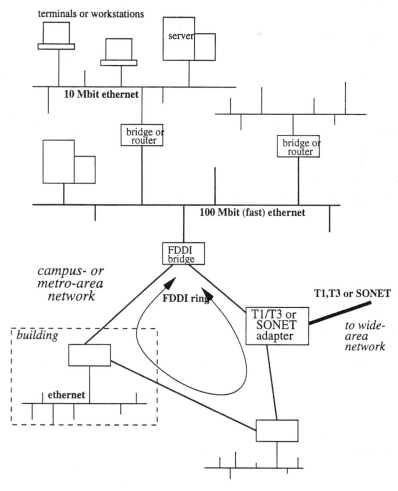

Fig. 11.2 Topology of a typical LAN.

11.1.3.   BROADBAND INTEGRATED SERVICES DIGITAL
NETWORK REQUIREMENTS

The need for integrating data- and telecommunications functions is apparent from several trends:

- Data communication has long made use of digital telephone switching networks for wide-area data transport.
- The use of facsimile and personal computer modems has increased phenomenally during the past several years, using local analog phone lines for digital services.
- New applications in consumer and business fields are emerging that share some of the characteristics of both traditional datacom and telecom applications, such as teleconferencing, advanced television and video-on-demand, interactive multimedia training and education, and telecommuting.

Moreover, as both the bandwidth requirement per user and the number of users connected continue to increase, networks must make increasing use of fiber optics as opposed to electronics at the physical layer.

The requirements on integrated services digital networks (ISDNs) are thus:

- Support for connectionless and packet-oriented data services, in particular TCP/IP.
- Efficient support for connection-oriented "constant bit rate" (CBR) services such as telephony or (uncompressed) digital video.
- Support for new classes of traffic such as "variable" (but steady) bit rate (VBR) services, such as compressed video or voice; or "bursty" VBR/CBR traffic from interactive multimedia applications.
- Efficient multicasting support, both for media multicasting and for TCP/IP functionality.

11.1.4.   ASYNCHRONOUS TRANSFER MODE AND
SYNCHRONOUS OPTICAL NETWORK OVERVIEW

ATM has been proposed as an enabling network technology to support broadband integrated services. It is not a complete, stand-alone networking standard; rather, ATM defines a common layer of interoperability called the ATM layer, on which various services ranging from telephony and video conferencing to TCP/IP data networking and multimedia can be

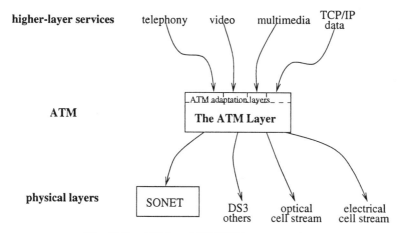

Fig. 11.3   ATM and SONET in perspective.

delivered. The ATM layer defines a common format used for switching and multiplexing bit streams from one end of an ATM network to another.

The ATM layer, in turn, uses the hardware facilities of lower layers to deliver the bits across individual links in a network. A variety of such physical layers have been defined, most of which are based on existing standards in order to maximally leverage existing technologies and installed bases. These relationships are summarized in Fig. 11.3.

One family of ATM physical layers is based on SONET. SONET is a synchronous, time-division multiplexing standard based on transmission over optical media (actually, a family of standards at a variety of bit rates). It was primarily designed to support telecommunications and long-haul, broadband services.

11.2. SONET

11.2.1. *Historical Perspective*

The SONET standards were developed in the mid-1980s to take advantage of low-cost transmission over optical fibers. SONET defines a hierarchy of data rates, formats for framing and multiplexing the payload data, as well as optical signal specifications (wavelength and dispersion), allowing multi-vendor interoperability. SONET was originally proposed by Bellcore in

Table 11.3 **Basic SONET/SDH Data Rates**

| SONET | | | |
Electrical	Optical	SDH	Rate
STS-1	OC-1	—	51.840 Mb/s
STS-3	OC-3	STM-1	155.520 Mb/s
STS-9	OC-9	—	466.560 Mb/s
STS-12	OC-12	STM-4	622.080 Mb/s
STS-18	OC-18	—	933.120 Mb/s
STS-24	OC-24	—	1.244160 Gb/s
STS-36	OC-36	—	1.866240 Gb/s
STS-48	OC-48	STM-16	2.488320 Gb/s

1985 and later standardized by ANSI and the CCITT [synchronous digital hierarchy (SDH) is a compatible set of standards in Europe] [1–3].

SONET is a synchronous, time-division multiplexing scheme designed to support existing telephone network trunk traffic and also designed with broadband ISDN (BISDN) services in mind. Its TDM basis readily supports fixed-rate services such as telephony. Its synchronous nature is designed to accept traffic at fixed multiples of a basic rate, without requiring variable stuff bits or complex rate adaptation. The SONET data transmission format is based on a 125-μs frame consisting of 810 octets, of which 36 are overhead and 774 are payload data. The basic SONET signal, whose electrical and optical versions are referred to as STS-1 and OC-1, respectively, is thus a 51.84 Mb/s data stream that readily accommodates TDM channels in multiples of 8 kb/s.

SONET defines a hierarchy of signals at multiples of the basic STS-1 rate. The SONET rates currently standardized are shown in Table 11.3. SDH is a compatible European counterpart to SONET. Due to compatibility issues with European switching equipment, the basic SDH rate, called STM-1, is three times the STS-1 rate (i.e., STS-3), or 155.52 Mb/s.

11.2.2. STS DATA RATES AND FRAMING

To efficiently support telephony, SONET bit rates rest fundamentally on voice-quality audio sampling rates, i.e., 8000 samples per second at 8 bits per sample. The SONET data transmission format is therefore based on a

125-μs frame illustrated in Fig. 11.4. This figure shows the basic STS-1 frame. Higher rates are achieved by byte-interleaving multiple STS-1 frames. The 125-μs frame contains 6480 bit periods, or 810 octets (bytes). This can be viewed as a two-dimensional arrangement of nine rows by 90 columns (of bytes) that is scanned row-wise from the upper left.

Thus, a single voice channel occupies a single octet in each 125-μs frame, and after leaving room for various "overhead" octets (see below), 774 64 kb/s voice channels can be time-division multiplexed into a single STS-1 frame. The bit rate for an OC-N link is thus given by

$$\begin{aligned} \text{OC-}N \text{ bitrate} &= N \cdot 8000 \text{ Hz} \cdot 90 \text{ columns} \cdot 9 \text{ rows} \cdot 8 \text{ bits/octet} \\ &= 51.84N \text{ Mb/s}, \end{aligned} \tag{11.1}$$

and the payload capacity (after accounting for four overhead columns per frame) is

$$\text{OC-}N \text{ capacity} = N \cdot 8000 \text{ Hz} \cdot 86 \cdot 9 \cdot 8 = 49.536\, N \text{ Mb/s}. \tag{11.2}$$

The first three columns in each frame (i.e., the first three of every 90 octets) are reserved for various overhead bytes. Overhead information is organized into section, line, and path overhead.

SONET can be thought of as following a layered model. At the lowest layer (the physical layer), SONET specifies characteristics of the optical signal, such as maximum dispersion, etc. The lowest level physical link between two pieces of SONET equipment (i.e., an optical fiber pair) is called a section (Fig. 11.5). Multiple sections may be linked together via signal repeaters (regenerators) to form a line. The two ends of a line attach to line termination equipment. At the next level up, a physical line may be used by one or more paths, which are connected on both ends to path

column

		1	2	3	4	...	90
	1	*0*	*1*	*2*	*3*	...	*89*
	2	*90*	*91*	*92*	*93*	...	*179*
	3	*180*	*181*	*182*	*183*	...	*etc.*
	4	section					
row	5	and				Payload	
	6	line					
	7	overhead					
	8						
	9	*720*	*721*	*722*	...		*809*

Fig. 11.4 SONET framing format.

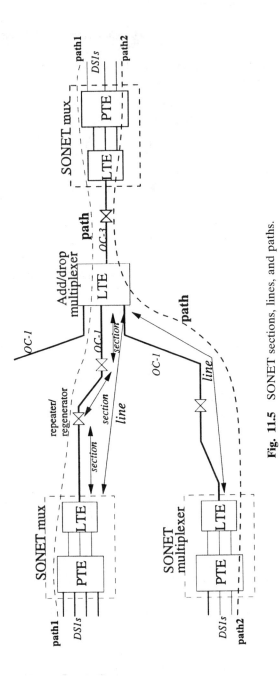

Fig. 11.5 SONET sections, lines, and paths.

Services	DS1, DS3, ATM cells...
Path Layer	Envelope
Line Layer	STS-N blocks
Section Layer	Frame
Physical Layer	Photons

Fig. 11.6 SONET layered architecture.

termination equipment. It is at the path termination equipment that SONET frames are assembled and disassembled. The layered approach allows the use of equipment for handling functions related to one or the other layer individually, keeping costs down by not requiring all layers to be handled at once. Figure 11.6 illustrates the SONET layers.

The first three columns of each frame contain the section and line overhead bytes. The first three rows of this are for the section overhead, and the last six rows are for the line overhead. This is illustrated in Fig. 11.7. The remaining 87 columns contain the synchronous payload envelope (SPE). The SPE contains the actual payload data as well as a single column of path overhead bytes.

Note that the SPE need not be exactly aligned in the payload frame. In fact, the first byte of the SPE may reside (and usually does) anywhere

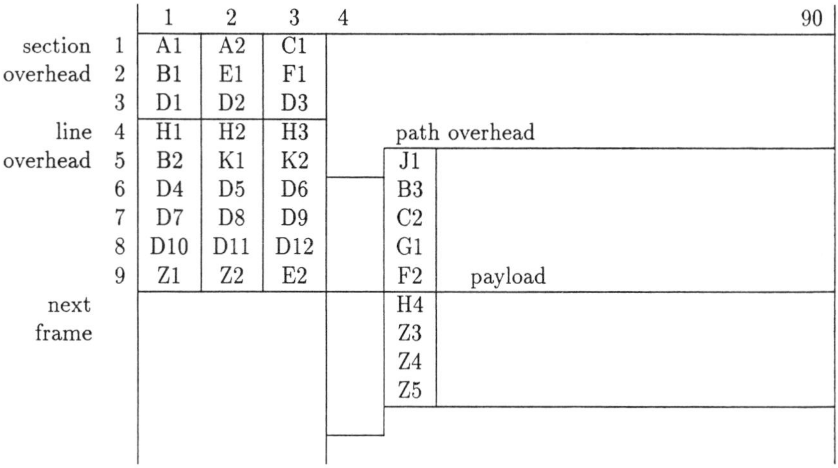

		1	2	3	4			90
section	1	A1	A2	C1				
overhead	2	B1	E1	F1				
	3	D1	D2	D3				
line	4	H1	H2	H3	path overhead			
overhead	5	B2	K1	K2		J1		
	6	D4	D5	D6		B3		
	7	D7	D8	D9		C2		
	8	D10	D11	D12		G1		
	9	Z1	Z2	E2		F2	payload	
next						H4		
frame						Z3		
						Z4		
						Z5		

Fig. 11.7 SONET frame overhead bytes.

within the frame; hence, the path overhead is not always in column 4. Overhead octets H1 and H2 form a pointer to the location of the first SPE octet. This feature is useful in connecting two lines whose bit clocks differ slightly, as they do in practice. This allows the SPE to "slip" slightly with respect to the frame. A stuff byte is provided in H3 to make up the bandwidth deficit in the case in which the signal to transmit is faster than the line clock. This scheme separates the synchronization of data payload frames from the generation of the framing signals, which can be done from a transmitter's local clock.

11.2.2.1. Section Overhead Octets

The first three rows of the first three columns in each frame are used for section-related functions. The functions of these bytes, which include framing, identification, section error monitoring, and auxiliary data channels, are summarized in Table 11.4.

11.2.2.2. Line Overhead Octets

The last six rows in the first three columns of each frame are used for line-related functions, as summarized in Table 11.5.

11.2.2.3. Path Overhead Octets

The first column in the SPE of an STS-1 signal is used for various path-related functions, as summarized in Table 11.6. In an OC-N signal, which carries N byte-interleaved STS-1 SPEs, the first column in each STS-1 is used for path-related overhead. By contrast, in a "concatenated" OC-Nc signal, there is only a single column of path overhead, with the remaining $87N-1$ columns available for payload data.

11.2.3. Payload Envelope Pointer

The SPE of a SONET frame need not be perfectly aligned with the framing overhead. Pointer octets H1 and H2 are used to locate the SPE within the frame. The lower 10 bits of H1 and H2 are an offset to the beginning of the SPE, i.e., the number of octets between H3 and J1, the first octet in the SPE. This feature makes it easier to synchronize multiple signals and multiple pieces of equipment, while allowing each signal source to generate its own framing structure based on a local clock.

Table 11.4   **SONET Section Overhead Octets**

Symbol	Bits	Name	Description
A1, A2	16	Framing	F628 Hex (1111011000101000 binary); provided in all STS-1 signals within an STS-N signal
C1	8	STS-1 identification	Unique number assigned just prior to interleaving that stays with STS-1 until deinterleaving
B1	8	Section BIP-8	Allocated in each STS-1 for a section error monitoring function
E1	8	Orderwire	Used as a local orderwire channel; reserved for communications between regenerators, hubs, and remote terminal locations
F1	8	Section user channel	This byte is set aside for the network provider's purpose; it is passed from one section level entity to another and is terminated at all section-level equipment
D1–D3	24	Section datacomm	A 192 kb/s channel for alarms, maintenance, control, etc. between section terminating equipment

The upper 4 bits of H1 and H2 are used to signal changes in the pointer value: A value of 0110 signals an increment or decrement by 1, and a value of 1001 signals some larger change. In the frame in which the pointer is incremented by 1, the lower 10 (H1,H2) bits do not contain the new pointer value but rather the old pointer value, with all the even bits (including the LSB) inverted; on a decrement by 1, the odd bits are inverted. Once the pointer stabilizes the true new value is used in the lower 10 (H1,H2) bits.

Because the frequency deviation imposed by the standard is small, pointer adjustments will take place infrequently in practice. If an upstream clock is too slow, the downstream equipment will have to periodically increment its pointer and delay outgoing SPEs. When eventually the pointer overflows the maximum value of 809, an entire frame will be skipped. If the upstream clock is too fast, the pointer will have to be decremented periodically. When this happens, the missing byte is put in the H3 octet to

Table 11.5   **SONET Line Overhead Octets**

Symbol	Bits	Name	Description
H1, H2	16	Pointer	Indicates the offset in bytes between the pointer and the first byte in the STS SPE
H3	8	Pointer action	Stuff byte for downstream frame advancement
B2	8	Line BIP-8	Allocated in each STS-1 for a line error monitoring function; Used as a local orderwire channel; reserved for communications between regenerators, hubs, and remote terminal locations
K1, K2	16	APS channel	Allocated for APS signaling between two line-level entities; also carries other management signals
D4–D12	72	Line datacomm	Nine bytes (576 kb/s) allocated for line data communication for alarms, maintenance, control, etc.
Z1, Z2	16	Growth	Future expansion
E2	8	Orderwire	Express orderwire between line entities

compensate. Essentially, the H3 stuff byte provides the extra bandwidth needed for slow-running clocks to keep up with the required data rate.

11.2.4. Multiplexing

Higher speed transmission than STS-1 rates is achieved by byte-interleaving N STS-1 signals to obtain an STS-N signal (which is then converted to an optical OC-N signal). This allows, for example, several STS-1 signals to be multiplexed for tranmission over an OC-3 (or higher) link.

Alternately, higher speed channels can be obtained using concatenated STS-1s to achieve a single channel with N times the capacity of an STS-1. In this case, N STS-1 frames are again byte-interleaved to obtain the STS-Nc framing structure. In the STS-Nc frame, there are $3N$ columns for transport (section and line) overhead, with $87N$ columns remaining for the payload. However, this payload is multiplexed, switched, and transported through the network as a single entity. Hence, only a single column of path overhead is needed (leaving slightly more bandwidth available for data capacity compared to noncontatenated STS-N).

Table 11.6   **SONET Path Overhead Octets**

Symbol	Bits	Name	Description
J1	8	STS path trace	Used by path-terminating equipment to verify its connection to the source, which continuously sends a fixed 64-byte pattern
B3	8	Path BIP-8	Path error monitoring
C2	8	STS path signal label	Indication of valid construction of SPE
G1	8	Path status	Path-terminating status and performance, back to an originating path
F2	8	Path user channel	For network provider
H4	8	Multiframe	A192 kb/s channel for alarms, maintenance, control, etc. between section terminating equipment
Z3–Z5	24	Growth	Future expansion

11.2.5. Virtual Tributaries

In order to directly support services with lower bandwidth requirements than the basic STS-1 payload, several standard "virtual tributary" formats have been defined for SONET. These are summarized in Table 11.7. The VT1.5, for example, allows a DS1 or T1 signal to be carried end to end on a SONET path without having to remultiplex the 24 DS0 (voice) channels contained therein.

Each virtual tributary format is defined as some integral number of columns of the SONET SPE, which includes room for the carried signal as well as any VT-related overhead octets.

Table 11.7   **Virtual Tributaries**

Name	Service	Data Rate	No. Columns
VT1.5	DS1	1.544	3
VT2	CEPT	2.048	4
VT3	DS1C	3.088	6
VT6	DS2	6.176	12

11.2.6. *International Interoperability*

SONET is compatible with an international set of standards called the SDH. SDH was developed based on SONET, but with the additional goal of providing compatibility between North American and European telecom carriers. Whereas SONET starts with a 51.84 Mb/s signal consisting of nine rows by 90 columns every 125 μs (STS-1), SDH starts with a 9 $\times$ 270 frame every 125 μs, or a 155.52 Mb/s signal.

The basic 155.52 Mb/s SDH signal, called STM-1, is similar and can be made compatible with SONET STS-3. There are some differences in the usage of section and line overhead octets between SONET and SDH. For a more detailed discussion of the differences, the reader is referred to Minoli [3]. See also Table 11.3 for SDH data rates.

11.2.7. *SONET Physical Specifications*

Specifications for the transmitter, receiver, and optical signal path characteristics for various SONET line rates are given in Table 11.8 [4].

11.3. ATM

11.3.1. *CELL VS PACKET SWITCHING*

ATM is designed for high-speed transport of a variety of traffic types. Due to its high-speed nature, it is believed that using fixed-size cells will allow efficient hardware implementations of various multiplexing and routing functions. Unlike LAN environments using Ethernet, fast Ethernet, or fiber distributed data interface (FDDI), and capable of tens to hundreds of megabits of throughput on variable-sized packet traffic, ATM is designed to work into the gigabit per second range [2, 3, 5–7]. Moreover, ATM is designed to be able to support both switched- and packet-oriented applications. Finally, ATM allows the quality of service to be specified within a range of parameters during call setup time.

The use of small, fixed-sized cells has several advantages over larger variable-sized packets as used in Ethernet or FDDI:

- Cell boundaries can be easily recognized at high speed in hardware, should loss of framing occur.
- Individual packets cannot monopolize the bandwidth of the channel.

Table 11.8 SONET Physical Layer Optical Specifications

Parameter	Units	OC-1	OC-3	OC-9	OC-12	OC-18	OC-24	OC-36	OC-48
Data rate									
Bit rate	Mb/s	51.84	155.52	466.56	622.08	933.12	1244.16	1866.24	2488.32
Tolerance	ppm			100					
Transmitter									
Type		MLM/LED	MLM/LED	MLM/LED	MLM/LED	MLM	MLM	MLM	MLM
$\lambda_{W\min}$	nm	1260	1260	1260	1260	1260	1260	1260	1265
$\lambda_{W\max}$	nm	1360	1360	1360	1360	1360	1360	1360	1360
$\Delta\lambda_{\max}$	nm	80	40/80	19/45	14.5/35	9.5	7	4.8	4
$P_{T\max}$	dBm	−14	−8	−8	−8	−8	−5	−3	−3
$P_{T\min}$	dBm	−23	−15	−15	−15	−15	−12	−10	−10
$r_{e\min}$	dB	8.2	8.2	8.2	8.2	8.2	8.2	8.2	8.2
Optical path System									
$ORL_{\max}$	dB	na	na	na	na	20	20	24	24
$D_{S\,R\max}$	ps/nm	na	na	31/na	13/na	13	13	13	12
Max sndr. to rcvr. reflectance	dB	na	na	na	na	−25	−25	−27	−27
Receiver									
$P_{R\max}$	dBm	−14	−8	−8	−8	−8	−5	−3	−3
$P_{R\min}$	dBm	−31	−23	−23	−23	−23	−20	−18	−18
P_0	dBm	1	1	1	1	1	1	1	1

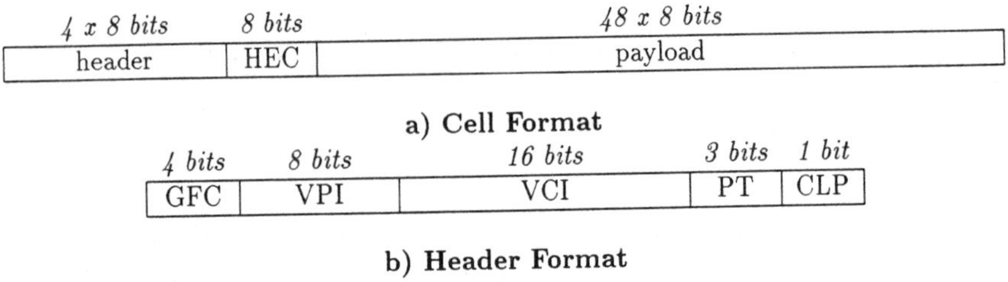

4 x 8 bits	*8 bits*	*48 x 8 bits*
header	HEC	payload

a) Cell Format

4 bits	*8 bits*	*16 bits*	*3 bits*	*1 bit*
GFC	VPI	VCI	PT	CLP

b) Header Format

Fig. 11.8   ATM cell format.

- Cell-handling decisions (e.g., during congestion or for traffic policing of individual connections) can be easily made based solely on the number of cells, without having to examine their headers for packet size information.
- Cell-buffering hardware in switches and other equipment is simplified.
- Circuit-like switching of cells replaces store-and-forward routing of packets, with much lower latency over multihop paths.

The disadvantages are that header information may consume a larger fraction of available bandwidth than for large packets, and that sending very small amounts of information is less efficient than it is for small packets (although both are inefficient).

The structure of ATM cells is shown in Fig. 11.8 [8]. It consists of a 5-byte header followed by 48 bytes of payload data. The header contains the following fields: Generic flow control (GFC), virtual path identifier (VPI), virtual channel identifier (VCI), payload type (PT), a cell loss priority bit (CLP), and header error correction (HEC). A brief description of each of these fields is given in Table 11.9.

11.3.2.   CELL VS CIRCUIT SWITCHING

Another key feature of ATM is its ability to transport "constant bit rate" data such as (uncompressed) telephony or video over virtual circuits with guaranteed bandwidth and latency characteristics. In other words, ATM provides a service that mimics a point-to-point, synchronous connection normally provided by a TDM network. Features of ATM that enable this include the following:

- Cell size is kept small, because cell size directly affects latency at the source and destination associated with packing and unpacking a

Table 11.9 **ATM Cell Fields**

Field	Bits	Name	Description
GFC	4	Generic flow control	
VPI	8	Virtual path identifier	Identifies 1 of 256 possible paths out of the current switch or device. Used with VCI to distinguish and locally route different cell streams
VCI	16	Virtual channel identifier	Identifies 1 of 65,536 possible channels in the given path out of the current switch or device. Used with VPI to distinguish and locally route different cell streams
PT	3	Payload type	Differentiates control vs data cells, etc.
CLP	1	Cell loss priority	Used to mark low-priority cells that may be discarded if network traffic is high
HEC	8	Header error correction	A CRC checksum on the first four header octets, using the generator polynomial $x^8 + x^2 + x + 1$. The resulting code is also XORed with 01010101 to get the HEC bits
Data	48×8	Payload	User data and headers/trailers from higher network layers

bit stream into cells and, to some extent, affects latency of cell handling at network-switching elements.

- Keeping cell size fixed makes it feasible to allocate link bandwidth to individual connections, and reducing the cell size increases the bandwidth resolution at which this can be done.

- Fixed-size cells make scheduling of periodic or pseudoperiodic traffic at switching elements feasible in principle.

However, the main justification for ATM is its ability to mix synchronous with other types of traffic such as variable bit rate or connectionless and "bursty" data: By using cells, rather than TDM time slots, the channel bandwidth can be reallocated to different "virtual connections" on a cell-by-cell basis instead of requiring a TDM time slot to be allocated (requiring an end-to-end call setup) as short-lived connections come and go or as the bandwidth requirement of a single channel waxes and wanes. This makes efficient statistical multiplexing possible, where a large number of variable-bandwidth connections can be supported over a broadband channel with capacity for the sum of the connections' average bandwidth requirement, even though the sum of maximum instantaneous bandwidth requirements may exceed the channel's capacity.

Paradoxically, one of the physical layers used for ATM is the SONET. Here, SONET frames are simply used to transport ATM cells across a SONET path. The payload carried by the ATM cells need not be synchronous, however, because from frame to frame (and from cell to cell), the payload carried by the ATM cells can come from completely different ATM connections. The ATM cells in the SONET payload are opaque to the SONET layer. Allocation of the ATM cell bandwidth to CBR, VBR, and connectionless data channels is handled entirely at the ATM layer and above.

11.3.3.   ATM LAYERED ARCHITECTURE

ATM is based on a layered architecture (Fig. 11.9). The major layers are the physical layer, the ATM layer, and the ATM adaptation layer (AAL). Above the AAL reside the data source layers, corresponding approximately to open systems interconnection (OSI) layers 3–7. The physical layer is further divided into a lower physical medium-dependent sublayer (PMD) and the transmission convergence sublayer (TC). The adaptation layer is also divided into the segmentation and reassembly sublayer (SAR) and the convergence sublayer (CS).

ATM model			OSI model
			Transport
	source layers (3-7)		Network
AAL	Adaptation Layer	*Segmentation and Reassembly* (SAR)	Data
		Convergence Sublayer (CS)	
ATM	ATM Layer		Link
PHY	Physical Layer	*Transmission Convergence* (TC)	
		Physical Medium Dependent (PMD)	Physical

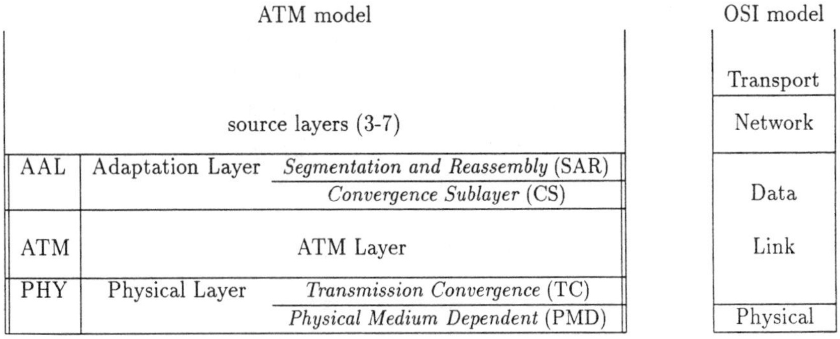

Fig. 11.9   ATM reference model.

Table 11.10   **ATM Service Classes**

			ABR	
	CBR *(class 1)*	**VBR** *(class 2)*	**Class 3**	**Class 4**
Timing	Synchronous	Synchronous	Asynchronous	Asynchronous
Bit rate	Constant	Variable	Variable	Variable
Connection mode	Connection oriented	Connection oriented	Connection oriented	Connectionless

ATM layers do not correspond to the standard seven-layer OSI model, although an approximate correspondence is shown in Fig. 11.9. In most applications, AAL, ATM, and the TC sublayer of PHY can be thought of as providing the functionality of the OSI data link layer, i.e., the error-free transmission of bits from one end of a link to another. Although this may involve the traversal of several switches (which in turn uses routing information in cells' VPI and VCI fields), the actual network layer function of establishing these routes on call setup is left to higher layers.

The ATM layer is fixed, but a variety of adaptation layers and physical layers have been defined. The services provided by the adaptation layer depend on the traffic type being supported. Traffic types vary in their data rate characteristics (constant data rate versus variable or bursty data traffic), connectionless (datagram) versus connection-orientedness, allowing ATM to support the spectrum of services including voice, video, and computer data services and interactive multimedia. Four basic traffic classes have been defined as shown in Table 11.10, and an adaptation layer has been defined for each (AAL-1–AAL-4). (The adaptation layers for class 3 and class 4 available bit rate traffic has been combined into a single layer, AAL-3/4.)

A fifth adaptation layer, AAL-5 (originally called SEAL, the simple and efficient adaptation layer), has also been defined to serve as a convenient application programmer interface (API) for computer applications to build directly on top of ATM services.

11.3.4. *ATM PHYSICAL LAYERS*

ATM is a switching and multiplexing scheme for BISDN, but it is not necessarily tied to a particular physical layer. Fiber optic as well as electronic physical layers are possible at a variety of data rates [8–11]. At the time

Table 11.11   **Standardized ATM Physical Layers**

Rate (Mb/s)	Media	Framing	UNI Specification
1.544	Twisted pair	DS1	Public
2.048	Twisted pair, coax	E1	Public
6.312	Coax	J2	Public
25.6	UTP-3	Cell stream, 32 Mbaud	Private
34.368	Coax	E3	Public
44.736	Coax	DS3	Public
51.84	UTP-3	SONET STS-1	Private
100	MMF	Cell stream, 125 Mbaud	Private
155.52	SMF	SONET OC-3c	Public/private
155.52	UTP-3, coax	SONET STS-3c	Private
155.52	MMF	Cell stream, 194.4 Mbaud	Private
155.52	STP	Cell stream, 194.4 Mbaud	Private
622.08	SMF, MMF	SONET OC-12	Private

of this writing, the ATM Forum Technical Committee has standardized the following physical layers for ATM: 155 and 622 Mb/s fiber optic layers based on SONET; 100 and 155 Mb/s cell-stream fiber optic layers; 155 and 25 Mb/s layers for twisted-pair connections; and a DS1 (1.5 Mb/s) layer based on T1. The physical layers standardized for ATM are summarized in Table 11.11 [12].

The ATM user–network interface (UNI) specification [8] includes two kinds of interfaces; public and private. Public ATM service providers and any equipment connecting to public ATM networks must adhere to the public UNI specification, whereas the less stringent private UNI specification is suitable for use in local-area networking equipment. The private UNI does not need the operation and maintenance complexity or the link distance provided by the public UNI for telecom lines.

11.3.4.1. SONET/SDH

SONET-based fiber optic physical layers for ATM have been defined at 155 Mb/s (OC-3) and 622 Mb/s (OC-12) rates. In both these cases, the PMD sublayer is essentially identical to the corresponding SONET–SDH specification. The TC sublayer makes use of SONET framing by encapsulating ATM cells into the SONET SPE. The relationship between the ATM physical layer and the SONET layered architecture is shown in Fig. 11.10.

The SONET payload envelope presents a bandwidth resource that is used by the TC sublayer to carry ATM cells. However, because the ATM cell size (53 octets) does not evenly divide the size of either the STS-3c or STS-12c payload envelopes, no synchronization between ATM cells and the SONET framing structure is implied (i.e., cells may cross SONET frame boundaries). In the STS-3c UNI, the available capacity for ATM cells is nine rows by 260 columns (the payload envelope minus one column of path overhead), or 149.760 Mb/s. In the 622 Mb/s interface, there are three fixed stuff columns following the path overhead, so the available cell carrying capacity is $9 \times (1044 - 4)/125 \, \mu s = 599.04$ Mb/s. In both cases, the available capacity is packed with ATM cells, and any rate decoupling between the ATM and PHY layers is accomplished by inserting empty cells into the stream.

Because of the asynchrony, the TC sublayer is also responsible for cell delineation. This is accomplished via the HEC bits in the cell headers. If cell synchronization is lost, the TC sublayer receiver continuously scans the SONET payload, testing whether each new octet starts a 5-octet ATM header with a valid HEC field. If so, it enters a presynch state, and if several valid cells are detected in a row, the synch state is entered and cell synchronization is assumed. HEC checking continues during normal transmission to verify that cell synchronization is not lost.

As long as cell synchronization is maintained in steady state, the HEC field is also used to correct any single-bit errors found in individual cell headers. The HEC field uses a polynomial code as indicated in Table 11.9 to perform single-bit correction and multiple-bit error detection on the header portion of each cell.

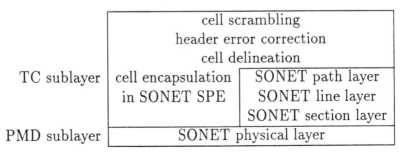

Fig. 11.10   OC-3 and OC-12 SONET-based ATM physical layer.

Finally, prior to insertion in the cell stream (and after removal on the receive side), the TC sublayer scrambles the payload portion of ATM cells to avoid any problems with DC levels or repeated bit patterns in the SONET payload envelope. This uses a self-synchronizing scrambler polynomial described in ITU recommendation I.432 [8].

11.3.4.2. Cell Stream

Alternately, cells may be sent directly over optical media, without using SONET framing. Several such physical layers have been defined for ATM at 100 and 155.52 Mb/s data rates. In order to maintain DC levels on the line and aid in synchronization, a 4B/5B (100 Mb/s) or 8B/10B (155 Mb/s) line code is used; hence, the actual required baud rate on the line is 125 or 194.4 Mb/s, respectively. Similarly, an electronic cell-stream physical layer has also been defined for twisted-pair cable, with a data rate of 25.6 Mb/s (32 Mbaud, due to 4B/5B coding) [10]. The 100 Mb/s PHY layer uses the same symbol coding and physical fiber and signal specifications as the FDDI LAN standard.

   In the cell-stream interfaces, the TC sublayer is responsible for the functions of cell delineation and HEC verification and for the 155 Mb/s UNI, 125-μs clock recovery. In both of these private UNIs, the HEC is used for detection of errored cells only and not for correction because the use of a 4B/5B (or 8B/10B) code means that any line bit errors result in multiple data bit errors. ATM cells are simply discarded from the stream sent to the ATM layer.

11.3.4.2.1. 100 Mb/s TC Sublayer

In the 100 Mb/s interface (also called TAXI) there is no framing structure; therefore, when no cells are being transmitted, a special 8-bit symbol "JK" (Sync) is continuously sent (not the FDDI "idle line" code). Three special 8-bit symbols are used by the ATM PHY layer, each of which is a pair of FDDI line symbols (5-line bits each), as shown below:

FDDI Code Pair	*Symbol/Function*
JK (Sync)	Idle
TT	Start of cell
QQ	Loss of signal

The start of a new cell is indicated by the "TT" byte. This is followed immediately by the 53 bytes of the cell.

Thus, although ATM cells are available from the ATM layer, they are transmitted on the line continuously as 54 FDDI symbol pairs each (a start of cell followed by 53 data bytes). However, the JK Sync code is also used to gain byte alignment; therefore, as a minimum, 1 JK symbol pair must be inserted by the TC sublayer on the line every 0.5 s.

11.3.4.2.2. 155-Mb/s TC Sublayer

The 155 Mb/s interface, on the other hand, does have a framing structure consisting of 1 ATM cell used as a physical layer overhead unit (PLOU) followed by 26 cells of data. All 27 cells consist of 53 bytes each, and each byte is coded on the line as a single 8B/10B symbol as specified in the Fibre Channel physical layer document. Synchronization is maintained by using special codes as the PLOU cell's header (four "K28.5" symbols followed by one "K28.7" symbol).

Unlike the 100 Mb/s interface, cell rate decoupling is performed by inserting idle cells rather than some idle line symbols. Thus, as in the case of the SONET-based PHYs, the available bandwidth is packed with a contiguous stream of whole cells (some of which may be "empty").

An additional function performed by the 155 Mb/s TC sublayer is the delivery of a 125-μs clock across the link. This is done by using another special line code (K28.2), which can be inserted anywhere within the symbol stream. On the receive end, this symbol is used as a 125-μs strobe and removed from the symbol stream prior to extracting ATM cells.

11.3.4.3. Physical Media Requirements

The optical ATM physical layers include those based on SONET–SDH and direct cell-stream physical layers. Currently, all public UNI specifications for fiber optic transmission are based on SONET–SDH; hence, they are suitable for long-distance links. The optical specifications can be found in Table 11.8.

The non-SONET cell-stream PHYs have been approved only for the private UNI. These are intended for shorter distance links (up to 2 km) in LANs. The 100 Mb/s layer is based on the physical specifications for FDDI (see Chapter 14). The optical specifications for the cell-stream ATM PHYs are summarized in Table 11.12 [8].

Table 11.12   **ATM Optical Specifications for the Cell-Stream Private UNIs**

Parameter	Units	100 Mb/s TAXI	155 Mb/s
Data rate			
Data rate	Mb/s	100	155.52
Line rate	Mb/s	125	194.4
Tolerance	ppm		100
Transmitter			
$\lambda_{W_{min}}$	nm		1270
$\lambda_{W_{min}}$	nm		1380
Spectral width (FWHM)	nm		<200
$P_{T_{max}}$	dBm		−14
$P_{T_{min}}$	dBm		−20
Optical path			
Type		Multimode fiber	Multimode fiber, graded index
Diameter	μm	62.5	62.5/125 or 50
Modal bandwidth	MHz/km		500
Zero dispersion Wavelength, λ_0	nm		1295–1365
Max dispersion Slope, s_0	ps/nm^2 − km		0.110
Max distance	km	2	2
Receiver			
$P_{R_{max}}$	dBm		−14
$P_{R_{min}}$	dBm		−29

11.3.4.4. Payload Capacity Comparison

It is instructive to compare the delivered payload capacity of various SONET and ATM formats. Table 11.13 gives the baud rate (line data rate), the bit rate (nominal symbol rate), and delivered payload capacity of raw SONET and several ATM PHY layers in the nominal 100–155 Mb/s range. In the ATM case, the payload capacity listed is the total data payload (no cell headers or HEC) presented to the ATM layer. In the case of SONET, it is the synchronous payload envelope minus any path overhead bytes.

    In terms of overall efficiency, raw SONET requires only approximately 4% overhead but delivers only synchronous data. ATM incurs an additional

Table 11.13   **Effective Payload Capacity Comparison**

	SONET		ATM		
	OC-3	**OC-3c**	**OC-3c**	**155 Mb/s**	**100 Mb/s TAXI**
Baud rate	155.52	155.52	155.52	194.4	125
Bit rate	155.52	155.52	155.52	155.52	100
Payload capacity	148.608	149.760	135.6317	135.6317	88.889
Total efficiency	95.56%	96.30%	87.21%	69.77%	71.11%

10% of overhead (5 cell header bytes per 48 bytes of data) — the price for the added flexibility of cell versus synchronous TDM switching. The ATM-cell-stream formats lose 20% in overhead due to the 4B/5B or 8B/10B line symbol coding (compared to 4% for some of the same functionality provided by SONET). Note that the SONET OC-3c and 155 Mb/s cell-stream ATM PHYs have the same effective payload capacity by design, although the forms of overhead are different (SONET framing versus one PLOU per 26 cells).

11.3.5. ATM LAYER

As discussed previously, ATM can be implemented atop a variety of physical layers. On the other hand, ATM supports a variety of different services and traffic classes by providing different adaptation layers to higher network levels. The ability to do so efficiently over a shared infrastructure is made possible by a common middle layer, called the ATM layer.

ATM does not use the standard OSI seven-layer reference model, but the ATM layer performs many of the functions of OSI level 2, the data link layer. For example, the ATM layer and AAL-5 together provide data link layer functionality similar to OSI layer 2, i.e., the ability to transmit error-free frames of size up to 64,000 bytes from a source to a destination ATM entity [5].

The ATM layer is responsible for switching and multiplexing of ATM cells. Because ATM is based on switched point-to-point links, as opposed to a broadcast medium, ATM functionality is basically connection oriented (although connectionless services are supported through an adaptation layer). Although the ATM layer performs the basic operations that transmit cells along multilink paths from source to destination, the establishment of these paths [i.e., network-layer routing using, for example, an Internet protocol address] is left to higher layers. The ATM layer is designed for simplicity and for ease of high-speed hardware implementation.

11.3.5.1. Virtual Channels and Paths

A virtual channel is a contiguous stream of cells transmitted between two points in an ATM network (e.g., a single user's data stream). To reach the destination from the source, this data stream must traverse a set of ATM switches, going out a particular port on each switch to reach the next switch. This constitutes a virtual path, and many virtual channels may share the same virtual path. Virtual channels can be thought of as being contained inside virtual paths (Fig. 11.11).

Each ATM cell header has 24 bits for identifying the VC and VP that a cell belongs to, the VCI and VPI fields, respectively. This information ultimately is used to route the cell to the correct output port on each switch in its intended path. The generic local routing procedure at each switch is as follows: When a cell enters an input port, its header is examined and the VCI and/or VPI field is extracted. This information is used to index a look-up table, which (i) identifies the output port to send the cell to and (ii) provides new values to be placed into the outgoing cell's VCI and/or VPI fields.

Hence, the VCI and VPI fields are not constant, but rather change on each hop through the network. A virtual path (channel) is a string of VPI (VCI) values stored as values in the switch look-up tables, forming a linked

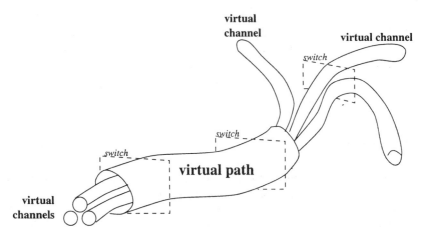

Fig. 11.11   Virtual channels and virtual paths.

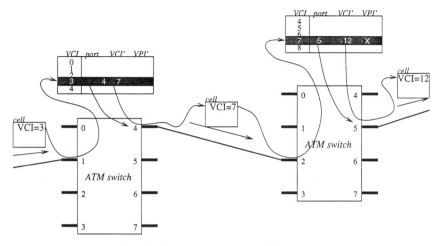

Fig. 11.12   Virtual path establishment.

list of table entries along the path (Fig. 11.12). A connection-establishment procedure is responsible for setting up the proper look-up table values whenever a new channel or path is initiated.

Depending on the kind of switch, either just the VPI information is used in routing or both VPI and VCI fields are used. Because logically VCs are viewed as being contained inside VPs, VP-only switches are not allowed to substitute VCI fields of outgoing cells; VP-only routing is considered a lower sublayer than VC routing [3].

Note also that cells entering a particular switch destined for different output ports must have distinct VPI values. Within a given VP, different VCs are distinguished by differing VCI values. However, different channels may share the same VCI value if their VPI fields differ (they belong to a different virtual path).

References

1. Hac, A., and H. B. Mutlu. 1989 (November). Synchronous optical network and broadband ISDN protocols. *Computer* 22(11):26–34.
2. Stallings, W. 1992. *ISDN and broadband ISDN*, 2nd ed. New York: Macmillan.
3. Minoli, D. 1993. *Enterprise networking: Fractional T1 to SONET, frame relay to BISDN*. Boston: Artech House.

4. DeCusatis, C. 1995. Data processing systems for optoelectronics. In *Optoelectronics for data communication,* ed. R. Lasky, U. Osterberg, and D. Stigliani, chap. 6. New York: Academic Press.

5. Miller, A. 1994 (June). From here to ATM. *IEEE Spectrum* 31(6):20–24.

6. Jungkok Bae, J., and T. Suda. 1991 (February). Survey of traffic control schemes and protocols in ATM networks. *Proc. IEEE* 79(2):170–189.

7. Roohalamini, R., V. Cherkassky, and M. Garver. 1994 (April). Finding the right ATM switch for the market. *Computer* 27(4):16–28.

8. ATM Forum Inc. 1995. *ATM user network interface (UNI) specification Version 3.1, 1/e.* New York: Prentice Hall.

9. The ATM Forum Technical Committee. 1994. DS1 physical layer specification. Technical Report AF-PHY-0016.000, The ATM Forum, September.

10. The ATM Forum Technical Committee. 1995. Physical interface specification for 25.6 Mb/s over twisted pair cable. Technical Report AF-PHY-0040.000, The ATM Forum, November 7.

11. The ATM Forum Technical Committee. 1996. 622.08 Mb/s physical layer specification. Technical Report AF-PHY-0046.000, The ATM Forum, January.

12. Klessig, B. 1995 (July). Status of ATM specifications. http://www.3com.com/0files/mktg/pubs/3tech/795atmst.html.

Chapter 12 | Fibre Channel Standard

Scholto Van Doren

Siemens Corporation, Santa Clara, California 95054

Casimer DeCusatis

IBM Corporation, Poughkeepsie, New York 12601

12.1. Introduction

This chapter provides an overview of the American National Standards Institute (ANSI) Fibre Channel Standard for data communications (this standard applies to both copper and fiber optic media and uses the term "fibre" to denote both types of physical layers). We will also take the opportunity in this chapter to mention Gigabit Ethernet, an emerging standard under the governance of ANSI that is similar in some ways to Fibre Channel and that could become very important in the next few years. Before going into the details of this standard, we will provide some background and motivation for this subject by describing the international standardization process in general, and the International Standards Organization (ISO)/ ANSI standards in particular. The goal of standards is to write a minimal set of requirements to achieve a maximum amount of interoperability. A minimum set of requirements is important to encourage innovation and multiple solutions to a design problem. This in turn gives the end user a choice among multiple products from different suppliers that will interoperate. Standards are documented agreements containing technical specifications or other precise criteria to be used consistently as rules, guidelines, or definitions of characteristics to ensure that materials, products, processes, and services are fit for their purpose. However, it should be noted that a standard usually describes only the minimum, high-level requirements for interoperability and thus it is not the same as a product specification.

415

HANDBOOK OF FIBER OPTIC
DATA COMMUNICATION

Indeed, the intent of a standard is to allow for many different design implementations, each of which may have its own specification and may differ in detail that are important to the end user, such as reliability, higher level functionality, etc. We can see that although multiple products may claim compliance with a given standard, they may have component or functional specifications that are different.

Standardization is a condition existing within a particular industrial sector when the large majority of products or services conform to the same standards. It results from consensus agreements reached between all economic players in that industrial sector — suppliers, users, and often governments. They agree on specifications and criteria to be applied consistently in the choice and classification of materials, the manufacture of products, and the provision of services. The aim is to facilitate trade, exchange, and technology transfer through the following:

- Enhanced product quality and reliability at a reasonable price
- Improved health, safety and environmental protection, and reduction of waste
- Greater compatibility and interoperability of goods and services
- Simplification for improved usability
- Reduction in the number of models, and thus reduction in costs
- Increased distribution efficiency and ease of maintenance

Users have more confidence in products and services that conform to international standards. Assurance of conformity can be provided by manufacturers' declarations or by audits carried out by independent bodies.

12.2. Why Are International Standards Needed?

The existence of nonharmonized standards for similar technologies in different countries or regions can contribute to so-called "technical barriers to trade." Export-minded industries have long sensed the need to agree on world standards to help rationalize the international trading process. This was the origin of the establishment of ISO. International standardization is now well established for a wide variety of technologies in such diverse fields as information processing and communications, packaging, distribution of goods, energy production and utilization, shipbuilding, banking, and financial services. It will continue to grow in importance for all sectors of industrial activity for the foreseeable future. The main reasons for this are the following:

- Worldwide progress in trade liberalization: Today's free-market economies increasingly encourage diverse sources of supply and provide opportunities for expanding markets. On the technology side, fair competition needs to be based on identifiable, clearly defined common references that are recognized from one country to the next and from one region to the other. An industrywide standard, internationally recognized, developed by consensus among trading partners, serves as the language of trade.

- Interpenetration of sectors: No industry in today's world can truly claim to be completely independent of components, products, rules of application, etc. that have been developed in other sectors. Bolts are used in aviation and for agricultural machinery; welding plays a role in mechanical and nuclear engineering, and electronic data processing has penetrated all industries. Environmentally friendly products and processes and recyclable or biodegradable packaging are pervasive concerns.

- Worldwide communications systems: The computer industry offers a good example of technology that needs to be standardized quickly and progressively at a global level. ISO's open systems interconnection (OSI) is the best known series of international standards in this area. Full compatibility among open systems fosters healthy competition among producers and offers real options to users because it is a powerful catalyst for innovation, improved productivity, and cost-cutting.

- Global standards are needed for emerging technologies: Standardization programs in completely new fields are being developed. Such fields include advanced materials, the environment, life sciences, urbanization, and construction. In the very early stages of new technology development, applications can be imagined but functional prototypes do not exist. Here, the need for standardization is in defining terminology and accumulating databases of quantitative information.

- Developing countries: Development agencies are increasingly recognizing that a standardization infrastructure is a basic condition for the success of economic policies aimed at achieving sustainable development. Creating such an infrastructure in developing countries is essential for improving productivity, market competitiveness, and export capability.

12.3. ISO

The technical work of ISO is highly decentralized, carried out in a hierarchy of approximately 2700 technical committees, subcommittees, and working groups. In these committees, qualified representatives of industry, research institutes, government authorities, consumer bodies, and international organizations from all over the world come together as equal partners in the resolution of global standardization problems.

The major responsibility for administrating a standards committee is accepted by one of the national standards bodies that make up the ISO membership — AFNOR, ANSI, BIS, CSBTS, DIN, SIS, etc. The member body holding the secretariat of a standards committee normally appoints one or two persons to do the technical and administrative work. A committee chairperson assists committee members in reaching consensus. Generally, a consensus will mean that a particular solution to the problem at hand is the best possible one for international application at that time.

12.4. An Introduction to ANSI

ANSI has served in its capacity as administrator and coordinator of the United States private-sector voluntary standardization system for 78 years. Founded in 1918 by five engineering societies and three government agencies, the institute remains a private, nonprofit membership organization supported by a diverse constituency of private- and public-sector organizations. Throughout its history, the ANSI federation has maintained as its primary goal the enhancement of global competitiveness of U.S. business and the American quality of life by promoting and facilitating voluntary consensus standards and conformity assessment systems and promoting their integrity. The institute represents the interests of its nearly 1400 company, organization, government agency, institutional, and international members through its headquarters in New York City and its satellite office in Washington, DC.

ANSI does not itself develop American National Standards (ANSs); rather, it facilitates development by establishing consensus among qualified groups. The institute ensures that its guiding principles — consensus, due process, and openness — are followed by the more than 175 distinct entities currently accredited under one of the federation's three methods of accreditation (organization, committee, or canvass). In 1995, the number of ANSs increased by 10% to a total of 11,500. ANSI-accredited developers are committed to supporting the development of national and, in many cases, international standards, and addressing the critical trends of technological innovation, marketplace globalization, and regulatory reform. ANSI promotes the use of U.S. standards internationally, advocates U.S. policy and technical positions in international and regional standards organizations, and encourages the adoption of international standards as national standards where these meet the needs of the user community. ANSI is the sole U.S. representative and dues-paying member of the two major nontreaty international standards organizations, ISO and, via the U.S. National Committee (USNC), the International Electrotechnical Commission (IEC).

ANSI was a founding member of the ISO and plays an active role in its governance. ANSI is one of five permanent members to the governing ISO council and one of four permanent members of ISO's Technical Management Board. United States participation, through the USNC, is equally strong in the IEC. The USNC is one of 12 members on the IEC's governing Committee of Action and the current president of the IEC is from the United States. Through ANSI, the United States has immediate access to the ISO and IEC standards development processes. ANSI participates in almost the entire technical program of both the ISO (78% of all ISO technical committees) and the IEC (91% of all IEC technical committees) and administers many key committees and subgroups (16% in the ISO and 17% in the IEC). As part of its responsibilities as the U.S. member body to the ISO and the IEC, ANSI accredits U.S. Technical Advisory Groups (U.S. TAGs) or USNC Technical Advisors (TAs). The U.S. TAGs' (or TAs') primary purpose is to develop and transmit, via ANSI, U.S. positions on activities and ballots of the international technical committee. In many instances, U.S. standards are taken forward, through ANSI or USNC, to the ISO or IEC, where they are adopted in whole or in part as international standards. Because the work of international technical committees is carried out by volunteers from industry and government, the success of these efforts often is dependent on the willingness of U.S. industry and the U.S. government to commit the resources required to ensure strong U.S. technical participation in the international standards process.

Conformity assessment, the term used to describe steps taken by both manufacturers and independent third parties to assess conformance to standards, also remains a high priority for the institute. ANSI's program for accrediting third-party product certification has experienced significant growth in recent years, and the institute continues its efforts to obtain worldwide acceptance of product certifications performed in the United States and the promotion of reciprocal agreements between U.S. accreditors and certifiers. One of the best indicators of the strength of the U.S. system is the government's extensive reliance on, and use of, private-sector voluntary standards. Pursuant to Office of Management and Budget (OMB) Circular A119, federal government agencies are required to use voluntary standards for regulatory and procurement purposes when appropriate. State and local governments and agencies have formally adopted thousands of voluntary standards produced by the ANSI federation, and the process appears to be accelerating. In summary, the ANSI federation continues to be fully involved in its support of the goals of U.S. and global standardization and committed to enhancing of the quality of life for all global citizens.

12.4.1. X3 COMMITTEE

The ANSI Accredited Standards Committee X3, more commonly referred to as simply X3, was established in 1961. It is accredited by ANSI to develop voluntary standards. The X3 secretariat is held by Information Technology Industry Council. As a standards developer in information technology, X3 is an active participant in the U.S. TAG to ISO/IEC JTC 1 (JTC1 TAG). X3 provides its 2000 members with countless opportunities to know and influence the key issues, activities, and people that drive national and international information technology standards in dynamic areas of commerce, technology, and society. X3 members have early and ready access to a wealth of technical information affecting the future of information technology on a global basis. Members actively participate in a deliberative consensus-building process that is both national and international in scope. The X3 Operational Management Committee (OMC) is an advisory committee to X3 on new national and international standards development projects, standards requirements, and review of proposed standards. OMC also manages the standardization process within the X3 technical committees (TCs). OMC considers the functional and economic, rather than the detailed technical aspects of standardization. X3's standards are developed by its TCs. A brief description of the TCs relevant to this book follows:

X3T10: I/O interface — Lower level (SCSI)

X3T11: I/O interface — Device level (IPI, HIPPI, FC)

The current chairman of X3T11:
  Roger Cummings
  Distributed Processing Technology, Inc.
  140 Candace Drive
  Maitland, FL 32751
Phone: (407) 830-5522 x348
Fax: (407) 260-5366
Vice chair:
  Carl Zeitler
  IBM Austin, MS 9250
  11400 Burnet Road
  Austin, TX 78758, USA
  Phone: (512) 838-1797

X3T12: I/O interface — Distributed data (FDDI)

X3T13: I/O interface — AT attachment

You may order completed ANSI publications by calling
  The American National Standards
  11 West 42nd Street
  New York, New York 10036
  Phone: (212) 642-4900 Agents are available from 8:45 AM to
4:45 PM (EST)

For standards in progress:
  Global Engineering Documents
  15 Inverness Way East
  Englewood, CO 80112 USA
  Phone: 303-397-2715; 800-854-7179 (United States and Canada)

The Fibre Channel Association can be reached on the World Wide
Web at http://www.amdahl.com/ext/CARP/FCA/FCA.html

12.5. IEEE 802.3z (Gigabit Ethernet)

Before we discuss Fibre Channel Standard, we will describe the emerging
gigabit Ethernet standard. This standard is not yet finalized and may undergo many changes before its final adoption sometime in 1997 or 1998.
The set of objectives for IEEE P802.3z (gigabit Ethernet) standard include
the following:

  1. Speed of 1000 Mb/s at the MAC/PLS service interface

2. Use 802.3/ethernet frame format

3. Meet 802 FR, with the possible exception of Hamming distance

4. Simple forwarding between 1000, 100, and 10 Mb/s

5. Preserve min and max frame size of current 802.2 Std

6. Full- and half-duplex operation

7. Support star-wired topologies

8. Use CSMA/CD access method with support for at least one repeater/collision domain

9. Support fiber media and, if possible, copper media

10. Use ANSI Fibre Channel FC-1 and FC-0 as basis for work

11. Provide a family of physical layer specifications that support a link distance of

    At least 500 m on multimode fiber

    At least 25 m on copper (100 m preferred)

    At least 2 km on single-mode fiber

12. Decide between collision domain diameter of $\geq$50 m or $\geq$ 200 m

13. Support media selected from ISO/IEC 11801

14. Accommodate proposed P802.3x flow control

There are five criteria motivating the development of gigabit Ethernet technology; its market potential, compatibility with existing Ethernet standards, its distinguishing features from other standards, and its technical and economic feasibility. We will review each of these criteria in turn to give some insight into the forces guiding the standards committee because the final specification has not yet been written.

12.5.1. BROAD MARKET POTENTIAL

- Broad set(s) of applications
- Multiple vendors, multiple users
- Balance cost, local area network (LAN) vs attached stations

The fast growth of CPU speed is forcing the development of new LANs with higher bandwidth. Applications that can benefit from this capability include backbone, server, and gateway connectivity. Higher bandwidth for multimedia, distributed processing, imaging, medical, CAD/CAM, and other applications is also motivating the aggregation of 100 Mb/s switches and the potentially large upgrade base for installed 10/100 Ethernet. Multiple vendors and users have demonstrated interest by attending the gigabit Ethernet tutorial (more than 200 participants), attending the preliminary study group meeting (more than 120 participants), and enrolling in the higher speed e-mail reflector (more than 210 participants). Eighty-one participants representing at least 54 companies indicate that they plan to participate in the standardization of 1000 Mb/s 802.3. This level of commitment indicates that a standard will be supported by a large group of vendors. This in turn will ensure that there will be a wide variety of equipment to support a multitude of applications. Higher speed 802.3 solutions, which include scaled-up versions of existing 802.3 topologies, have balanced cost. Prior experience with scaling 802.3 across the range of 1 to 100 Mb/s indicates that the cost balance between adapters, cabling, and hubs remains roughly constant, provided that the operating speed can be achieved within the limits of current technology.

12.5.2. COMPATIBILITY WITH IEEE STANDARD 802.3

- Conformance with CSMA/CD media access control (MAC), PLS
- Conformance with 802.2
- Conformance with 802 FR

The proposed standard will conform to the CSMA/CD MAC, with currently authorized extensions appropriately adapted for 1000 Mb/s operation. In a fashion similar to the 100 BASE-T standard, the current physical layers will be replaced with new physical layers (PHY) as appropriate for 1000 Mb/s operation. The proposed standard will conform to the 802.2 LLC interface and the 802 Functional Requirements Document (with the possible exception of Hamming distance). The CSMA/CD access method will not support a 2-km network diameter at this speed while maintaining the current values in the MAC parameter table. This portion of the application space will be addressed at 1000 Mb/s with the full-duplex operating mode of 802.3.

12.5.3. DISTINCT IDENTITY

- Substantially different from other 802.3 specs/solutions
- Unique solution for problem (not two alternatives/problem)
- Easy for document reader to select relevant spec

The proposed standard is an upgrade for 802.3 users, based on the 802.3 CSMA/CD MAC, running at 1000 Mb/s. Maximum compatibility with the installed base of more than 60 million CSMA/CD nodes is maintained by adapting the existing CSMA/CD MAC protocol for use at 1000 Mb/s. Established benefits of CSMA/CD and the 802.3 MAC include the following:

- Optimistic transmit access method
- High efficiency in full-duplex operating mode
- Well-characterized and understood operating behavior
- Broad base of expertise in suppliers and customers
- Straightforward bridging between networks at different data rates

The management information base (MIB) for 1000 Mb/s 802.3 will maintain consistency with the 802.3 MIB for 10/100 Mb/s operation. Therefore, network managers, installers, and administrators will see a consistent management model across all operating speeds. The proposed standard will encompass one physical layer solution for each specific type of network media (e.g., single-mode fiber, multimode fiber, coaxial cable, and balanced pair cable). It will be a supplement to the existing 802.3 standard, formatted as a collection of new clauses, making it easy for the reader to select the relevant specification.

12.5.4. TECHNICAL FEASIBILITY

- Demonstrated feasibility; reports and working models
- Proven technology and reasonable testing
- Confidence in reliability

Technical presentations, given to 802.3, have demonstrated the feasibility of using the CSMA/CD MAC in useful network topologies at a rate of 1000 Mb/s. Technical presentations given to 802.3 from multiple current vendors of full-speed Fibre Channel components have demonstrated the feasibility of physical layer signaling at a rate of 1.06 Gbaud on both fiber optic and copper media. Many of these vendors have expressed support for an increase in the signaling rate to 1.25 Gbaud, which would support a MAC data rate of 1000 Mb/s. The principle of scaling the CSMA/CD MAC to higher speeds has been well established by previous work within 802.3. The 1000 Mb/s work will build on this experience. The principle of building bridging equipment that performs rate adaptation between 802.3 networks operating at different speeds has been amply demonstrated by the broad set of product offerings that bridge between 10 and 100 Mb/s.

Vendors of full-speed Fibre Channel components and systems are building reliable products that operate at 1.06 Gbaud and that meet worldwide regulatory and operational requirements.

The FC-1 layer is the encode/decode layer and defines the DC-balanced 8B/10B code, byte synchronization, and character-level error control. In the FCS standard, 8-bit data bytes rate converted to 10-bit transmission characters with an error-checking code; the 8-bit bytes are recovered at the other end of the link. Using 10 bits rather than 8 provides for 1024 possible encoded values rather than only 256. Not all these values are used; to maintain the DC balance of the code, only characters that contain 4 zeros and 6 ones, 6 zeros and 4 ones, or 5 zeros and 5 ones are used. Some of the extra characters are reserved for lwo-level link control, and one special character called a comma or K28.5 is used for byte synchronization.

The FC-2 layer is the framing protocol layer; it defines the signaling protocol including frame and byte structure. This level includes framing protocol, flow control, 32-bit CRC generation, and various classes of service definitions. A frame is the smallest unit of data that can be transmitted on a Fibre Channel link. Frame size depends on the implementation and is variable up to a maximum of 2148 bytes. Each frame contains a 4-byte start of frame delimiter, a 24-byte header, up to 2112 bytes of FC-4 payload, a 4-byte CRC, and a 4-byte end of frame delimiter. The FC-4 payload consists of 0–64 bytes of optional headers and 0–2048 bytes of data. Related frames are grouped together to form a sequence, and multiple sequences can be grouped to form an exchange [comparable to a SCSI input/output (I/O) process]. If an error is detected that requires a recovery operation, then the sequence containing that frame may be retransmitted. The FC-2 layer also defines three classes of service. Class 1 establishes a dedicated connection between two devices (a circuit-switched connection), guarantees delivery of frames in the order in which they were transmitted, and provides confirmation of delivery. Class 2 is a multiplexed, connectionless, frame-switched link; delivery is guaranteed with acknowledgment of receipt but not necessarily in the order of transmission. Class 3 does not guarantee either delivery or order of receipt. If no fabric is present, class 2 and 3 become special cases of point-to-point interconnection.

The FC-3 or common services layer defines a set of services common across multiple ports of a node. This layer is still under discussion and no functions have been formally defined at this time.

The FC-4 or protocol mappings layer is the topmost layer and defines mapping of the upper level protocol (ULP) layers to Fibre Channel. Each

ULP supported by FCS is specified in a separate FC-4 document; for example, the Fibre Channel protocol for SCSI is called FCP; it supports SCSI without requiring an SCSI bus because the upper layer protocols are not tied to a particular physical medium or interface.

There are many FCS-compatible products available from workstations to transceivers and chip sets. In 1994, IBM announced the use of FC-0 links as part of its parallel coupling facility for large servers [1]. These intersystem channels offer two options: a shortwave, 531 Mb/s link on multimode fiber with a maximum distance of 1 km and a longwave, 1062.5 Mb/s link on single-mode fiber with a maximum distance of 3 km [1 km longer than the maximum distance in the standard, achieved by using nonstandard open fiber control (OFC) timing]. Both links use OFC, although the standard does not specify timing for gigabit links; this implementation uses timing for the 266 Mb/s OFC links and thus achieves a slightly longer maximum distance. These links have been used on three generations of large, CMOS-based servers to interconnect parallel processors, including the Multiprise 2000 and Generation 3 machines announced in 1996.

12.5.5. *ECONOMIC FEASIBILITY*

- Cost factors known and reliable data
- Reasonable cost for performance expected
- Total installation costs considered

Cost factors are derived from the current full-speed Fibre Channel component supplier base. A reasonable cost increase (three times that of 100BASE-FX) with a 10-fold increase in available bandwidth in the full-duplex operating mode will result in an improvement in the cost/performance ratio by a factor of 3.33 for multimode fiber applications. The provision for a half-duplex operating mode using the 802.3 CSMA/CD MAC will permit the construction of very inexpensive repeating hubs. Customers will in many cases be able to re-use their existing fiber that has been installed in accordance with ISO/IEC 11801. Installation costs for new fiber runs based on established standards are well known and reasonable. Costs for coaxial-based short-run copper links are well established for full-speed Fibre Channel. Although the cost model for the horizontal copper cabling is well established, the cost model for 1000 Mb/s physical layers that will operate on horizontal copper cabling has not yet been firmly established. Presentations have been given to the high speed study group (HSSG), that suggest a cost multiple of two times relative to 100BASE-T2.

12.6.   Fibre Channel Standard

Fibre Channel Standard is a data transfer protocol intended to meet the emerging needs of high-bandwidth communication links. Development of this standard began in 1988 under the auspices of the X3T9.3 Working Group, as an outgrowth of the Intelligent Physical Protocol (IPI) Enhanced Physical Project. The motivation behind Fibre Channel was to develop an approach that encompassed the desirable features of many existing protocols, such as small computer interface (SCI) and high-performance parallel interface (HIPPI), while providing a scalable method for transferring data at the fastest speeds currently achievable. The standard provides for the attachment of both networking and peripheral I/O communication over a single channel using the same drivers, ports, and adapters. In October 1994, the Fibre Channel physical and signaling interface standard FC-PH was approved as ANSI standard X3.230-1994. Many vendors have stated their support for Fibre Channel Standard (FCS) attachments; in particular, leading workstation vendors, such as IBM, SUN, and HP, have formed the Common Optics Group to develop a common communications profile for workstations based on FCS [2].

Fibre Channel is logically a bidirectional point-to-point serial interface, using IBM's 8B/10B encode/decode scheme [3]. Physically, it can be implemented in three different topologies to interconnect different numbers of devices, or nodes [1; 4–9]. These topologies are called point to point, arbitrated loop, and cross-point switched or fabric. Each node contains one or more ports through which the topologies communicate over the FCS network; the standard refers to a generic node point as an N port, and the connections between ports are called links.

The point-to-point topology is simply a direct link between two devices, such as a processor and a device controller. This is the simplest default topology; because only two devices are connected together, there is no need for more complex functions such as name mapping or switching.

Fibre Channel arbitrated loop, or FC-AL [5], provides for the interconnection of between 2 and 126 devices through attachments called L ports in a loop configuration. An L port may consist of I/O devices or processor systems; because this interconnection loop does not require hubs or switches, it is intended to be a low-cost solution for small to medium-sized applications. The loop provides an upward growth path for interconnections to a switched fabric, which we will discuss shortly. All devices on the loop share the loop bandwidth and contribute to management of the loop; there is no dedicated loop master. Any node can become loop master; this is selected when the loop is initialized. After initialization, each node arbitrates for temporary control of the loop using a fairness algorithm; each node has equal opportunity to communicate with any other node on the loop. Only one communication path, or loop circuit, can be active at any time between two devices on the loop. After completing communications and relinquishing control of the loop, a given port cannot win arbitration again until all other ports have had their turn. There are several different kinds of loops. A private loop does not connect with a fabric but rather only to other private nodes via attachment points called *NL* ports. A public loop does attach to a fabric through an *FL* port. A disk loop interconnects a number of high-performance storage disks [for example, a redundant array of inexpensive disks (RAID) architecture].

The third FCS topology is an interconnected switched network or fabric. A switchable fabric, as defined by FCS, refers to a netwrok in which all station management functions are controlled by switching point rather than by each node. An analogy is the telephone network, in which users specify an address (phone number) of the device with which they want to communicate and the switched network provides them with an interconnection path. This approach removes the need for complex switching algorithms at each node. A fabric appears as a single entity to attached nodes, for *F* ports. In principle, there is no limit to the number of nodes that can be interconnected in a fabric; practically, addressing space limits the maximum to approximately 16 million unique addresses and a typical switch may have between 4 and 16 *F* ports attached. An example of an FCS fabric is shown in Fig. 12.1, with a bridge providing interconnection to other protocols such as asynchronous transfer mode (ATM), fiber distributed data interface (FDDI), and HIPPI [1]. Development of FCS is expected to compliment current efforts in ATM development of high-speed packet switching.

Fibre Channel is structured as a set of five hierarchical layers, denoted as FC-0–FC-4. FCS supports a number of existing protocols such as SCSI, Internet protocol, or IPI (known as upper-level protocols); these are outside the scope of FCS, which is not a high-level protocol like SCSI. FCS does not have a native command set, and it is not aware of the content of the data being transported. Networking protocols are mapped to Fibre Channel constructs, encapsulated in FCS frames, and transported across the FCS fabric in a transparent fashion. The main goal of FCS is to provide the means for a large number of existing protocols to operate over a variety of physical media, as shown in Fig. 12.2. We now provide a brief overview of each layer of the Fibre Channel Standard.

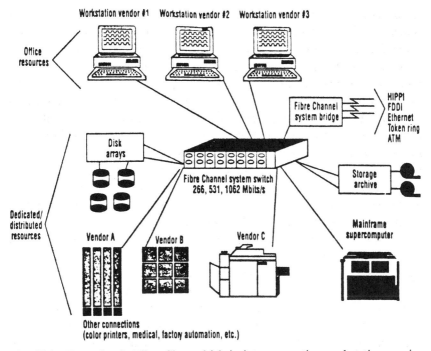

Fig. 12.1 Example of a Fibre Channel fabric, interconnecting workstations, mainframes, and storage through a circuit or packet-switched network. A bridge provides connections to other protocols such as ATM, FDDI, and HIPPI (from Ref. [1]).

The lowest of the five levels is FC-0, the physical layer. It defines the physical characteristics of the media (either optical fiber or copper), connectors, drivers, transmission rates, and other parameters. Table 12.1 summarizes the various supported media and speeds. The standard provides for a minimum bit error rate of 10e-12 in all cases, and additional physical parameters such as jitter are provided in an appendix to the standard. The standard provides for longwave laser (1300 nm) single-mode data links at 265–625, Mb/s, 531.25, and 1062.5 Mb/s over distances from 2 to 10 km. Multimode links are specified as well, with an additional data rate of 132.8125 Mb/s. The two slowest multimode speeds use longwave light-emitting diode (LED) sources, whereas the higher speeds on multimode fiber use shortwave lasers (780–850 nm). For each possible combination, the transmitter output, receiver sensitivity, and other factors are specified (see Tables 12.2 through 12.8); this guarantees operation of the link at the required distances and data rates if all components meet this specification. For optical fiber, FCS uses the SC duplex connector, keyed to prevent

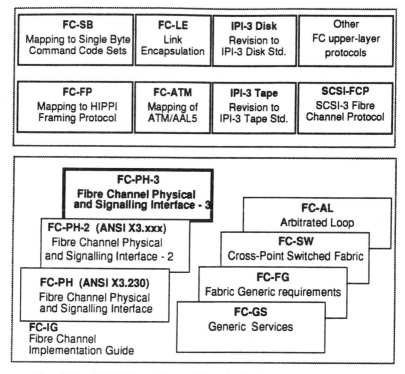

Fig. 12.2 ANSI Fibre Channel Standard document relationships.

insertion of a multimode fiber into a single-mode receptacle (see Chapter 1 for the SC duplex connector and Chapter 13 for the FCS single-mode connector design taken from the Fibre Channel Standard). Note that in the early 1980s, before the standard was finalized, a version of the SC duplex connector was produced in which the keys were rotated 90° to one side of the connector housing. The SC connector with rotated keys is not compatible with the SC connector adopted by Fibre Channel Standard, which has led to some confusion. Early literature refers to the "SC connector" and "SC duplex" connectors, meaning the version with the rotated keys; although this version is not in wide use today, some applications continue to use this connector type. Some literature attempts to resolve the confusion by referring to the nonrotated keys as the "FCS connector" or "FCS duplex" connector. However, most applications have now stan-

Table 12.1   **Fibre Channel Media Performance**

Media type	Data rate (Mb/s)	Max. distance	Signaling rate (Mbaud)	Transmitter
Single-mode fiber	100	10 km	1062.5	LW laser
	50	10 km	1062.5	LW laser
	25	10 km	1062.5	LW laser
50 μm multimode fiber	100	500 m	1062.5	SW laser
	50	1 km	531.25	SW laser
	25	2 km	265.625	SW laser
	12.5	10 km	132.8125	LW LED
62.5 μm multimode fiber	100	300 m	1062.5	SW laser
	50	600 m	531.25	SW laser
	25	1 km	265.625	LW LED
	12.5	2 km	132.8125	LW LED
105-ohm type-1 shielded twisted-pair electrical	25	50 m	265.125	ECL
	12.5	100 m	132.8125	ECL
75-ohm mini coax	100	10 m	1062.5	ECL
	50	20 m	531.25	ECL
	25	30 m	265.625	ECL
	12.5	40 m	132.8125	ECL
75-ohm video coax	100	25 m	1062.5	ECL
	50	50 m	531.25	ECL
	25	75 m	265.625	ECL
	12.5	100 m	132.8125	ECL
150-ohm twinax or STP	100	30 m	1062.5	ECL
	50	60 m	531.25	ECL
	25	100 m	265.625	ECL

dardized on the shorthand notation "SC duplex" for the nonrotated keys, while simply avoiding references to the earlier connector versions.

All FCS links must meet international class 1 laser safety standards (see Chapter 7). This can be achieved by using conventional mechanical shutters, by limiting the maximum launch power into the fiber, or by using OFC, which is specified in the standard for multimode links only. Using OFC, the link transceivers perform a handshake to initialize the channel (implemented in hardware); if the full-duplex link is interrupted

Table 12.2 **Optical Cable Plant Overview**

Single mode				
100-SM-LL-L	100-SM-LL-I	50-SM-LL-L	25-SM-LL-L	25-SM-LL-I
Clause 6.1	Clause 6.1	Clause 6.1	Clause 6.1	Clause 6.1
SM	SM	SM	SM	SM
1300 nm	1300 nm	1300 nm	1300 nm	1300 nm
2 m–10 km	2 m–2 km	2 m–10 km	2 m–10 km	2 m–2 km
Multimode (62.5 μm)				
100-M6-SL-I	50-M6-SL-I	25-M6-SL-I	25-M6-LE-I	12-M6-LE-I
Clause 6.2	Clause 6.2	Clause 6.2	Clause 6.3	Clause 6.3
MM	MM	MM	MM (LED)	MM (LED)
780 nm	780 nm	780 nm	1300 nm	1300 nm
2–175 m	2–350 m	2–700m	2 m–1.5 km	2 m–1.5 km
Multimode (50 μm)				
100-M5-SL-I	50-M5-SL-I	25-M5-SL-I	25-M5-LE-I	12-M5-LE-I
Clause 6.2	Clause 6.2	Clause 6.2	Clause 6.3	Clause 6.3
MM	MM	MM	MM (LED)	MM (LED)
780 nm	780 nm	780 nm	1300 nm	1300 nm
2–500 m	2 m–1.5 km	2 m–2 km	*	*

* Alternate fiber cable plant.

Table 12.3 **Electrical Cable Plant Overview**

Video Coax			
100-TV-EL-S	50-TV-EL-S	25-TV-EL-S	12-TV-EL-S
Clause 7.2	Clause 7.2	Clause 7.2	Clause 7.2
0–25 m	0–50 m	0–75 m	0–100 m
Miniature Coax			
100-MI-EL-S	50-MI-EL-S	25-MI-EL-S	12-MI-EL-S
Clause 7.2	Clause 7.2	Clause 7.2	Clause 7.2
0–10 m	0–15 m	0–25 m	0–35 m
Shielded twisted pair			
25-TP-EL-S	12-TP-EL-S		
Clause 7.3	Clause 7.3		
0–50 m	0–100 m		

Table 12.4A   **Single-Mode Cable Plant**

FC-0	100-SM-LL-L	100-SM-LL-I	50-SM-LL-L	25-SM-LL-L	25-SM-LL-I
Clause	6.1	6.1	6.1	6.1	6.1
Operating range	2 m–10 km	2 m–2 km	2 m–10 km	2 m–10 km	2 m–2 km
Cable plant dispersion (ps/nm)	35	12	60	60	12
Dispersion-related penalty (dB)	1	1	1	1	1
Reflection-related penalty (dB)	1	1	1	1	1
Loss budget (dB)	14	6	14	14	6

Table 12.4B   **62.5-μm Multimode Cable Plant**

FC-0	25-M6-LE-I	12-M6-LE-I
Clause	6.3	6.3
Data rate	25 Mb/s	12 Mb/s
Operating range	2 m–1.5 km	2 m–1.5 km
Loss budget	6 dB	6 dB

Table 12.4C   **62.5-μm Fiber Type**

Nominal Core Diameter (FOTP-58)	Cladding Diameter (FOTP-45 and FOTP-176 or FOTP-48)	Nominal Numerical Aperture (FOTP-177)
62.5 μm	125 μm	0.275

Table 12.5   62.5-μm Multimode Bandwidth

Wavelength (nm)	Modal Bandwidth (· 3 dB optical min) (MHz · km)	Test Per
850	160	FOTP-30 or FOTP-51 with FOTP-54
1300	500	FOTP-30 or FOTP-51 with FOTP-54

(because of a broken fiber or failed transmitter, for example) then the link automatically detects this condition and shuts down transmitters at both ends of the link. This occurs quickly enough so that there is no danger of human exposure to unsafe optical power levels; the OFC timing is specified as part of the standard. When the link enters this state, both transmitters emit a low duty cycle pulse train until the link is closed again; when this occurs, the condition is detected and the link attempts to reestablish itself. An FCS transceiver contains identifying information to ensure that a module with OFC capability will only interoperate with another module of the same type; the module type and speed are called out by 2 bits implemented in hardware and passed on to the rest of the system, either as a serial or parallel ID.

Table 12.6   50-μm Multimode Cable Plant

FC-0	100-M5-SL-I	50-M5-SL-I	25-M5-SL-I
Clause	6.2	6.2	6.2
Data rate (Mb/s)	100	50	25
Operating range	2–500 m	2 m–1 km	2 m–2 km
Loss budget (dB)	6	8	12

Table 12.7 FC-0 Physical Links for Single-Mode Classes

FC-0	Units	100-SM-LL-L	100-SM-LL-I	50-SM-LL-L	25-SM-LL-L	25-SM-LL-I
Clause		6.1	6.1	6.1	6.1	6.1
Data rate	Mb/s	100	100	50	25	25
Nominal bit rate	Mbaud	1062.5	1062.5	531.25	265.625	265.625
Tolerance	ppm	±100	±100	±100	±100	±100
Operating range (typ)	m	2 m–10 km	2 m–2 km	2 m–10 km	2 m–10 km	2 m–2 km
Fiber core diameter	μm	9	9	9	9	9
Transmitter (S)						
Type		Laser	Laser	Laser	Laser	Laser
Spectral center wavelength	nm (min)	1285	1270	1270	1270	1270
	nm (max)	1330	1355	1355	1355	1355
RMS spectral width	nm (max)	3	6	3	6	30
Launched power, max	dBm (ave)	−3	−3	−3	−3	−3
Launched power, min	dBm (ave)	−9	−12	−9	−9	−12
Extinction ratio	db (min)	9	9	9	6	6
RIN_{12} (max)	dB/Hz	−116	−116	−114	−112	−112
Eye opening @ BER = 10^{-12}	% (min)	57	57	61	63	63
Deterministic jitter	% (p-p)	20	20	20	20	20
Receiver (R)						
Received power, min	dBm (ave)	−25	−20	−25	−25	−20
Received power, max.	dBm (ave)	−3	−3	−3	−3	−3
Optical path power penalty	db (max)	2	2	2	2	2
Return loss of receiver	dB (min)	12	12	12	12	12

Table 12.8 FC-0 Physical Links for Multimode Classes

FC-0	Units	100-M5-SL-I	50-M5-SL-I	25-M5-SL-L	25-M6-LE-I	12-M6-LE-I
Clause		6.2	6.2	6.2	6.3	6.3
Data rate	Mb/s	100	50	25	25	12.5
Nominal bit rate	Mbaud	1062.5	531.25	265.625	265.625	132.8125
Tolerance	ppm	±100	±100	±100	±100	±100
Operating range (typ)	m	2–500	2 m–1 km	2 m–2 km	2 m–1.5 km	2 m–1.5 km
Fiber core diameter	μm	50	50	50	62.5	62.5
Transmitter (S)						
Type		Laser	Laser	Laser	LED	LED
Spectral center wavelength	nm (min)	770	770	770	1280	1270
	nm (max)	850	850	850	1380	1380
Spectral width	nm RMS (max)	4	4	4	NA	NA
	nm FWHM (max)	NA	NA	NA	—	250
Launched power, max	dBm (ave)	1.3	1.3	0	−14	−14

436

Launched power, min	dBm (ave)	−7	−7	−5	−20	−22
Extinction ratio	dB (min)	6	6	6	9	10
RIN$_{12}$ (max)	dB/Hz	−116	−114	−112	NA	NA
Eye opening @ BER = 10^{-12}	% (min)	57	61	63	NA	NA
Deterministic jitter	% (p-p)	20	20	20	16	24
Random jitter	% (p-p)	NA	NA	NA	9	12
Optical rise fall time	ns (max)	NA	NA	NA	2.0/2.2	4.0
Receiver (R)						
Received power, min	dBm (ave)	−13	−15	−17	−26	−28
Received power, max	dBm (ave)	+1.3	+1.3	0	−14	−14
Return loss of receiver	dB (min)	12	12	12	NA	NA
Deterministic jitter	% (p-p)	NA	NA	NA	19	24
Random jitter	% (p-p)	NA	NA	NA	9	12
Optical rise/fall time	ns (max)	NA	NA	NA	2.5	4.3

437

References

1. DeCusatis, C. 1995. Data processing systems for optoelectronics. In *Optoelectronics for data communication*, eds. R. Lasky, U. Osterberg, and D. Stigliani, 219–283. New York: Academic Press.
2. Allan, I. D., and E. Frymayer. 1993 (May). FCS can break the datacom gridlock. *Electron. Design*, 87.
3. Widmer, A. X., and P. A. Franaszek. 1983. A DC balanced, partition block 8B/10B transmission code. *IBM J. Res. Dev.* 27:440–451.
4. American National Standards Institute. 1994. Fibre channel — Physical and signaling interface, (FC-PH), X3.230-1994, rev. 4.3.
5. American National Standards Institute. 1995 (June). Fibre Channel — Arbitrated loop (FC-AL), X3.272-199x, rev. 4.5.
6. American National Standards Institute. 1995 (May 30). Fibre Channel protocol for SCI (FCP), X3.269-199x, rev. 0.12.
7. The Fibre Channel Association. 1994. *Fibre Channel: Connection to the future.*
8. IBM Corporation. 1994. Coupling facility channel I/O interface physical layer document. IBM Document Number SA23-0395, rev. 1.0, 1994. Mechanicsburg, PA: IBM Corp.
9. Primmer, M. 1996 (October). An introduction to Fibre Channel. *Hewlett Packard J.* 47:94–98.

Chapter 13 | Enterprise Systems Connection Fiber Optic Link

Daniel J. Stigliani, Jr.

IBM Corporation, Poughkeepsie, New York 12601

13.1. Introduction

The information technology (IT) industry has been evolving at a tremendous rate since the advent of computers in the business world in the early 1950s. This "explosion" of IT (formerly data processing) in the commercial world since the late 1980s has been fueled by the following trends:

- Desire of business to achieve competitive advantage through the use of IT
- Development and exploitation of large-scale integration and complementary metal oxide semiconductor (CMOS) technology
- Desire to have ubiquitous information and processing capability
- Desire to realize efficiencies in the business process and reduce product cost
- Development of new application (e.g., multimedia and real-time video and audio at the desktop) and business communication models (e.g., World Wide Web)

All of these items and more have put tremendous emphasis on dissemination of data from server (host computer) to server, server to client, and client to client. The information transfer is likely to start with a client or server request for data from a large database and end with the storage of new or revised data in a remote database or stored locally, for further processing in the future (e.g., data mining). The key factors in a successful data transfer link or network interconnection are

- High aggregate throughput
- Low response latency

HANDBOOK OF FIBER OPTIC
DATA COMMUNICATION

- Easy to use
- Allowance for future growth

This environment has placed tremendous demands on International Business Machines Corporation (IBM) System/390 large servers to improve not only data processing and server capability but also the system interconnection capability. In the early 1990s IBM introduced the first in a series of new large-scale servers that provided a new system structure and architecture (Enterprise Systems Architecture/390) for coupling multiple data processing systems together and Enterprise System Connection (ESCON) architecture to provide high bandwidth interconnection capability for System/390 products and attachments. This was the beginning of the large-server interconnection network evolution into the modern information technology paradigm.

This chapter provides an understanding of the ESCON interconnection from a system perspective and design consideration. Detailed design parameters and implementation requirements of the ESCON fiber optic link are discussed. Section 13.2 discusses the system perspective, interconnection architecture and topology, and link-level protocol. The link design consideration and trade-offs are discussed in Section 13.3. Sections 13.4 and 13.5 present the ESCON physical layer specification for multimode and single-mode links, respectively. Section 13.6 discusses the planning and installation considerations and recommendations of an ESCON link. The link optical loss budget analysis is presented in Section 13.7, whereas link troubleshooting is reviewed in Section 13.8. The final section (13.9) discusses ESCON link eye safety, link qualification, and the impending single byte connectivity (SBCON) American National Standards Industry (ANSI) standard.

This chapter makes extensive use of illustrations and tables published in documents copyrighted by International Business Machines Corporation, which are listed in the References. The illustrations and figures are reprinted by permission from International Business Machines Corporation.

13.2. ESCON System Overview

ESCON systems architecture is a total network interconnection system for large server complexes [1, 2]. ESCON encompasses fiber optic technology links, serial data transfer, new link-level protocols, data encoding/decoding, new system transport architecture, and a new topology. The application for ESCON is intended as the backbone network that spans a customer premise. In some cases it may be a machine room, whereas in other cases it could be a large multibuilding campus that may span 20 km or more.

The objective of ESCON is to provide a versatile, robust, expandable, high-bandwidth interconnection system for IBM System/390 servers. Some of the key attributes of ESCON include the following:

- Provide high-speed data transfer capability (200 Mb/s) on a media with the ability to support enhanced performance
- Allow interconnection of ESCON-compatible devices at both short (4 m) and long distance (20 km)
- Allow the addition and deletion of components on the network (including servers) without disruption of the network and its operation
- Provide improved connectivity (multipath) between servers and attached devices
- Reduce the quantity and bulk of interconnection cables
- Ease configuration and installation complexity

13.2.1. DESIGN CONSIDERATIONS AND ATTRIBUTES

A key element of achieving the attributes described previously was the choice of a link technology that would result in

- High-speed data capability
- Long-distance capability
- Small size and weight
- Highly reliable
- Capable of significant performance improvement

The original copper implementation of the System/390 OEMI channel utilized a parallel data transfer protocol and was severely limited in distance capability by skew on the cable. This would not satisfy the previously mentioned requirements and a serial data transfer approach was adopted. This decision further exacerbated the need for a technology that supports very high-speed data transfer. After significant investigation (both theoretical and experimental) in both copper (e.g., coaxial cable) and fiber optic technology transport capability, it was decided that fiber optics technology could meet all the system requirements and support potential future growth. The inherent property of electromagnetic compatibility for fiber optics was also an important consideration.

In the mid-1980s fiber optic technology was being successfully implemented in the telecom environment but there had been no significant use in the datacom market. IBM System/390 division initiated an exhaustive technology investigation of fiber optic technology to define an application component set that provided the following:

- Long-distance transmission (several kilometers)
- High-speed capability (greater than 100 Mb/s)
- Proven capability for high reliability (both soft and hard failure)
- Application robustness in an uncontrolled building environment

The technology set of 1300-nm long-wavelength light-emitting diode (LW LED), 62.5/125 μm multimode fiber (MMF), and unidirectional transmission per fiber was chosen. To facilitate ease of use and avoidance of fiber misplugging, a push-on duplex optic fiber connection was developed. This is the base ESCON link technology. As the need for distances greater than 2 or 3 km became apparent, the ESCON Extended Distance Feature (XDF) was developed. This extended the ESCON distance to 20 km using a long-wavelength laser source and single-mode fiber (SMF). This capability completed the ESCON fiber technology set. Both of these fiber optic technologies are in use today and will be discussed in detail later in this chapter.

The use of fiber optic technology and long-distance transmission changed the System/390 interconnection paradigm from a parallel asynchronous data connection approach to a high-speed serial data communication link. The properties (e.g., long-distance transmission and extremely small number of random bit errors) of such a link fostered a new architecture. In order to facilitate recovering the data, with minimal latency and error, the architecture was designed such that information "idle characters" are continually transmitted to ensure the receiver is continually synchronized to the transmitter. During a data transfer, the receiver is locked to the incoming data by design and can transmit the data to the processing circuits with minimum delay. The data, which are in byte format, are aggregated into groups of bytes that is called a frame. Also, to enable the receiver to sample the data at the optimum point (usually in the middle of the data pulse) the transmit clock is encoded within the data (in-band signaling). Several techniques are common (e.g., Manchester encoding) in the industry. ESCON uses an 8B/10B code where the 8-bit byte is mapped into a 10-bit character.

The physical topology of ESCON was also a critical factor because it had to ensure connectivity to all devices, maintain high throughput, minimize the number of connections (cables), and offer network growth in a realistic and nondisruptive manner. Several network topologies and infrastructures were investigated (e.g., Ethernet CDMA Ring, Dual Insertion Ring, and Host Star). The topology chosen for ESCON is "switched point-to-point." It offered the highest throughput, excellent connectivity with minimal number of links, and the ability to grow the network in a nondisruptive manner.

13.2.2. ESCON TOPOLOGY

The switched point-to-point topology utilizes a central switch (director) to direct the network traffic to the various elements of the network [3]. The use of a director allows the connectivity of any unit on the network to any other unit on the network. The physical connections are point-to-point links that are ideally suited to fiber optic technology. This topology enables the ability to isolate links in the network for failure analysis and repair. An n-port nonblocking director can accommodate $\frac{n}{2}$ simultaneous conversations between endpoints in the switched point-to-point network.

This type of network offers some very nice attributes that have been implemented in ESCON [2]:

- Dynamic reconfiguration of data exchanges by terminating one connection and establishing a new connection to a different endpoint
- Multipathing of links between endpoints to ensure 100% availability in all circumstances
- The addition and deletion of physical links without disruption of the existing network
- Dedicated high-priority links between endpoints
- Ability to isolate (fencing) portions of the network
- Cascading of directors for enhanced connectivity

13.2.2.1. Typical Configuration

The application of a switched point-to-point topology in ESCON is straightforward and provides the primary interconnection configuration for System/390 platforms. Most computing complexes today contain multiple servers and devices (any attached unit, e.g., DASD controller or communication element). A typical example is shown in Fig. 13.1. An important availability element of this configuration is that all servers and devices have two paths

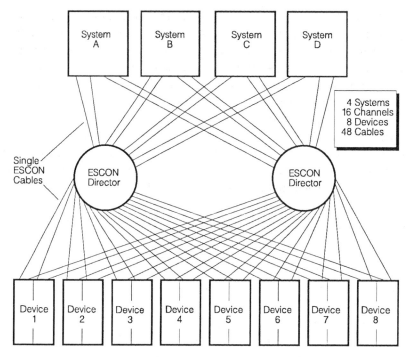

Fig. 13.1 Examples of an IBM System/390 ESCON configuration [2]. (Copyright 1992 by International Business Machines Corporation, reprinted with permission.)

to each director. This configuration provides not only connectivity between servers and devices but also full redundancy and multipathing. For example, if any one of the links or directors becomes inoperative, there is an alternate path between the system and device. Also, by adding four links (two to each director) a new server can be included in the network nondisruptively with immediate full connectivity.

13.2.3. ESCON ARCHITECTURE AND CHANNEL

The IBM ESCON architecture establishes the rules and syntax used by the server to communicate to attached devices [4]. The architecture was defined to provide efficient transmission of data over long distances via a communication channel with a bit error rate (BER) of 10^{-10} (1 error in 10^{10} bits) or less. The architecture can be divided into two fundamental categories: device level and link level. The device level defines the rules for communication of a large server to an attached device using the facilities of the physical link. It defines data and control messages and the protocol to implement the server input/output (I/O) functions. The link-level architecture defines the actual transmission of information across the physical path. It defines the frame structure, type of frames, link initialization, exchange setup, data and control messages, address structure, and link error recovery. These functions are implemented in the ESCON channel [5] hardware, which interfaces with the fiber optic link hardware.

The ESCON link is a synchronous link with information transferred continually to ensure both transmitter and receive side are in lock step. Data and control information will be sent as required by the server or attached device synchronous to the signal on the line.

13.2.3.1. Link Protocol

All information on the ESCON link is transferred within a frame structure or a sequence of special characters [4]. The ESCON frame is used to transport control and data information and is structured as shown in Fig. 13.2 [6]. The ESCON frame is delimited by a start of frame (SOF) and end of frame (EOF) ordered set of characters, respectively. The SOF and EOF are unique sets that are also used by the director to establish a connection, continue a connection, or disconnect after completion of the

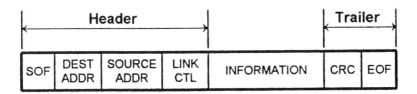

SOF :	Two character start-of-frame delimiter.
DEST ADDR :	Two byte destination address of frame.
SOURCE ADDR :	Two byte source address of frame.
LINK CTL :	One byte of link control information.
INFORMATION :	Zero to 1028 bytes of data.
CRC :	Two byte cyclic-redundancy-check information.
EOF :	Three character end-of-frame delimiter.

Fig. 13.2 ESCON frame structure.

frame transmission. The SOF delimiter is composed of two characters (20 bits). The next 16 bits (before encoding) are reserved for the destination address, the next 16 bits (before encoding) contain the source address, and the next 8 bits (before encoding) are a link control field. The link control field indicates the type and format of the frame. The four fields above are known, as a group, as the link header.

The next field following the header (Fig. 13.2) is the information field, which may contain data or system information and can vary from 0 to 1028 bytes. The link trailer consists of two fields, cyclic-redundancy-check field (CRC) and the EOF field. In order to ensure the data are received correctly, a CRC is generated at the transmitter and included in the frame as a 16-bit CRC field. The receiving device uses the CRC to verify the information field. The use of fiber optic technology has ensured that link errors from external stimulus are extremely low and the random bit error rate of the optical link due to receiver noise is less than 10^{-15}. Based on these low error rates, the recovery approach is to retransmit the frame if an error has occurred. Because this happens so seldom, the system performance is not affected by this recovery approach. The EOF field is a three-character (30-bit) field that signifies the end of the frame. The data between the SOF and EOF delimiters are modulo of 8 bits before transmission and encoded into 10-bit characters for transmission on the link.

The architecture also defines an ordered set of sequences that can be transmitted over the link in the presence of a very high error rate condition (in which frames cannot be transmitted correctly). Each sequence contains a continuous repetition of an ordered set to maximize the likelihood that a sequence will be correctly recognized. Some typical sequences are not operational sequence, in which a link-level facility (at the server or device) cannot interpret a received signal, or offline sequence, in which the appropriate link-level facility is indicating that it is off-line with respect to sending any information. These and other sequences are interpreted at a level above the link layer and appropriate action is taken by the server.

An idle character is always sent on the link when no frames or control sequences are being sent. The idle character is a special ordered set of bits (named K28.5) [7]. Also, idles are sent between frames as well. The idle sequence ensures that the receiver is both in bit and character synchronization with the transmitter. If the receiver becomes out of synchronization with the transmitter, the architecture has defined a set of rules and procedures whereby synchronization can be reacquired [7].

13.2.3.2. Data Encoding/Decoding

High-speed fiber optic receivers perform best over environmental and man-ufacturing variation when they are AC coupled. The ESCON optical receiver is designed in this manner. In order to prevent DC baseline wander, it is important to ensure that the information on the link is encoded from the normal nonreturn to zero (NRZ) computer code to a DC-balanced code. Several codes (e.g., Manchester and 4B/5B) were investigated and an 8B/10B code was chosen for ESCON. This technique was chosen because it provides the most robust code and a minimum bandwidth overhead (25%). For example, the 8B/10B code contains special control characters that will not degrade into a another valid character with single bit errors.

The 8B/10B encoding transforms a byte (8 bit) of information at a time into a 10-bit transmission character. The 10-bit character is sent serially by bit over the fiber optic link and decoded at the receiver into the original 8-bit byte. Conceptually, the 256-bit combinations of the 8-bit byte are mapped into a subset of the 1024 10-bit characters such that the maximum run length of 1's or 0's is 5. Special control characters and sequences (e.g., idle, SOF, and EOF) are defined that are not derived from the 8-bit original but are meaningful only as architected control and definition characters. For example, the + K28.5 idle character (0011111010) is unique and there is no valid data character with this 10-bit sequence [7]. A single bit error will not result in a valid 10-bit character. Only 536 of the 1024 possible characters are valid. All others will cause an architected error condition.

The running disparity (difference between the number of 1's and 0's in a character) is continually monitored to ensure a DC balance. If disparity exceeds the bounds, an error condition occurs. The 8B/10B code is well behaved with regard to DC balance and the number of transmissions between 0's and 1's is sufficient to ensure that the receiver, retiming, and character recognition circuits can reliably perform the required fuctions.

13.2.3.3. Bit Error Rate Thesholding

The architecture is tolerant of bit errors on the link that may be detected as code violation, sequencing, or CRC errors [8]. A code violation occurs when an invalid transmission character is received. A sequencing error occurs when a sufficient number of consecutive special ordered sets (discussed earlier) cannot be transmitted without error. Finally, a CRC error occurs when the CRC result of the received frame contents is not equal to the expected value.

For the link design, the number of retries due to link errors has a negligible effect on link performance. However, as the rate of retries increases beyond a threshold value the degradation of the link may be noticeable. A report is generated when the specified threshold is reached on a link for further analysis and maintenance. The threshold for ESCON is set a 1 error in 10^{10} bits. At this level the link performance is still tolerable and maintenance can be deferred until a convenient time. Beyond this level, the server will begin to realize degraded performance on that link.

The actual measurement is done by counting the number of code violation events within a specified time. A bit error will likely cause more than one code violation. Consequently the concept of an error burst has been developed. To prevent a single bit error causing multiple error counts, one or multiple code violations within a 1.5-s period are considered as one error burst for the threshold count. Fifteen or more error bursts within a 5-min period will result in a threshold error recorded by the server. The threshold count is reset when the threshold is reached or every 5 min, whichever occurs first. Detailed information is given in Ref. [8].

13.3.   ESCON Link Design

The transition of computer interconnection from parallel copper technology to a radically different technology ("serial" fiber optics) generated many questions and concerns. Most of the concerns centered around the reliability of the link in a computer data center environment. Can the technology meet the stringent reliability requirement for both bit errors on the link and hardware failures? The fiber optic link must perform equal to or better than the copper links it replaces. The ESCON link design [9] and component selection were made to achieve both high data rate and reliability.

13.3.1   MULTIMODE DESIGN CONSIDERATIONS

The multimode ESCON link replaced a parallel (8-bit wide) copper coaxial cable link that had proven reliability and performance. Any replacement of the copper link must be easier to use, offer higher data rate and distance performance, be smaller in size, be lighter in weight, and be equal or better in reliability. To achieve those requirements, a technology investigation was initiated to determine the best technology set for a communication link. Fiber optics was the clear choice with both multimode and single-mode fiber. The technology allows for significant future performance enhancements as the need for higher data rates continue. Figure 13.3 depicts the size reduction of the interconnection cable and connector of ESCON compared to the equivalent (two) parallel copper cables and connectors it replaces.

The optical link must extend throughout a campus environment (typically 2 or 3 km) and achieve very reliable data transfer. A optical link BER design of 10^{-15} for the worst-case link (longest length) was chosen.

The optical cable must be easy to handle and install without major complications. The optical cables (jumper cables) that connected to the server units were chosen to be robust, flexible, and contain no metal. The absence of any metal in the cable allowed the link design to take advantage of the inherent electromagnetic compatibility (EMC) of the technology. The optical cable had to have the capability to be pulled under the raised floor, through ceiling conducts/raceways, etc. and withstand this application environment without any functional or reliability impact. The connector was designed with polarization to ensure the cable could not be misplugged.

Fig. 13.3 Parallel copper and ESCON channel cables [9]. (Copyright 1992 by International Business Machines Corporation, reprinted with permission.)

13.3.1.1. Major Components

The major components of the optical link are illustrated in Fig. 13.4. The serializer (typically implemented in CMOS technology) takes the 10 parallel bits of 8B/10B encoded data and serializes the data into a 200 Mb/s rate serial bit stream, whereas the deserializer performs the complementary function. The deserializer also includes the retiming function, which extracts the clock from the serial data. The derived clock is used to latch and reshape the serial data prior to deserialization.

The transceiver contains the electrooptic transducer that converts the electrical signal to an optic signal and vise versa. It also includes the drive and receive electronics that interface to the transducers. The transmitter uses a light-emitting diode (LED) operating at 1300 nm and the receiver uses a positive–intrinsic–negative (PIN) photodiode. Both devices are made of InGaAsP quaternary material. The 1300-nm LED was chosen because this wavelength is at the optimum attenuation and bandwidth of multimode fiber and has excellent reliability and low cost.

For a System/390 CMOS server four channels are implemented on a single printed circuit card as depicted in Fig. 13.5. Individual channels on the card are configured horizontally with the electrooptic transceivers on the right card edge. The CMOS channel logic module is the 25-mm capless module to the left of the transceiver. This module contains the retiming, serializer, deserializer, 8B/10B encoding/decoding, and all channel logic functions associated with the ESCON link.

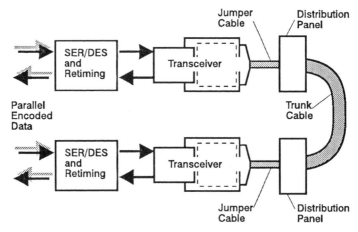

Fig. 13.4   Block diagram of fiber optic link elements [9]. (Copyright 1992 by International Business Machines Corporation, reprinted with permission.)

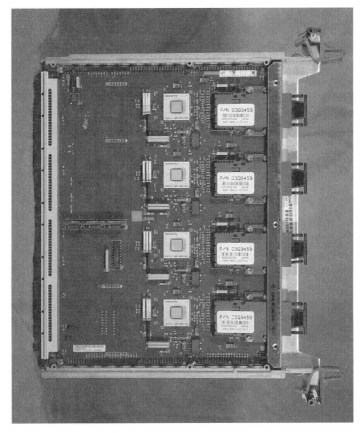

Fig. 13.5 An IBM ESCON channel card (courtesy of IBM Corp.).

The jumper cable is a two-fiber (one inbound and one outbound), rugged, yet flexible, cable assembly that uses an aramid fiber strength member. The ESCON connector is a low-profile, polarized, push-on connector that latches into a transmitter receiver subassembly (TRS) or coupler assembly. The fiber used in the jumper is multimode 62.5/125 μm and the ferrules are made of zirconia ceramic material. The ESCON link is designed to be used with either 62.5/125 or 50/125 μm multimode trunk fiber. The use of 62.5/125 μm trunk supports a link length of 3 km, whereas the 50/125 μm trunk fiber supports a 2-km link distance. The difference in distance capability is due to the additional loss associated with connecting a 62.5/125 μm jumper fiber to a 50/125 μm trunk fiber.

13.3.2. SINGLE-MODE DESIGN CONSIDERATIONS

In general, ESCON multimode links satisfy the majority of the distance requirements for a large server complex. However, there has been a significant need to extend the ESCON distance beyond the campus environment. To meet these requirements, it was necessary to develop an alternate technology set that is complementary to the telecommunications infrastructure being established in fiber optics. However, the link signaling, protocol, and architecture remain the same as with the base ESCON link.

The new long-distance ESCON link, called ESCON XDF, uses a long-wavelength laser as the source and single-mode optical fiber (SMF). The single-mode fiber chosen is the same as that used by the telecommunications industry and generally available. This is an important consideration because these long-distances typically will traverse right-of-ways and likely the fiber is owned by another company (e.g., posts, telephone, and telegraphs; local telephone provider; and power company). In general, the computer customer is not interested in fiber optics as an entity but only as a means of efficient communication within his or her network. To ensure ease of use and not require of the customer anything more than the base link requirements, the XDF must be an international class 1 laser safety product. This category allows unrestricted access by uncertified laser personnel because the product conforms with "eye safe" government and industry criteria.

13.3.2.1. Major Components

The single-mode link is very similar in structure to the multimode link illustrated in Fig. 13.4. The major differences include the following:

1. Use of 1300-nm laser source
2. Transceiver uses the Fibre Channel Standard (FCS) duplex SC connector
3. Single-mode fiber jumpers and trunk cables

The single-mode fiber was chosen for its high bandwidth and low loss characteristics. The jumper cables use 9/125 μm fiber, whereas the trunk can use either 9- or 10-μm core fiber. There is no distance penalty associated with the use of 10-μm core trunk fibers.

The XDF feature provides a 20-km link capability at 200 Mb/s without the use of repeaters. The link distance is a function of the optical loss budget and is a trade-off of laser transceiver cost and complexity versus distance. The laser power output is maintained at a low enough power level to ensure compliance with class 1 laser safety standards (see Section 13.9.1).

The laser transceiver discussed in this chapter is a second-generation transceiver that utilizes the single-mode asynchronous transfer mode industry standard module package with the FCS connector. The prior version was a single-mode ESCON connectorized module that is no longer in production. It was designed and produced by IBM because there was no industry product at that time that could meet the requirements of System/390 servers. The new and original laser transceivers are fully compatible and have similar specifications.

13.3.3. OPTICAL LINK DESIGN

The ESCON optical link must perform reliably while allowing maximum user flexibility and performance (data rate and distance). The data rate is fixed at 200 Mb/s and cannot be altered or compromised. The optical link parameters (e.g., transmitter optical coupled power, LED spectral width, and receiver sensitivity) can be optimized to result in the longest possible link using state-of-the-art technology.

The link system design must consider the effects of all the parameters that contribute to an optical loss either directly or indirectly via a bandwidth limitation [10]. The fiber optic cable bandwidth will decrease the achievable link length (for a given data rate) unless this phenomenon is compensated for with an increase in the light signal level. This compensation methodology can be used, generally, for all link parameters provided the individual parameter limitation effects are maintained at modest levels (<3–5 dB). If the effect is very pronounced or extreme, then the use of optical power to compensate for bandwidth effects is very difficult to use because of the nonlinearity of the correlation function. For example, the slope of the optical signal bandwidth transfer function becomes very steep beyond the classical knee of the curve. At this point an unrealistic power increase is needed to compensate for bandwidth roll-off.

The design of the ESCON link is based on each parameter being represented by an optical loss or an equivalent optical loss (simulation of those parameter effects that are not direct optical loss characterized). Each of the parameters can then be characterized as a statistical parameter element (mean and standard deviation). The individual elements are summed statistically to arrive at a a total optical loss. The total optical loss must be less than the optical link power available from the respective transmitter and receiver pair. The statistical summation of the loss elements has the advantage of more precisely representing the link components in the field. It allows the designer to more accurately describe the component elements in the form of distributions and improve focus on cost/performance trade-offs as well as design optimization.

13.3.3.1. Multimode Link Design and Specification

The ESCON link budget elements are grouped into two major categories:

1. Cable plant
   The cable plant loss includes connector loss, fiber attenuation, higher order mode loss, and splices.
2. Available power
   The available power is the resultant optical power available for the link after the optical budget associated with the transmitter and receiver is adjusted for link losses such as

   - Fiber dispersion penalty (modal and spectral)
   - Retiming penalty
   - BER specification conversion from 10^{-12} to 10^{-15}
   - LED end-of-life degradation
   - Transceiver coupling variation
   - Data dependency.

ink parameters are defined into these categories to allow maximum
ility over the elements that can be controlled by the user (e.g., fiber
ation) and incorporate into the available power those elements that
ifficult or cannot be controlled (e.g., fiber dispersion) by the user.

e elements of the available power budget are statistically summed to
a resultant available power as a distribution with a mean and standard
viation. The following condition must be satisfied for the link to meet
its design criteria as follows,

$$U_{\mathrm{av}} - n\sigma_{\mathrm{av}} \geq C_t, \tag{13.1}$$

where U_{av} is the available power, n is the number of standard deviations, σ_{av} is the standard deviation of the available power, and C_t is the total cable plant optical loss. For an ESCON link $n = 3$ (3 σ design) for the longest link allowed in the configuration at a BER of 10^{-15}. The resultant mean and standard deviation for the available power is determined using a Monte Carlo technique to sum the various elements. This was done

Table 13.1   ESCON Maximum Link Loss (at 1300-nm Wavelength)

Maximum link length (km)	Maximum link loss (dB)	Trunk fiber core size (μm)	Minimum trunk modal bandwidth (MHz/km)
2.0	8.0	62.5	500
2.0	8.0	50.0	800
2.0–3.0	8.0	62.5	800

Note. From Ref. [11]. The maximum link length includes both jumper and trunk cables. The maximum total jumper cable length cannot exceed 244 m when using either 50/125 μm trunk fiber or when a 62.5/125 μm link exceeds 2 km.

because all the parameter distributions are not necessarily Gaussian and in fact the transmitter output power and receiver sensitivity are truncated distributions. The use of a 3 σ design point for the worst-case link (3 km for 62.5-μm trunk and 2 km for 50-μm trunk) ensures that all shorter links are designed conservatively and the risk of an install link budget failure is extremely remote.

Table 13.1 illustrates the resultant specification of the cable plant to ensure the multimode link operates in accordance with the link design requirements. The maximum link loss is established at 8 dB independent of link configuration. The loss budget was maintained at 8 dB by adjusting the fiber bandwidth and in turn the dispersion penalty. The standard 2-km 62.5/125 μm link uses 500 MHz-km fiber, whereas the 2-km 50/125 μm and 3-km 62.5/125 μm link use a higher bandwidth (800 MHz/km) grade of fiber. This allows the customer maximum flexibility to adjust his or her configuration to the environment. The user can trade off number and connector quality with fiber attenuation and length to achieve an optimized installation.

13.3.3.2. Single-Mode Link Design and Specification

The single-mode link design follows the same approach used for the multimode design. The jumper fiber is 9/125 μm. The XDF link supports both 9- or 10-μm core fiber without any effect on distance. The excess loss (approximately 0.2 dB) associated with the coupling of a 9-μm core jumper to a 10-μm core trunk fiber is included in the available power category and transparent to the overall link budget. The dispersion penalty of the fiber due to spectral width of the laser is small and has also been accounted for

Table 13.2   **ESCON XDF Maximum Link Loss (at 1300-nm Wavelength)**

Maximum link length (km)	Maximum link loss (dB)	Trunk fiber core size (μm)
20.0	14.0	9–10

Note. From Ref. [11]. The maximum link length includes both jumper and trunk cables. The minimum length of a single-mode jumper cable is 4 m. In a single-mode trunk cable, distance between connectors or splices must be sufficient to ensure only the lowest order bound mode propagates.
Single-mode connectors and splices must meet a minimum return loss specification of 28 dB. The minimum return loss of a single-mode link must be 13.7 dB.

in the available power budget along with any effects due to laser mode hopping and relative intensity noise [10]. All these time domain effects are relatively small for a 200 Mb/s single-mode link and are included as a 1.5-dB fixed (no distribution) "AC optical path" penalty. The single-mode link specification is given in Table 13.2. A maximum link length of 20 km can be achieved with a maximum optical cable plant loss budget of 14 dB for the cable plant.

In order to ensure the laser is well behaved under all operating conditions it is important to minimize any optical reflections occurring in the cable plant. This is done by specifying that all connections and splices in the link have a minimum return loss of 28 dB. Mode partition noise in the XDF link is alleviated by specifying that no jumper less than 4 m may be used. The minimum length in conjunction with the specified cutoff wavelength of the fiber ensures only the lowest order bound mode propagates in the jumper. Likewise, the trunk installer must ensure any connectors or splices in the trunk meet the return loss specification and all connectors or splices are placed sufficiently apart to ensure only the lowest order mode is propagating prior to any connectors, splices, or other optical discontinuities.

13.4.   Multimode Physical Layer

The multimode physical layer [11] defines a common set of specifications at the device interface that ensures compatibility between the link transmitter and receiver through a defined cable plant. The physical layer provides a design point that allows any server or attached device manufacturer to build an ESCON-compatible link. Transceivers by multiple manufacturers

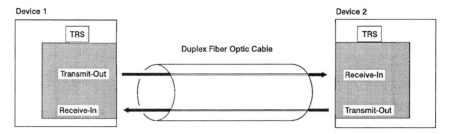

Fig. 13.6   Bidirectional fiber optic ESCON link [12].

can be intermixed on a link and still conform to the ESCON performance requirements.

The ESCON link is composed of two unidirectional point-to-point links as shown in Fig. 13.6. Information (light pulses) is transferred from the server to a device over one fiber and from the device to the server over the other fiber. Both fibers are encased in a common cable assembly. The link is full duplex—both links may be active simultaneously.

The multimode physical interface specifies the optical and mechanical requirements of the optical output and input interface to the jumper cable and the performance requirements of the cable plant. Data are transmitted over the link at a transmission rate of 200 ± 0.04 Mb/s. The encoding and link protocols were discussed in Section 13.2. A "1" bit corresponds to light on and a "0" bit to light off. All specifications are for worst-case operating conditions, including end-of-life effects.

13.4.1.   MULTIMODE OUTPUT OPTICAL INTERFACE

The optical coupled light specifications required for an ESCON link are given in Table 13.3. The parameters specified will allow the maximum distance requirements and loss budget, as specified in Table 13.1 with a BER of 10^{-15}. The light source is an incoherent light-emitting diode.

13.4.2.   MULTIMODE INPUT OPTICAL INTERFACE

The input optical interface specifications are given in Table 13.4. A loss-of-light function and operation is specified for link failure indication and diagnostic use. The design of the machine receiving this information determines how this state change information is utilized.

Table 13.3   **Multimode Optical Output Interface Specifications**

Parameter	Minimum	Maximum	Unit
Average power[a,b]	−20.5	−15.0	dBm
Center wavelength	1280	1380	nm
Spectral width (FWHM)		175.0	nm
Rise time (t_r) (20–80%)[a,c]		1.7	ns
Fall time (t_f) (80–20%)[a,c]		1.7	ns
Eye window[a]	3.4		ns
Optical output jitter[d]		0.8	ns
Extinction ratio[a,e]	8		dB
t_r, t_f at optical path output[c,f]		2.8	ns

Note. From Ref. [11].

[a] Based on any valid 8B/10B code. The length of jumper cable between the output interface and the instrumentation is 3 m.

[b] The output power shall be greater than −29 dBm through a worst-case link as specified in Table 13.1. Higher order mode loss (HOML) is the difference in link loss measured using the device transmitter compared to the loss measured with a source conditioned to achieve an equilibrium mode distribution in the fiber. The transmitter shall compensate for any excess HOML occurring in the link (e.g., HOML in excess of 1 dB for a 62.5-μm link).

[c] The minimum frequency response bandwidth range of the optical waveform detector shall be 100 kHz to 1 GHz.

[d] The optical output jitter includes both deterministic and random jitter. It is defined as the peak-to-peak time-histogram oscilloscope value (minimum of 3000 samples) using a 2^7-1 pseudo-random pattern or worst-case 8B/10B code pattern. The transmitter output light is coupled to a PIN photodiode O/E converter (e.g., Tektronix P6703A or equivalent) via a 3-m cable and jitter measured with a digital sampling oscilloscope [13].

[e] Measurement shall be made with a DC-coupled optical waveform detector that has a minimum bandwidth of 600 MHz and whose gain flatness and linearity over the range of optical power being measured provide an accurate measurement of the high and low optical power levels.

[f] The maximum rise or fall time (from, e.g., chromatic, modal dispersion, etc.) at the output of a worst-case link as specified in Table 13.1. The 0 and 100% levels are set where the optical signal has at least 10 ns to settle. The spectral width of the transmitter shall be controlled to meet this specification.

13.4.3.   MULTIMODE FIBER OPTIC CABLE SPECIFICATION

The two optical fibers are assembled into a duplex optical cable assembly for the jumper and assembled into pairs for the trunk. The jumper cable assembly is terminated in the ESCON duplex fiber optic connector. The trunk cable, however, is usually installed in high-count configurations (e.g., 12, 24, 36, 72, and 144 fiber counts) by professionals skilled in the art of fiber optic installation. The planning and installation of the trunk is reviewed in Section 13.6. The two fibers in a jumper cable are assembled as illustrated

Table 13.4 **Multimode Optical Input Interface Specifications**

Parameter	Minimum	Maximum	Unit
Sensitivity[a,b]		−29.0	dBm
Saturation level[a]	−14.0		dBm
Acquisition time[c]		100	ns
LOL threshold[d]	−45	−36	dB
LOL hysteresis[d,e]	0.5		dB
Reaction time for LOL state change	3	500	μs

Note. From Ref. [11].

[a] Based on any valid 8B/10B code pattern measured at, or extrapolated to, 10^{-15} BER measured at center of eye. This specification shall be met with worst-case conditions as specified in Table 13.3 for the output interface and Table 13.1 for the fiber optic link. This value allows for a 0.5-dB retiming penalty.

[b] A minimum receiver output eye opening of 1.4 ns at 10^{-12} should be achieved with a penalty not exceeding 1 dB.

[c] The acquisition time is the time to reach synchronization after the removal of the condition that caused the loss of synchronization. The pattern sent for synchronization is either the idle character or an alternating sequence of idle and data characters.

[d] In direction of decreasing power: If power $>$ −36 dBm, LOL state is inactive; if power $<$ −45 dBm, LOL state is active. In direction of increasing power: If power $<$ −44.5 dBm, LOL state is active; if power $>$ −35.5 dBm, LOL state is inactive.

[e] Required to avoid random transitions between LOL being active and inactive when input power is near threshold level.

in the cable cross section (Fig. 13.7). The cable assembly is nonmetallic and uses aramid fiber as the strength member. All the elements are encased in a flexible polyvinyl chloride (PVC) jacket.

The optical specifications in this section are primarily associated with the fiber and are necessary to ensure the link meets its performance objectives. They also ensure consistency among various ESCON-compatible devices.

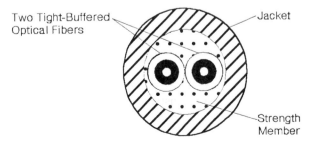

Fig. 13.7 Multimode jumper cable construction [12].

13.4.3.1. Multimode Jumper Cable Assembly

The MMF jumper cable is only offered in a 62.5/125 μm fiber configuration and the optical specifications are given in Table 13.5. The cable jacket color is orange. All the parameters are specified and measured in accordance with the applicable industry standards as indicated.

13.4.3.2. Multimode Trunk Fiber Specification

Two multimode fiber types are supported for the trunk. The required optical parameters of both trunk fibers are specified in Table 13.6. Both fiber types conform to applicable European and United States industry standards [14–16]. All fiber parameters are specified and measured in accordance with the applicable industry standards as indicated.

13.4.4. ESCON CONNECTOR (MULTIMODE)

The ESCON connector (illustrated in Fig. 13.8) is a ruggedized, two-ferrule connector that is polarized to prevent misplugging. The polarization is accomplished by beveling two corners of the connector as shown in Fig. 13.8. The connector is a push-on type that is retained in the receptacle by two latches. The ferrules are protected from handling and dirt by a protective spring-loaded cap that retracts upon insertion in a receptacle, allowing

Table 13.5   **Multimode (62.5/125 μm) Jumper Cable Specifications**

Parameter	*Specification*
Fiber type	Graded index with glass core and cladding
Operating wavelength	1300 nm
Core diameter[a]	62.5 ± 3.0 μm
Cladding diameter[b]	125 ± 3.0 μm
Numerical aperture[c]	0.275 ± 0.015
Minimum modal bandwidth[d]	500 MHz-km
Attenuation	1.75 dB/km at 1300 nm (maximum)

Note. From Ref. [11].
[a] Measured in accordance with EIA 455 FOTP 58, 164, 167, or equivalent.
[b] Measured in accordance with EIA 455 FOTP 27, 45, 48, or equivalent.
[c] Measured in accordance with EIA 455 FOTP 47 or equivalent.
[d] Measured in accordance with EIA 455 FOTP 51 or equivalent.

Table 13.6   **Multimode Trunk Fiber Specifications**

Parameter	Specification
62.5/125 μm multimode fiber	
Fiber type	Graded index with glass core and cladding
Operating wavelength	1300 nm
Core diameter[a]	62.5 ± 3.0 μm
Core noncircularity	6% maximum
Cladding diameter[b]	125 ± 3.0 μm
Cladding noncircularity	2% maximum
Core and cladding offset	3.0 μm maximum
Numerical aperture[c]	0.275 ± 0.015
Minimum modal bandwidth[d]	500 MHz-km at ≤ 2 km
	800 MHz-km at >2 km and ≤3 km
Attenuation[e]	1.0 dB/km at 1300 nm
50/125 μm multimode fiber	
Fiber type	Graded index with glass core and cladding
Operating wavelength	1300 nm
Core diameter[a]	50 ± 3.0 μm
Core noncircularity	6% maximum
Cladding diameter[b]	125 ± 3.0 μm
Cladding noncircularity	2% maximum
Core and cladding offset	3.0 μm maximum
Numerical aperture[c]	0.200 ± 0.015
Minimum modal bandwidth	800 MHz-km
Attenuation[e]	0.9 dB/km at 1300 nm

Note. From Ref. [11].
[a] Measured in accordance with EIA 455 FOTP 58, 164, 167, or equivalent.
[b] Measured in accordance with EIA 455 FOTP 27, 45, 48, or equivalent.
[c] Measured in accordance with EIA 455 FOTP 47 or equivalent.
[d] Measured in accordance with EIA 455 FOTP 51 or equivalent.
[e] The attenuation is a typical value rather than a specification. Table 13.1 is the specification.

the ferrules to be inserted into alignment sleeves. The ferrules are made of high-precision, stabilized zirconia ceramic. The ferrule ensures accurate alignment of the fiber to an LED, photodiode (PD), or fiber receptacle.

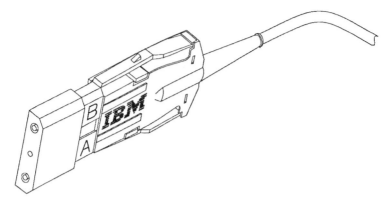

Fig. 13.8   ESCON multimode duplex connector [12].

The ends of the ceramic are polished into a convex shape to ensure that glass physical contact is always achieved in fiber-to-fiber connections. This feature ensures low loss connections (typically less than 0.4 dB) and high return loss from the connection interface.

The color of the multimode connector is black. The connector also has bend radius-limiting boot at the connection of the fiber cable. This ensures that excessive bend stresses cannot be applied to the fiber at the connector.

13.4.4.1.   Multimode Duplex Receptacle Specification

The duplex receptacle for the ESCON multimode connector is specified in detail in Fig. 13.9. All the necessary dimensions and specifications are illustrated to build a transceiver or fiber (coupler) receptacle while maintaining full functional and dimensional compliance with the ESCON multimode connector.

13.5.   Single-Mode Physical Layer

The single-mode physical layer follows the same approach as the multimode physical layer and defines a common set of specifications at the device interface that ensure interoperability [11]. The ESCON XDF link is composed of two simplex, point-to-point counterdirectional links encased in a duplex configuration. Both the optical and mechanical properties of the interface are specified as well as the cable plant.

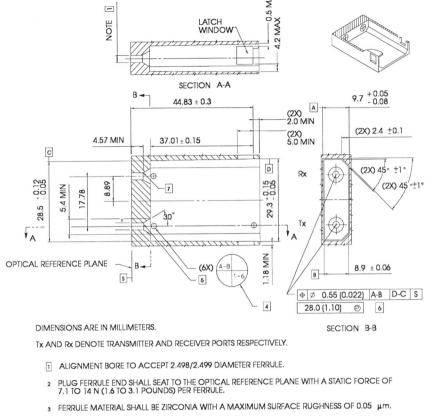

SECTION A-A

SECTION B-B

DIMENSIONS ARE IN MILLIMETERS.

Tx AND Rx DENOTE TRANSMITTER AND RECEIVER PORTS RESPECTIVELY.

1. ALIGNMENT BORE TO ACCEPT 2.498/2.499 DIAMETER FERRULE.

2. PLUG FERRULE END SHALL SEAT TO THE OPTICAL REFERENCE PLANE WITH A STATIC FORCE OF 7.1 TO 14 N (1.6 TO 3.1 POUNDS) PER FERRULE.

3. FERRULE MATERIAL SHALL BE ZIRCONIA WITH A MAXIMUM SURFACE RUGHNESS OF 0.05 μm.

4. DATUM TARGETS SHOWN APPLY TO TOP AND BOTTOM SURFACES. THEY ARE TO BE USED IN PAIRS TO ESTABLISH CENTERLINES.

5. THE MODULE BORE MUST BE ABLE TO WITHSTAND A FORCE OF 2.5 N (0.55 POUNDS) APPLIED PERPENDICULAR TO THE OPENING AT THE OUTERMOST CONTACT POINT OF THE SLEEVE.

6. AS SPECIFIED FOR A SOLID MODULE SLEEVE. FOR A SPLIT SLEEVE,

7. BLEND/TAPER BORE ENTRANCE.

Fig. 13.9 Diagram of the multimode duplex receptacle [11].

The physical layer defines a set of specifications that ensure link performance up to a 20-km distance, without retransmission, using dispersion unshifted single-mode fiber. The data rate specification of 200 ± 0.04 Mb/s is the same as multimode. A "1" bit corresponds to a light on condition. All specifications are for worst-case operating conditions, including end-of-life conditions.

13.5.1. SINGLE-MODE OUTPUT OPTICAL INTERFACE

The parameter specifications in Table 13.7 define the optical output interface for the light coupled into the single-mode fiber. The parameter specifications ensure the BER does not exceed 10^{-15}. The light source is a 1300-nm semiconductor laser.

13.5.2. SINGLE-MODE INPUT OPTICAL INTERFACE

The single-mode optical input interface requirements are given in Table 13.8. A loss-of-light function and operation is specified for link failure indication and diagnostic use. The design of the machine receiv-

Table 13.7 **Single-Mode Optical Output Interface Specifications**

Parameter	Minimum	Maximum	Unit
Average power into SMF[a]	−8.0	−3.0	dBm
Center wavelength[a]	1261	1360	nm
Spectral width (rms)		7.7	nm
Rise time (t_r) (20–80%)[a,b]		1.5	ns
Fall time (t_f) (80–20%)[a,b]		1.5	ns
Eye window[a]	3.5		ns
Optical output jitter[c]		0.8	ns
Extinction ratio[a,d]	8		dB
Relative intensity noise (RIN_{12})[e]		−112.0	dB/Hz
t_r, t_f at optical path output[c,f]		2.8	ns

Note. From Ref. [11].

[a] Based on any valid 8B/10B code pattern, The measurement is made using at 4-m single-mode jumper cable and only includes the power in the lowest order fundamental mode.

[b] The minimum frequency response bandwidth range of the optical waveform detector shall be 100 kHz to 1 GHz.

[c] The optical output jitter includes both deterministic and random jitter. It is defined as the peak-to-peak time-histogram oscilloscope value (minimum of 3000 samples) using a 2^7-1 pseudo-random pattern or worst-case 8B/10B code pattern. The transmitter output light is coupled to a PIN photodiode O/E converter (e.g., Tektronix P6703A or equivalent) via a 3-m cable and jitter is measured with a digital sampling oscilloscope [13].

[d] Measurement shall be made with a DC-coupled optical waveform detector that has a minimum bandwidth of 600 MHz and whose gain flatness and linearity over the range of the high and low optical power levels.

[e] The relative intensity noise is measured with a 12-dB optical return loss into the output interface.

[f] The maximum degradation in input interface sensitivity (from, e.g., jitter, mode hopping, and intersymbol interference) that can occur by using a worst-case link as specified in Table 13.2. The spectral width of the transmitter shall be controlled to meet this specification.

Table 13.8 **Single-Mode Optical Input Interface Specifications**

Parameter	Minimum	Maximum	Unit
Sensitivity[a,b]		−28.0	dBm
Saturation level[a]	−3.0		dBm
Return loss[c]	12.5		dB
Acquisition time[c]		100	ns
LOL threshold[d]	−45	−32	dB
LOL hysteresis[e]	1.5		dB
Reaction time for LOL state change	0.25	5000	μs

Note. From Ref. [11].

[a] Based on any valid 8B/10B code pattern measured at, or extrapolated to, 10^{-15} BER measured at center of eye. This specification shall be met with worst-case conditions as specified in Table 13.7 for the output interface and Table 13.2 for the fiber optic link. This value allows for a 0.5-dB retiming penalty.

[b] A minimum receiver output eye opening of 1.4 ns at 10^{-12} should be achieved with a penalty not exceeding 1 dB.

[c] The measurement is made using a 4-m single-mode jumper cable and only includes the power in the lowest order fundamental mode.

[d] The acquisition time is the time to reach synchronization after the removal of the condition that caused the loss of synchronization. The pattern sent for synchronization is either the idle character or an alternating sequence of idle and data characters.

[e] Required to avoid random transitions between LOL being active and inactive when input power is near threshold level.

ing this information determines how this state change information is utilized.

13.5.3. *SINGLE-MODE FIBER OPTIC CABLE SPECIFICATION*

The ESCON XDF link is similar to the multimode link. It is composed of two unidirectional links assembled in a common cable and connector housing. The cable is terminated in a polarized duplex connector. The trunk cable is usually installed by professionals in high-count cables and physically distributed in distribution panels (see Section 13.6).

13.5.3.1. **Single-Mode Jumper Cable Assembly**

The SMF jumper cable assembly is a second-generation product that conforms to the ANSI Fibre Channel Standard [17]. The cable assembly is nonmetallic and uses aramid fiber as a strength member. The cable construction is illustrated in Fig. 13.10.

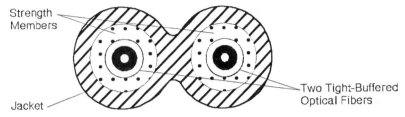

Fig. 13.10   Single-mode jumper cable construction [12].

The single-mode jumper is only offered in 9/125 μm nondispersion-shifted fiber with the optical and mechanical specifications given in Table 13.9. The color of the cable jacket is yellow. All parameters are specified and measured in accordance with the applicable industry standards as indicated.

13.5.3.2.   Single-Mode Trunk Fiber Specification

The allowable single-mode trunk fiber optical specifications are given in Table 13.10. The nominal mode field diameter of the fiber can vary from 9 to 10 μm to account for all the standard single-mode fiber types. The specifications are consistent with the applicable European and United States industry standards [14, 18, 19]. All fiber parameters are specified and measured in accordance with the applicable industry standards as indicated.

Table 13.9   **Single-Mode Jumper Cable Specifications**

Parameter	*Specification*
Fiber type	Dispersion unshifted
Operating wavelength	1261–1360 nm
Mode field diameter[a]	9.0 ± 1.0 μm
Zero dispersion wavelength[b]	1310 ± 10 nm
Dispersion[b]	6.0 ps/nm-km maximum
Cutoff wavelength (λ_c)[c]	1260 nm maximum
Attenuation[d]	0.8 dB/km maximum

Note. From Ref. [11].
[a] Measured in accordance with EIA 455 FOTP 164, 167, or equivalent.
[b] Measured in accordance with EIA 455 FOTP 168 or equivalent.
[c] Measured in accordance with EIA 455 FOTP 80 or equivalent.
[d] Measured in accordance with EIA 455 FOTP 78 or equivalent.

Table 13.10   **Single-Mode Trunk Fiber Specifications**

Parameter	Specification
Fiber type	Dispersion unshifted
Operating wavelength	1261–1360 nm
Mode field diameter[a]	9.0–10.0 ± 10%
Core concentricity error[b]	1 μm maximum
Cladding diameter[b]	125 ± 2.0 μm
Cladding noncircularity[b]	2% maximum
Zero dispersion wavelength[c]	1295–1322 nm (1310 nm nominal)
Zero dispersion slope[c]	0.095 ps/nm^2-km maximum
Cutoff wavelength (λ_c)[d]	1280 nm maximum
Cutoff wavelength (λ_{cc})[e]	1260 nm maximum
Attenuation above nominal[f, g]	0.6 dB/km maximum
Attenuation (nominal)[g]	0.5 dB/km at 1310 nm

Note. From Ref. [11].

[a] Measured in accordance with EIA 455 FOTP 164, 167, or equivalent.

[b] Measured in accordance with EIA 455 FOTP 45, 48, or equivalent.

[c] Measured in accordance with EIA 455 FOTP 168 or equivalent.

[d] Measured in accordance with EIA 455 FOTP 80 or equivalent.

[e] Measured in accordance with EIA 455 FOTP 170 or equivalent.

[f] The maximum attenuation for wavelengths from 1261 to 1360 nm shall not exceed the attenuation at 1310 nm by more than 0.06 dB/km. Typically, this specification can be met by fiber with 1383-nm water absorption peaks below 2 dB/km.

[g] The attenuation is a typical value rather than a specification. Table 13.2 is the specification.

13.5.4. *ESCON XDF DUPLEX CONNECTOR*

The single-mode ESCON XDF connector is the same as the FCS SC push-on, polarized (via keying) duplex fiber optic connector (Fig. 13.11) [17]. The ferrules are also made of zirconia ceramic. The ends of the ferrules are polished to ensure physical contact when mated to another connector. For the single-mode version, one key is narrower than the other. It is mechanically retained in the duplex receptacle by latch arms that engage the connector upon plugging. Each fiber has its own subassembly that is compatible with the industry standard SC connector. The mating, external dimension, and interface requirements of the connector conform to the ANSI Fibre Channel Standard [17]. In order to improve usability, the connector housing is gray in color. This helps to differentiate the single-

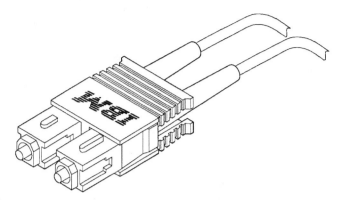

Fig. 13.11   ESCON single-mode duplex connector [12].

mode link from the multimode link. The connector also uses bend radius-limiting boots to minimize bending stress at the cable/connector interface.

13.5.4.1.   Single-Mode Duplex Receptacle Specifications

The duplex receptacle for the XDF single-mode connector is specified in detail in Fig. 13.12. The specification is the same as specified in the Fibre Channel Standard FC-PH document. Any single-mode connector that is compliant with FCS will interoperate with the XDF single-mode receptacle.

13.6.   Planning and Installation of an ESCON Link

This section discusses the planning and installation considerations for fiber optic cabling in a large data processing environment [20]. The size, layout, and number of buildings along with the physical location of the equipment to be interconnected determine the cable plant layout. The equipment may be located

- in a single floor within a building;
- or multiple floors or in multiple rooms within a building;
- between two or more buildings within a campus; or
- between two or more campuses.

The distance between ESCON-compatible equipment may vary from 4 m to 20 km and beyond when equipment (e.g., director) is used to retransmit the optical signal.

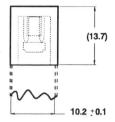

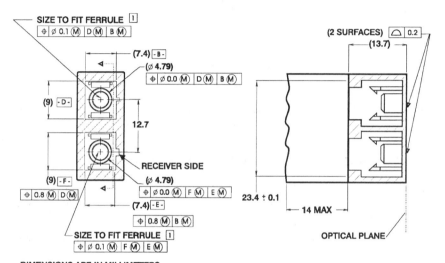

DIMENSIONS ARE IN MILLIMETERS

① ALIGNMENT BORE TO ACCEPT 2.499/2.4985 DIAMETER FERRULE.

2. PLUG FERRULE END SHALL SEAT TO THE OPTICAL REFERENCE PLANE WITH A STATIC FORCE OF 6.7 TO 13.1 N PER FERRULE.

3. FERRULE MATERIAL SHALL BE ZIRCONIA WITH A MAXIMUM SURFACE ROUGHNESS 0.05 μ m.

Fig. 13.12 Diagram of ESCON XDF duplex receptacle [17].

Fiber optic cable planning is very similar to planning for a copper cable plant. The cables should be installed in troughs or conduits, which may be run in the ceiling or under the floor. For a small installation, where the quantity of equipment is small and the equipment pieces are located close to each other, the cable plant will likely consist of jumper cables only.

In a more complex configuration, where equipment is numerous and widely dispersed, the cable plant will consist of a network of distribution panels and trunk cables. It is generally recommended to place the distribution panel at several strategic locations, which minimize the length of the jumper cables. The distribution panels are interconnected with large-count (12, 24, 36, 72, or 144 fibers) trunk cables. This is the more common configuration and requires a significant investment in the procurement and installation of the cable plant. The trunk cables are run throughout the buildings and between buildings. Generally, the planning, purchase, and installation of the cable plant is handled by professionals under contract from the end user. For example, IBM Availability Services will provide a total turnkey design and installation of the cable plant (see Section 13.6.5).

In general, the cable plant is a considerable investment and should be designed with the following considerations:

- Potential future reconfiguration, expansion, and growth
- Future applications
- Local laws and ordinances that would affect the type of cable (e.g., required use of plenum or halogen-free cables in place)
- Adherence to all link specifications with margin for growth (if possible)
- The number and size of the cables required
- Obtaining right-of-way access if cables are to be installed through or over property owned by others
- Potential data rate or dark fiber services that may be available for right-of-way crossings

13.6.1. CABLE PLANT CONFIGURATION CONSIDERATIONS

As mentioned earlier, the physical location of present and future (if possible) equipment is a key factor in determining an optimized cable plant. Establish an equipment location layout of all the equipment to be interconnected. Include any right-of-way requirements in the layout. The layout should also include the number of links to each equipment, to and from locations, and potential future locations. Based-on this layout, a judicious choice of trunk cables, fiber counts, and distribution panel locations can be established.

13.6.1.1. Link Requirements

After determining the optimum link configuration:

1. Use a floor plan to assess the possible link paths, and determine the length and type of each link (for example, jumper only or

jumper and trunk, and multimode or single-mode) and identify and locate each connector, splice, and distribution panel.

2. Ensure that link specifications agree with the fiber cable types, calculated link loss is within specifications (see Section 13.7), and links conform to national, state, and local building codes.

3. Establish a labeling/identification technique for ease of installation and future cable management.

13.6.1.2. Installation Planning

Before installing a link, consider the following:

1. Cable routing diagrams
   - Location and length of each link
   - Type, location, and identification of connectors, adapters, and couplers
   - Locations of splices and distribution panels
2. Manufacturer data sheets
   - Cables
   - Bend radius control
   - Strain relief
   - Splices
   - Distribution panels
   - Attached equipment

The link must be compatible with the equipment that attaches to it. Section 13.6.2 contains cable physical specifications required to support ESCON-capable equipment.

13.6.1.3. Trunk Cables

The trunk cable should meet all the requirements in Sections 13.4.3.2 and 13.5.3.2. The connectors used to terminate the trunk should be of high quality, physical contact, and low loss. Industry standard connector (e.g., ST and FC/PC) meet these requirements and are readily available. Both types are available for multimode and single-mode fiber. Splices are often part of the trunk cable (particularly for lengths greater than 2 km). They should also be high quality, low loss, and high return loss (>28 dB).

Consideration should be given to adding extra fibers in the trunk cable to allow for future growth and expansion and/or additional conduits for the addition of future trunks. In those cases in which the trunk cable communication link is critical to the operation of the data processing complex, and right-of-ways are traversed, an alternate trunk cable route is strongly suggested to ensure uninterrupted service. System/390 servers can be set up to automatically reroute traffic if a trunk cable fails for whatever reason.

13.6.1.4. Distribution Panels

The jumper cable and trunk cable are connected at the distribution panel. In the distribution panel, the trunk cable is split into "two-fiber" groups, connectorized, and connected to an ESCON coupler (e.g., trunk cable terminated with ST connectors and connected to an ESCON/ST coupler). Distribution panels should be located to provide the following [20]:

- An easy entry, exit, and connection point for trunk and jumper cables
- A concentration point for jumper cables to allow manual I/O reconfiguration
- A suitable access point to service an area of processors and equipment

Because of fiber optic channel link distance capabilities, panels can also be located away from the processors and attached equipment, thereby saving floor space.

Distribution panels should contain an adapter and coupler panel, a splice tray, and a storage area for jumper cable slack [20]. The panel should also provide strain relief and radius control for jumper cables:

- Adapter and coupler panels provide the hardware to join identical (coupler) or different (adapter) cable connectors. They can be used separately or in combination within the panel.
- Splice trays should have the same cable capacity as the adapter and coupler panels and should be equipped with hardware to provide strain relief for all cables entering and exiting the tray.
- Jumper cable slack storage provides adequate organizing hardware and storage space for excess jumper cable. The hardware should allow identification and removal of an individual jumper cable, without having to disconnect or remove other jumper cables within the panel.

- Jumper cable strain relief and radius control provides hardware that prevents damage to the jumper cables caused by excessive bending or pulling of the cables.

13.6.2. CABLE ENVIRONMENTAL REQUIREMENTS

In many instances the fiber optic cables will be routed in troughs in the ceiling or under raised floors along with copper cables. The fiber cables are very durable, rugged, and will withstandard significant stress and strain due to rough handling. Nevertheless, it is good practice not to intermingle fiber optic jumper cables and large copper cables (e.g., System/390 bus and tag cables). Trunk cables in general are designed for extremely rugged environments and would be the recommended cable for outdoor, underground, and harsh building environments. When cables are routed properly and the cable environment is organized, no additional fiber cable protection is required.

As a guide to determine the suitability of the jumper cable for the intended environment, the specification, the specifications in Tables 13.11 and 13.12 for multimode and single-mode jumper cables, respectively, govern. Each data processing complex must be carefully analyzed to determine the amount of cable activity or disruption that could occur under the raised floor. If that activity creates a significant exposure to the fiber cables, do any or all of the following:

- Consider installing a direct attached trunk system (e.g., IBM's Fiber Transport System, described in Section 13.6.5).
- Educate personnel about proper installation and handling procedures for fiber cables.
- Bundle the fiber cables. For example, use tie wraps to reduce the possibility of damage.
- Route the fiber cables to minimize the exposure caused by sharp edges, excessive bends, or congestion caused by bus and tag cables.

Also, fiber cables can be installed in conduits, tubings, raceways, troughs, or other protective devices.

13.6.3. LINK CONTINUITY

Light propagation occurs in a single direction within a fiber. In an ESCON jumper cable (for either multimode or single mode), light enters the "B"-labeled port and exits the "A"-labeled port of the cable assembly (Fig.

Table 13.11   **Multimode Jumper Physical and Environmental Specifications**

Parameter	Specification
Physical	
Jacket material	PVC
Jacket outside diameter	4.8 mm
Weight (for information only)[a]	20 g per meter
Installation tensile strength	1000 N maximum
Minimum bend radius (during installation)	4.0 mm; 5 s maximum at 400 N
Minimum installed bend radius	
No load	12 mm
Long-term residual	25 mm at 89 N
Flammability	Underwriters laboratory UL-1666 ONFR (Optical Fiber Nonconductive; Plenum UL-910 is also acceptable)
Crush resistance	500 N/cm maximum
Maximum unsupported vertical rise	100 m
Environmental	
Operating environment	Inside buildings only
Operating temperature	10–60°C
Operating relative humidity	5–95%
Storage and shipping	−40 to 60°C
Lightning protection	None required
Grounding	None required

Note. From Ref. [21].
[a] Cable only; connectors not included.

13.13). The transceiver output is always coupled to the B port and the transceiver input is always coupled to the A port. The fiber crossover is designed into the jumper cable assembly and, by virtue of the polarized connector, the transceiver output is always connected to the corresponding transceiver input at the other end of the cable.

The design of the trunk cable plant must be done to account for the fiber crossover. Only an odd number of crossovers will result in a viable link. In general, a link contains two jumper cables, which requires the trunk fiber to be crossed in order to ensure link continuity. This configuration is illustrated in Fig. 13.14.

Table 13.12   **Single-Mode Jumper Physical and Environmental Specifications**

Parameter	Specification
Physical	
Jacket material	PVC
Jacket outside diameter	Zip cord
Weight (for information only)[a]	20 g per meter
Installation tensile strength	1000 N maximum
Minimum bend radius (during installation)	50 mm; 5 s maximum at 100 N
Minimum installed bend radius	
No load	30 mm
Long-term residual	50 mm at 80 N
Flammability	Underwriters laboratory UL-1666 ONFR (Optical Fiber Nonconductive; Plenum UL-910 is also acceptable)
Crush resistance	889 N/cm maximum
Maximum unsupported vertical rise	100 m
Environmental	
Operating environment	Inside buildings only
Operating temperature	0–60°C
Operating relative humidity	8–95%
Storage and shipping	−40 to 60°C
Lightning protection	None required
Grounding	None required

Note. From Ref. [21].

[a] Cable only; connectors not included.

13.6.4.   *CAMPUS LAYOUT EXAMPLE*

The physical location and quantity of equipment are the primary factors affecting the cable layout. The application environment may vary from a single-room configuration with limited equipment to a multicampus environment with multiple buildings and floors within a building.

The example illustrated in Fig. 13.15 is a single campus environment with data processing equipment in two buildings. The point of egress of each building is at a "building interface panel." This panel provides a common access point for each building. The trunk cable (multimode or single mode) between the buildings will be determined by the actual

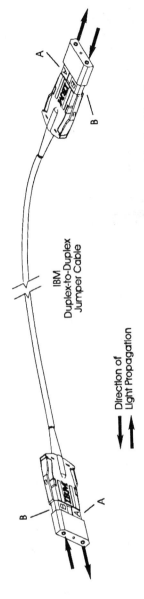

Fig. 13.13 Direction of light propagation in an IBM jumper cable (multimode depicted) [20].

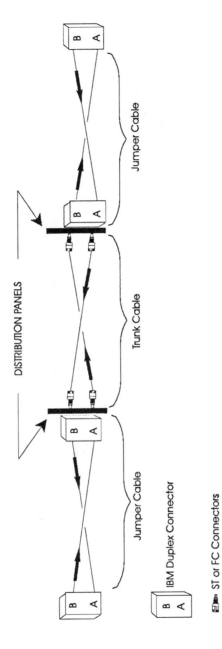

Fig. 13.14 Direction of light propagation in a link with a trunk [20].

477

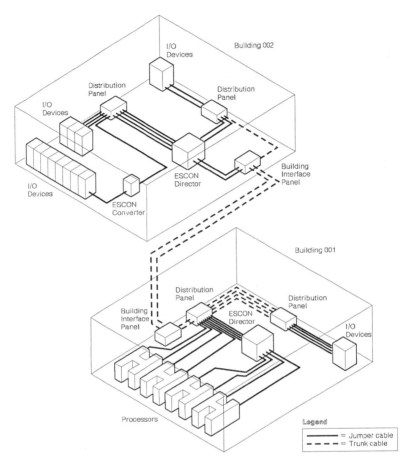

Fig. 13.15 Example of a single campus link environment between two buildings [20].

"run" distance of the intended cable. In general, single mode is chosen for distance greater than 3 km. The trunk cable should be a rugged outdoor cable type that can be installed in an underground conduit (preferred) or strung on utility poles. Extra count trunks should be considered for expansion.

Within a building, distribution panels are optimally located. Both building trunks and jumpers are used to interconnect distribution panels. In general, jumpers connecting to distribution panels should be kept as short as possible. Distribution panels should be optimally located with multifiber trunks interconnecting the distribution panels.

The flexibility and congestion relief inherent with distribution panels requires additional termination hardware and increased link loss due to the inclusion of additional connections and splices. This must be considered in the loss budget analysis (Section 13.7) to ensure the loss budget requirements are not violated.

If the two buildings were located on different campuses, then right-of-way access is required from the owner of the land. Leasing of "dark fiber" may also be a possibility. The negotiation, however, for right-of-way, leasing, etc. may greatly extend the total planning and installation time. This must be considered in the overall project schedule.

For distance requirements beyond 20 km (XDF feature maximum), channel extenders may be used such as the IBM 9036 ESCON Remote Channel Extender or the IBM 9729 Fiber Optic Multiplexer. The IBM 9036 is a repeater that retransmits the optical signal to achieve an additional 20-km distance. The Fiber Optic Multiplexer utilizes wavelength division multiplexing to transmit up to 10 individual ESCON channels on a single pair of optical fibers. Each channel is converted to a different wavelength and transmitted simultaneously on the fiber. Distances of 50 km can be achieved with the IBM 9729.

13.6.5. DIRECT ATTACH TRUNKS

For large installations, in which the amount of equipment in a data processing complex is large, the use of direct attach trunk cables has become popular [20]. Multifiber ribbon (12 fibers per ribbon) trunk cables are directly attached to an IBM System/390 server as illustrated in Fig. 13.16. For large ESCON channel count servers, direct-attached trunks offer the least congested cabling approach.

The trunk may contain as many as 12 ribbons (144 fibers total). Each ribbon is terminated in a 12-fiber multifiber terminated push-on (MTP) connector. The small size of the connectors easily enables six MTPs to be terminated inside the server frame in a coupler bracket. Several brackets can be installed in a machine. A harness is used inside the machine to convert the 12 fibers (representing six ESCON channels) of a ribbon to six ESCON connectors for insertion into the appropriate transceiver.

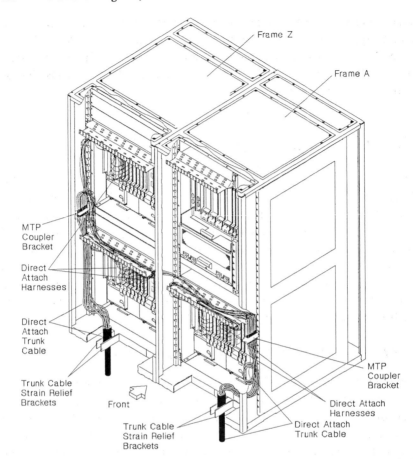

Fig. 13.16 IBM's FTS-III direct-attach trunk system in a System/390 server [20].

Once the harnesses are installed in the machine, all fiber connects and disconnects can be done using the trunk cable MTP connectors. The harnesses remain inside the machine, plugged into the individual ESCON or coupling facility ports, whereas the trunk cables can be quickly removed. The machine can then be relocated within the data complex because the trunk cables are easily rerouted to the new location and replugged into the harnesses. This greatly reduces the time spent unplugging, rerouting, and plugging the individual fiber optic jumper cables back into the machine ports.

MTP-to-MTP multiribbon trunks can be used throughout the complex to directly connect equipment together as well as equipment to distribution panels. For these installations the use of jumper cables is minimized. This cabling approach is used for both multimode and single-mode versions.

IBM Availability Services offers this approach as Fiber Transport System III (FTS-III) and is also available to do the planning, design, and installation of this interconnection system.

13.7. Loss Budget Analysis

The cable plant loss (C_t) of an ESCON link is a critical parameter and is determined by the link length, fiber attenuation, number of connections and jumper cable loss, jumper/trunk core diameter mismatch splices, etc. [22]. These parameters are greatly influenced by the cable plant layout and quality of components used. As discussed in Section 13.3.3, the required optical cable plant loss is established by the available power as determined by Eq. (13.1). The cable plant optical link loss must be less than or equal to the available power to ensure the link design performance. This is true for both multimode and single-mode links.

13.7.1. MULTIMODE CABLE PLANT LINK LOSS

The multimode link loss is determined by statistically adding all the individual loss elements in the cable plant. A good approximation is that the individual loss distributions are Gaussian and can be described with a mean value and variance. The means and variances of all the individual elements are summed together to yield an overall cable plant mean (M_c) and variance (σ_c^2), where σ_c is the cable plant loss standard deviation. The total cable plant loss is given by

$$C_t = M_c + 3\sigma_c. \tag{13.2}$$

Representative loss values for the individual elements can be obtained from component manufacturers or from a professional fiber optic network planner. Table 13.13 contains typical loss distributions for the elements in a cable plant. Table 13.13 can be used as a guide if more accurate data are not available; however, it is preferred to use data from the cable plant installer or component manufacturer.

In general, component losses (e.g., fiber attenuation, and connection loss) are measured using an equilibrium mode distribution in the fiber. In order to account for the optical loss of a link associated with mode

Table 13.13   **Typical Optical Component Loss Values**

Component	Description	Fiber core size (μm)	Mean loss (dB)	Variance (dB²)
Connector[a]	Physical contact	62.5–62.5	0.40	0.02
		50.0–50.0	0.40	0.02
		9.0–9.0[b]	0.35	0.02
		62.5–50.0	2.10	0.12
		50.0–62.5	0.00	0.01
Splice	Mechanical	62.5–62.5	0.15	0.01
		50.0–50.0	0.15	0.01
		9.0–9.0[b]	0.15	0.01
Splice	Fusion	62.5–62.5	0.40	0.01
		50.0–50.0	0.40	0.01
		9.0–9.0[b]	0.40	0.01
Cable	IBM MMF jumper	62.5	1.75[c]	NA
	IBM SMF jumper	9.0	0.8[c]	NA
	Trunk	62.5	1.00[c]	NA
	Trunk	50.0	0.90[c]	NA
	Trunk	9.0	0.50[c]	NA

Note. From Ref. [21].

[a] The connector loss value is typical when attaching identical connector types. The loss can vary significantly if attaching different connector types.

[b] Single-mode connectors and splices must meet a minimum return loss of 28 dB.

[c] Actual loss value in dB/km.

redistribution in multimode fibers, the following loss should be included in a link:

- 1.5 dB for 50/125 μm trunk fiber, or
- 1.0 dB for 62.5/125 μm trunk fiber

A typical example of a link is illustrated in Fig. 13.17. As a multimode link, a typical configuration might be two jumper cables with a combined length of 90 m, 1.5 km of 50/125 μm trunk (800 MHz/km bandwidth), two physical contact connectors (one 62.5–50 μm and the other 50–62.5 μm), and six splices. A completed loss worksheet is shown in Fig. 13.18. The element loss values used are from Table 13.13. In this case the total cable plant link loss is 7.3 dB.

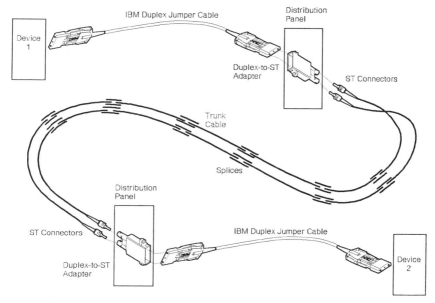

Fig. 13.17   Example of a generic ESCON link configuration [22].

13.7.2.   SINGLE-MODE CABLE PLANT LINK LOSS

This section describes the calculation of the loss in a single-mode link. The approach and procedure are the same as those in the multimode link. The individual element losses are statistically added to yield a total link loss mean and variance value. Equation (13.2) is used to calculate the cable plant loss. In the case of a single-mode fiber, no equilibrium mode distribution loss occurs but there is a small loss associated with mode field diameter mismatches (excess connector loss). This loss adds to a 0.5-dB total (two connections) and is small enough to be included in the loss budget for all configurations (whether present or not). Using the generic configuration in Fig. 13.17, a typical single-mode link could consist of two jumper cables with a combined length of 210 m, 19.76 km of 9/125 μm trunk, two physical contact connections, and two mechanical splices. Using the typical values in Table 13.13, Fig. 13.19 illustrates a completed single-mode cable plant loss analysis. In this case the loss is 12.86 dB.

A. Calculating the Multimode Component Mean Loss

Connection loss multiplied by the
number of connections in the link:

_62.5_ -μm-to-_50.0_ -μm connection:	_2.10_ dB	×	_1_	= _2.10_ dB
_50.0_ -μm-to-_62.5_ -μm connection:	_0_ dB	×	_1_	= _0_ dB
_____-μm-to-_____-μm connection:	____ dB	×	____	= ____ dB

Splice loss multiplied by the total
number of splices in the link: _0.15_ dB × _6_ = _0.90_ dB

Jumper cable loss multiplied by the
combined length of the jumper cables: _1.75_ dB/km × _0.09_ km = _0.16_ dB

Trunk loss per kilometer multiplied
by the total trunk length (in km): _0.90_ dB/km × _1.5_ km = _1.35_ dB

(+) ―――――――

Total Component   _4.51_ dB
Mean Loss

B. Calculating the Multimode Component Variance Loss

Connection variance multiplied by the
number of connections in the link:

_62.5_ -μm-to-_50.0_ -μm connection:	_0.12_ dB^2	×	_1_	= _0.12_ dB^2
_50.0_ -μm-to-_62.5_ -μm connection:	_0.01_ dB^2	×	_1_	= _0.01_ dB^2
_____-μm-to-_____-μm connection:	____ dB^2	×	____	= ____ dB^2

Splice variance multiplied by the
number of splices in the link: _0.01_ dB^2 × _6_ = _0.06_ dB^2

(+) ―――――――

Total Component   _0.19_ dB^2
Variance Loss

C. Calculating the Total Multimode Link Loss

Total component mean loss: = _4.51_ dB

Square root of total component
variance loss multiplied by 3: $\sqrt{\_0.19\_ dB^2}$ = _0.44_ dB × 3 = _1.32_ dB

Higher-order mode loss (ESCON only): = _1.5_ dB

50.0-μm trunk = 1.5 dB
62.5-μm trunk = 1.0 dB   (+) ―――――――

Calculated Link Loss   _7.3_ dB

Fig. 13.18   Example of a completed multimode link loss worksheet [22].

13.7.3.   SPECIFICATION VERIFICATION

Upon completion of the link loss analysis, the resultant values should be
checked against the requirements given in Tables 13.1 and 13.2 for multimode and single-mode links, respectively. Both of the examples illustrated
earlier fulfill this requirement.

A. Calculating the Single-Mode Component Mean Loss

Connection loss multiplied by the number of connections in the link:	_0.35_	dB	×	_2_		=	_0.70_	dB
Splice loss multiplied by the total number of splices in the link:	_0.15_	dB	×	_2_		=	_0.30_	dB
Jumper cable loss multiplied by the combined length of the jumper cables:	_0.8_	dB/km	×	_0.21_	km	=	_0.17_	dB
Trunk loss per kilometer multiplied by the total trunk length (in km):	_0.5_	dB/km	×	_19.76_	km	=	_9.88_	dB

(+) ‾‾‾‾‾‾‾

Total Component Mean Loss   _11.05_ dB

B. Calculating the Single-Mode Component Variance Loss

Connection variance multiplied by the number of connections in the link:	_0.06_	dB^2	×	_2_		=	_0.12_	dB^2
Splice variance multiplied by the number of splices in the link:	_0.01_	dB^2	×	_2_		=	_0.02_	dB^2

(+) ‾‾‾‾‾‾‾

Total Component Variance Loss   _0.14_ dB^2

C. Calculating the Total Single-Mode Link Loss

Total component mean loss:	=	_11.05_	dB
Square root of total component variance loss plus jumper assembly variance loss multiplied by 3:	$\sqrt{\underline{0.14} + 0.05\ dB^2}$ = _0.436_ dB × 3 =	_1.31_	dB
Jumper assembly loss plus excess connector loss:	=	0.50	dB

(+) ‾‾‾‾‾‾‾

Calculated Link Loss   _12.86_ dB

Fig. 13.19   Example of a completed single-mode link loss worksheet [22].

13.7.4. CABLE PLANT LINK VERIFICATION

Verification of the cable plant ensures that a link meets IBM specifications before equipment is attached. This is usually done by the cable plant installer prior to delivering the installation to the end user. Verifying a link requires using an optical source to transmit a signal through the link and a power meter to measure this signal at the other end. For multimode links, an optical mode conditioner tool [23], to achieve a equilibrium mode distribution, must be used to provide consistent power loss measurements.

Before you begin link verification, you need to review the link documentation, e.g., link configuration, layout, and distribution panels, to ensure you understand what loss value to expect based on the analysis given earlier in this section. Link verification consists of (i) calibrating the test equipment, (ii) obtaining reference power levels, and (iii) measuring the total loss of the links under verification. The amount of loss introduced by the link depends on the jumper cable lengths, the trunk cable length and specifications, and the number and type of link splices and connectors. This measured loss should be less than the maximum allowable cable plant loss.

13.8. Link Troubleshooting

This section summarizes the link troubleshooting strategy and process if a failure of the link has occurred after initial installation, verification, and operation. A link failure has occurred if the link has failed completely (no intelligible signal can be transmitted over the link) or the bit error rate is higher than the allowable threshold and the processor has declared the link defective. In either case, the processor end of the link will report a malfunction and cause a repair call. Depending on the severity (e.g., one path of multiple paths to the equipment) of the problem at the system level, the channel will be required immediately or at a convenient maintenance interval. The latter is generally the case for System/390 servers. The servers will also tell the system operator or repair technician exactly which link is defective.

The link troubleshooting process consists of verification of the output interface operation, cable loss, and input optical interface, usually in this order [24]. The generic link configurations that may be found in an installation are the following:

- Links consisting of one jumper cable between two devices
- Links consisting of two jumper cable connected together in a distribution panel; In this case, the duplex-to-duplex coupler performs a fiber crossover
- Links consisting of two jumper cables, two distribution panels, and a trunk cable connecting the distribution panels

The various link configuration are treated similarly in the link troubelshooting, fault location process.

The most common cause of link failure, after the link was verified upon installation, is the "no-light" condition on the link. This condition is most likely caused by

1. Reconfiguration of link causing a light continuity error
2. Dust or dirt at the interfaces blocking the light
3. Jumper failure
4. Trunk failure
5. Transmitter operation failure
6. Receiver operation failure

Items 1 and 2 can be investigated by obtaining the link diagram and ensuring that there are an odd number of crossovers in the link (Section 13.6.3), the link connects the correct equipment, and by inspecting and cleaning all connections. The remaining items will be discussed in the following sections.

Fiber optic bandwidth measurements and output optical sensitivity measurements are very difficult to measure in the field. Specially trained technicians or technical people are needed to perform these types of measurements with sophisticated test equipment. These elements are verified at the factory and are not likely to fail in the field. They are rarely measured in the field.

13.8.1. LINK (MMF OR SMF) ANALYSIS

The link failure analysis consists primarily of optical power measurements at strategic locations throughout the link. The appropriate test tools (e.g., optical power source and optical mode conditioner) will be needed to perform the measurements. A tool kit is available from IBM or your local cable installer to perform these measurements. Figure 13.20 illustrates the four key measurements points (P_T, P_1, P_2, and P_R) in an ESCON link. The four measurements should be made on both paths of the link or until the problem has been found. A typical procedure follows:

Step 1: Measure the average output power in dB, P_T, of the transmitter by using an ESCON duplex-to-simplex (ST or Biconic) test jumper and a power meter. The test jumper should be made with 62.5/125 μm fiber for multimode link and 9/125 μm fiber for a single-mode link. Ensure that all connections are clean and the power meter is properly set and calibrated. The average optical power should be within the range specified in Table 13.3 for a multimode link or Table 13.7 for a single-mode link. If the specification is satisfied go to step 2; if not, replace the channel card and retest the new port for average optical power. If the specification is achieved, have the system restart the ESCON channel and verify correct operation.

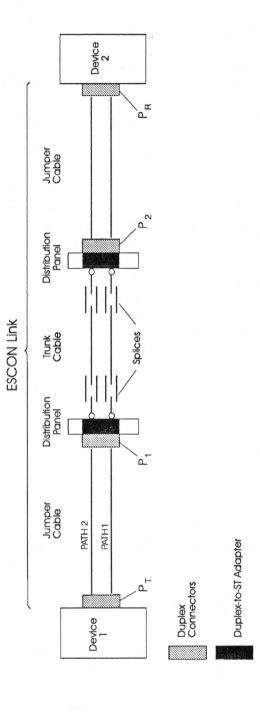

Fig. 13.20 Measurement positions in an ESCON link.

Step 2: Measure the average power in dB, P_1, exiting the jumper cable by using a duplex-to-duplex coupler, a large core (100/140 μm) duplex-to-simplex (ST or Biconic) bucket fiber jumper, and the same power meter. Ensure all surfaces are clean. The P_1 reading should be between the saturation and sensitivity levels are specified in Table 13.4 for multimode and Table 13.8 for single-mode. Also, the jumper loss, obtained by subtracting P_1 from P_T, should not exceed 2 dB. If these requirements are satisfied go to step 3, or if this is a single jumper link, repeat steps 1 and 2 for the return path. If any condition is not met, the jumper is defective and should be replaced. Remeasure P_1 with the new jumper and verify the requirements are satisfied. If the specification is achieved have the system restart the ESCON channel and verify correct operation.

Step 3: This measurement is made only if a trunk cable is present. If the link is composed of only two jumper cables, proceed to step 4. Measure the output power in dB, P_2, of the output of the trunk cable at the distribution panel using a 62.5/125 μm duplex-to-simplex (ST or Biconic) fiber jumper and the same power meter. Ensure all services are clean. The P_2 reading should be between the saturation and sensitivity levels as specified in Table 13.4 for a multimode link and Table 13.8 for a single-mode link. Also, the sublink loss, obtained by subtracting P_2 from P_T, should not exceed 8 dB for a multimode link (specified in Table 13.1) or 14 dB for a single-mode link (specified in Table 13.2). If these conditions are satisfied, proceed to step 4. If the requirements are not satisfied, proceed to Section 13.8.2.

Step 4: This procedure measures the amount of light available to the receiver. Measure the output power (dB), P_R, of the total link at the receiver input using a duplex-to-duplex coupler, a large core (100/140 μm) duplex-to-simplex (ST or Biconic) bucket fiber jumper, and the same power meter. Ensure all surfaces are clean. The P_R reading should be between the saturation and sensitivity levels as specified in Table 13.4 for a multimode link and Table 13.8 for a single-mode link. Also, the jumper cable loss, obtained by subtracting measurement P_R from P_2, should be less than 2 dB. If the jumper cable loss requirement is not satisfied, replace the jumper cable and retest the link. If the two conditions are now satisfied, restart the channel and perform a system test. If the average power does not meet the receiver requirements then the trunk cable is defective and proceed to

Section 13.8.2. If the specifications in Table 13.4 or Table 13.8 are satisfied and the link still does not function properly, then the receive channel card is likely defective. Replace the receive channel card and restart the channel from the server and perform a system test.

If all the optical conditions on the cable and transmitters are satisfied (e.g., P_T, P_1, P_2, and P_R) and the link still does not function after replacing the receive channel card, then the transmitter is defective (e.g., LED "stuck on") and the transmitter channel card should be replaced. In most cases, the system will be able to identify which path in the channel link is defective. However, in the case in which the defective path of the channel link is not known, repeat the four steps for the return path.

13.8.2.   TRUNK CABLE ANALYSIS

If it is determined that the trunk cable or its terminating connection is suspected of being defective, then the trunk cable should be tested with a "conditioned" light source. The trunk loss should be measured from the first distribution panel nearest device 1 of the link to the last distribution panel nearest device 2 (see Fig. 13.17). The trunk cable may be a long link (up to 3 km for multimode and 20 km for single mode) and may interconnect buildings or extend over right-of-ways. The trunk cable test procedure is intended to determine if the total trunk meets requirements and does not attempt to isolate the defective element.

The test equipment required for a multimode trunk include an LED optical source with a mode conditioner, two 62.5/125 μm duplex-to-simplex (ST or Biconic) test jumpers, a large-core (100/140 μm) duplex-to-simplex (ST or Biconic) bucket fiber jumper, a duplex-to-duplex coupler and other couplers as appropriate, and a calibrated power meter. The mode conditioner is placed between the LED optical source and the first test jumper. The purpose of the condition is to ensure the light exiting the "conditioned" optical source is equivalent to an equilibrium mode distribution. This can also be achieved by using 1 km of fiber or wrapping a length of fiber several times about a mandrel.

The test equipment for a single-mode trunk cable is a laser power source, two simplex-to-duplex 9/125 μm test jumpers at least 2 m long, a large-core (100/140 μm) duplex-to-simplex (ST or Biconic) bucket fiber jumper, duplex-to-duplex coupler, and a calibrated power meter. The conditioning tool is not needed because the 2-m length of test jumper ensures single-mode operation.

The same procedure is used for multimode or single-mode trunk:

Step 1: The intent of this step is to assemble and measure an optical source whose optical output is at the end of a duplex test jumper. Assemble the optical source to the mode conditioner (not required for SMF), and connect the simplex side of the appropriate (MMF or SMF) test jumper to the conditioner (MMF only) output. For single-mode setup connect the SMF test jumper directly to the laser power source. Measure the output optical power of the assembled source by coupling the large-core bucket jumper to the duplex connector end of the test jumper via the appropriate duplex-to-duplex coupler. Insert the appropriate simplex end into the power meter and measure the average output power, P_1, which is the average optical power inserted into the duplex trunk connector at the first distribution panel.

Step 2: With the optical source connected to the trunk cable (do not disconnect cable after step 1 measurement), connect the appropriate (MMF or SMF) test jumper to the last distribution panel in the link. Insert the appropriate simplex end of the test jumper to the same power meter and measure the average output power, P_2, of the trunk cable. The trunk cable loss is obtain by subtracting P_2 from P_1. This value includes all the losses in the trunk, including the connections at the first and last distribution panel. The measurement should be less than 8 dB for a multimode link and 14 dB for a single-mode link. The measurement should also be compared to the trunk loss measured when the trunk was installed and verified.

If the cable loss criteria in step 2 is not satisfied or the loss is significantly different from the initial installation measurement, the trunk cable installer or a qualified fiber optic professional should be called to reevaluate and repair the trunk. Once the link is corrected, restart the channel from the server and perform a system test.

13.9. Additional Aspects

This section discusses some additional considerations for the ESCON fiber optical link that are very important but do not conform to the theme of the previous sections. The IBM ESCON link, in conjunction with the ANSI standard discussed in Section 19.9.3, will likely continue to evolve and improve in the future with technology advances. Nevertheless, any evolution of both ESCON and ESCON XDF will be accomplished in a manner such that it is compatible with the existing installed based.

13.9.1. LASER COMPLIANCE

Both ESCON and ESCON XDF links are designed to be class 1 optical products. Class 1 optical products are not considered hazardous to any untrained personnel who are using or servicing the links. A general requirement imposed on transceiver manufacturers is that class 1 certification (if applicable) must be inherent to the transceiver. No requirements should be placed on the system hardware or software by the transceiver manufacturer to achieve class 1 compliance.

The LED transceiver used in ESCON, by virtue of its low maximum output power (-15 dBm) and incoherence, is inherently compliant to the class 1 criteria used by the Department of Health Services (DHHS) in the United States the International Electrotechnical Community (IEC) in Europe. Certification is not required. For customer awareness, however, optical radiation warning labels should be placed inside the host equipment containing ESCON and class 1 compliance should be discussed in appropriate literature.

The ESCON XDF transceiver contains a 1300-nm semiconductor laser and must be certified by the appropriate agencies to be considered a class 1 laser product. The certification consists of submitting a written report of device operation, maximum conditions, fail-safe mechanisms, manufacturer's quality control, etc. to both U.S. and European certification agencies. The IBM ESCON XDF transceiver modules are certified in the United States to conform to the requirements of DHHS-21 CFR Subchapter J for class 1 laser products. Elsewhere, they are certified to be in compliance with IEC 825 and CENELEC HD 482 S1 as a class 1 laser product. Appropriate labels should be placed inside the host equipment and discussed in the appropriate literature.

13.9.2. QUALIFICATION

The ESCON fiber optic link has undergone an extensive functional and environmental test program to ensure design robustness and that the product meets all the application requirements. The qualification consists of the following major elements:

1. Component functional and environmental testing

2. Link optical and mechanical performance testing over the application environment

3. System-level testing in simulated and actual application environments.

Component testing consists of extensive environmental testing (e.g., temperature, humidity, shock, vibration, and accelerated life testing) and ensuring all functional specification parameters are in conformance during the testing. A measure of reliability is obtained from the data. The components tested include the transceivers, jumper cables, couplers, and channel logic modules.

The link performance testing involves a parametric analysis of link performance over worst-case conditions (functional and environmental) using a representative sample of manufactured components. Link margins are determined and conformance to the link model is assessed utilizing simulated data transmission of all possible combinations. All link configurations are evaluated.

The system-level testing involves utilizing a large sample of links in a real data processing environment. The BER is evaluated with simulated and real workloads, all link configurations and operational environmental extremes. Electromagnetic compatibility (EMC) testing is completed at this time.

13.9.3. SBCON ANSI STANDARD

ESCON has proven to be an excellent communication link for the data processing environments. IBM, with the support and encouragement of other companies in the information industry, has requested that ANSI sanction an official work effort to make ESCON an industry standard. The activity was formally approved by ANSI in 1995 and a draft standard SBCON has been established [25]. The SBCON standard was approved in 1997. The information contained in this chapter is consistent with this standard.

ACKNOWLEDGMENTS

I thank the International Business Machines Corporation for allowing the extensive use of copyrighted material in this chapter. I also thank the many individuals (too numerous to list here) both within and outside of IBM whose hard work and dedication helped make ESCON a reality.

References

1. IBM Corp. Introducing Enterprise Systems Connection (IBM Document No. GA23-383). Mechanicsburg, PA: IBM Corp.
2. Calta, S. A., deVeer, J. A., Loizides, E., and R. N. Strangwayes. 1992. Enterprise Systems Connection Architecture—System Overview. *IBM J. Res. Dev.* 36:535–551.
3. Georgiou, C. J., Larsen, T. A., Oakhill, P. W., and Salimi, B. 1992. The IBM Enterprise Systems Connection (ESCON) director: A dynamic switch for 200 Mb/s fiber optic links. *IBM J. Res. Dev.* 36:593–616.
4. Elliott, J. C., and Sachs, M. W. 1992. The IBM Enterprise Systems Connection Architecture. *IBM J. Res. Dev.* 36:577–592.
5. Flanagan, J. R., Gregg, T. A., and Casper, D. F., 1992. The IBM Enterprise Systems Connection (ESCON) channel—A versatile building block. 1992. *IBM J. Res. Dev.* 36:617–632.
6. IBM Corp. Enterprise Systems Architecture/390 ESCON I/O interface (IBM Document No. SA22-7202, Chap. 2). Mechanicsburg, PA: IBM Corp.
7. IBM Corp. Enterprise Systems Architecture/390 ESCON I/O interface (IBM Document No. SA22-7202, Chap. 8). Mechanicsburg, PA: IBM Corp.
8. IBM Corp. Enterprise Systems Architecture/390 ESCON I/O interface (IBM Document No. SA22-7202, Appendix). Mechanicsburg, PA: IBM Corp.
9. Aulet, N. R., Boerstler, D. W., DeMario, G., Ferraiolo, F. D., Hayward, C. E., Heath, C. D., Huffman, A. L., Kelley, W. R., Peterson, G. W., and Stigliani, D. J., Jr. 1992. IBM Enterprise Systems multimode fiber optic technology. *IBM J. Res. Dev.* 36:553–577.
10. DeCusatis, C. 1995. Data processing systems and optoelectronics. In *Optoelectronics for Data Communication,* eds. R. C. Lasky, U. L. Osterberg, and D. J. Stigliani, Chap. 6. New York: Academic Press.
11. IBM Corp. Enterprise Systems Architecture/390 ESCON I/O interface physical layer (IBM Document No. SA23-0394). Mechanicsburg, PA: IBM Corp, courtesy of International Business Machines Corporation.
12. IBM Corp. Planning for fiber optic channel links (IBM Document No. GA23-0367, Chap. 1). Mechanicsburg, PA: IBM Corp, courtesy of International Business Machines Corporation.
13. Gregurick, V. 1992. Fiber optic transmitter measurement procedure, IBM Engineering Specification 49G3489. East Fishkill, NY: IBM Corp.
14. Electronics Industry Association/Telecommunications Industry Association Commercial Building Telecommunications Cabling Standard (EIA/TIA-568-A).
15. Electronics Industry Association. Detail specification for 62.5 μm core diameter/125 μm cladding diameter class 1a multimode. Graded index optical wave-

guide fibers (EIA/TIA-492AAAA). New York: Electronics Industry Association.

16. International Telecommunications Union. Characteristics of a 50/125 μm multimode graded index optical fibre cable (CCITT Recommendation G.651).

17. Fiber Channel Standard. Physical layer and signaling specification. American National Standards Institute X3T9/91-062, X3T9.3/92-001 (rev. 4.2).

18. Electronics Industry Association Detail specification for class IVA dispersion unshifted single-mode optical wavelength fibers used in communications systems (EIA/TIA-492BAAA). New York: Electronics Industry Association.

19. International Telecommunications Union. Characteristics of single-mode optical fibre cable (CCITT Recommendation G.652).

20. IBM Corp. Planning for fiber optic channel links (IBM Document No. GA23-0367, Appendix A). Mechanicsburg, PA: IBM Corp, courtesy of International Business Machines Corporation.

21. IBM Corp. Planning for fiber optic channel links (IBM Document No. GA23-0367, Chap. 2). Mechanicsburg, PA: IBM Corp, courtesy of International Business Machines Corporation.

22. IBM Corp. Planning for fiber optic channel links (IBM Document No. GA23-0367, Chaps. 3 and 4). Mechanicsburg, PA: IBM Corp, courtesy of International Business Machines Corporation.

23. Light launch conditions for long-length graded-index optical fiber spectral attenuation measurements (EIA/TIA-455-50A), available from Electronics Industry Association, New York.

24. IBM Corp. Maintenance information for enterprise systems connection links (IBM Document No. SY27-2597). Mechanicsburg, PA: IBM Corp.

25. American National Standards Institute. Single-byte command code sets CONnection architecture (SBCON). Draft ANSI Standard X3T11/95-469 (rev. 2.2).

Chapter 14 | Local Area Networks and Fiber Distributed Data Interface

Rakesh Thapar

InterOperability Laboratory, University of New Hampshire, Durham, New Hampshire 03824

14.1. Introduction

Local area networks (LANs) are commonly used in many applications and environments. As the technology is changing so are the LANs. Tomorrow's LANs are promising much higher speeds and reach. Unfortunately, the competing technologies being developed are based on different philosophies, thereby making them incompatible. This chapter introduces the modern fiber-based LAN technologies that either exist today or are expected in the market during the next few years. In particular, the chapter discusses Fast Ethernet, gigabit Ethernet, VG-AnyLan, Internet protocol (IP) over asynchronous transfer mode (ATM), ATM LAN emulation, and fiber distributed data interface (FDDI).

Several proposed technologies promise better performance for tomorrow's LANs. The promises create difficult choices for LAN owners who need better performance today but do not want to buy equipment that will be nonstandard or obsolete when tomorrow's technologies are standardized. In this chapter, I give an overview of the current LAN technologies, some of which are still undergoing an active standardization process.

14.2. Fast Ethernet

Fast Ethernet [1–3] or the 100BASE-T, is a 100 Mb/s networking technology based on IEEE 802.3u standard. It uses the same media access control (MAC) protocol, the carrier sense multiple access/collision detection

496

HANDBOOK OF FIBER OPTIC
DATA COMMUNICATION

(CSMA/CD) method (running 10 times faster) that is used in the existing Ethernet networks [International Standards Organization/International Electrotechnical Community (ISO/IEC) 8802-3] such as 10BASE-T, connected through a media-independent interface (MII) to the physical layer device (PHY) running at 100 Mb/s. The supported PHY sublayers are 100BASE-T4, 100BASE-TX, and 100BASE-FX. The general name of the PHY group is 100BASE-T. One of the most important advantages of Fast Ethernet is that it uses the standard 802.3 MAC. This allows data to be interchanged between a 10BASE-T and 100BASE-T nodes without protocol translation, thereby allowing the possibility of a phased introduction of 100BASE-T in the existing 10BASE-T networks and a cost-effective migration path that maximizes use of existing cabling and network management systems. Autonegotiation is another key element of the 100BASE-T Fast Ethernet IEEE 802.3u standard. Autonegotiation allows an adapter, hub, or switch capable of data transfer at both 10 and 100 Mb/s rates to automatically use the fastest rate supported by the device at the other end. Autonegotiation signals the capabilities it has available, detects the technology that exist in the device it is being connected to, and automatically configures to the highest common performance mode of operation. The relationship of Fast Ethernet model to the open systems interconnection model is shown in Fig. 14.1.

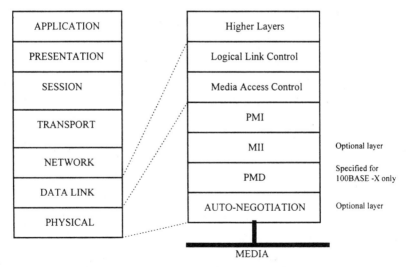

Fig. 14.1 Fast Ethernet (IEEE 802.3u) model.

A. THE PHYSICAL LAYER

1. 100BASE-T4

This physical layer defines the specification for 100BASE-T Ethernet over four pairs of category 3, 4, or 5 unshielded twisted-pair (UTP) wire. This is aimed at those users who want to retain the use of voice-grade twisted-pair cable. Additionally, it does not transmit a continuous signal between packets, which makes it useful in battery-powered applications.

With this signaling method, one pair is used for carrier detection and collision detection in each direction and the other two are bidirectional. This allows for a half-duplex communication using three pairs for data transmission. The unidirectional pairs are the same ones used in 10BASE-T (it uses only two pairs) for consistency and interoperability. Because three pairs are used for transmission, to achieve an overall 100 Mb/s each pair needs only to transmit at 33.33 Mb/s. If Manchester encoding was to be used at the physical layer, as in 10BASE-T, the 30-MHz limit for the category 3 (cat 3) cable would be exceeded. To reduce the rate, a 8B6T block code is used that converts a block of 8 bits into six ternary symbols, which are then transmitted out to three independent channels (pairs). The effective data rate per pair thus becomes $(6/8) \cdot 33.33 = 25$ MHz, which is well within the cat 3 specifications of 30 MHz.

2. 100BASE-X

Two physical layer implementations, 100BASE-TX and 100BASE-FX, are collectively called 100BASE-X when referring to issues common to both. 100BASE-X uses the physical layer standard of FDDI by using its physical media-dependent sublayer (PMD) and medium-dependent interfaces (MDI). The 125 Mb/s full-duplex signaling system for a twisted pair, defined in FDDI, forms the basis for 100BASE-TX and the system defined for transmission on optical fiber forms the basis for 100BASE-FX. Basically, 100BASE-X maps the characteristics of FDDI PMD and MDI to the services expected by the CSMA/CD MAC. It defines a full-duplex signaling standard of 125 Mb/s for multimode fiber, shielded twisted-pair and unshielded twisted-pair wiring. The physical sublayer maps 4 bits from MII into 5-bit code blocks and vice versa using the 4B/5B encoding scheme (same as FDDI).

3. 100BASE-TX

This physical layer defines the specifications for 100BASE-T Ethernet over two pairs of category 5 UTP wire or two pairs of shielded twisted-pair (STP) wire. With one pair for transmit and the other for receive, the wiring scheme is identical to that used for 10BASE-T Ethernet (see Table 14.1 for the connector details).

4. 100BASE-FX

This physical layer defines the specification for 100BASE-T Ethernet over two strands of multimode (62.5/125 μm) fiber cable. One strand is used for transmit, whereas the other is used for receive.

100BASE-T Fast Ethernet and 10BASE-T Ethernet differ in their topology rules. 100BASE-T preserves 10BASE-T's 100-m maximum UTP cable runs from hub to desktop. The basic rules revolve around two factors: the network diameter and the class of the repeater (or hub). The network diameter is defined as the distance between two end stations in the same collision domain. Fast Ethernet specifications limit the network diameter to approximately 205 m (using UTP cabling), whereas traditional Ethernet could have a diameter up to 500 m.

Table 14.1   **Fast Ethernet Copper Connector Pinouts**

	100BASE-T4		100BASE-TX	
Contact	**PHY without internal crossover**	**PHY with internal crossover**	**PHY without internal crossover**	**PHY with internal crossover**
1	Transmit +	Receive +	Transmit +	Receive +
2	Transmit −	Receive −	Transmit −	Receive −
3	Receive +	Transmit +	Receive +	Transmit +
4	Bidirectional-1 +	Bidirectional-2 +	Not used	Not used
5	Bidirectional-1 −	Bidirectional-2 −	Not used	Not used
6	Receive −	Transmit −	Receive −	Transmit −
7	Bidirectional-2 +	Bidirectional-1 +	Not used	Not used
8	Bidirectional-2 −	Bidirectional-1 −	Not used	Not used

Repeaters are the means used to connect segments of a network medium together in a single collision domain. Different physical signaling systems can be joined into a common collision domain using a repeater. Bridges can also be used to connect different signaling systems; however, each system connected to the bridge will have a different collision domain.

In traditional Ethernet, all hubs are considered to be functionally identical. In Fast Ethernet, however, there are two classes of repeaters: class I and class II. Class I repeaters perform translations when transmitting, enabling different types of physical signaling systems to be connected to the same collision domain. Because they have internal delays, only one class I repeater can be used within a single collision domain when maximum cable lengths are used, i.e., this type of repeater cannot be cascaded and the maximum network diameter is 200 m. On the other hand, class II repeaters simply repeat the incoming signals with no translations, i.e., provide ports for only one physical signaling system type. Class II repeaters have smaller internal delays and can be cascaded using a 5-m cable with a maximum of two repeaters in a single collision domain if maximum cable lengths are used. Cable lengths can always be sacrificed to get additional repeaters in a collision domain. This is summarized in Table 14.2.

With traditional 10BASE-T Ethernet, networks are designed using three basic guidelines: maximum UTP cable runs of 100 m, four repeaters in a single collision domain, and a maximum network diameter of 500 m. With 100BASE-T Fast Ethernet, the maximum UTP cable length remains 100 m. However, the repeater count drops to two and the network diameter drops to 205 m. On the surface, the 100BASE-T Fast Ethernet rules seem restrictive, but with use of repeaters, bridges, and switches, Fast Ethernet can be easily implemented in a network.

Table 14.2   **Maximum Collision Domain Diameter**

	All UTP cable (m)	**UTP (T4) and fiber (m)**	**UTP (TX) and fiber (m)**	**All fiber (m)**
Point-to-point (no repeater)	100	N/A	N/A	412
One class I repeater	200	231	260	272
One class II repeater	200	304	308	320
Two class II repeaters	205	236	216	228

B. AUTONEGOTIATION

Autonegotiation is the process that enables two devices, sharing a link segment, to communicate necessary information with one another in order to interoperate taking maximum advantage of their abilities. The basis of autonegotiation is a modified 10BASE-T link integrity pulse sequence as defined in Section 28 of the Institution of Electrical and Electronic Engineers (IEEE) 802.3u standard. The technologies currently supported by autonegotiation are 10Base-T half duplex, 10Base-T full duplex, 100Base-TX half duplex, 100Base-TX full duplex, and 100Base-T4.

The foundation for all of autonegotiation's functionality is the fast link pulse (FLP) burst, which is simply a sequence of 10Base-T normal link pulses, also known as link test pulses (in 10Base-T technology) that come together to form a "word" (link code word) that identifies supported operational modes. On power-on or command from management entity, a device capable of autonegotiation issues a FLP burst.

Each FLP is composed of 33 pulse positions, with the 17 odd-numbered positions corresponding to clock pulses and the 16 even-numbered positions corresponding to data pulses. The time between pulse positions is 62.5 $\pm$ 7 μs, and therefore 125 $\pm$ 14 μs between each clock pulse. All clock positions are required to contain a link pulse. On the other hand, if there is a link pulse present in a data position, it is representative of a logic 1, whereas the lack of a link pulse is representative of a logic 0. The amount of time between FLP bursts is 16 $\pm$ 8 ms, which corresponds to the time between consecutive link test pulses produced by a 10Base-T device to allow interoperation with fixed-speed 10BASE-T devices.

The 16 data positions in an FLP burst come together to form a link code word that is typically encoded as shown in Fig. 14.2.

The 5-bit selector field contains 32 possible combinations, only 2 of which are currently defined and allowed to be sent (Table 14.3). The next 8 bits, which make up the technology ability field, are used by a device to advertise its abilities to support various IEEE 802.3 technologies. The abilities are ad-

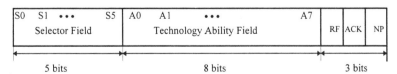

Fig. 14.2   Fields of a 802.3u link word. Reprinted from [1], IEEE Std 802.3u-1995, Copyright © 1995. IEEE. All right reserved.

Table 14.3   **Selector Field Values in Link Word**

Bit position					
S0	**S1**	**S2**	**S3**	**S4**	*Description*
1	0	0	0	0	IEEE 802.3
0	1	0	0	0	IEEE 802.9
Other combinations					Reserved for future use

vertised in parallel, i.e., a single selector field value will advertise all the supported technologies. A logic 1 in any of these positions symbolizes that the device is capable of a particular IEEE 802.3 technology (as shown in Table 14.4). The device should advertise only the technologies that it supports.

The devices at the two ends of the link segment may have an ability to support multiple technologies; the highest common denominator ability is always chosen, i.e., the technology with the highest priority that both sides can support. The relative priorities of the technologies supported by the IEEE 802.3 selector field value are as follows (from highest to lowest):

1. 100BASE-TX full duplex
2. 100BASE-T4
3. 100BASE-TX

Table 14.4   **Ability Field Values in Link Word**

Bit	*Technology represented*
A0	10BASE-T
A1	10BASE-T full duplex
A2	100BASE-TX
A3	100BASE-TX full duplex
A4	100BASE-T4
A5	Reserved
A6	Reserved
A7	Reserved

4. 10BASE-T full duplex

5. 10BASE-T

Remote fault: This bit can be set to inform a station that a remote fault has occurred.

Acknowledge: This bit is set to confirm the receipt of at least three complete, consecutive, and consistent FLP bursts from a station. This functionality will be discussed in detail later.

Next page: This is a means by which devices can transmit additional information beyond their link code words. This bit simply indicates whether or not there are more Next Pages to come. When set to logic 0, it indicates that additional pages will follow, whereas a logic 1 indicates that there are no remaining pages.

C. BASIC OPERATION

When a Fast Ethernet device is brought up, it performs autonegotiation if capable. The objective, as described previously is to agree to operate in the highest common denominator mode. Once the operating parameters are determined, the operations are exactly like the traditional Ethernet or IEEE 802.3 standard MAC operation. The MAC frame formats, back-off algorithms, and the collision detection scheme work exactly as in traditional CSMA/CD (in IEEE 802.3 standard).

14.3. Gigabit Ethernet

The purpose of the gigabit Ethernet is to extend the 802.3 protocol to an operating speed of 1000 Mb/s in order to provide a significant increase in bandwidth while maintaining maximum compatibility with the installed base of 10/100 CSMA/CD nodes, research and development, network operation, and management. It is being studied by an IEEE working group aiming to standardize this technology as IEEE 802.3z [4] and expected to be in place by early 1998.

The following objectives have been adopted by the IEEE 802.3 High-Speed Study Group for this technology:

1. Speed of 1000 Mb/s at the MAC–PHY service interface

2. Use 802.3/Ethernet frame format

3. Meet 802 FR, with the possible exception of Hamming distance

4. Simple forwarding between 1000, 100, and 10

5. Preserve minimum and maximum frame size of the current 802.3 standard

6. Full- and half-duplex operation

7. Support star-wired topologies

8. Use CSMA/CD access method with support for at least one repeater per collision domain

9. Support fiber media and, if possible, copper media

10. Use American National Standards Institute (ANSI) Fibre Channel FC-1 and FC-0 as basis for work

11. Provide a family of physical layer specifications that support
    - A link distance of at least 2 km on single-mode fiber
    - A link distance of at least 500 m on multimode fiber
    - A link distance of at least 25 m on copper (100 m preferred)

12. Decide between collision domain diameter of $\geq$50 m or $\geq$200 m

13. Support media selected from ISO/IEC 11801

14. Accommodate proposed P802.3x flow control

14.4. VG-Any LAN

The demand priority protocol (IEEE 802.12) [2, 5, 6], commonly referred to as 100VG-Any LAN [5, 6], is a local area network standard that seeks to provide a high-speed, shared-media LAN utilizing the existing copper media. The fundamental components of VG-Any LAN are end nodes and repeaters (Fig. 14.3).

Typically, the end nodes are user workstations, bridges, routers, switches, or file servers. They are connected to the network through an interface card in the end node. End nodes have two modes of operation: private and promiscuous. Promiscuous nodes receive all frames on the network, whereas private nodes receive only the messages specifically sent to them.

Repeaters are devices that interconnect end nodes and other components in demand priority. They manage the network access by performing "round-robin" scans of network port requests while keeping track of addresses and making sure that the link is up and the end station functional. Each repeater has a special uplink port through which another repeater can be cascaded. Each port on the repeater can be configured to operate in normal

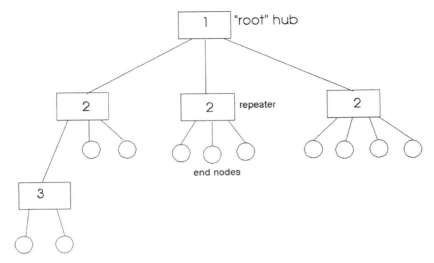

Fig. 14.3   A general VG-Any LAN topology.

mode or promiscuous mode. In the normal configuration, a port receives only packets addressed to the attached node. In the promiscuous mode, the port receives all packets. The repeater may be configured to handle IEEE 802.3 CSMA/CD or IEEE 802.5 Token Ring frame formats. However, all repeaters in the same segment must use the same format.

A demand priority network employs a physical star-wired tree topology. A minimum network of this type will consist of one repeater, two end nodes, and two network links. All the end nodes in the network are connected to the repeaters. A usual small network consists of between 6 and 32 end nodes connected to a single repeater. A larger network can be constructed by cascading several repeaters using the uplink. A maximum of five levels of cascading are allowed by the protocol because of the timing issues. The layered model for demand priority is shown in Fig. 14.4.

A. *PHYSICAL LAYER*

The physical layer is made of the physical media-independent (PMI) sublayer, the MII sublayer, and the physical media-dependent (PMD) sublayer. The network links are made of either copper or fiber media (see Table 14.5). The bundled cable (25 pair) is allowed in end node-to-repeater links provided that the repeater supports privacy mode and store-and-forward capability. Bundled cable is not allowed in repeater-to-repeater,

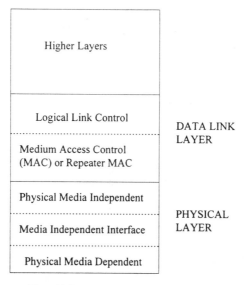

Fig. 14.4   Demand priority model.

repeater-to-bridge, or repeater-to-promiscuous end node connections. No loops or branches are allowed in the topology. Flat cable is prohibited. The maximum number of nodes in one segment is limited to 1024.

PMI sublayer, which is link independent, is responsible for taking a frame from the MAC layer and dividing it into 5-bit data streams (called quintets) and then sequentially distributing them among four channels. If four-pair UTP is used, then channel 0 is transmitted on leads 1 and 2, channel 1 uses leads 3 and 6, channel 2 uses leads 4 and 5, and channel 3 uses leads 7 and 8. When fiber optic or the two-pair STP cable is used, the PMD sublayer uses a multiplexing scheme to convert the four channels into one or two channels. A data scrambling algorithm is used to randomize

Table 14.5   **Cable Type vs Distance for Demand Priority**

Cable type	Category	Maximum distance (m)
4 pair UTP	3,4	100
2 pair STP	3,4	100
4 pair UTP	5	200
Fiber	Single or multimode	2000

bit patterns on each transmission pair to minimize interference and cross-talk. After scrambling, the quintets are converted to 6-bit sextets using the 5B/6B encoding scheme. Finally, a preamble, start delimiter, and end frame delimiter are added to each data channel along with padding, if needed. A reverse role is played when PMI receives a frame to be handed to MAC or repeater MAC (RMAC) layers.

The MII is an interface between the PMI and the PMD. PMD service primitives are passed across this interface and provide for interchangeability of PMDs, easy reconfiguration of end nodes, and ports to support multiple link media.

The PMD sublayer is link dependent and provides control signal generation and recognition, data stream signal conditioning, clock recovery, and channel multiplexing appropriate to the link medium. The transmission over the physical medium is half duplex. The actual signaling of data uses non-return-to-zero (NRZ) encoding scheme. The clock rate for transmitting is 30 MHz for each wire pair; therefore, 30 Mb/s are transmitted on each pair, with a total data rate of 120 Mb/s. Because every 5 data bits are encoded as 6 bits, the actual data transmission rate is 100 Mb/s.

B. MEDIA ACCESS CONTROL

The MAC layer is the control center for data transfer for a specific node. It is responsible for handling data transfer requests to the network and encapsulation of data into frames, on receipt removing data from frames and error checking.

The end nodes are continually polled by the repeater for any data it needs to send. The transmitted data can be either normal priority or high priority. The repeater first takes the high-priority requests and then the normal priority data while arbitrating between the ports in round-robin. Starvation of normal data by high-priority data is avoided by changing normal priority data to high priority if it has been waiting for longer than the threshold. Data packet transmission consists of a series of handshaking sequences in which the sending side of the end node–repeater local port on a point-to-point connection makes a request and the other side acknowledges the request. The sequence of sending a data packet transmission is requested by an end node and controlled by the repeater. The sequence of events proceeds as follows:

1. If an end node has a data packet ready to send, it transmits either a *Request_Normal* or *Request_high* control signal. Otherwise, the end node transmits the *Idle_Up* control signal.

2. The repeater polls all local ports to determine which end nodes are requesting to send a data packet and at what priority level (normal or high).

3. The repeater selects the next end node with a high-priority request pending. Ports are selected in port order. If no high-priority requests are pending, then the next normal-priority port is selected (in port order). This selection causes the selected port to receive the *Grant* signal. Packet transmission begins when the end node detects the *Grant* signal.

4. The repeater then sends the *Incoming* signal to all other end nodes, alerting them to the possibility of an incoming packet. The repeater decodes the destination address from the frame being transmitted as it is being received.

5. When an end node receives the *Incoming* control signal, it prepares to receive a packet by stopping the transmission of requests and listening on the media for the data packet.

6. Once the repeater has decoded the destination address, the packet is delivered to the addressed end node or end nodes and to any promiscuous nodes. Those nodes not receiving the data packet receive the *Idle_Down* signal from the repeater.

7. When the end node(s) receives the data packet, it returns to its state prior to the reception of the data packet, either sending an *Idle_Up* signal or making a request to send a data packet.

This process is used throughout the demand priority protocol to allow end nodes to transmit data packets to other end nodes.

Demand priority is designed to operate in compatibility modes for both Ethernet and Token Ring frame formats. This means that software and protocols above the logical link control (LLC) need only to know that they are operating on either an Ethernet or Token Ring network with regard to frame formats. The LLC provides the MAC with the primitives containing the information used to construct the Ethernet or Token Ring frame. The MAC then constructs a demand priority frame containing the proper elements. The supported demand priority frame formats are as follows:

- ***802.12 training frames:*** These frames are used during link training to establish validity of a link. They are initially constructed at the lower end of the link. These frames are forwarded to all network repeaters.

- *802.12 void frames:*   These frames are generated when a repeater detects that the acknowledged device did not send a packet within an expected time span or because the packet has errors.
- *802.3 frame:*   This frame format allows the demand priority network to communicate with existing Ethernet networks.
- *802.5 frame:*   This frame format enables the demand priority network to communicate with existing Token Ring networks.

C.   REPEATER MAC

An RMAC is to a repeater what a MAC is to an end node. The RMAC accepts transmit requests from the end nodes, arbitrates transfer orders, and routes the incoming packets to the proper outbound ports. The two major tasks of RMAC are round-robin polling and link training.

In order to provide each node fair access to demand priority network, repeaters perform round-robin polling. Each repeater maintains two requested pointers to keep track of the next port to be serviced: one for the next normal-priority port and one for the next high-priority port. All repeater ports are polled at least once per packet transmission to determine which have requests pending. High-priority requests exist so that time-sensitive transmissions, such as video, sound, etc., will not be caught in normal traffic.

High-priority requests are serviced before normal-priority requests. No requests are interrupted once data are being transmitted. To avoid starvation, any normal-priority traffic that has been waiting to transmit for 200–300 ms is automatically elevated to high-priority status and is serviced in port order by the high-priority pointer.

The repeater–end node link is initialized using link training. The actual training involves the transmission of training frames in each direction between the lower entity and the repeater. Such frames are always considered as normal-priority frames. In order to join the network, the lower entity must send and receive 24 consecutive, uncorrupted training frames. This ensures the cable validity. Within the training frame is the field that indicates the "requested configuration" of the lower entity. When the repeater responds, the "allowed configuration" field instructs the lower entity of the allowed configuration. The lower entities' address must be acknowledged by the upper repeater.

Link training is used for many purposes, such as verification of cable quality for data transmission, allowing the receiver to adapt to the link, establishing the MAC address of the end node, determining the link node type — 802.3 or 802.5, etc. The link training is performed each time a link is established, such as power-up, or at cable connections or during certain error conditions and is always initiated by the lower entity device desiring to connect to the network, which can be either an end node or a repeater.

D.   CLASSICAL IP OVER ATM

Classical IP over ATM (RFC 1577) [7, 8] had been proposed by the Internet Engineering Task Force (IETF) as a way of connecting IP-based work-stations on ATM. RFC 1577 basically emulates the IP layer (network layer) over ATM to provide end-to-end connectivity to the higher layers. In this approach, the IP end stations connected to the ATM cloud are divided into logical IP subnets (LISs). The subnets are administered in the same manner as the conventional subnets. Hosts connected to the same subnet-work can communicate directly; however, communication between two hosts on different LISs is only possible through an IP router, regardless of whether direct ATM connectivity is possible between the two hosts.

Implementation of classical IP over ATM requires a mapping between IP and ATM addresses. IP addresses are resolved to ATM addresses using the ATM address resolution protocol (ATMARP) and vice versa using the inverse ATMARP (InATMARP) within a subnet. ATMARP is used for finding the ATM address of a device given the IP address. It is analogous to the IP–ARP associated with IP protocol. Just like conventional ARP, it has a quintuple associated with it: source IP address, source ATM address, destination IP address, and destination ATM address. On the other hand, InATMARP is used to find the IP address of a station given the ATM address [almost equivalent to conventional reverse address resolution protocol (RARP)]. Typical use of InATMARP is by the ATMARP server to find out the IP address of the station connected to the other end of an open virtual circuit. This information is used to update database entries and to refresh the entry on time-outs.

Every end station is configured statically with the address of the ATMARP server. On initialization, the end station opens a virtual channel connection (VCC) to the ATMARP server. The ATMARP server, on detecting a new VCC, performs an InATMARP on it to find the IP address of the end station connected at the other end of the VCC. This information is stored in the tables of the ATMARP server for further use. Each end station maintains a local ARP cache that acts as the primary cache and the

APR server acts as a secondary cache. If the ATM end station wants to contact another station, it will query its local cache first for the ATM address for a given IP address. If that fails, it queries the ARP server for the ATM address. Once it has the ATM address of the destination, the end station proceeds to open a direct VCC to the destination.

The basic drawback to this approach is that this work only for IP-type traffic. It does not support multicast or broadcast. It requires static assignment of ARP server address and the ARP server becomes the single point of failure. The IETF has recently removed some of these drawbacks by introducing a new concept to enhance RFC 1577, i.e., multicast address resolution server. Work on it has been going on since late 1994.

E. ATM LAN EMULATION

LAN emulation (LANE) [9–11] has been proposed by ATM Forum and has been widely accepted by the ATM industry as a way to emulate conventional LANs. The necessity of defining LANE arose because most of the existing customer premises networks use LANs such as IEEE 802.3/802.5 (Ethernet and Token Ring) and customers expect to keep using existing LAN applications as they migrate toward ATM. To use the vast repertoire of LAN application software, it became necessary to define a service on ATM that could emulate LANs. The idea is that the traditional end-system applications should interact as if they are connected to traditional LANs. This service should also allow the traditional (legacy) LANs to interconnect to ATM networks using today's bridging methods.

LANE has been defined as a MAC service emulation, including encapsulation of MAC frames (user data frames). This approach, as per ATM Forum, provides support for maximum number of existing applications. This is not easy because there are some key differences between legacy LANs and ATM networks (Table 14.6).

The main objective of LAN emulation service is to enable existing applications to access an ATM network via protocol stacks, such as NetBIOS, IPX, IP, AppleTalk, etc., as if they were running over traditional LANs. In many cases, there is a need to configure multiple separate domains within a single network. This objective is fulfilled by defining an emulated LAN (ELAN) that comprises a group of ATM-attached devices. It appears as a group of stations connected to a IEEE 802.3 or 802.5 LAN segment. Several ELANs could be configured and membership in an ELAN is independent of the physical location of the end station. An end station could belong to multiple ELANs.

Table 14.6   **LAN vs ATM**

LANs	ATM
Connectionless	Connection oriented
Broadcast and multicast are easily accomplished through shared medium of LAN	ATM does not support broadcast or multicast
LAN MAC addresses based on manufacturers organizationally unique identifiers and serial numbers are independent of topology	ATM addresses depend on topology, i.e., where the station is plugged in (i.e., network prefix)

1. Components

There are four basic components in LANE: LAN emulation client (LEC), LAN emulation configuration server (LECS), LAN emulation server (LES), and broadcast and unknown server (BUS).

LEC is an entity in the ATM workstation or ATM bridges that performs data forwarding, address resolution, and other control functions. This provides a MAC-level emulated Ethernet/IEEE 802.3 or IEEE 802.5 service interface to applications running on top. It implements the LANE user–network interface (LUNI) when communicating with other entities within the emulated LAN.

The LES implements the control coordinating function for the ELAN. It provides a facility for registering and resolving MAC addresses or route descriptors to ATM addresses. LECs register the LAN destinations they represent with the LES. A client can also query the LES when the client wishes to resolve a MAC address to an ATM address. A LES will either respond directly or forward the query to other clients so they may respond.

BUS handles data sent by an LEC to the broadcast MAC address ("FFFFFFFFFFFF" hex), all multicast traffic, and, as an option, some initial unicast frames sent before the target ATM address is resolved.

2. LEC Initialization Phases

The basic states that a LEC goes through before it is operational are shown in Fig. 14.5 and described as following:

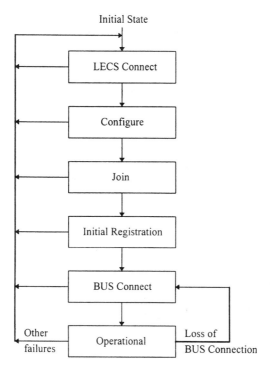

Fig. 14.5 Initialization phases of LANE [9]. Copyright 1995 The ATM Forum.

Initial state: In this state LES and LEC know certain parameters (such as address, ELAN name, maximum frame size, etc.) about themselves.

LECS connect phase: LEC sets up a call to LECS. The VCC that is opened is referred to as configuration-direct VCC. At the end of configuration, this VCC may be closed by the LEC.

Configuration phase: LEC discovers LES in preparation for join phase.

Join phase: During the join phase, LEC establishes its control connections to the LES. Once this phase is complete, the LEC has been assigned a unique LEC identifier (LECID), knows the emulated LAN's maximum frame size and its LAN type, and has established the control VCC with the LES.

Initial registration: After joining, an LEC may register any number of MAC addresses in addition to the one registered during the join phase.

BUS connect: In this phase a connection is set up to the BUS. The address of the BUS is found by issuing an *LE_ARP* for an ATM address with all 1's. The BUS then establishes a multicast-forward VCC to the LEC.

3. Connections

A LEC has separate VCCs for control traffic and for data traffic. Each VCC carries traffic for only one ELAN. The VCCs form a mesh of connections between the LECs and other LANE components such as LECS, LES, and BUS.

4. Control Connections

A control VCC links the LEC to the LECS and LEC to the LES. The control VCCs never carry data frames and are set up as a part of the LEC initialization phase. The control connection terminology is listed below:

Configuration-direct VCC is a bidirectional point-to-point VCC set up by a LEC as part of the LECS connect phase and is used to obtain configuration information, including the address of LES. This connection may be closed after this phase is over.

Control-direct VCC is a bidirectional point-to-point VCC to the LES set up by a LEC for sending control traffic. This is set up during the initialization process. Because LES has the option of using the return path to send control data to the LEC, this requires the LEC to accept control traffic on this VCC. This VCC must be maintained open by both LES and LEC while participating in the ELAN.

Control-distribute VCC is a unidirectional point-to-multipoint or point-to-point VCC from LES to the LEC to distribute control traffic. This is optional and LES, at its discretion, may or may not set this up. This VCC is also set up during the initialization phase. This VCC, if set up, must be maintained while participating in the ELAN.

5. Data Connections

Data VCCs connect the LECs to each other and to the BUS. These carry either Ethernet or Token Ring data frames and under special conditions a flush message (optional). Apart from flush messages, data VCCs never carry control traffic:

Data-direct VCC is a bidirectional point-to-point VCC established between LECs that want to exchange unicast data traffic.

Multicast-send VCC is a bidirectional point-to-point VCC from LEC to BUS. It is used for sending multicast data to the BUS and for sending initial unicast data. The BUS may use the return path on this VCC to send data to the LEC, so this requires the LEC to accept traffic from this VCC. The LEC must maintain this VCC while participating in the ELAN.

Multicast-forward VCC is either a point-to-multipoint VCC or a unidirectional point-to-point VCC from the BUS to the LEC after the LEC sets up a multicast-send VCC. It is used for distributing data from the bus. The LEC must attempt to maintain this VCC while participating in the ELAN.

Figure 14.6 shows all type of connections. LEC1 maintains its configuration directly though LEC2 and LEC3 has torn down that connection after becoming operational. Here, all three LECS are operational.

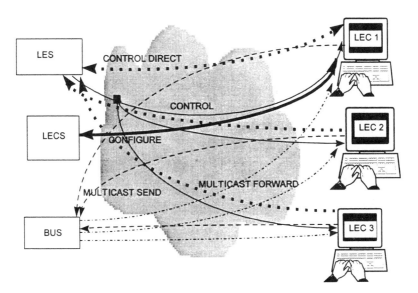

Fig. 14.6 Various lane connections.

6. Operation

To get to the operational state, i.e., the state at which the LEC can start exchanging information with other LECs, it has to go through an initialization process that consists of several phases. First, if required, it must contact the LECS. This phase is optional and may not exist if a preconfigured switched virtual circuit or permanent virtual circuit (PVC) to LES are used. The LEC will locate the LECS by using the following mechanisms to be tried in the following order: (i) Get the LECS address via interim local management interface (ILMI) using the ILMI Get or ILMI Get Next to obtain the ATM address of the LECS for the UNI; (ii) using the well-known LECS address: If LECS address cannot be obtained via ILMI or if LEC is unable to establish a configuration direct VCC to that address, then an ATM Forum specified well-known address "47.00.79.00.00.00.00.00.00.00.00.00.00-00.A0.3E. 00.00.01-00" hex must be used to open a configuration direct VCC; (iii) using a well-known PVC: If VCC could not be established to the well-known address in the previous step then the well-known PVC of virtual path identifier = 0 and virtual channel identifier = 17 (decimal) must be used.

The configuration phase prepares the LEC for join phase by providing the necessary operating parameters for the emulated LAN that the LEC will join. Once the LECS is found then LEC sends a *LE_Configure_Request* and waits for a *LE_Configure_Response,* which is a part of the LE configuration protocol. All control frames have the structure shown in Fig. 14.7.

Marker is always a fixed 2-byte value "FF00" hex. The op-code determines the type of control frame, e.g., "0001" for *LE_Configure_Request* "0101" for *LE_Configure_Response,* etc. Status is used in the responses to inform about reasons of denial for the requests or to indicate success. Type length values are used to exchange specific information in the control frames such as timer values, retry counts, etc.

During the configuration, the LECS provides the LEC with the ATM address of LES and also provides all kinds of timers values, time-out periods, and retry counts.

Armed with this information, LEC enters the join phase. Here, the LEC establishes its connection with the LES and determines the operating parameters of the emulated LAN. The LEC implicity registers one MAC address with the LES as a part of the joining process. LEC must initiate the UNI signaling to establish a control-direct VCC (or use a control-direct PVC) and then send a *LE_JOIN_Request* over this VCC to the LES. The

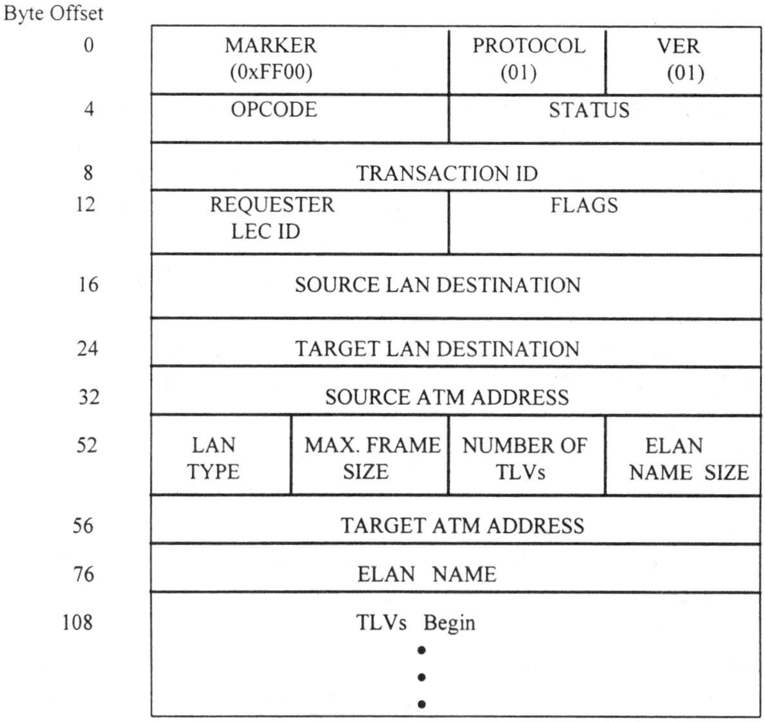

Byte Offset

0	MARKER (0xFF00)		PROTOCOL (01)	VER (01)
4	OPCODE		STATUS	
8	TRANSACTION ID			
12	REQUESTER LEC ID		FLAGS	
16	SOURCE LAN DESTINATION			
24	TARGET LAN DESTINATION			
32	SOURCE ATM ADDRESS			
52	LAN TYPE	MAX. FRAME SIZE	NUMBER OF TLVs	ELAN NAME SIZE
56	TARGET ATM ADDRESS			
76	ELAN NAME			
108	TLVs Begin • • •			

Fig. 14.7   LANE frame format [9]. Copyright 1995 The ATM Forum.

LES, may optionally establish a control-distribute VCC back to the LEC. After that the LES will send back a *LE_JOIN_Response* that may be sent on either control direct or control distribute (if created). To each LEC that joins, the LES assigns a unique LECID.

If the join phase is successful then the LEC is allowed to register additional MAC addresses, which it represents with the LES. This is called the registration phase. However, this can happen any time and is not restricted to this phase. However, additional registrations cannot be done before joining the ELAN.

This is followed by the BUS connect phase in which LEC has to establish connection to the BUS. For this purpose, the LEC needs to find out the address of the BUS. This is accomplished by the ARP. In this procedure whenever a LEC is presented with a frame for transmission whose LAN destination is unknown to the client, it must issue LANE ARP (*LE_ARP*) request frames to the LES over its control-direct VCC. The LES may issue an *LE_ARP* reply on behalf of a LEC that had registered the LAN destination earlier with the LES or alternatively can forward the request to the appropriate client(s) using the control-distribute VCC or one or more control-direct VCCs and then the *LE_ARP_Reply* from the appropriate LEC will be relayed back over the control VCCs to the original requester. Each LEC also maintains a local cache of addresses.

For connecting to the BUS, LEC first issues an *LE_ARP_Request* to the LES for the broadcast MAC address, i.e., all 1's ("FFFFFFFFFFFF" hex). The *LE_ARP_Response* gives the ATM address of the BUS. The LEC then proceeds to set up a multicast-send VCC to the BUS, which then immediately opens a multicast-forward VCC back to the LEC. At this point the LEC is considered operational.

Now, if the LEC wants to exchange information with another LEC, it can use the address resolution procedure to get its address and then set up a data-direct VCC to the other LEC and transfer information. However, to save time, if the target LEC's address is not known, then the originating LEC issues a *LE_ARP_Request* and starts sending frames through the BUS. Once the *LE_ARP_Reply* is received the LEC is required to stop using the BUS and open a data-direct VCC. Despite all this, ATM guarantees in-order delivery and therefore a flush message is sent to BUS that ensures that no frames are transmitted on the data-direct VCC until all the previous ones routed through the BUS are delivered. Flush request message is a way to inform the other side that following that request, data will be transmitted on a different channel, e.g., switching from multicast-send to data-direct VCC. The flush request needs to be responded by flush response so that the side issuing the flush request understands that all the previously sent messages have been delivered on the old channel and it is safe to switch channels and still maintain in-order delivery of messages.

14.5. FDDI

A. *STANDARD*

FDDI is an ANSI standard and has an ISO approval. The basic FDDI standard [12–14] can be categorized as shown in Fig. 14.8.

It is based on a dual-ring topology composed of two counterrotating rings that operate at a data rate of 100 Mb/s. It uses dual counterrotating rings to enhance reliability: the primary ring and the secondary ring. Second-

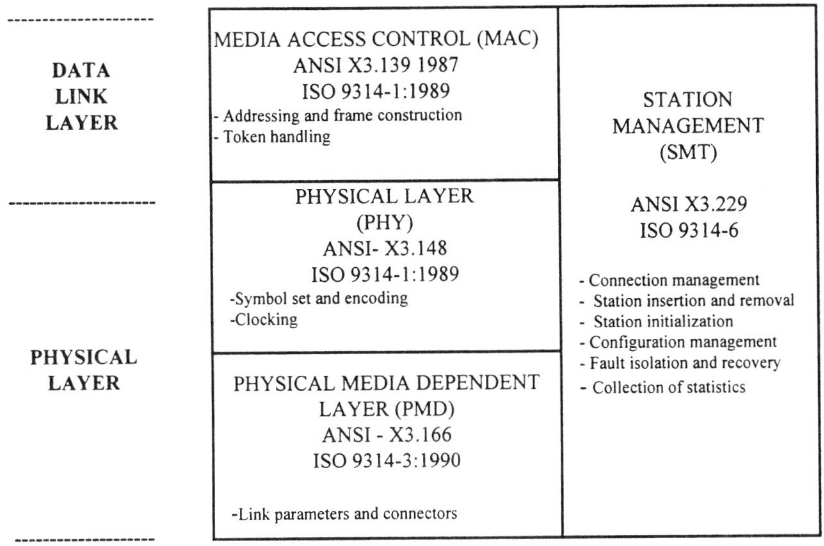

DATA LINK LAYER	MEDIA ACCESS CONTROL (MAC) ANSI X3.139 1987 ISO 9314-1:1989 - Addressing and frame construction - Token handling	STATION MANAGEMENT (SMT) ANSI X3.229 ISO 9314-6 - Connection management - Station insertion and removal - Station initialization - Configuration management - Fault isolation and recovery - Collection of statistics
PHYSICAL LAYER	PHYSICAL LAYER (PHY) ANSI- X3.148 ISO 9314-1:1989 -Symbol set and encoding -Clocking	
	PHYSICAL MEDIA DEPENDENT LAYER (PMD) ANSI - X3.166 ISO 9314-3:1990 -Link parameters and connectors	

Fig. 14.8 FDDI model.

ary ring is used in case of failures. Multiple failures, however, result in multiple rings. Multimode fiber connects each station together and the total ring can be 100 km in length. A maximum of 500 stations can be connected in the ring and hence it forms an ideal backbone network.

FDDI consists of a set of stations serially connected by a transmission medium to form a closed loop. Information is transmitted as symbols (equivalent of 4 information bits) from one active station to the next. Each station, including the destination, simply regenerates each symbol on the ring in the direction of rotation. The destination, however, makes a copy as the information passes it. The frame is removed by the station that originated the transmission.

B. STATION TYPES

There are two types of FDDI stations: a dual-attach station (DAS), which is connected to both rings, and a single-attached station (SAS), which is attached only to the primary ring. A dual-attached station has at least two ports — an A port, where the primary ring comes in and the secondary ring goes out, and a B port, where the secondary ring comes in and the primary goes out. A station may also have a number of M ports, which are attachments for single-attached stations. Stations with at least one M port are called concentrators. SASs have S ports, which can be connected to the FDDI ring only through concentrators using M ports. A dual ring of DAS devices cannot be readily moved, changed, or added into. A single user disconnecting a dual-attached workstation causes a break in the ring. Dual-attach stations require twice the number of connectors and cables.

C. PHYSICAL LAYER SPECIFICATIONS

FDDI supports two basic media types — copper and fiber. With copper it works with STP cable and allows data at 100 Mbs for up to 100 m. When using STP cabling, we get FDDI over copper commonly known as copper distributed data interface. Copper cables will be used to connect workstations to hub only. UTP has also been considered by the X3T9.5 committee. When using fiber, there are three alternatives: single-mode fiber, multimode fiber, and the plastic or low-cost fiber. The single-mode fiber can be used to work at FDDI speed over a distance of 50 km but is expensive. Multimode fiber is the most commonly used media. Plastic or low-cost fiber were on the market for a while but, due to the nature of plastic, the distance was limited to 100 m and interest in them was short lived.

An FDDI station attaches to the fiber optic cable by a media interface connector (MIC) whose main function is to mechanically align the fibers. MIC plugs have mechanical latches that mate with the latch points in the MIC receptacle. As an option, the MIC receptacles of a station can be keyed to prevent improper attachment of input and output. Keys are used to determine the PMD entries contained in the FDDI node. Four keys are defined: type A or red key/label (primary in/secondary out) and type B or blue key/label (secondary in/primary out) are used to connect DAS into an FDDI ring. Type M or green key/label (master) is used on the concentrator side to connect a SAS to a concentrator. Type S or no key/white label (slave) is used on the station side to connect a SAS to the concentrator (i.e., to connect to a port M). Table 14.7 summarizes the general MIC connector key usage and Table 14.8 gives the FDDI interface power levels.

1.   Coding and Symbol Set

FDDI uses 4B/5B encoding, i.e., each 4 bits are encoded into a 5-bit symbol. FDDI communicates all of its information using symbols. This 5-bit encoding provides 16 data symbols (normal hex range: 0–F), eight control symbols (Q, H, I, J, K, T, R, and S), and eight code violation symbols (V) as shown

Table 14.7 **MIC Connector Key Usage**

FDDI connection type	Key type
Workstation to wall	S/S
Distribution panel to concentrator M port	M/M
Distribution panel to concentrator A port	A/A
Distribution panel to concentrator B port	B/B
Concentrator A to concentrator M port	A/M
Concentrator B to concentrator M port	B/M
Concentrator 1A to concentrator 2B port	A/B
Concentrator 1B to concentrator 2A port	B/A
Workstation to concentration M port	S/M

in Table 14.9. After the data are 4B/5B encoded, they are further encoded using non-return-to-zero invert coding before the bits are let out on the media.

2. Line States

Line states are used by the physical layer to communicate information. A line state is a continuous stream of a certain symbol(s) sent by transmitter (physical layer) that, upon receipt of another station, will uniquely identify the state of the communication line. The following are the commonly encountered line states:

- Quiet line state: A continuous stream of "Q" symbols. This line state is used to break a link and restart a connection.
- Halt line state: A continuous stream of "H" symbols. This line state is used during connection management signaling.

Table 14.8 **FDDI Interface Power Levels**

	Output interface		Input interface	
	Min	Max	Min	Max
Center wavelength	1270 nm	1380 nm	1270 nm	1380 nm
Average power	−20.0 dBm	−14.0 dBm	−31.0 dBm	−14.0 dBm

Table 14.9   **FDDI Symbol Set**

Data symbols			Control symbols		
Symbol	Binary	Bit stream (4B/5B encoded)	Symbol		Bit stream (4B/5B encoded)
0	0000	11110	Q	QUIET	00000
1	0001	01001	I	IDLE	11111
2	0010	10100	H	HALT	00100
3	0011	10101	J	Used in Starting Del.	11000
4	0100	01010	K	Used in Starting Del.	10001
5	0101	01011	T	Used in Ending Del.	01101
6	0110	01110	R	RESET (logical 0)	00111
7	0111	01111	S	SET (logical 1)	11001
8	1000	10010		Code violations	
9	1001	10011	V or H	To be taken as HALT	00001
A	1010	10110	V or H	To be taken as HALT	00010
B	1011	10111	V		00011
C	1100	11010	V		00101
D	1101	11011	V		00110
E	1110	11100	V or H	To be taken as HALT	01000
F	1111	11101	V		01100
			V or H	To be taken as HALT	10000

- Master line state: An alternating series of "H" and "Q" symbols. This line state is used to indicate a trace and also used during physical connection management (PCM) signaling.

- Idle line state: A continuous stream of "I" symbols. This line state is used to separate information bits and to provide clock synchronization.

- Active line state: This line state is asserted by a "JK" pair and continues to be asserted until it receives "R," "S," or "T" symbols. It is used by MAC layer to transmit data frames.

- Link state unknown: This is an indication by the physical layer if the receiver is unable to determine the line state.

- Noise line state: This is an indication by the physical layer if the receiver has received enough symbols but is unable to uniquely recognize the line state (i.e., the receiver is getting garbled information on the line).

D. MAC SPECIFICATIONS

FDDI is a controlled-access network. The access to transmission is through a token. The token is a particular unique combination of symbols that circulates on the ring in one direction from station to station. A station receives the token in its entirety on its input port and then regenerates the token on the output port to continue circulation. In order to transmit, a station waits for the token and on receipt of the token removes it from the ring, thereby capturing the token. It then puts out the frame(s) it wants to transmit and then at the end introduces a new token. A station can hold on to a token for a finite time defined as the token-holding time. The station that transmits a frame is also responsible for removing the frame from the ring. The basic two protocol data units that are used by the FDDI MAC are the token and the frames.

1. Token

The general format of a token is shown in Fig. 14.9. The preamble (PA) is a minimum of 16 symbol of idles. Because of ring latency and timing constraints, subsequent stations may see a smaller length of preamble. Tokens are recognized when received with a preamble of zero or greater length. The start delimiter is made up of two symbols, J and K, which will not be seen anywhere except to mark the beginning of a new token or a frame. Similarly, the end delimiter is made of two symbols, which are both a T. For a token there are two possible values (as shown previously) for the frame control field depending on whether it is a nonrestricted or restricted token. The nonrestricted mode is the normal mode of operation in which the bandwidth is shared equally among all requesters. The restricted mode is used to dedicate bandwidth to a single extended dialog between specific requesters (these are described later in this section).

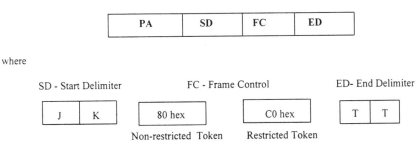

Fig. 14.9 FDDI token format.

2. Frame

FDDI defines three basic types of frames: (i) the MAC frames, which carry MAC control data; (ii) logical link control frames (LLC), which carry user–user information; and (iii) Station management frames (SMT), which carry management information between stations and layers. MAC frames include frames used in ring initialization and beacon frames used in the process of ring fault isolation. These frames never leave the FDDI ring (i.e., they never go across a bridge or a router) because they are specific for a ring only. LCC frames carry user information between nodes on a network. These frames do cross bridges or routers. SMT frames, like MAC frames, do not cross bridges or routers because they are used to control the operation of a ring only.

Figure 14.10 shows the general format used to transmit the LLC, MAC, and SMT frames. The SMT frames are discussed later and this section will focus on MAC and LLC frames: The preamble, start delimiter, and the end delimiter are the same as in a token. The frame control is used to distinguish between a token, MAC frame, LLC frame, or SMT frame. The Fibre Channel Standard (FCS) is a 32-bit polynomial checksum.

The frame status consists of three indicators that may have one or two values. The indicators can either be set (S) or reset (R). Every frame is originally transmitted with all the indicators set to R. The indicators can be set by intermediate stations when they retransmit the frame. The three indicators are error, address recognized, and copy. Error is set when a station determines that the frame is in error. This may be due to a format error or failure of cyclic redundancy checksum (CRC). The address recognized bit is set when a station receives the frame and determines that the destination address is its own MAC address (or if it was broadcast, then the first recipient sets it). Copy indicator is set when the destination station is able to copy the contents into its buffers.

Fig. 14.10   FDDI LLC and MAC frame format. SFS, start of frame sequence; SD, start delimiter (2 symbols); PA, preamble ($\geq$4 symbols); FC, frame control (2 symbols); DA, destination address (4 or 12 symbols); SA, source address (4 or 12 symbols); INFO, information; FCS, frame check sequence (8 symbols); EFS, end of frame sequence; ED, ending delimiter (1 or 2 symbols); and FS, frame status (3 symbols).

3. Ring Operation

Access to the ring is controlled by passing a token around the ring. The token gives the downstream station (relative to the one passing the token) an opportunity to transmit a frame or a sequence of frames. If a station wants to transmit, it must first capture the token, i.e., strip the token off the ring and then begin transmitting its eligible queued frames. After finishing transmission, the station immediately issues a new token for use by downstream stations, without waiting for frames to return on the ring. Optionally, only for various station management functions, the station may wait to see one or more or all the transmitted frames return before it issues the token. If a station has nothing to transmit, it acts as a simple repeater, i.e., it repeats the incoming symbol stream. While repeating, the station must determine if the information is destined for it. This is done by matching the destination address (DA) to its own address or a relevant group address. If a match occurs, then the subsequent symbols up to FCS are processed by the MAC (for MAC frames) or sent to LLC (for LLC frames).

Each transmitting station is responsible for stripping its own transmitted frames from the ring. This is done by stripping the remainder of each frame whose source address (SA) matches the station's address and replacing it with IDLE symbols. This process leaves remnants of frames, consisting of PA, start delimiter (SD), frame control (FC), DA, and SA fields, followed by IDLE symbols, because the decision to remove the frame is based on matching of the SA field and that cannot occur until after the initial part of the frame is repeated. Remnants are easily distinguishable because they are always followed by IDLE symbols. They are removed from the ring when they encounter a transmitting station.

4. Ring Scheduling

Transmission of normal data on the ring is controlled by a timed token protocol. This protocol supports two types of transfer: (i) asynchronous — dynamic bandwidth sharing and (ii) synchronous — guaranteed bandwidth and response time. The asynchronous transfer is typically used for those applications that are bursty and whose response time is not critical. The bandwidth is instantaneously allocated from the pool of remaining unallocated or unused bandwidth. On the other hand, synchronous transfer is the choice for those applications that are predictable enough to permit a preallocation of bandwidth [via station management (SMT)].

Each station maintains a token-rotation timer (TRT) to control the ring scheduling. A target token rotation time (TTRT) is negotiated during ring initialization via the claim token bidding process. Initially, TRT is set to TTRT. A token arriving before the TRT reaches zero, called an early token, causes the TRT to be reset, and can be used for transmitting both the synchronous and asynchronous traffic. On the other hand, a token arriving after TRT reaches zero, called a late token, can be used only for synchronous transmissions. Each station also maintains a counter called the late counter (LC), which records the number of times the token was late and a token holding timer (THT), which dictates how long a station can hold the token.

This protocol guarantees an average TRT (or average synchronous response time) not greater than TTRT, and a maximum TRT (or maximum synchronous response time) not greater than twice TTRT.

When a station receives an early token, i.e., LC = 0, it saves the remaining time of TRT in THT and then resets and enables TRT for countdown, i.e.,

THT = TRT

TRT = TTRT

Enable TRT countdown

Transmit synchronous frames (if any)

Enable THT countdown

Transmit asynchronous frames as long as THT > 0

If the received token is a late token, i.e., LC > 0, then

LC = 0

TRT not changed and continues to run

Transmit only synchronous frames

Each station has a known allocation of synchronous bandwidth, i.e., the maximum time that the station may hold the token without THT being enabled. Synchronous bandwidth is expressed as a percentage of TTRT, i.e., a station would require a 100% allocation to transmit for a time equal to TTRT before issuing a token. The sum of current allocations of all stations should not exceed the maximum useable synchronous bandwidth of the ring (like successive expiration of TRT with late counter in the MAC or invalid frames on the ring or a logical/physical break in the ring).

5. Claim Token Process

Any station detecting a requirement for ring initialization shall initiate a claim token process. In this process one or more stations bid (TTRT) for the right to initialize the ring by continuously transmitting claim frames. Each station also looks for incoming claim frames and compares the received bid with the station's own bid. Any station receiving a lower bid shall enter or reenter the bidding, whereas any station receiving a higher bid shall yield. The claim token process is completed when one station receives its own claim frames after the frames have passed around the ring. At this point all stations around the ring have yielded to this station's claim frame. The winning station proceeds to initialize the ring.

6. Beacon Process

When a station detects that claim token process has failed or upon a request from SMT, it initiates the beacon process. This happens when the ring has, most probably, been physically interrupted or reconfigured. The purpose of the beacon process is to signal to all remaining stations that a significant logical break has occurred and to provide diagnostic or other assistance to the restoration process. In this process, the station begins continuous transmission of beacon frames. If a station receives a beacon frame from its upstream neighbor, it repeats that beacon and stops its own. If it receives its own beacon, it assumes that the logical ring is restored and it stops beaconing and starts the claim token process. If a station receives no beacon and it has been continuously beaconing for at least the time indicated by stuck beacon timer, then ring management begins the stuck beacon recovery procedure. The recovery procedure starts with a directed beacon and eventually traces the stations around the fault, at which time all stations around the fault remove themselves from the ring, perform a self-test, and can rejoin the ring if they pass those tests.

E. STATION MANAGEMENT

Station management provides the control necessary at the station (node) level to manage the processes under way in the various FDDI layers such that a station may work cooperatively as a part of an FDDI network. SMT provides services such as connection management, station insertion and removal, station initialization, configuration management, fault isolation

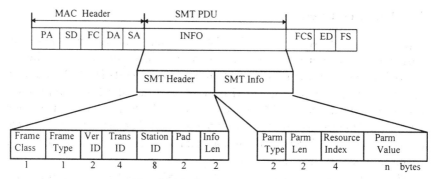

Fig. 14.11   SMT frame format.

and recovery, scheduling policies, and collection of statistics. A variety of internal node configurations are possible. However, a node shall have one, and only one, SMT entity. It may, however, have multiple instances of MACs, PHYs, and PMDs. For a SMT frame, the information field in the general frame format is occupied by a SMT header and a SMT information portion. The SMT header is the protocol header for all SMT frames. SMT information is the information indicated by the SMT header. An SMT frame has subfields as shown in Fig. 14.11.

1.   SMT Header

SMT frames are identified by their frame class, e.g., neighbor information frame [NIF(01 hex)], configuration status information frame [SIF-Cfg (02 hex)], operation status information frame [SIF-Oper (03 hex)], Echo frame [ECF (04 hex)], and so on. The frame type is an indicator of whether this frame is an announcement (01 hex), a request (02 hex), or a response (03 hex). The *Version_ID* field indicates the structure of SMT into field. At the moment only two possible values are acceptable — 0001 hex for stations using a version lower than 7.x and 0002 hex for stations using version 7.x of SMT. To ensure backward compatibility, NIF, SIF, and ECF frames will have a constant *Version_ID* of 0001 hex. *Transaction_ID*'s sole purpose is to match the request and response frames. The *Station_ID* is the unique identifier of the FDDI station transmitting the SMT frame. The pad field is two bytes of all zeros. This is used to make the header an even 32 bytes. The length of the information field is reported in the info length field. This value does not include the length of the SMT header or this field. Its value can be between 0 and 4458 bytes.

2. SMT Information

The SMT info field consists of a list of parameters. If more than one parameter is present in the frame, they will be listed one after another.

NIFs are used by a station for periodic announcement of its MAC address and basic station description. NIFs are used in the neighbor notification protocol that allows a MAC to determine its logical upstream neighbor address and logical downstream neighbor address. The protocol also detects duplicated MAC addresses on an operational ring. NIFs can be used by a monitoring station to build a ring map.

SIFs are used, in conjunction with the SIF protocol, to request and provide in response to a station's configuration and operation information. There are two classes of SIFs to provide this function: the SIF configuration and the SIF operation request and response frames. Potential uses are fault isolation and statistics monitoring.

ECFs are defined for SMT-to-SMT loopback testing on an FDDI ring using the ECHO protocol. The ECH frames may contain any amount of echo data (up to the maximum frame size). Potential uses include confirming that a station's port, MAC, and SMT are operational.

Resource allocation frames are defined to support a variety of network policies such as allocation of synchronous bandwidth.

Request-denied frames are used to notify a requester that its request has been denied because of version or protocol problems, e.g., if a station receives a request with a *Version_ID* that is not supported or if the request frame has a length error.

Status report frames are used by a station to announce the station status, which may be of interest to the manager of the ring. These are sent when certain conditions become (in)active. The conditions that are reported include frame errors, link errors, duplicate address, peer wrap condition, and not-copied condition.

Parameter management frames (PMFs) provide the means for remote access to station attributes via the parameter management protocol. They operate on all SMT MIB attributes. There are two types of PMFs: PMF-Get frames to look up a value and PMF-Set frames to change a value. Not all stations support the PMF-Set protocol.

3. Physical Connection Management

Within each FDDI station there is one SMT entity per port, called the physical connection management (PCM) entity. The number of PCM entities within a station is exactly equal to the number of ports of that station.

PCM entities are a part of SMT that control the operations of the ports. In order to make a connection, two ports must be physically connected to each other by means of fiber optic or copper cable. When this happens, the PCMs that are responsible for those ports can recognize each other's existence and begin communicating. They do this by sending line states out of the port onto the cable. The PCM at the other end of the connection will recognize the line state and respond accordingly.

When the PCM sees another PCM on the other end of the connection, they will synchronize and communicate with each other. During this communication, the following important things happen: (i) the PCMs figure out the type of each port and determine if they are compatible and (ii) the PCMs perform a link confidence test (LCT). The LCT determines if the quality of the link is good enough to establish a connection. If not, then a connection will not be made (allowed); otherwise, PCM will establish a connection and place the ports on a token path that goes through that station. At this point the ports become a part of the network and data can be sent through these ports.

14.6. FDDI-II

FDDI-II is an enhancement to the original FDDI being developed by the X3T9.5 committee. Both FDDI and FDDI-II run at 100 Mb/s on the fiber. FDDI can transport both asynchronous and synchronous types of frames. FDDI-II has a new mode of operation called hybrid mode. Hybrid mode uses a 125-μs cycle structure to transport isochronous traffic in addition to synchronous/asynchronous frames. FDDI-II supports integrated voice, video, and data capabilities and therefore expands the range of applications of FDDI. FDDI and FDDI-II stations can be operated in the same ring only in the basic mode.

FDDI-II has the ability to carry the isochronous (constant bit rate) and multimedia data, which is sometimes not possible in the original FDDI due to variation in delay that can be obtained with the token access control method.

FDDI-II nodes can run in basic mode. If all nodes on the ring are FDDI-II nodes, then the ring can switch to the hybrid mode in which isochronous service is provided in addition to basic mode services. In the basic mode on FDDI-II, synchronous and asynchronous traffic is transmitted in a manner identical to that on FDDI. In the basic mode, FDDI-II behaves in the same way as FDDI, i.e., all ring accesses for transmission are controlled by a token and the available bandwidth is time shared.

In the alternative mode, also called the hybrid mode, the bandwidth is divided into channels using time-division multiplexing, whereas some of the channels can be reserved for isochronous data. The establishment of multiple channels is done by a station known as the "cycle master," which also controls the utilization of the channels. To provide isochronous service, FDDI-II deploys a special frame called a cycle. A cycle is generated every 125 μs. At 100 Mbs, 1262.5 bytes can be transmitted in 125 μs. The bytes of the cycles are preallocated to various channels on the ring. The 1560 bytes of the cycle are divided into 16 wideband channels (WBCs) that can carry either circuit-switched or packet-switched traffic. All packet-switched WBCs are concatenated together to form a single channel, which is operated with the FDDI-timed token protocol.

Despite the strengths and the characteristics of FDDI-II, it never caught on in the market. Very few vendors have products supporting this standard and the interest seems to be dying.

References

1. IEEE Std 802.3. Local and metropolitan area networks, supplement—Media access control (MAC) parameters, physical layer, medium attachment units and repeater for 100 Mb/s operation, type 100BASE-T (clauses 21–30)" Copyright © 1995 by the Institute of Electrical and Electronics Engineers, Inc. The IEEE disclaims any responsibility or liability resulting from the placement and use in the described manner. Information is reprinted with the permission of the IEEE.
2. Molle, M., and G. Watson. 1996. 100Base-T/IEEE 802.12/packet switching. *IEEE Commun. Mag.* 34(8):64–73.
3. F. Halsall. 1995. *Data communications, computer networks and open systems.* 4th ed. Reading, MA: Addison-Wesley.
4. IEEE 802.3z, http://stdbbs.ieee.org/pub/802_main/802.3/gigabit.
5. IEEE Std 802.12. Demand-priority access method, physical layer and repeater specifications for 100Mb/s operation. Institute of Electrical and Electronic Engineers, 345 East 47th Street, New York, NY 10017.
6. Watson, G., A. Albrecht, J. Curcio, D. Dove, S. Goody, J. Grinham, M. P. Spratt, and P. A. Thaler. 1995. The demand priority MAC protocol. *IEEE Network* 9(1):28–34.
7. RFC 1577. Classical IP over ATM, request for comments. Internet Engineering Task Force.

8. D. E. Comer. 1995. *Internetworking with TCP/IP — Volume 1,* 3rd ed. Englewood Cliffs, NJ: Prentice Hall.

9. ATM Forum. 1995. *LAN Emulation over ATM Version 1.0,* af-lane-0021.000. ATM Forum, 303 Vintage Park Drive, Foster City, CA 94404-1138.

10. Siu, K., and R. Jain. 1995. A brief overview of ATM: Protocol layers, LAN emulation and traffic management. *ACM SIGCOMM* Comput. *Commun. Rev.* 25(2):6–20.

11. Finn, N., and T. Mason. 1996. ATM LAN emulation. *IEEE Commun. Mag.* 34(6):96–100.

12. ANSI X3T9.5. FDDI — Physical layer medium dependent (PMD). American National Standards Institute, 1430 Broadway, New York, NY 10018.

13. ANSI X3.148. FDDI — Token ring physical layer protocol. American National Standards Institute, 1430 Broadway, New York, NY 10018.

14. ANSI X3.139. FDDI — Token ring media access control (MAC). American National Standards Institute, 1430 Broadway, New York, NY 10018.

Part 4 | The Manufacturing Technology

Chapter 15 | Semiconductor Laser and Light-Emitting Diode Fabrication

Wenbin Jiang
Michael S. Lebby

*Phoenix Applied Research Center, Motorola, Incorporated, Tempe,
Arizona 85284*

15.1. Introduction

In Chapter 2, we briefly described the technology behind light-emitting diodes (LEDs), edge-emitting lasers, and vertical cavity surface-emitting laser (VCSEL) sources. We will now focus on manufacturing issues for these optical sources, such as material growth and device fabrication. We will concentrate on the fabrication of laser sources because manufacturing LEDs uses most of the same principles as those in the fabrication of edge-emitting semiconductor laser diodes, albeit with simpler manufacturing in general. VCSEL manufacturing will be described in detail because this is an emerging field that will potentially have a great impact on the data communication industry.

15.2. LED Fabrication

Figure 15.1 shows a cross-sectional and top view of an LED. In manufacturing an LED, a substrate such as GaAs or GaP is grown as an ingot, using a seed crystal of appropriate crystal orientation. During the crystal growing process, the ingot is doped *n* type and possibly with other dopants such as sulfur (S) consistent with the bandgap associated with the desired wavelength of emitted light, as described in Chapter 2. The ingot is sawed to make wafers or starting substrates.

Epitaxial (epi) layers are deposited using a chemical vapor deposition

535

HANDBOOK OF FIBER OPTIC
DATA COMMUNICATION

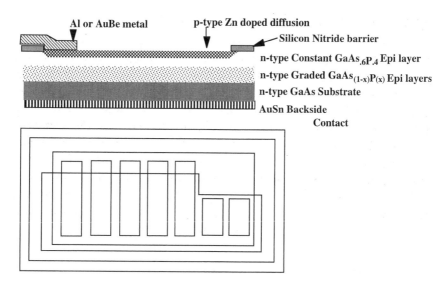

Fig. 15.1   Cross-sectional and top view of an LED.

process, with the composition of the epi layers gradually changed to maintain a perfect crystal lattice. Appropriate dopants, again consistent with the desired bandgap and *n*-type doping, are incorporated during the epitaxial deposition process. The wafers are then coated with a layer of a dielectric, such as silicon nitride, using chemical vapor deposition.

The silicon nitride is patterned using photolithography, and the silicon nitride is selectively etched. A *p*-type dopant such as zinc is introduced to the surface of the wafer, with the silicon nitride acting as a barrier against the diffusion of *p*-type dopant wherever it remains after the selective etch process. The diffusion of zinc into the opening forms the *p–n* junction for the light-emitting diode.

Metal such as AuBe or aluminum is deposited using a process such as sputter deposition and patterned using photolithography and metal etch processing so that it contacts the *p*-type diffused area and also extends over the silicon nitride-capped region, providing a bonding pad.

The back of the wafer is exposed, and a metal such as AuGe or AuSn is deposited by sputter or evaporative deposition onto the back of the wafer, forming the backside contact.

15.3.   Material Growth — Lasers

Edge-emitting semiconductor lasers and vertical cavity surface-emitting lasers are based on the same material systems but with very different epitaxy growth control requirements. VCSEL technology is in essence an epitaxy growth technology because a VCSEL structure may consist of more than 100 epitaxy layers. The thickness and the doping level of each layer must be precisely controlled. Any slight deviation to the designed parameters may affect the VCSEL performance, which will end up affecting the final manufacturing yield. Fortunately, modern epitaxy techniques have made the VCSEL growth possible after many years of development. In contrast, edge-emitting semiconductor laser epitaxy growth involves much fewer steps and is more tolerant to the parameter variation.

15.3.1.   EDGE-EMITTING LASERS

Three major epitaxial growth techniques are used for edge-emitting laser diodes: liquid phase epitaxy (LPE), metal organic chemical vapor deposition (MOCVD), and molecular beam epitaxy (MBE). LPE was the dominant growth technique in the early stage of the semiconductor laser development [1] but has gradually been replaced by MBE and MOCVD techniques due to the requirements for high quality, high uniformity, and high reproducibility in heteroepitaxial material growth, especially in the growth of quantum well structures. Today, the vast majority of the commercial semiconductor lasers are grown by either MBE or MOCVD.

Molecular beam epitaxy offers an advantage to grow very thin active layers down to only one monolayer because of its slow growth rate. It is therefore an excellent technique to grow single-quantum well or multiquantum well lasers. In practice, MBE has been used for the growth of commercial GaAs/AlGaAs laser structures. In GaAsP/InP and InGaAlP/InGaP structures have also been demonstrated with MBE [2, 3]. Figure 15.2 is a schematic diagram of a MBE system. The MBE growth is conducted in an ultrahigh vacuum (UHV) of approximately 10^{-10} torr with heated effusion cells filled with Ga, Al, As, and doping sources to produce an adequate beam flux onto the substrate. The effusion cells are enclosed in liquid nitrogen-cooled shrouds. The Al content of $Al_xGa_{1-x}As$ is controlled by varying the partial gas pressures of Al and Ga. The typical doping sources are Si or Sn for n type and Mg or Be for p type. During the growth, the

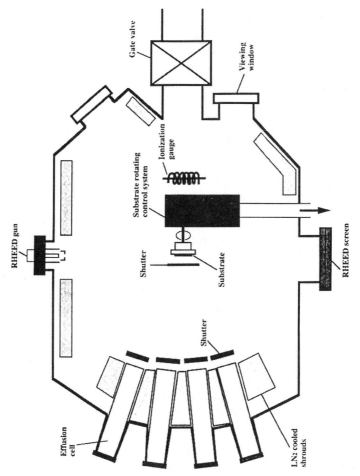

Fig. 15.2 Schematic diagram of a MBE system.

Gate valve

Viewing window

Ionization gauge

Substrate rotating control system

RHEED gun

Shutter

Substrate

RHEED screen

Shutter

Effusion cell

LN₂ cooled shrouds

substrate can be rotated to achieve the thickness uniformity across the whole wafer. Reflection high-energy electron diffraction (RHEED) is usually used for *in situ* growth monitoring. The detailed description of MBE growth can be found in Ref. [4]. Recently, MBE techniques have also been used for GaN-based material growth [5, 6], but the crystal quality has not reached that grown by MOCVD techniques.

MOCVD, also known as metal organic vapor-phase epitaxy, offers excellent growth uniformity across a wafer and run-to-run reproducibility. Because it has the advantages of high growth rate and multiwafer growth capability, MOCVD has become the dominant technique in the growth of commercial semiconductor lasers.

Figure 15.3 is a schematic diagram of a typical MOCVD system. The growth sources used for MOCVD are arsine (AsH_3) or phosphine (PH_3) for group V elements and trimethylgallium (TMGa), trimethylaluminum (TMAl), and trimethylindium (TMIn) for group III elements. Some reactors also use triethylgallium (TEGa), triethylaluminum (TEAl), and triethylindium (TEIn) sources as group III elements. During the growth, liquid-state group V sources having high vapor pressure at room temperature and group III AsH_3 source are transferred to the reactor by bubbling with hydrogen (H_2) as a carrier gas. The sources are decomposed at temperatures ranging from 500 to 800°C in the reactor to form III–V compound crystal structures. Dopants for AlGaAs-based materials are silicon (Si) from disilane (Si_2H_6) or tellurium (Te) from diethyltellerium (DETe) for *n*-type materials, and carbon (C) from CCl_4 or zinc (Zn) from diethylzinc (DEZn) for *p*-type materials. Because phosphine is not easily decomposed at room temperature, it is first thermally decomposed in a special furnace before being transferred into the reactor. In the growth of InGaAsP/InP structures, lattice matching of InGaAsP material to the InP substrate is achieved by adjusting the growth source ratio. The group III to group V source ratio adjustment is also used for bandgap control. Selenium (Se) from H_2Se or S from H_2S is used for *n*-type doping, and Zn from dimethylzinc (DMZn) is used for *p*-type doping. The growth rate is controlled by the amount of group V sources introduced into the reactor, i.e., by the flow rate of H_2 carrier gas for group V elements.

Metal organic chemical vapor deposition techniques have been used to grow high-quality InGaN/AlGaN/GaN-based crystal structure for blue visible LED. Visible blue edge-emitting laser has also been demonstrated based on the III-nitride materials grown by MOCVD. The material growth temperature in the MOCVD reactor is between 750 and 1050°C. Ammonia

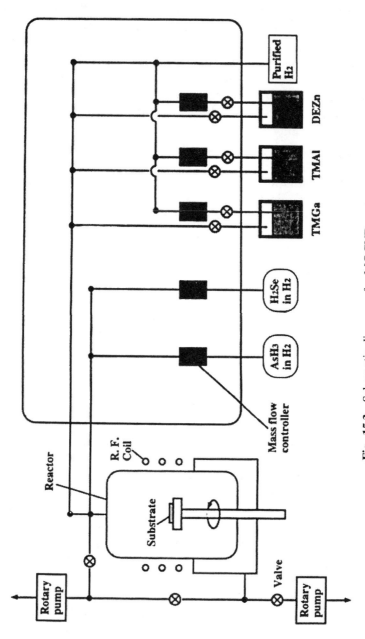

Fig. 15.3 Schematic diagram of a MOCVD system.

(NH$_3$) is the source of nitride to replace either AsH$_3$ or PH$_3$ sources in the MOCVD reactor for the growth. Magnesium (Mg) from *bis*-cyclopentadienyl magnesium (CP$_2$Mg) is the *p*-type dopant, and Si from silane (SiH$_4$) is the *n*-type dopant. Postgrowth thermal annealing at 700°C for 10 min in N$_2$ atmosphere [7] or low-energy electron beam irradiation [8] have been effective in activating the *p*-type dopants and reducing the *p*-type GaN film resistance.

15.3.2.   *VERTICAL CAVITY SURFACE-EMITTING LASERS*

Although the growth of edge-emitting semiconductor lasers is straightforward and involves minimum calibration because only a few layers are needed for the positive–intrinsic–negative (PIN) structure, the growth of VCSEL structures poses a real challenge in growth uniformity and reproducibility. A typical VCSEL consists of more than 100 quarter-wavelength distributed Bragg reflector (DBR) mirror and multiple quantum well (MQW) active layers. The total variation of each layer thickness across the wafer cannot exceed 1% during the growth. Such a precise thickness control is needed for two reasons. First, the net optical gain in the VCSEL cavity is very small because of the thin active layer. Highly reflective DBR mirrors, typically in excess of 99.5%, are needed to ensure that the laser operates effectively. In order to achieve such a high reflectivity, each repetitive layer in the DBR mirror has to be exactly the same, i.e., one quarter wavelength for each layer, to retain appropriate constructive interferences among those layers. Second, the cavity resonance peaks of a VCSEL are usually spaced widely apart due to the typical one-λ cavity. The requirement that the spectral position of this resonance fall within the optical gain region of the active MQWs places a severe constraint on the optical thicknesses of the layers making up the cavity.

Figure 15.4 shows the simulation results of net reflectivity of a completed 850-nm GaAs VCSEL structure grown on a GaAs substrate that consists of a top DBR, a bottom DBR, and an active region sandwiched between, as shown in Fig. 2.14 in Chapter 2. Figure 15.4a is the reflectivity of an idealy grown structure. A deep Fabry–Perot resonate peak sits in the center of the reflective band (850 nm). Figure 15.4b illustrates the reflectivity of a structure with the top DBR wavelength being longer than the bottom DBR wavelength by 1%. The resonate peak shifts toward the longer wavelength and becomes less pronounced, and a small glitch appears on the long-wavelength end of the high-reflectivity band. Figure 15.4c illustrates

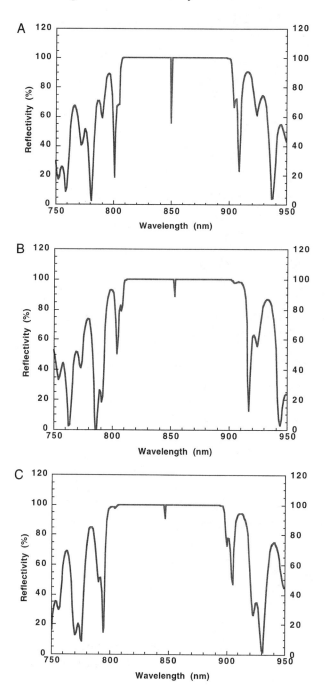

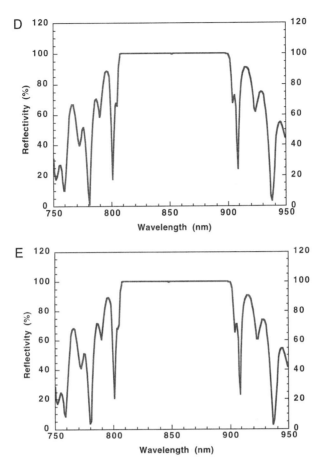

Fig. 15.4 Net reflectivity simulation results of a complete GaAs VCSEL structure grown on a GaAS substrate for (A) a perfectly grown structure with identical top and bottom DBR mirrors, (B) the top DBR center wavelength being 1% longer than the bottom DBR center wavelength, (C) the top DBR center wavelength being 1% shorter than the bottom DBR center wavelength, (D) the VCSEL cavity being 1% longer than the designed wavelength, and (E) the VCSEL cavity being 1% shorter than the designed wavelength.

the reflectivity of a structure with the top DBR wavelength being shorter than the bottom DBR wavelength by 1%. The resonate peak shifts toward the shorter wavelength and becomes less pronounced, and a small glitch appears on the short-wavelength end of the high-reflectivity band. Figure 15.4d illustrates the laser cavity being 1% longer than the designed wavelength of 850 nm. The center resonate peak moves toward the longer wavelength, and the depth of the resonate peak is reduced. Figure 15.4e illustrates the laser cavity being 1% shorter than the designed wavelength of 850 nm. Accordingly, the center resonate peak moves toward the shorter wavelength, and the depth of the resonate peak is also reduced. In order to achieve better than 1% growth control for the VCSEL structure, not only is the exact knowledge of the refractive index of the required alloy concentration is needed but also the precise growth rate has to be determined and maintained throughout the growth. Typically, the simulation results as shown in Fig. 15.4 are used as references to judge if the growth is successful or how much deviation there is from a designed structure. The information drawn from the comparison will be used to make corrections for the subsequent growth runs. A successful growth requires an extremely stable epitaxial growth reactor that has the necessary long-term reproducibility and controllability. Similar to the growth of edge-emitting semiconductor laser materials, two types of state-of-the-art epitaxial techniques are typically used to grow VCSEL structures: MBE and MOCVD.

A modern MBE growth system can usually take a 2-in. or a 3-in. GaAs wafer. The growth process takes 8–12 h to complete the entire VCSEL structure. The standard high-temperature effusion cells provide group III sources, such as Al, Ga, and In, and a group V source such as As. The MBE system is also equipped with cells for n-type and p-type dopants, which are usually Si and Be, respectively. Modifications are usually made to the MBE system to reduce the growth rate transients to less than 1% during the repetitive switching of constituents for the VCSEL multiple layers. When substrate rotation is employed during the growth, 1% lateral thickness uniformity across entire 2-in. wafers can be achieved. This will lead to high yield and high uniformity across the wafer, as is required for manufacturing product.

RHEED oscillation measurements can be used to count the atomic layers being deposited, thus precisely monitoring the growth thickness [9, 10]. However, RHEED can only be used to monitor one particular spot on a wafer. When RHEED is used for *in situ* growth monitoring, the wafer is not allowed to rotate, leading to approximately 5%/cm linear variation in layer thickness across the wafer. Although such a variation may be advantageous in the research environment, it is not acceptable to the commercial application due to wafer growth nonuniformity affecting production yield. Therefore, RHEED oscillation calibrations are usually performed prior to the actual growth on a small test sample centrally mounted on a different sample block. The reevaporation effect of group II species can be a problem that will affect the crystal quality grown by MBE. This effect is strongly substrate temperature dependent. To achive good crystal quality, the GaAs-AlGaAs-AlAs layers are usually grown at a substrate temperature of 580–630°C and InGaAs at 520–530°C.

The growth rate calibration acquired just before the actual growth is only accurate to approximately 1%. It is not only time consuming but also subject to drift. A modified RHEED *in situ* monitoring scheme is based on discrete substrate rotations [11]. Following the growth of each pair of quarter-wavelength Bragg layers, the growth is paused for a few seconds and the substrate is rotated 180°. Any growth nonuniformity across the wafer will be compensated during the subsequent growth after the substrate rotation. The net effect of the discrete rotation will be equivalent to that of the continuous substrate rotation.

Optical reflectometry measurement is another effective *in situ* monitoring technique in which the bottom DBR mirror and most of the VCSEL cavity, ranging between 94 and 100% of the cavity, is grown first. The partially finished wafer is then taken out to evaluate the reflectivity spectrum. A comparison with the simulation will reveal information about how to complete the rest of the cavity and the top DBR [12].

A more rigorous *in situ* reflectivity measurement apparatus is shown in Fig. 15.5 [13], in which the measurement is conducted at room temperature through the viewpoint of the MBE transfer tube. The wafer remains in the UHV environment for the measurement so that any potential oxidation-induced defects can be avoided at the interface between the portions grown and the subsequent growth. In order to minimize the weight and bulk of the equipment attached to the viewport, a light pipe is used to direct the white light toward and receive the reflected light from the wafer. A chart recorder synchronized with the scan rate of the spectrometer records the unnormalized reflectivity spectra. During the growth, the reflectivity spectrum is first measured after only a few periods of the bottom DBR mirror have been grown. Any relative error in the center wavelength determined from the measurement will be used to correct the subsequent growth. The second interruption and reflectivity measurement can be made just after the growth of the cavity. However, because the bottom DBR mirror reflectivity is very high and cavity absorption is low, the Fabry–Perot resonant dip is very shallow and sometimes difficult to identify. To enhance the resonant dip visibility, two or three periods of the top DBR mirror can be grown before the second reflectivity spectrum measurement. Any cavity

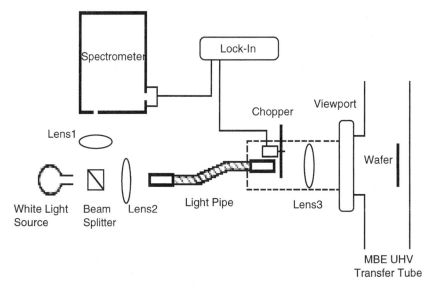

Fig. 15.5 An *in situ* reflectivity measurement system for growing VCSELs by a MBE system. (Reprinted with permission from Ref. [13]. Copyright 1992 American Institute of Physics.)

thickness correction can then be made by adjusting the thickness of the immediate one layer in the top DBR mirror when the growth resumes.

Other methods, such as multiwavelength pyrometric interferometry measurements [14] and single-wavelength pyrometric interferometric measurements [15, 16], have also demonstrated that they are capable of *in situ* monitoring the VCSEL growth precisely to within 1%.

Although the predominant number of research publications on near-infrared VCSELs (780–980 nm) are based on MBE, MOCVD has its own advantages in VCSEL applications due to its high growth rate, wide range of host and dopant species, and capability of continuously varying material composition.

In a high-volume manufacturing environment, VCSEL growth time of 8–12 h by a MBE reactor is feasible but may suffer from a high cost of ownership and low-throughput arguments. Conversely, a MOCVD reactor can achieve growth rates approximately three times as high as MBE [17] while maintaining ±1% thickness uniformity across the entire wafer. Reports of good yields have also been demonstrated with multiwafer MOCVD reactors, which can hold up to 15 × 2-in. wafer loads [18].

In order to maintain sharp doping profiles and prevent the dopant outdiffusion into the active region of the device, dopants with low diffusion coefficients must be used. Zinc is the most commonly used p dopant in MOCVD growth [19] and is the dominant dopant in GaAs-based LEDs. It possesses, however, a high diffusion coefficient and is therefore not recommended for the VCSEL p dopant from device performance consideration [20]. The other commonly used p-doping alternative is carbon (C), which has proven to be a reliable p dopant with very low diffusivity and high solubility [21–24]. The most commonly used carbon precursor for p doping is carbon tetrachloride (CCl_4). The growth rate of C-doped AlGaAs is usually lower than that of undoped or n-doped AlGaAs with the effect of slow etching of AlGaAs growth surface while doping with CCl_4. It should be noted that CCl_4 is an ozone-depleting material and its use has been discouraged from the environmental consideration. Therefore, the search for alternative C-doping method has been ongoing within the research community. One candidate is the use of organometallic As sources such as tertiarybutylarsine (tBAs). Tertiarybutylarsine allows the incorporation of intrinsic carbon into the grown material due to enhancement of background carbon levels from organometallic decomposition and has the advantage of being approximately on an order of magnitude less toxic than arsine. This has encouraged research centers such as universities to install MOCVD reactors in addition to MBE reactors [25–27].

Low series resistance in the p-doped DBR mirror ensures high VCSEL performance, which is usually achieved by grading the heterointerfaces between the alternative DBR mirror pairs, as discussed in Chapter 2. In MBE, the continuous interface grading can be accomplished by varying source cell temperature [28, 29] or by repeatedly switching the cell shutters to create a superlattice grading [30]. Even though the techniques have been demonstrated with high reproducibility, the processes are not easy to implement from the manufacturing standpoint. In comparison, the continuous grading of an arbitrary composition profile can be readily achieved with MOCVD, and it is therefore expected that within the near future MOCVD will dominate the growth process.

The MOCVD growth is typically carried out in a commercially available horizontal quartz tube reactor with graphite susceptor and fast-switching run-vent manifold. Sources used are TMIn, TEGa or TMGa, TMAl, AsH_3, and PH_3. Dopants are Si from Si_2H_6 or Te from DETe for n-type materials, C from CCl_4 for p-type AlGaAs alloys, Zn from DEZn for p-type GaAs cap, and Zn or Mg from bis-cyclopentadienyl magnesium (Cp_2Mg) for *p*-type AlGaInP. Growth pressure inside the reactor is in the range of 80–200 mbar, and growth temperatures are typically in the range of 725–775°C. With AlGaAs DBR mirrors, the growth rate is approximately 3 μm/h, V to III ratio is 50:1, and the interface grading, whether it is linear or parabolic, can be accomplished by simultaneous ramping of TMGa and TMAl flows in very short increments, such as 0.1 s, controlled by the computer. Good yield and uniformity have been reported with 3-in, wafers [31, 32] and with 4-in, wafers [33] by MOCVD.

The growth reproducibility of VCSEL structures is of major concern because the DBR structures are very sensitive to growth fluctuations. The statistics obtained for 150 VCSEL wafers grown by MOCVD and operating at 850 nm are presented in Fig. 15.6, in which the variation of the reflectivity center wavelength as well as bandwidth of the reflectivity spectrum are shown in Fig. 15.6a and 15.6b, respectively. The standard deviation is 11 and 6 nm for the center wavelength and the bandwidth, respectively [34]. The application of *in situ* reflectometry growth monitoring [35–38] has allowed a better run-to-run thickness control across a 3-in. wafer using MOCVD [39].

15.4.   Device Processing

Similar to standard silicon device fabrication, semiconductor laser processing includes photolithography, dopant diffusion, ion implantation, oxide deposition, chemical etching, metalization, etc. Unlike standard silicon device fabrication, however, a compound semiconductor wafer that contains epitaxial growth is used to begin the process. Each fabrication step used in the building of lasers may be functionally like that used in the silicon industry but is customized for compound semiconductor material systems.

15.4.1.   EDGE-EMITTING LASER FABRICATION

The process procedures of edge-emitting semiconductor lasers vary with device structures. Edge-emitting semiconductor lasers can be catagorized into gain-guided lasers and index-guided lasers based on how current and optical modes are confined. Gain-guided lasers include broad-area lasers and stripe lasers, and index-guided lasers include weak index-guided lasers and strong index-guided lasers.

Broad-area semiconductor lasers do not employ any current confinement scheme, and their fabrication involves only a few process steps. First,

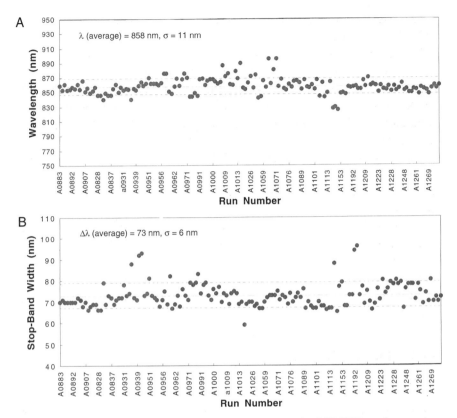

Fig. 15.6 The statistics for 150 VCSEL wafers grown by MOCVD and operating at 850 nm. (A) The variation of the reflectivity center wavelength and (B) the bandwidth of the reflectivity spectrum (after Ref. [34]).

p-type contact metal, usually AuZn/Au or Ti/Pt/Au, is deposited onto the epitaxy side of an as-grown wafer. AuZn/Au is an alloy contact and the alloying is carried out at approximately 400–450°C for 30 s. Ti/Pt/Au is a nonalloy contact. After the wafer is thinned down to approximately 100 μm, n-type contact metal, usually Au/Sn or Ni/AuGe, is deposited onto the back side of the substrate. The alloying for the n-type contact metal is carried out at 350–400°C for approximately 30 s. The wafer is then cleaved into bars of appropriate width, which determine the laser cavity length. The bars are finally sawed into individual chips, as shown in Fig. 15.7. Broad-area lasers are frequently used to characterize the laser material quality.

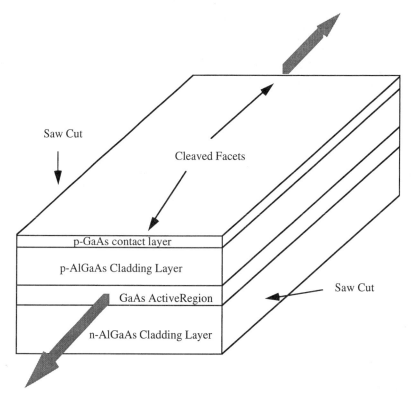

Saw Cut

Cleaved Facets

p-GaAs contact layer

p-AlGaAs Cladding Layer

GaAs ActiveRegion

Saw Cut

n-AlGaAs Cladding Layer

Fig. 15.7   Schematic diagram of an broad-area GaAs semiconductor laser chip.

Because the current in broad-area semiconductor lasers is not confined, the broad-area lasers have high threshold current and seldom lase continuous wave (CW) at room temperature. A gain-guided stripe laser has a current restriction stripe built along the junction plane to enhance the carrier confinement and is used to reduce the threshold current. The stripe lasers allow CW operation [40] and exhibit fundamental transverse mode [41]. Three types of typical gain-guided stripe lasers based on AlGaAs/ GaAs material systems are shown in Fig. 15.8. The first type is an oxide-stripe laser as shown in Fig. 15.8a. A SiO$_2$ layer of the p-GaAs contact layer is used to force the injected current flowing into a small region through an opening in the dielectric film [40]. The SiO$_2$ dielectric film can be deposited by plasma-enhanced chemical vapor deposition (PECVD). Photolithography is used to define a stripe width of 15 μm or less. Either wet chemical etching or reactive ion etching is used to remove the dielectric

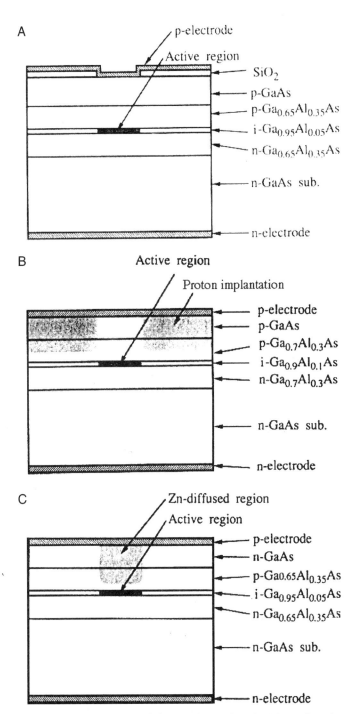

Fig. 15.8 Schematic diagram of (A) an oxide stripe GaAS semiconductor laser, (B) a proton implantation stripe GaAs semiconductor laser, and (C) a junction stripe GaAs semiconductor laser with Zn diffusion.

material within the stripe. After the removal of the photoresist, p-contact metal is deposited onto the entire wafer. The wafer is then thinned to approximately 100 μm and n-contact metal is deposited onto the back side of the substrate. Finally, the wafer is cleaved into individual laser chips. The second type uses proton implantation to define a region of high resistivity that restricts the current flowing into an opening in the implanted region [41], as shown in Fig. 15.8b. A thick photoresist stripe can be used as a mask for proton implantation to generate high-resistive region outside of the stripe region covered by the photoresist. After the removal of the photoresist, metal contacts are deposited onto both sides of the wafer to finish the device fabrication. The third type is a junction-stripe laser [42], as shown in Fig. 15.8c. Zinc diffusion is used to convert a small region of the top n-type GaAs layer into p type, thus providing a current path. The reverse-biased junction over the remaining region provides current confinement. Typically, a SiO_2 mask is used to define a stripe window on the wafer for the Zn diffusion, which is performed at 700°C for 30 min using $ZnAs_2$ as a diffusion source.

Although the current confinement has been improved with a gain-guided stripe laser, the lateral optical mode confinement is poor because of the carrier-induced antiguiding effect in the active layer. In order to further reduce the laser threshold and increase the differential quantum efficiency, the laser structure is modified to introduce an index step to guide the optical mode.

A weakly index-guided laser relies on layer thickness nonuniformity in the cladding region. The effective index difference between the optical mode guiding region and the region outside of the waveguide is on the order of 10^{-2}, larger than the carrier-induced antiguiding effect. There are a number of weakly index-guided structures that can be grouped into two categories, ridge waveguide devices [43] and rib waveguide devices [44] (Fig. 15.9).

The ridge waveguide laser fabrication (Fig. 15.9a) starts with an epitaxial wafer grown by LPE, MBE, or MOCVD techniques. The wafer is etched to form a ridge 3–5 μm wide. Dry etching is typically used to ensure a better scale control, but wet etching will result in less crystal damage. The etching stops in the p-cladding layer approximately 0.5 μm above the active layer. The etching depth is controlled *in situ* by a reflective laser interferometer [45]. Alternatively, an etch stop layer can be used for precise etching depth control [46]. A layer of SiO_2 dielectric material is then deposited onto the wafer. Photolithography and etching are used to remove the SiO_2

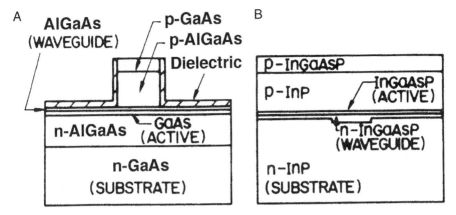

Fig. 15.9   Schematic diagram of (A) a ridge waveguide GaAs semiconductor laser and (B) a rib waveguide InGaAsP semiconductor laser.

on the ridge for *p*-type metalization, followed by substrate thinning and *n*-type metalization on the back side of the substrate. Finally, the wafer is cleaved into laser chips. During the laser operation, a fraction of the optical mode overlaps with the dielectric, which has a considerably lower refractive index than the semiconductor cladding layer. The optical mode overlap introduces an effective lateral index step more significant than that induced by the carrier effect, thus eliminating the antiguiding effect and improving the optical mode confinement.

A channeled substrate planar (CSP) laser [44], one of several basic rib waveguide lasers, is shown in Fig. 15.9b. Epitaxial layers, including *n*-InGaAsP cladding layer, InGaAsP active layer, *p*-InP cladding layer, and *n*-InGaAsP contact layer, are grown by either LPE or MOCVD on a patterned *n*-InP substrate with etched-stripe structures. Zinc diffusion is used to convert the *n*-InGaAsP contact layer in the channeled structure region into *p* type for current injection. Metal contact of *p* type is then deposited onto the epitaxy side of the wafer. After the wafer thinning, a *n*-type metal contact is deposited onto the back side of the substrate. The wafer is finally cleaved into laser chips. When this CSP laser is forward biased, current can only flow through the Zn-diffused region because the remainder is reverse biased. A fraction of the optical mode overlaps with the lower index InP substrate on either side of the active region. This overlap introduces a lateral index step for optical mode guiding. The degree of guiding in effect depends on the width of the active region, the thickness of the n-cladding layer, and the depth of the channel stripes.

Strong index-guided semiconductor lasers have active regions buried in the higher bandgap materials (Fig. 15.10). The index differences between the guided region and the region outside is larger than 0.1. These types of lasers are called buried heterostructure (BH) semiconductor lasers. There are a number of different types of BH lasers, such as etched-mesa heterostructure (EMBH) lasers [47], double-channel planar buried heterostructure lasers [48, 49], planar buried heterostructure lasers [50], strip buried heterostructure lasers [51, 52], V-grooved substrate buried heterostructure lasers (which are also known as channel-substrate buried heterostructure lasers [53]), mesa-substrate buried heterostructure lasers [54], and buried-crescent buried heterostructure lasers [55].

Buried heterostructure semiconductor lasers rely generally on high-quality epitaxial regrowth techniques. For example, the fabrication of an EMBH laser (Fig. 15.10) starts with the first growth by LPE or MOCVD of four epitaxial layers, including a n-InP buffer layer of 3 or 4 μm doped at 2×10^{18} cm^{-3}, an undoped InGaAsP active layer of 0.2 μm at a wavelength of 1.3 μm, a p-InP cladding layer of 2 μm doped at 10^{18} cm^{-3}, and a p-InGaAsP contact layer of 0.5 μm doped at 10^{19} cm^{-3} with a bandgap corresponding to a wavelength of 1.1 μm [47]. Mesas of 2 or 3 μm wide

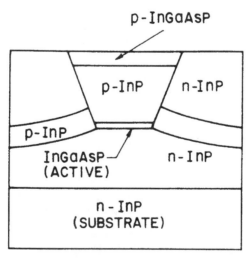

Fig. 15.10   Schematic diagram of an etched-buried mesa heterostructure laser.

are then formed by etching on the as-grown wafer using a solution of 1% Br_2 and CH_3OH with plasma-deposited SO_2 as an etching mask. After the etching, the etching mask SiO_2 is removed, and p-InP and n-InP blocking layers are grown using the second LPE or MOCVD growth. After the regrowth, a p-type metal contact is deposited onto the epitaxial side of the wafer. The wafer is then thinned to approximately 100 μm before a n-type metal contact is deposited on the back side of the substrate. Finally, the wafer is cleaved to produce semiconductor lasers. Regrowth techniques are generally used for InP-based material systems, although good-quality GaAs-based regrowth has also been demonstrated [56, 57].

Semiconductor lasers are required to operate at a single longitudinal mode for long-distance optical communications. Distributed feedback (DFB) lasers are the most widely used single longitudinal mode device in commercial applications. A typical long-wavelength DFB laser is shown in Fig. 15.11 [58]. To fabricate the DFB laser, a grating is first formed on a n-GaAs or n-InP substrate with the grating period determined by the required lasing wavelength. The optical length of the grating period should

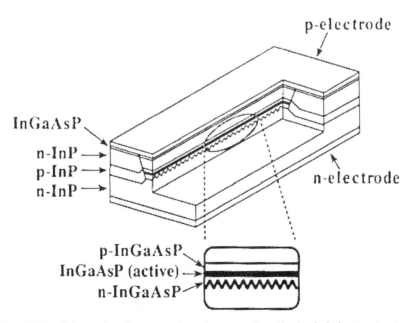

Fig. 15.11 Schematic diagram of a long-wavelength buried heterostructure DFB laser.

be equal to the half wavelength or its integer multiplication. The first-order grating is used for an edge-emitting laser, and the second-order grating is used for a grating surface-emitting laser that emits light perpendicular to the surface of the substrate [59].

A two-beam holographic exposure method is generally used to fabricate the uniform gratings [60]. A laser beam is first expanded and collimated by a pinhole and a pair of lenses. The collimated beam is then divided by a beam splitter into two beams that meet again in the sample stage. Grating patterns generated by the interference between the two collimated beams are transferred photolithographically onto the surface of a semiconductor substrate that is spin-coated with a photoresist to a thickness of approximately 100 nm. The grating patterns of the developed photoresist are transferred to the substrate by selective wet etching or dry etching. The grating constant is determined by the writing laser wavelength and the angle of the two crossed laser beams. A $\lambda/4$-shifted grating can be fabricated by multilayer photoresist processing [61] or by using a phase mask [62], and a chirped grating can be fabricated by using a spherical beam method that uses divergent beams instead of collimated beams [63]. The $\lambda/4$-shifted grating or the chirped grating will result in more stable single longitudinal mode operation. There are other ways to make gratings, such as a photomask self-interference method [64] or an electron beam lithography technique [65].

After the grating structure is fabricated on the InP substrate, multiple epitaxial layers, including n-InGaAsP lower-guide layer, InGaAsP active layer, p-InGaAsP upper-guide layer, p-InP cladding layer, and p-InGaAsP cap layer, are grown by LPE or MOCVD. Either a wet or dry etching technique is used to form mesa structures approximately 1 μm wide. A small mesa width is used to ensure a fundamental transverse mode. The mesa structures are then embedded epitaxially with p-InP and n-InP blocking layers or with semi-insulating blocking layer. After the wafer thinning and metalization, the wafer is cleaved into bars 200–400 μm wide. The laser emission facets are then coated with a thin dielectric film by either sputtering or electron beam evaporation. Stable operation of a longitudinal single mode relies strongly on the facet reflectivity and is therefore sensitive to the thickness of the coating film. Typically, a $\lambda/4$-shifted DFB laser should have a facet reflectivity of less than 1%. The facet coating also acts as a passivation means to protect the exposed active layer at the facets, improving the device reliability. Finally, the laser bars are separated into individual laser chips.

A DBR laser can also be used to achieve longitudinal single-mode operation [66–68]. The DBR laser grating is formed outside of the active region (Fig. 15.12). Laser oscillation takes place in a resonator formed by two grating mirrors or by one grating mirror and one cleaved facet. The wavelength selectivity of the grating mirrors enables the single-mode oscillation in a DBR laser. Because there are many modes in the laser resonator, temperature variation can cause the DBR laser mode jumping between the adjacent modes. DBR laser fabrication is similar to DFB laser fabrication except that the corrugated grating structure in a DBR laser is in the passive waveguide region instead of above or below the active layer, as in an index-coupled DFB laser.

15.4.2. *VCSEL FABRICATION*

From the emission direction standpoint, there are top-emitting VCSELs and bottom-emitting VCSELs. Any VCSELs based on GaAs substrate will not emit toward the bottom substrate direction if the VCSEL wavelength is shorter than 870 nm because of the substrate absorption, unless the substrate is removed. In this section, we will concentrate on top-emitting GaAs VCSEL processing. The same fabrication technology can easily be extended to bottom-emitting VCSELs, such as strained InGaAs VCSELs at approximately 980 nm.

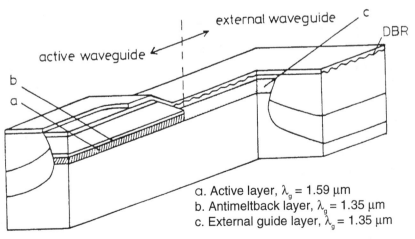

a. Active layer, $\lambda_g = 1.59\ \mu m$
b. Antimeltback layer, $\lambda_g = 1.35\ \mu m$
c. External guide layer, $\lambda_g = 1.35\ \mu m$

Fig. 15.12 Schematic diagram of a GaInAsP/InP buried heterostructure DBR laser.

Similar to edge-emitting laser diodes, VCSELs are either gain guided or index guided. The gain-guided VCSEL has a planar top surface as shown in Fig. 15.13. The current path of the gain-guided VCSEL is surrounded by the proton (H⁺) implant buried in the top p-doped DBR mirror, which forms a high-resistance enclosure right above the active region. The optical mode is determined by the active region where the current flows through.

The fabrication of the gain-guided VCSEL is straightforward. Immediately after the photoresist patterning on top of the wafer that defines the VCSEL emission aperture window, BeAu/Ti/Au three-layer p-contact metal is deposited by thermal evaporation [69] for liftoff. The total contact metal thickness is approximately 1500–3000 Å. The emission window diameter typically varies from 5 to 50 μm, depending on the requirements for optical output power and transverse mode design. A thick layer of photoresist is patterned to cover the laser emission window with the diameter slightly larger than that of the emission window as a proton (H⁺) implant mask. The photoresist implant mask thickness varies from 6 to 12 μm, depending on the proton implant energy. The implant energy is such that

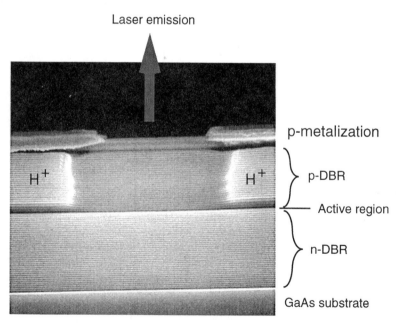

Fig. 15.13 Cross-section SEM photo of a plannar GaAs VCSEL structure, where proton (H⁺) implantation region is delineated by stain etching.

the majority of the protons rest at 0.5 μm above the active region to create a highly resistive layer, leaving the topmost portion of the DBR mirror relatively conductive. For the top DBR mirrors of 18–30 quarter-wavelength stack pairs, the required implant energy varies from 300 to 380 kV. The proton implant dose is approximately 10^{14} or 10^{15} cm^{-2}. The implant straggle width is approximately 0.5 μm. The proton implant defines an annular high-resistance region underneath the p-metal contact and right above the active region in the p-DBR mirror as that the injected current can funnel into the central active region underneath the laser emission aperture window. The diameter of the implant mask is chosen to be larger than the laser emission window defined by the p-metal contact. This helps lower the p-metal contact resistance because of a ring contact with an area unaffected by the implant. The smaller emission window also behaves like a filter that enhances the lowest order transverse mode [69, 70].

After the photoresist mask is stripped off, a new photoresist mask pattern is deposited for a shallow proton implantation to electrically isolate individual VCSELs between each other. The second implantation can be replaced by etched trenches if the requirement for planar surface is not essential [71]. Subsequently, a n-metal contact layer consisting of GeAu/Ni/Au is deposited by either thermal evaporation or e-beam evaporation on the bottom of the GaAs substrate. The wafer is then annealed at 450°C for 30 s. The anneal serves to alloy both contacts and to further shape the resistance profile due to proton implantation by further reducing the surface resistance while retaining a high-resistance layer under the metallized region [72]. A metal interconnect deposition on the top surface completes the device fabrication process. As an option, a layer of dielectric material, such as SiN or SiO$_2$, can be deposited by PECVD for the purposes of laser device surface passivation.

The gain-guided VCSEL by proton implantation has been proven manufacturable with exceptionally high yield [31, 32]. There are, however, concerns regarding the devices' turn-on delay due to thermal lensing effect [73] and turn-on delay-induced timing jitter [74, 75]. These effects give rise to problems in high-speed fiber optical applications. Adverse effects of the turn-on delay have been alleviated via careful biasing of the laser at or above threshold [76, 77]. To further address this issue, index-guided VCSELs or ridge waveguide VCSELs have been developed and have achieved better performances than the gain-guided VCSELs [32].

A typical index-guided VCSEL structure is shown in Fig. 2.15 of Chapter 2 [78]. The fabrication process starts with the deposition of 4000 Å TiW metal alloy by sputtering and 4000 Å SiN by PECVD. A photoresist mask is used to define the index-guided VCSEL mesa diameter, which usually ranges from 8 to 50 μm. The SiN and TiW outside of the photoresist mask is dry etched by fluorinated chemistry in a reactive ion-etching (RIE) reactor to expose the AlGaAs semiconductor epitaxial layer. After the removal of the photoresist, the remaining SiN serves as a hard mask for the following mesa etching by chlorine chemistry in a RIE reactor. The mesa etching depth is *in situ* monitored using laser interferometry to a level above the active layer. Protons are then implanted into the remaining *p*-DBR mirror to further electrically isolate the mesa region. The implantation energy is 50 keV and the implantation dose is in the range of 10^{14} to 10^{15} cm^{-2}. After the implantation, a layer of 4000 Å SiN is deposited by PECVD onto the entire wafer. A photoresist mask is then spun onto the SiN with the mesa top exposed for dry etching using fluorine chemistry to remove the SiN on the mesa. This etching process is not selective, so some TiW will be etched as well. After the photoresist is removed, 4000 Å TiW will be sputtered onto the entire wafer. A new photoresist mask is then used to define a VCSEL emission aperture on the mesa. The diameter of the emission aperture is smaller than the mesa diameter and fluorine chemistry is used to selectively remove the TiW in the emission aperture. Gold interconnect pads can then be deposited on top of the TiW by a photoresist liftoff process. This completes the device fabrication on the *p*-side. A broad-area *n*-metal contact formed by Ni/AuGe/Au is deposited on the bottom surface of the GaAs substrate. The wafer is then annealed in a rapid thermal annealing chamber for 30 s at 420°C. Finally, a SiN passivation layer can be deposited on the VCSEL surface for protection. The wafer is then sawed into individual chips as opposed to cleaving, which is usually done for edge-emitting lasers. This fabrication process is highly manufacturable and is currently being used in production for Motorola's trademarked Optobus optical interconnect product (Fig. 15.14) [32].

When the VCSEL is driven at a current well above threshold, multitransverse modes develop partially due to spatial hole burning in the active region. The number of allowable transverse modes of a regular index-guided VCSEL is largely determined by the waveguide, which is composed of etched mesa and the materials surrounding the mesa, such as SiN, metal, or air. As a consequence of the large index difference between the semiconductor material and the surrounding materials, the diameter of the mesa must be very small for the waveguide to support only the fundamental/single transverse mode. Single-mode VCSELs are preferred for an optical

A

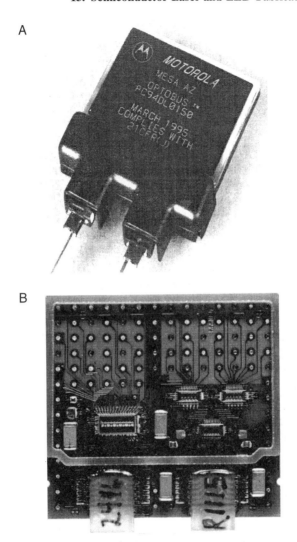

B

Fig. 15.14 Motorola Optobus module, a 10-channel data link at 400 MHz/channel.

interconnect based on single-mode fibers, laser printing, compact disk data storage, bar code scanning, etc. Reducing the mesa diameter also helps reduce the VCSEL threshold. Super low threshold VCSEL designs are appealing for some high-speed applications as well as for some basic physics studies. However, the mesa-type index-guided VCSEL cannot be too small in diameter because of the optical scattering loss from the etched sidewall.

The mesa sidewall damage due to exposure to the ion beam [79] may also become nonradiative recombination centers that limit the minimum achievable threshold and cause potential reliability problem.

A selective lateral oxidation defined index-guided VCSEL (shown in Fig. 2.27 in Chapter 2) is a promising alternative to the mesa-defined index-guided VCSEL [80–83]. It has been reported that native oxide of Al_xO_y can be stable enough to be utilized in the device fabrication for current confinement [84–87]. The effective index difference between the semiconductor and its native oxide also provides for an excellent waveguide for optical mode confinement. The diameter of the lateral oxide VCSEL waveguide can be made relatively larger for supporting only the single transverse mode, owing to the relatively moderate effective index difference. In addition, the p metal has a larger area to make contact with for the oxide VCSEL, reducing the series resistance significantly, even for a very small active area. Lower series resistance is effective in reducing the active junction temperature during operating; therefore, it is more desirable from a reliability standpoint.

For example, to fabricate the device structure shown in Fig. 2.27a, a VCSEL epitaxial structure is grown that includes $Al_{0.85}Ga_{0.15}As/Al_{0.16}Ga_{0.84}As$ DBR mirrors and a triple GaAs quantum well active region in a graded AlGaAs cavity, with the aluminum mole fraction of the AlGaAs mirror layer closest to the cavity increased from $x = 0.85$ to $x = 0.98$ on the p side. Devices are formed in mesas created by etching away all the surrounding p-side mirror stacks and part of the cavity materials above the active layer to expose the edge of the cavity layers. A portion of the AlGaAs layers starting from the exposed edge is then converted to an electrically insulating aluminum oxide (Al_xO_y) by wet thermal oxidation at 425°C for a few hours. The aluminum oxide has a refractive index of ~1.55. The lateral oxidation is conducted in a furnace supplied with a flow of N_2 bubbled through deionized water heated to a temperature of 95°C. The $x = 0.98$ layer oxidizes more rapidly [88] and the resultant oxide protrudes further into the mesa to provide electrical and optical confinement about the active region. A back-side substrate contact and an annular contact on the mesa complete the VCSEL [83]. This selective lateral oxidation is the most effective method in achieving record low-threshold VCSELs [89, 90]. Single transverse mode operation of up to 2.7 mW has also been demonstrated [91].

The same oxidation technology described previously can be used to oxidize through the high aluminum mole fraction layer to create DBR mirrors with one layer of semiconductor material and another layer of a native oxide Al_xO_y. The large index difference between the two alternative quarter-wavelenght-thick layers enables the demonstration of a high-reflective and wide-bandwidth DBR mirror with only 4 or 5 mirror pairs [92, 93].

Even though the oxide VCSEL has superb reported performances in single-mode power operation [91] and low threshold [89, 90], its long-term reliability, manufacturing uniformity across the wafer, and yield are still unknown. Other techniques for high-power single-mode emission include the leaky mode solution [94], which uses the regrowth technique to deposit higher index materials to surround the etched-mesa pillar [95, 96]. A more simple leaky mode technique involves using a planar proton-implanted VCSEL with smaller current isolation diameter and emission aperture. This technique has allowed the generation of a single-mode output power up to 4.4 mW at 980 nm [97].

In systems that are polarization sensitive, such as some optical data storage systems, the precise knowledge of the VCSEL polarization is required. VCSELs generally polarize along |110> and |1h0> crystalline directions. In order to stabilize the VCSEL polarization along a predetermined direction, certain asymmetries must be introduced into the laser structure. The most straightforward technique is to utilize the anisotropic transverse cavity geometry to select and maintain a dominant polarization state [98, 99]. Typically, the airpost or ridge waveguide mesa structure is elongated into a rectangular, rhombus, or dumbbell shape. This technique works successfully in stabilizing the VCSEL polarization along the elongated mesa direction. Alternatively, the polarization can be stabilized by using a differential loss method obtained from coatings on the sides of the upper DBR [100] or by applying biaxial stress to the active layer [101]. A grating formed on the very top layer of the p-DBR, with metalization on the sidewalls of the grating structure, has also been demonstrated with good polarization extinguishing ratio [102].

Even though the previously mentioned fabrication procedures use GaAs VCSELs for the example, the techniques should apply to InGaAs/GaAs VCSELs at 980 nm and InAlGaP/GaAs VCSELs at red visible wavelengths. Long-wavelength VCSELs at 1.3 and 1.55 μm are of interest to the telecommunication industries, but their fabrication has met some difficulty due to Auger nonradiative recombination-induced loss and the low reflectivity of the monolithic InGaAsP/InP DBR mirrors. Dielectric mirrors with 8.5 pairs of MgO/Si multilayers and Au/Ni/Au on the p side and 6 pairs of SiO2/Si on the *n* side have been used instead of the semiconductor DBR mirrors [103]. Continuous-wave VCSEL operation at 14°C has therefore been demonstrated at a wavelength of 1.3 μm.

A more rigorous technique in the long-wavelength VCSEL fabrication is to use a wafer-fusion or -bonding process to unite separately made GaAs/AlAs mirrors formed on a GaAs substrate to the InGaAsP double-heterostructure (DH) laser diode [104]. The higher reflectivity of the GaAs system mirrors at either 1.3 or 1.55 μm make them more advantageous than InP system mirrors. The wafer-fused VCSEL fabrication begins by fusing the GaAs/AlAs mirror wafer with the top of the InP DH wafer at 650°C for 30 min in a H_2 ambient environment under a graphite load [105]. After the wafers have been fused together, the InP substrate is removed by a selective wet chemical etchant such as HCL : H_2O (3 : 1). The resulting fused structure, an InP DH active layer with a GaAs/AlAs DBR mirror on a GaAs substrate, is mechanically robust and undergoes typical fabrication procedures without any special handling. The completed mesa structure VCSEL has a Pd/Zn/Pb/Au intracavity ring contact and a Si/SiO$_2$ dielectric mirror on the *p* side and an AuGe/Ni/Au contact on the n-type GaAs substrate [106]. This device operates pulsed at room temperature and CW at −45°C at a wavelength of 1.3 μm. When both sides of the InP DH wafer are fused with the GaAs/AlAs DBR mirrors, the VCSEL performances are further improved [107]. A double-fused VCSEL operating CW up to 64°C has been demonstrated at 1.55 μm by combining with the lateral DBR mirror oxidation technique [108]. This result represents the best long-wavelength VCSEL performance achieved to date.

15.4.3. CLEAVED-MIRROR FORMATION AND DICING

Before an edge-emitting laser can be probe tested, a resonant cavity must be formed by cleaving the wafer into bars along the crystalline direction, typically |110> or |1h0> (Fig. 15.15). The cleaving facets function like laser cavity mirrors. Usually, the wafer to be cleaved is mounted on a sticky, thin, flexible tape, and small indentations are scribed near one edge of the wafer with a diamond scriber. Slight flexing will then result in cleavage along |110> crystalline perpendicular to the surface. This technique ensures perfectly aligned mirrors. For broad-area lasers, the cleaved bars can then be sawed perpendicular to the mirror faces with a diamond or string saw to separate them into rectangular device chips. The rough sawn sides and a cavity length-to-width ratio of at least 2.5 appear to prevent excitation

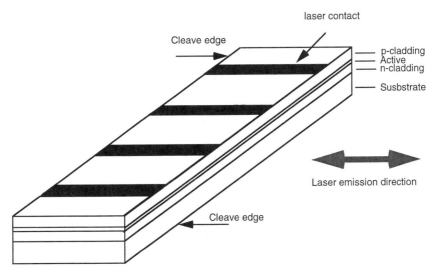

Fig. 15.15 Schematic diagram of a cleaved edge-emitting laser array bar.

of internally circulating modes [109]. The natural cleaved facets are usually mirror-like smooth with a reflectivity of approximately 30%, determined by the index difference between the semiconductor material and the air. The laser cavity length between the two cleaved facets usually ranges from 100 μm to 1 mm. When the device is biased above the threshold, the laser will be generated emitting toward both facets with an equal amount of output power. The emission from one of the laser facets can be used to monitor the laser output power from the other facet in applications in which constant output power is needed. When the monitoring photodiode detects the variation in laser output power, it will send a signal back into the driver circuitry to adjust the laser pump current until a predetermined laser output power is reached. This auto power control (APC) technique has been used in substantially all compact disk (CD) pickups and serial optical fiber data links to compensate for the power fluctuation due to ambient temperature change and/or device degradation.

The facets of edge-emitting lasers based on GaAs/AlGaAs are subject to oxidation when exposed to air [110, 111]. The catastrophic failure due to facet degradation is one of the most commonly seen failures occuring with edge-emitting lasers (Fig. 15.16). Facet passivation using SiO_2, Si_xN_y, or Al_2O_3 via deposition by chemical vapor deposition or sputtering is one

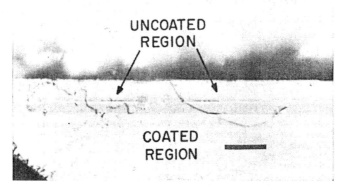

Fig. 15.16 Photo showing the facet erosion in the uncoated region of a partially coated laser diode after 600 h of operation. (Reprinted with permission from Ref. [136]. Copyright 1977 American Institute of Physics.)

of the many techniques used to prevent the facets from oxidation and has been proven effective in improving the device reliability.

When higher output power is needed from the laser, one of the laser facets can be antireflectively (AR) coated to reduce the facet reflectivity to only a few percent. More than 90% of the intracavity energy will then be coupled out from this coated facet. The reduced cavity Q by AR coating will cause the laser threshold to increase. A highly reflective (HR) dielectric mirror coating is usually deposited onto the other facet to increase the cavity Q back to or higher than the original level without the AR coating [112]. Some applications require an edge-emitting laser to operate at low threshold. This can be achieved by depositing HR dielectric mirror coatings on both facets, with one of the facets less reflective than the other for the laser to couple out. Edge-emitting lasers with sub-mA thresholds have been demonstrated in this way in the research environment [113].

Both HR coatings and AR coatings are deposited after the wafer cleaving. The individual edge-emitting laser chips, or laser bars, must be mounted one by one for the coatings. This is a high-cost manufacturing process that challenges the device yield and throughput. The high level of stress incorporated during the coating deposition also affects the device reliability. In contrast, the monolithic manufacturing process of VCSELs allows on-wafer probe test with a computer-controlled probe card as per the silicon industry. After testing, the wafer is separated into chips by a diamond saw similar to the dicing of silicon integrated circuits (IC) (Fig. 15.17). Known good dies will then be automatically picked out for packaging using com-

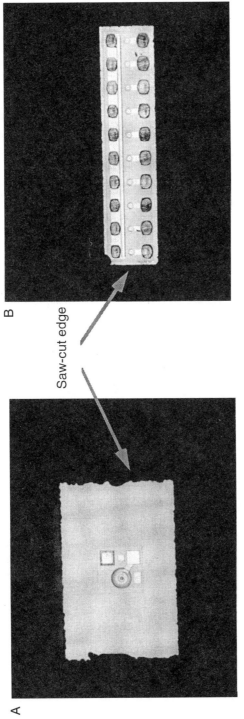

Saw-cut edge

Fig. 15.17 Photo showing (A) a discrete VCSEL and (B) a 1 × 10 VCSEL array.

A

B

puter control. This highly automated process is compatible with the existing silicon IC infrastructures and therefore preferred for low-cost manufacturing. Furthermore, the VCSEL active layers are not exposed to the air, and no facet-related degradation mode exists. Therefore, the GaAs VCSEL reliability improves dramatically compared to that of conventional GaAs edge-emitting lasers having unprotected facets. In addition, the VCSEL wafer can easily be sawed into chips of two-dimensional (2-D) arrays, which may either be individually addressed or matrix addressed depending on the interconnect metal deposited onto the wafer [114–117], whereas edge-emitting lasers can only be cleaved into one-dimensional array bars.

15.4.4. PACKAGING AND BONDING

The commercial semiconductor laser is usually packaged in either a can type or a dual-in-line pin (butterfly) type package. The can-type package is smaller in design and is currently used in high-volume manufacturing. It is less costly to manufacture and is therefore widely used for CD data storage, bar code scanners, pointers, laser printers, and serial data links. The 14-lead butterfly-type package costs more to manufacture and is typically reserved for long-wavelength semiconductor lasers.

Semiconductor lasers packaged in metal cans (TO cans) dominate today's commercial semiconductor laser market. The majority of earlier TO cans had a diameter of 9 mm, but the trend has been to use a smaller type TO-56 package with a diameter of 5.6 mm (Fig. 15.18). The laser chip is first

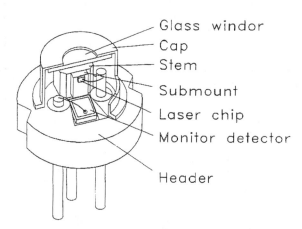

Fig. 15.18 Schematic diagram of an edge-emitting laser mounted on a TO-56 package with a power monitoring photodiode.

mounted upside down on a Si submount with Sn solder. Electrical transmission lines can be deposited on the submount for electrical contact. The submount with the laser chip turned on is then mounted sideways with In solder onto a copper heat sink that sits on a stem 5.6 mm in diameter so that the laser emission is vertical to the package. Sometimes the laser chip is directly mounted upside down onto the copper heat sink without any Si submount. The stem is based on copper, iron, or nickel, depending on the heat dissipation and cost requirements. Both the heat sink and the stem can be plated with gold for better heat dissipation.

If APC is needed, a silicon PIN detector is directly mounted onto the stem with a Sn or In solder to receive the laser emission from the back facet. The detector on the stem is tilted to avoid any reflected light from coupling directly back into the laser, thus destabilizing the laser operation. The detector can sometimes be made directly on the silicon submount that will detect one-half of the cone-shaped backward laser emission for the laser power monitoring. Gold wire bonding is used to connect electrodes on the semiconductor laser and the detector to the corresponding posts on the stem. Depending on the specific application, the semiconductor laser cathode can be either in common with the detector cathode, as shown in Fig. 15.19a, or in common with the detector anode, as shown in Fig. 15.19b. Finally, a lid is hermetically sealed onto the stem. Hermetic sealing is

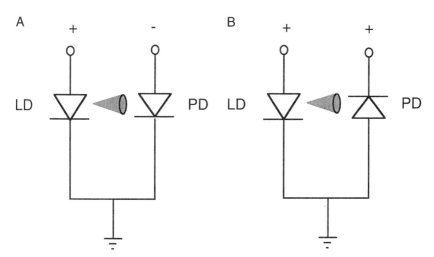

Fig. 15.19   Electrical contact diagram of a laser package with a monitoring photodiode with the laser cathode in common with the photodiode (A) and (B) anode.

needed to improve the laser reliability. The glass window on the lid is AR coated to reduce any backward reflection that is detrimental to the laser signal-to-noise ratio. A slanted lid is sometimes used to steer the backward reflection light away from entering the laser cavity (Fig. 15.20).

Dual in-line butterfly pigtail-type laser packages are typically used for optical communications. Laser light is required to couple into a single-mode fiber with a core diameter of approximately 10 μm. Butt-coupling is one of the commonly used active alignment techniques for fiber pigtailing. In the butt-coupling, a laser chip is first mounted onto a subcarrier and turned on. An optical fiber is brought close to the emission end of the laser to achieve the best coupling efficiency and then secured onto the subcarrier. The butt-coupling usually has a loss of more than 3 dB because of the poor matching in numerical aperture (NA) between the laser emission and the fiber. The coupling efficiency can be improved dramatically by inserting a lens between the laser and the fiber [118]. Many types of lenses can be used, but a gradient-index lens is among the best in performance [119]. An integrated ball lens can be made directly using the glass fiber tip and relatively good coupling has been achieved [120]. Many manufacturers use can-type packaged CD lasers at 780 nm because of their commercial availability at a substantially lower price. Typically, a lens between the can and the fiber is used to improve the coupling efficiency [121, 122].

To reduce the fiber pigtailing cost, especially when a laser array is needed to couple into a fiber array, a passive coupling alignment technique is

Fig. 15.20 A TO-56 package with slanted lid to avoid direct reflected light into the laser.

preferred in which the lasers are never turned on during the packaging. This passive coupling alignment can also increase the packaging throughput significantly. Figure 15.21 is an example of a four-channel fiber pigtail package with passive alignment [123]. A Si submount is first made with four V grooves uniformly spaced apart. The spacing is determined by the spacing of the lasers on the laser array chip. A four-channel laser chip is flip-chip mounted adjacent to the end of the V grooves with the position predetermined by the marks on the Si submount. A self-alignment technique can also be used to position the laser chip using solder reflow [124]. Four fiber cores are then placed into the V grooves on the Si submount with their ends leaning against the end of the V grooves. A Si cover carrier with V grooves and a space for the laser chip is then placed back onto the Si submount with optical glue to secure the whole package. No lenses are used between the lasers and the fibers because the laser emission facets are right against the fiber input ends and their relative positions are predetermined with very high placement resolution.

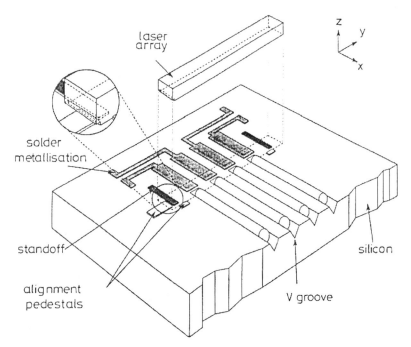

Fig. 15.21 A four-channel V-groove submount for passive alignment of laser array to fibers (after Ref. [123]).

The vertical emission feature of a VCSEL allows it to be packaged at a relatively lower cost. Because the VCSEL has a circular beam shape, its coupling efficiency into the fiber is high. A single VCSEL can be packaged in the same way as a surface-emitting LED on either a TO-type can or a molded leadframe (Fig. 15.22). A molded lens can be integrated to the leadframe surface mount-type package to adjust the laser beam divergence angle and thus to match the NA of the fiber for better coupling efficiency.

One-dimensional (1-D) or 2-D array capability is the unique advantage of VCSEL technology. The Optobus package is an example how a 1-D

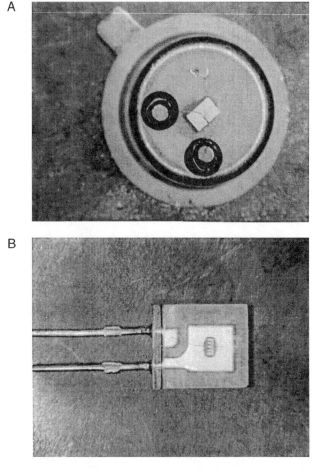

Fig. 15.22 Photos of VCSELs packaged with (A) TO-56 cans and (B) plastic molding.

VCSEL array is packaged to couple into multimode fibers [32] (Fig. 15.23). The package supports a 1×10 VCSEL array. Leadframes with 10 copper leads for the VCSEL anodes and two copper leads for the VCSEL cathodes are first embedded into a 10-channel polymer waveguide, the Guidecast. The 12 leads are bent toward the input end of the Guidecast for the direct chip attachment of the VCSEL array onto the leads. All the VCSEL anodes and cathodes are on the top side, as shown in Fig. 15.17b. The electrical contacts of the VCSEL array are made of plated Au with the same height above the top emission surface. The VCSEL array is flip-chip mounted onto the Guidecast with a robot. The robot has a placement accuracy of several micrometers — enough to guarantee an accurate alignment of the VCSEL with an emission aperture of 10–14 μm in diameter onto the Guidecast waveguide channels with dimensions on the order of 40 $\times$ 40 μm. The electrical contacts between the VCSELs and the Guidecast leads are connected by conductive epoxy. Underfill material with the same composition as the Guidecast is used to secure a robust attachment of the VCSEL array to the Guidecast. In addition to being electrical connectors, the Guidecast leads also function as thermal heat sinks to dissipate heat from the VCSELs. The output end of the Guidecast is attached to the standard MT-ferrules for light coupling into 10-channel multimode fiber ribbons. The whole concept of this approach is based on passive coupling alignment and polymer material molding to achieve low-cost packaging while utilizing the array function of VCSELs to achieve high-speed parallel data transmission.

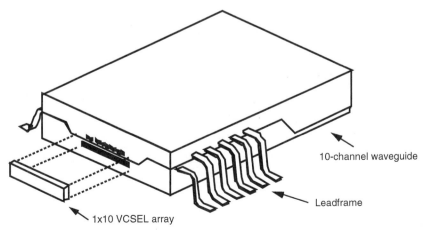

10-channel waveguide

Leadframe

1x10 VCSEL array

Fig. 15.23   Schematic diagram of a VCSEL GUIDECAST package.

The leadframe type Guidecast works properly for transmitting data at 155 Mb/channel. When a higher data transmission speed is needed, the electrical cross-talk between the leads becomes a limiting factor. To address this issue, a tape access-bonding (TAB) approach is used to make electrical contacts to the VCSEL array [33] (Fig. 15.24). The polymer Guidecast without the copper leads is then attached to the VCSEL array and functions as a waveguide to couple the VCSEL beams into the 10-channel fiber ribbon through the MT-ferrules. This approach has improved the Optobus data link speed to 400 Mb/channel and is expected to work well for the next-generation Optobus data link at 800 Mb/channel.

Vertical cavity surface-emitting laser array packaging for data links has also been demonstrated with direct fiber coupling [125–128] or through the use of flexible Polyguide [129]. Passive coupling alignment with the Polyguide approach has recently been demonstrated [130].

It should be noted that the efficiency to couple light into a fiber is determined by the beam transverse mode and the modes that the fiber can support. Edge-emitting semiconductor lasers, VCSELs, and LEDs have different beam profiles and therefore have different launch conditions for fiber coupling. Good coupling efficiency can be achieved by adjusting the light source beam profile with a lens to match the fiber NA. In telecommunication applications, long-wavelength semiconductor lasers are coupled into a single-mode fiber to achieve the optimum bandwidth–distance performance. Optical fibers are designed for applications at those wavelengths. The low-cost requirement for high-speed data communications pushes forward the applications of CD lasers operating on multimode fibers, and they

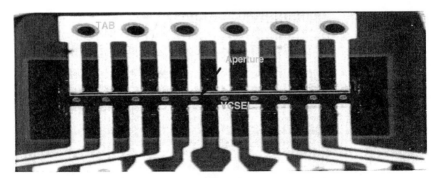

Fig. 15.24 Photo showing a tape-access-bonding (TAB) approach for VCSEL array packaging.

have very different launch conditions compared to long-wavelength laser sources due to the difference in optical modes that the multimode fiber can support at the CD wavelength region. Recently, some specifications for gigabit Ethernet were redrafted because the bandwidth and losses were originally calculated based on an incorrect LED launch model.

VCSELs have an advantage of low-cost manufacturing. In addition, GaAs VCSELs at 850 nm have demonstrated superb performance and reliability for applications in data communications. However, most standard graded-index 62.5-μm multimode fibers today are optimized for long-wave applications at either 1.3 or 1.55 μm and have much smaller bandwidth–distance performance at 850 nm. Therefore, today's VCSEL data link market is targeted for short-distance applications up to 300–500 m. As the VCSELs become more important and prevalent, it should be possible to buy fiber with higher bandwidth at shorter wavelengths. For example, a multimode fiber with a minimum of 800 MHz/km at 850 nm has been demonstrated that is achieved by optimizing the fiber manufacturing process at a slightly different point and tightly controlling the process variables. This cable is more expensive today than the standard long-wave fiber, but this trend could change soon if the market demand triggers sufficient volume production.

15.5. Conclusion

The semiconductor laser and LED manufacturing technologies have been discussed in this chapter with the emphasis on the lasers. Edge-emitting laser diodes have been commercialized for decades and are extensively used for CD/DVD data storage, laser printing, bar code scanners, pointers, optical communications, solid-state laser pumping sources, etc. VCSELs have just started to enter the commercial market and are currently used for parallel or serial optical fiber data links. It is expected that VCSELs will also see applications in areas currently dominated by edge-emitting laser diodes. In-plane surface-emitting lasers [59, 131] are another type of lasers that have been extensively studied by many research groups. Circular-grating surface-emitting lasers [132, 133] are considered to be in this category. Manufacturability and manufacturing cost relative to VCSELs are the two factors that have hindered their commercialization.

Edge-emitting lasers are grown by LPE, MBE, or MOCVD techniques, whereas VCSELs are grown by MBE or MOCVD techniques. Growth using LPE is not accurate enough for monolithic DBR structures of VCSELs. From a manufacturing standpoint, MOCVD offers the best growth throughput and excellent material quality; thus, it is predominantly used by all the commercial VCSEL manufacturers. Many edge-emitting laser manufacturers are also relying on the MOCVD technology for the high-volume production.

Currently, edge-emitting lasers are dominated by either DH-type or MQW-type structures. Index-guided structures are used to achieve low thresholds and single transverse mode operation. Distributed feedback lasers have been extensively studied to achieve single longitudinal-mode operation for long-distance optical communications. Precise wavelength control is critical for applications in wavelength-division multiplexing [134]. Vertical cavity surface-emitting lasers have only one longitudinal mode because of their extremely short laser cavity. Multitransverse mode operation, however, is common due to the large active area relative to the laser cavity length. There are three basic types of VCSEL structures: proton-implanted VCSELs, etched-mesa VCSELs, and lateral oxidation-confined VCSELs. Many applications require that the laser beam be diffraction limited. Single transverse-mode VCSELs can be achieved by reducing the laser emission aperture, but the single-mode output power becomes limited as well [91, 95–97]. Nevertheless, VCSEL beams have the advantage of being circular, and therefore offer a high coupling efficiency into an optical fiber.

Edge-emitting lasers can be tested only after the wafer is cleaved, and most of the time after being packaged. Vertical cavity surface-emitting lasers can be tested on the wafer, and only known good dies will be packaged after being sawed into chips. From a manufacturing standpoint, VCSELs are more suitable for low-cost mass production.

Edge-emitting lasers at 1.3 and 1.55 μm are still the only choice for long-distance optical communications. Tremendous progress has been made in VCSEL research at 1.3 and 1.55 μm using wafer-bonding techniques [106–108]. The device efficiency and yield sare still low and the manufacturability is unknown. Further development is needed before the long-wavelength VCSELs become practical for commercial applications.

Device reliability is one of the most important considerations for any commercial application. Edge-emitting lasers at 780 nm do not survive long, predominantly due to facet degradation. Facet passivation and hermetically sealed packages are effective in increasing the device lifetime. For applications requiring extreme device reliability, burn-ins and special screenings are necessary. Such procedures certainly affect the manufacturing cost. Vertical cavity surface-emitting lasers have been proven reliable with extrapolated mean time to failure of up to millions of hours at normal operation condition without any burn-in screening [32, 135]. Hermeticity may not be necessary for VCSEL packaging. Such an advantage will allow a VCSEL to be packaged, like a LED, using plastic molding and therefore substantially reducing the packaging cost.

References

1. Casey, H. C., Jr., and M. B. Panish. 1978. *Heterostructure lasers.* New York: Academic Press.
2. Temkin, H., M. B. Panish, R. A. Logan, and J. P. Van der Ziel. 1984. $\lambda \approx$ 1.5 μm InGaAsP ridge lasers grown by gas source molecular beam epitaxy. *Appl. Phys. Lett.* 45:330.
3. Asahi, H., Y. Kawamura, and H. Nagai. 1982. Molecular beam epitaxial growth of InGaAlP on (100) GaAs. *J. Appl. Phys.* 53:4928.
4. Herman, M. A. and H. Sitter. 1989. *Molecular beam epitaxy.* Berlin: Springer-Verlag.
5. Wang, C. and R. F. Davis. 1993. Deposition of highly resistive, undoped, and p-type, magnesium-doped gallium nitride films by modified gas source molecular beam epitaxy. *Appl. Phys. Lett.* 63:990.
6. Yung, K., J. Yee, J. Koo, M. Rubin, N. Newman, and J. Ross. 1994. Observation of stimulated emission in the near ultraviolet from a molecular beam epitaxy grown GaN film on sapphire in a vertical-cavity, single pass configuration. *Appl. Phys. Lett.* 64:1135.
7. Nakamura, S., T. Mukai, M. Senoh, and N. Iwasa. 1992. Thermal annealing effects on p-type Mg-dope GaN films. *Jpn. J. Appl. Phys.* 31:L139.
8. Amano, H., M. Kito, K. Hiramatsu, and I. Akasaki. 1989. P-type conduction in Mg-doped GaN treated with low-energy electron beam irradiation. *Jpn. J. Appl. Phys.* 28:L2112.
9. Neave, J. H., B. A. Joyce, P. J. Dobson, and N. Norton. 1983. Dynamics of film growth of GaAs by MBE from Rheed observations. *Appl. Phys. A.* 31:1.
10. Walker, J. D., D. M. Kuchta, and J. S. Smith. 1991. Vertical-cavity surface-emitting laser diodes fabricated by phase-locked epitaxy. *Appl. Phys. Lett.* 59:2079.
11. Walker, J. D., D. M. Kuchta, and J. S. Smith. 1993. Wafer-scale uniformity of vertical-cavity lasers grown by modified phase-locked epitaxy technique. *Electron. Lett.* 29:239.

12. Chalmers, S. A. and K. P. Killeen. 1993. Method for accurate growth of vertical-cavity surface-emitting lasers. *Appl. Phys. Lett.* 62:1182.
13. Bacher, K., B. Pezeshki, S. M. Lord, and J. S. Harris. 1992. Molecular beam epitaxy growth of vertical cavity optical devices with *in situ* corrections. *Appl. Phys. Lett.* 61:1388.
14. Grothe, H., and F. G. Boebel. 1993. *In situ* control of Ga(Al)As MBE layers by pyrometric interferometry. *J. Cryst. Growth* 127:1010.
15. Houng, Y. M., M. R. T. Tan, B. W. Liang, S. Y. Wang, L. Yang, and D. E. Mars. 1994. InGaAs(0.98 μm)/GaAs vertical cavity surface emitting laser grown by gas-source molecular beam epitaxy. *J. Cryst. Growth* 136:216.
16. Houng, Y. M., M. R. T. Tan, B. W. Liang, S. Y. Wang, and D. E. Mars. 1994. *In situ* thickness monitoring and control for highly reproducible growth of distributed Bragg reflectors. *J. Vacuum Sci. Tech.* B12:1221.
17. Lear, K. L., R. P. Schneider, K. D. Choquette, S. P. Kilcoyne, J. J. Figiel, and J. C. Zolper. 1994. Vertical cavity surface emitting lasers with 21-percent efficiency by metalorganic vapor phase epitaxy. *IEEE Photon. Tech. Lett.* 6:1053.
18. Hibbs-Brenner, M. K., R. P. Schneider, R. A. Morgan, R. A. Walterson, J. A. Lehman, E. L. Kalweit, J. A. Lott, K. L. Lear, K. D. Choquette, and H. Juergensen. 1994. Metalorganic vapour-phase epitaxial growth of red and infrared vertical-cavity surface-emitting laser diodes. *Microelectron. J.* 25:747.
19. Kawakami, T., Y. Kadota, Y. Kohama, and T. Tadokoro. 1992. Low-threshold current low-voltage vertical-cavity surface-emitting lasers with low-Al-content p-type mirrors grown by MOCVD. *IEEE Photon. Tech. Lett.* 4:1325.
20. Yoon, S. F. 1992. Observation of nonbiased degradation recovery in GaInAsP/InP laser diodes. *IEEE J. Lightwave Tech.* 10:194.
21. Cunningham, B. T., M. A. Haase, M. J. McCollum, J. E. Baker, and G. E. Stillman. 1989. Heavy carbon doping of metalorganic chemical vapor deposition grown GaAs using carbon tetrachloride. *Appl. Phys. Lett.* 54:1905.
22. Cunningham, B. T., J. E. Baker, and G. E. Stillman. 1990. Carbon tetrachloride doped $Al_xGa_{1-x}As$ grown by metalorganic chemical vapor deposition. *Appl. Phys. Lett.* 56:836.
23. de Lyon, T. J., N. I. Buchan, P. D. Kirchner, J. M. Woodall, G. J. Scilla, and F. Cardone. 1991. High carbon doping efficiency of bromomethanes in gas source molecular beam epitaxial growth of GaAs. *Appl. Phys. Lett.* 58:517.
24. Buchan, N. I., T. F. Kuech, G. Scilla, F. Cardone, and R. Potemski. 1989. Carbon incorporation in metal-organic vapor phase epitaxy grown GaAs from CH_xI_{4-x}, HI, and I_2. *J. Electron. Mater.* 19:277.
25. Stringfellow. 1989. Non-hydride group V sources for OMVPM. *Mater. Res. Soc. Symp. Proc.* 145:III–V Heterostructures for Electronic/Photonic Devices, 171.
26. Lum, R. M., J. K. Klingert, and F. A. Stevie. 1990. Controlled doping of GaAs films grown with tertiary-butylarsine. *J. Appl. Phys.* 67:6507.

27. Kuech, T. F., D. J. Wolford, E. Veuhoff, V. Deline, P. M. Mooney, R. Potemski, and J. Bradley. 1987. Properties of high-purity $Al_xGa_{1-x}As$ grown by the metal-organic vapor-phase-epitaxy technique using methyl precursors. *J. Appl. Phys.* 62:632.

28. Chalmers, S. A., K. L. Lear, and K. P. Killeen. 1993. Low resistance wavelength-reproducible p-type (Al, Ga)As distributed Bragg reflectors grown by molecular beam epitaxy. *Appl. Phys. Lett.* 62:1585.

29. Lear, K. L., S. A. Chalmers, and K. P. Killeen. 1993. Low threshold voltage vertical cavity surface-emitting laser. *Electron. Lett.* 29:584.

30. Peters, M. G., B. J. Thibeault, D. B. Young, J. W. Scott, F. H. Peters, A. C. Gossard, and L. A. Coldren. 1993. Band-gap engineered digital alloy interfaces for lower resistance vertical-cavity surface-emitting lasers. *Appl. Phys. Lett.* 63:3411.

31. Hibbs-Brenner, M. K., R. A. Morgan, R. A. Walterson, J. A. Lehman, E. L. Kalweit, S. Bounnak, T. Marta, and R. Gieske. 1996. Performance, uniformity, and yield of 850-nm VCSEL's deposited by MOVPE. *IEEE Photon. Tech. Lett.* 8:7.

32. Lebby, M., C. A. Gaw, W. B. Jiang, P. A. Kiely, C. L. Shieh, P. R. Claisse, J. Ramdani, D. H. Hartman, D. B. Schwartz, and J. Grula. 1996. Use of VCSEL arrays for parallel optical interconnects. *Proc. SPIE* 2683:81.

33. Lebby, M. L., C. A. Gaw, W. B. Jiang, P. A. Kiely, P. R. Claisse, and J. Grula. 1996 (November). Key challenges and results of VCSELs in data links. LEOS '96, WV2, Boston, MA.

34. Grodzinski, P., S. P. Denbaars, and H. C. Lee. 1995. From research to manufacture — The evolution of MOCVD. *JOM — J. Miner. Metal. Mater. Soc.* 47:25.

35. Kawai, H., S. Imanaga, K. Kaneko, and N. Watanabe. 1987. Complex refractive indices of AlGaAs at high temperature measured by *in situ* reflectometry during growth by metalorganic chemical vapor deposition. *J. Appl. Phys.* 61:328.

36. Breiland, W. G. and K. P. Killeen. 1995. A virtual interface method for extracting growth rates and high temperature optical constants from thin semiconductor films using *in situ* normal incidence reflectance. *J. Appl. Phys.* 78:6726.

37. Frateschi, N. C., S. G. Hummel, and P. D. Dapkus. 1991. *In situ* laser reflectometry applied to the growth of $Al_xGa_{1-x}As$ Bragg reflectors by metalorganic chemical vapour deposition. *Electron. Lett.* 27:155.

38. Azoulay, R., Y. Raffle, R. Kuszelewicz, G. Leroux, L. Dugrand, and J. C. Michel. 1994. *In situ* control of the growth of GaAs GaAlAs structures in a metalorganic vapour phase epitaxy reactor by laser reflectometry. *J. Cryst. Growth* 145:61.

39. Hou, H. Q., H. C. Chui, K. D. Choquette, B. E. Hammons, W. G. Breiland, and K. M. Geib. 1996. Highly uniform and reproducible vertical-cavity surface-

emitting lasers grown by metalorganic vapor phase epitaxy with *in situ* reflectometry. *IEEE Photon. Tech. Lett.* 8:1285.

40. Ripper, J. E., J. C. Dyment, L. A. D'Asaro, and T. L. Paoli. 1971. Stripe-geometry double heterostructure junction lasers: Mode structure and cw operation above room temperature. *Appl. Phys. Lett.* 18:155.

41. Dyment, J. C., L. A. D'Asaro, J. C. North, B. I. Miller, and J. E. Ripper. 1972. Proton-bombardment formation of stripe-geometry heterostructure lasers for 300 K CW operation. *Proc. IEEE* 60:726.

42. Yonezu, H., I. Sakuma, K. Kobayashi, T. Kamejima, M. Ueno, and Y. Nannichi. 1973. A GaAs-Al$_x$Ga$_{1-x}$As double heterostructure planar stripe laser. *Jpn. J. Appl. Phys.* 12:1585.

43. Kaminow, I. P., R. E. Nahory, M. A. Pollack, L. W. Stulz, and J. C. DeWinter. 1979. Single-mode CW ridge-waveguide laser emitting at 1.55 μm. *Electron. Lett.* 15:763. [Erratum, *Electron. Lett.* 16:75 (1980)].

44. Ueno, M., I. Sakuma, T. Furuse, Y. Matsumoto, H. Kawano, Y. Ide, and S. Matsumoto. 1981. Transverse mode stabilized InGaAsP/InP ($\lambda = 1.3$ μm) plano-convex waveguide lasers. *IEEE J. Quantum Electron.* QE-17:1930.

45. Chao, C. P., S. Y. Hu, P. Floyd, K. K. Law, S. W. Corzine, J. L. Merz, A. C. Gossard, and L. A. Coldren. 1991. Fabrication of low-threshold InGaAs GaAs ridge waveguide lasers by using in situ monitored reactive ion etching. *IEEE Photon. Tech. Lett.* 3:585.

46. Shieh, C. L. and D. E. Ackley. 1994. Top emitting VCSEL with etch stop layer. U.S. Patent No. 5,293,392.

47. Hirao, M., S. Tsuji, K. Mizuishi, A. Doi, and M. Nakamura. 1980. Long wavelength InGaAsP/InP lasers for optical fiber communication systems. *J. Opt. Commun.* 1:10.

48. Mito, I., M. Kitamura, K. Kobayashi, S. Murata, M. Seki, Y. Odagiri, H. Nishimoto, M. Yamaguchi, and K. Kobayashi. 1983. InGaAsP double-channel-planar-buried-heterostructure laser diode (DC-PBH LD) with effective current confinement. *IEEE J. Lightwave Tech.* LT-1:195.

49. Dutta, N. K., S. G. Napholtz, R. Yen, T. Wessel, T. M. Shen, and N. A. Olsson. 1985. Long wavelength InGaAsP ($\lambda \sim 1.3$ μm) modified multiquantum well laser. *Appl. Phys. Lett.* 46:1036.

50. Mito, I., M. Kitamura, K. Kaede, Y. Odagiri, M. Seki, M. Sugimoto, and K. Kobayashi. 1982. InGaAsP planar buried heterostructure laser diode (PBH-LD) with very low threshold current. *Electron. Lett.* 18:2.

51. Tsang, W. T. and R. A. Logan. 1979. GaAs-Al$_x$Ga$_{1-x}$As strip buried heterostructure lasers. *IEEE J. Quantum Electron* QE-15:451.

52. Nelson, R. J., P. D. Wright, P. A. Barnes, R. L. Brown, T. Cella, and R. G. Sobers. 1980. High-output power InGaAsP (1 = 1.3 μm) strip-buried heterostructure lasers. *Appl. Phys. Lett.* 36:358.

53. Ishikawa, H., H. Imai, T. Tanahashi, Y. Nishitani, M. Takusagawa, and K. Takahei. 1981. V-grooved substrate buried heterostructure InGaAsP/InP laser. *Electron. Lett.* 17:465.

54. Tamari, N. and H. Shtrikman. 1982. High-T_0 low-threshold crescent InGaAsP mesa-substrate buried-heterojunction lasers. *Electron. Lett.* 18:177.
55. Murotani, T., E. Oomura, H. Higuchi, H. Namizaki, and W. Susaki. 1980. InGaAsP/InP buried crescent laser emitting at 1.3 μm with very low threshold current. *Electron. Lett.* 16:566.
56. Ogura, M. 1995. In situ etching and regrowth process for edge- and surface-emitting laser diodes with an AlGaAs/GaAs buried heterostructure. *J. Vacuum Sci. Tech. B* 13:1529.
57. Rochus, S., M. Hauser, T. Röhr, H. Kratzer, G. Böhm, W. Klein, G. Tränkle, and G. Weimann. 1985. Submilliamp vertical-cavity surface-emitting lasers with buried lateral-current confinement. *IEEE Photon. Tech. Lett.* 7:968.
58. Itaya, Y., T. Matsuoka, Y. Nakano, Y. Suzuki, K. Kuroiwa, and T. Ikegami. 1982. New 1.5 μm wavelength GaInAsP/InP distributed feedback laser. *Electron. Lett.* 18:1006.
59. Zory, P. and L. D. Comerford. 1975. Grating-coupled double-heterostructure AlGaAs diode laser. *IEEE J. Quantum Electron.* QE-11:451.
60. Aiki, K., M. Nakamura, and J. Umeda. 1976. Lasing characteristics of distributed-feedback GaAs-GaAlAs diode lasers with separate optical and carrier confinement. *IEEE J. Quantum Electron.* QE-12:597.
61. Utaka, K., S. Akiba, K. Sakai, and Y. Matsushima. 1984. λ/4-shifted InGaAsP/InP DFB lasers by simultaneous holographic exposure of positive and negative photoresists. *Electron. Lett.* 20:1008.
62. Numai, T., M. Yamaguchi, I. Mito, and K. Kobayashi. 1987. A new grating fabrication method for phase-shifted DFB LDs. *Jpn. J. Appl. Phys.* 26:L1910.
63. Suzuki, A. and K. Tada. 1980. Fabrication of chirped gratings on GaAs optical waveguides. *Thin Solid Films* 72:419.
64. Okai, M., S. Tsuji, N. Chinone, and T. Harada. 1989. Novel method to fabricate corrugation for a λ/4-shifted distributed feedback laser using a grating photomask. *Appl. Phys. Lett.* 55:415.
65. Sekartedjo, K., N. Eda, K. Furuya, Y. Suematsu, F. Koyana, and T. Tanbun-Ek. 1984. 1.5 μm phase-shifted DFB lasers for single-mode operation. *Electron. Lett.* 20:80.
66. Reinhart, F. K., R. A. Logan, and C. V. Shank. 1975. GaAs-Al$_x$Ga$_{1-x}$As injection lasers with distributed Bragg reflectors. *Appl. Phys. Lett.* 27:45.
67. Abe, Y., K. Kishino, T. Tanbun-Kk, S. Arai, F. Koyana, K. Matsumoto, T. Watanabe, and Y. Suematsu. 1982. Room-temperature cw operation of 1.60 μm GaInAsP/InP buried-heterostructure integrated laser with Butt-jointed built-in distributed-Bragg-reflection waveguide. *Electron. Lett.* 18:410.
68. Mikami, O., T. Saitoh, and H. Nakagome. 1982. 1.5 μm GaInAsP/InP distributed Bragg reflector lasers with built-in optical waveguide. *Electron. Lett.* 18:458.
69. Hasnain, G., K. Tai, L. Yang, Y. H. Wang, R. J. Fischer, J. D. Wynn, B. Weir, N. K. Dutta, and A. Y. Cho. 1991. Performance of gain-guided surface emitting

lasers with semiconductor distributed Bragg reflectors. *IEEE J. Quantum Electron.* QE-27:1377.

70. Morgan, R. A., G. D. Guth, M. W. Focht, M. T. Asom, K. Kojima, L. E. Rogers, and S. E. Callis. 1993. Transverse mode control of vertical cavity top-surface emitting lasers. *IEEE Photon. Tech. Lett.* 5:374.

71. Lee, Y. H., B. Tell, K. Brown-Goebeler, and J. L. Jewell. 1990. Top-surface-emitting GaAs four-quantum-well lasers emitting at 0.85 μm. *Electron. Lett.* 26:710.

72. Tell, B., Y. H. Lee, K. F. Browngoebeler, J. L. Jewell, R. E. Leibenguth, M. T. Asom, G. Livescu, L. Luther, and V. D. Mattera. 1990. High-power cw vertical-cavity top surface-emitting gas quantum well lasers. *Appl. Phys. Lett.* 57:1855.

73. Hasnain, G., K. C. Tai, L. Yang, Y. H. Wang, R. J. Fischer, J. D. Wynn, B. Weir, N. K. Dutta, and A. Y. Cho. 1991. Performance of gain-guided surface emitting lasers with semiconductor distributed Bragg reflectors. *IEEE J. Quantum Electron.* QE-27:1377.

74. Ding, G., S. W. Corzine, M. R. T. Tan, S. Y. Wang, K. Hahn, K. L. Lear, and J. S. Harris, Jr. 1996 (October). Dynamic behavior of VCSELs under high speed modulation. OSA annual meeting, WR6, Rochester, NY.

75. Law, J., and G. P. Agrawal. 1996 (October). Effect of carrier diffusion on modulation and noise characteristics of vertical-cavity surface-emitting lasers. OSA annual meeting, WII8, Rochester, NY.

76. Schwartz, D. B., C. K. Y. Chun, B. M. Foley, D. H. Hartman, M. Lebby, H. C. Lee, C. L. Shieh, S. M. Kuo, S. G. Shook, and B. Webb. 1995. A low cost, high performance optical interconnect. *Proc. 45th Electron. Compon. Tech. Conf.,* 376.

77. Nordin, R. A., D. B. Buchholz, R. F. Huisman, N. R. Basavanhally, and A. F. J. Levi. 1993. High performance optical data link array technology. *Proc. 43rd Electron. Compon. Tech. Conf.,* 795.

78. Shieh, C. L., M. S. Lebby, and J. Lungo. 1995. Method of making a VCSEL. U.S. Patent No. 5,468,656.

79. Scherer,,A., H. G. Craighead, M. L. Roukes, and J. P. Harbison. 1988. Electrical damage induced by ion beam etching of GaAs. *J. Vacuum Sci. Tech.* B6:227.

80. Lebby, M. S., C. L. Shieh, and H. C. Lee. 1994. High efficiency VCSEL and method of fabrication. U.S. Patent No. 5,359,618.

81. Huffaker, D. L., D. G. Deppe, and K. Kumar. 1994. Native-oxide ring contact for low threshold vertical-cavity lasers. *Appl. Phys. Lett.* 65:97.

82. Huffaker, D. L., D. G. Deppe, and T. J. Rogers. 1994. Transverse mode behavior in native-oxide-defined low threshold vertical-cavity lasers. *Appl. Phys. Lett.* 65:1611.

83. Lear, K. L., K. D. Choquette, R. P. Schneider, and S. P. Kilcoyne. 1995. Modal analysis of a small surface emitting laser with a selectively oxidized waveguide. *Appl. Phys. Lett.* 66:2616.

84. Dallesasse, J. M., N. Holonyak, A. R. Sugg, T. A. Richard, and N. Elzein. 1990. Hydrolyzation oxidation of $Al_xGa_{1-x}As$-AlAs-GaAs quantum well heterostructures and superlattices. *Appl. Phys. Lett.* 57:2844.

85. Sugg, A. R., E. I. Chen, T. A. Richard, N. Holonyak, and K. C. Hsieh. 1993. Native oxide-embedded $Al_yGa_{1-y}As$-GaAs-$In_xGa_{1-x}As$ quantum well heterostructure lasers. *Appl. Phys. Lett.* 62:1259.

86. Sugg, A. R., E. I. Chen, T. A. Richard, N. Holonyak, and K. C. Hsieh. 1993. Photopumped room-temperature continuous operation of native-oxide-$Al_yGa_{1-y}As$-GaAs-$In_xGa_{1-x}As$ quantum-well-heterostructure lasers. *J. Appl. Phys.* 74:797.

87. Maranowski, S. A., A. R. Sugg, E. I. Chen, and N. Holonyak. 1993. Native oxide top-confined and botton-confined narrow stripe p-n $Al_yGa_{1-y}As$-GaAs-$In_xGa_{1-x}As$ quantum well heterostructure laser. *Appl. Phys. Lett.* 63:1660.

88. Choquette, K. D., R. P. Schneider, K. L. Lear, and K. M. Geib. 1994. Low threshold voltage vertical-cavity lasers fabricated by selective oxidation. *Electron. Lett.* 30:2043.

89. Huffaker, D. L., J. Shin, and D. G. Deppe. 1994. Low threshold halfwave vertical-cavity lasers. *Electron. Lett.* 30:1946.

90. Yang, G. M., M. H. MacDougal, and P. D. Dapkus. 1995. Ultralow threshold current vertical-cavity surface-emitting lasers obtained with selective oxidation. *Electron. Lett.* 31:886.

91. Weigl, B., M. Grabherr, R. Michalzik, G. Reiner, and K. J. Ebeling. 1996. High-power single-mode selectivity oxidized vertical-cavity surface-emitting lasers. *IEEE Photon. Tech. Lett.* 8:971.

92. M. H. MacDougal, P. Daniel Dapkus, V. Pudikov, H. M. Zhao, and G. M. Yang. 1995. Ultralow threshold current vertical-cavity surface-emitting lasers with AlAs oxide-GaAs distributed Bragg reflectors. *IEEE Photon. Tech. Lett.* 7:229.

93. M. H. MacDougal, G. M. Yang, A. E. Bond, C. K. Lin, D. Tishinin, and P. D. Dapkus. 1996. Electrically-pumped vertical-cavity lasers with Al_xO_y-GaAs reflectors. *IEEE Photon. Tech. Lett.* 8:310.

94. G. G. Hadley, K. D. Choquette, and K. L. Lear. 1996. Understanding waveguiding in vertical-cavity surface-emitting lasers. *CLEO Tech. Digest Ser.* 9:425.

95. Y. A. Wu, C. J. Chang-Hasnain, and R. Nabiev. 1993. Single mode emission from a passive-antiguide-region vertical-cavity surface-emitting laser. *Electron. Lett.* 29:1861.

96. Y. A. Wu, G. S. Li, R. F. Nabiev, K. D. Choquette, C. Caneau, and C. J. Chang-Hasnain. 1995. Single-mode, passive antiguide vertical cavity surface emitting laser. *IEEE J. Select. Top. Quantum Electron.* 1:629.

97. K. L. Lear, R. P. Schneider, K. D. Choquette, S. P. Kilcoyne, J. J. Figiel, and J. C. Zolper. 1994. Vertical cavity surface emitting lasers with 21% efficiency by metalorganic vapor phase epitaxy. *IEEE Photon. Tech. Lett.* 6:1053.

98. T. Yoshikawa, H. Kosaka, K. Kurihara, M. Kajita, Y. Sugimoto, and K. Kasahara. 1995. Complete polarization control of 8 × 8 vertical-cavity surface-emitting laser matrix arrays. *Appl. Phys. Lett.* 66:908.

99. K. D. Choquette and R. E. Leibenguth. 1994. Control of vertical-cavity laser polarization with anisotropic transverse cavity geometries. *IEEE Photon. Tech. Lett.* 6:40.

100. M. Shimizu, T. Mukaihara, T. Baba, F. Koyama, and K. Iga. 1991. A method of polarization stabilization in surface emitting lasers. *Jpn. J. Appl. Phys.* 30:L1015.

101. T. Mukaihara, F. Koyama, and K. Iga. 1992. Polarization control of surface emitting lasers by anisotropic biaxial strain. *Jpn. J. Appl. Phys.* 31:1389.

102. Ser, J. H., Y. G. Ju, J. H. Shin, and Y. H. Lee. 1995. Polarization stabilization of vertical-cavity top-surface-emitting lasers by inscription of fine metal-interlaced gratings. *Appl. Phys. Lett.* 66:2769.

103. Baba, T., Y. Yogo, K. Suzuki, F. Koyama, and K. Iga. 1993. Near room temperature continuous wave lasing characteristics of GaInAsP/InP surface emitting laser. *Electron. Lett.* 29:913.

104. Dudley, J. J., M. Ishikawa, B. I. Miller, D. I. Babic, R. Mirin, W. B. Jiang, J. E. Bowers, and E. L. Hu. 1992. 144°C operation of 1.3 μm InGaAsP vertical cavity laesrs on GaAs substrates. *Appl. Phys. Lett.* 61:3095.

105. Ram, R. J., L. Yang, K. Nauka, Y. M. Houng, M. Ludowise, D. E. Mars, J. J. Dudley, and S. Y. Wang. 1993. Analysis of water fusing for 1.3 μm vertical cavity surface emitting lasers. *Appl. Phys. Lett.* 62:2474.

106. Dudley, J. J., D. I. Babic, R. Mirin, L. Yang, B. I. Miller, R. J. Ram, T. Reynolds, E. L. Hu, and J. E. Bowers. 1994. Low threshold, wafer fused long wavelength vertical cavity lasers. *Appl. Phys. Lett.* 64:1463.

107. Babic, D. I., K. Streubel, R. P. Mirin, N. M. Margalit, J. E. Bowers, E. L. Hu, D. E. Mars, L. Yang, and K. Carey. 1995. Room-temperature continuous-wave operation of 1.54-μm vertical-cavity lasers. *IEEE Photon. Tech. Lett.* 7:1225.

108. Margalit, N. M., D. I. Babic, K. Streubel, R. P. Mirin, R. L. Naone, J. E. Bowers, and E. L. Hu. 1996. Submilliamp long wavelength vertical cavity lasers. *Electron. Lett.* 32:1675.

109. Ettenberg, M., H. F. Lockwood, and H. S. Sommers, Jr. 1972. *J. Appl. Phys.* 43:5047.

110. Imai, H., M. Morimoto, H. Sudo, T. Fujiwara, and M. Takusagawa. 1978. Catastrophic degradation of GaAlAs DH laser diodes. *Appl. Phys. Lett.* 33:1011.

111. Gfeller, F. R., and D. J. Webb. 1990. Degradation and lifetime studies of high-power single-quantum-well AlGaAs ridge lasers. *J. Appl. Phys.* 68:14.

112. Imafuji, O., T. Takayama, H. Sugiura, M. Yuri, H. Naito, M. Kume, and K. Itoh. 1993. 600 mW CW single-mode GaAlAs triple-quantum well laser with a new index guided structure. *IEEE J. Quantum Electron.* 29:1889.

113. Lau, K. Y., P. L. Derry, and A. Yariv. 1988. Ultimate limit in low threshold quantum well GaAlAs semiconductor lasers. *Appl. Phys. Lett.* 52:88.
114. Orenstein, M., A. C. Von Lehmen, C. Chang-Hasnain, N. G. Stoffel, J. P. Harbison, and L. T. Florez. 1991. Matrix addressable vertical cavity surface emitting laser array. *Electron. Lett.* 27:437.
115. Vakhshoori, D., J. D. Wynn, G. J. Zydik, and R. E. Leibenguth. 1993. 8 × 18 top emitting independently addressable surface emitting laser arrays with uniform threshold current and low threshold voltage. *Appl. Phys. Lett.* 62:1718.
116. Uchiyama, S. and K. Iga. 1985. Two-dimensional array of GaInAsP/InP surface-emitting lasers. *Electron. Lett.* 21:162.
117. Von Lehmen, A., C. Chang-Hasnain, J. Wullert, L. Carrion, N. Stoffel, L. T. Florez, and J. Harbison. 1991. Independently addressable InGaAs/GaAs vertical cavity surface emitting laser arrays. *Electron. Lett.* 27:583.
118. Khoe, G. K. and G. Kuyt. 1978. Realistic efficiency of coupling light from GaAs laser diode into parabolic-index optical fibers. *Electron. Lett.* 14:667.
119. Reith, L. A., P. W. Shumate, and Y. Koga. 1986. Laser coupling to single-mode fibre using graded-index lenses and compact disk 1.3 μm laser package. *Electron. Lett.* 22:836.
120. Mathyssek, K., J. Wittmann, and R. Keil. 1985. Fabrication and investigation of drawn fiber tapers with spherical microlenses. *J. Opt. Commun.* 6:142.
121. Cheng, W. H., and J. H. Bechtel. 1993. High-speed fibre optic links using 780 nm compact disc lasers. *Electron. Lett.* 29:2055.
122. Soderstrom, R. L., S. J. Baumgartner, B. L. Beukema, T. R. Block, and D. L. Karst. 1993 (June). CD lasers optical data links for workstations and midrange computers. ECTC'93, 505, Orlando, FL.
123. Armiento, C. A., M. Tabasky, C. Jagannath, T. W. Fitzgerald, C. L. Shieh, V. Barry, M. Rothman, A. Negri, P. O. Haugsjaa, and H. F. Lockwood. 1991. Passive coupling of InGaAsP/InP laser array and singlemode fibres using silicon waferboard. *Electron. Lett.* 27:1109.
124. Tan, Q. and Y. C. Lee. 1996. Soldering technology for optoelectronic packaging. *Proc. 46th Electron. Compon. Tech. Conf.*, 26.
125. Zeeb, E., B. Möller, G. Reiner, M. Ries, T. Hackbarth, and K. J. Ebeling. 1995. Planar proton implanted VCSEL's and fiber-coupled 2-D VCSEL arrays. *IEEE J. Select. Top. Quantum Electron.* 1:616.
126. Wong, Y. M., D. J. Muehlner, C. C. Faudskar, D. B. Buchholz, M. Fishteyn, J. L. Brandner, W. J. Parzygnat, R. A. Morgan, T. Mullally, R. E. Leibenguth, G. D. Guth, M. W. Focht, K. G. Glogovsky, J. L. Zilko, J. V. Gates, P. J. Anthony, B. H. Tyrone, Jr., T. J. Ireland, D. H. Lewis, Jr., D. F. Smith, S. F. Nati, D. K. Lewis, D. L. Rogers, H. A. Aispain, S. M. Gowda, S. G. Walker, Y. H. Kwark, R. J. S. Bates, D. M. Kuchta, and J. D. Crow. 1995. Technology development of a high-density 32-channel 16-Gb/s optical data link for optical interconnection applications for the optoelectronic technology consortium (OETC). *J. Lightwave Tech.* 13:995.

127. Kosaka, H., A. K. Dutta, K. Kurihara, Y. Sugimoto, and K. Kasahara. 1995. Gigabit-rate optical-signal transmission using vertical-cavity surface-emitting lasers with large-core plastic-cladding fibers. *IEEE Photon. Tech. Lett.* 7:926.

128. Matsuda, K., T. Yoshida, Y. Kobayashi, and T. Chino. 1996. A surface-emitting laser array with backside guiding holes for passive alignment to parallel optical fibers. *IEEE Photon. Tech. Lett.* 8:494.

129. Hahn, K. H. 1995 (May). POLO — Parallel optical links for gigabyte data communication. *Proc. 45th Electron. Compon. Tech. Conf.*, 368, Las Vegas, NV.

130. Rowlette, J. R., Jr., M. K. Kallen, J. Stack, and W. Lewis. 1996. Laser micromachining of polymer waveguides for low cost passive alignment to VCSELs. *Proc. LEOS '96* 2:169.

131. Ou, S. S., M. Jansen, J. J. Yang, J. J. Mawst, and T. J. Roth. 1991. High power cw operation of InGaAs/GaAs surface-emitting lasers with 45° intracavity micromirrors. *Appl. Phys. Lett.* 59:2085.

132. Fallahi, M., M. Dion, F. Chatenoud, I. M. Templeton, and R. Barber. 1993. High power emission from strained DQW circular-grating surface-emitting DBR lasers. *Electron. Lett.* 29:2117.

133. Fallahi, M., F. Chatenoud, M. Dion, I. Templeton, R. Barber, and J. Thompson. 1995. Circular-grating surface-emitting distributed Bragg reflector lasers on an InGaAs-GaAs structure for 0.98-μm applications. *IEEE J. Select. Top. Quantum Electron.* 1:382.

134. Zah, C. E., M. R. Amersfoort, B. Pathak, F. Favire, P. S. D. Lin, A. Rajhel, N. C. Andreadakis, R. Bhat, C. Caneau, and M. A. Koza. 1996. *IEEE Photon. Tech. Lett.* 8:864.

135. Guenter, J. K., R. A. Hawthorne, III, D. N. Granville, M. K. Hibbs-Brenner, and R. A. Morgan. 1996. Reliability of proton-implanted VCSELs for data communications. *Proc. SPIE* 2683:102.

136. Ladany, I., Ettenberg, M., Lockwood, H. F., and Kressel, H. 1977. Al_2O_3 half-wave films for long-life cw lasers. *Appl. Phys. Lett.* 30:87.

Chapter 16 | Receiver, Laser Driver, and Phase-Locked Loop Design Issues

Dave Siljenberg

IBM Corporation, Rochester, Minnesota 55901

16.1. Introduction

This chapter introduces receiver, laser driver, and phase-locked loop (PLL) designs used in high-speed fiber optic communications. The goal is not to explain the details of their design but to explore the real-world environment within which these designs must perform. I will also outline the strategies the designer must use to build a robust design.

16.2. Fiber Optic Receivers

A. DESCRIPTION

Fiber optic receivers must convert the optical signal into an electrical signal and amplify it enough to be treated as a digital signal. Although there are other topologies, I will use the transimpedance amplifier configuration to explain various design trade-offs. The transimpedance receiver has good sensitivity and dynamic range, making it one of the most popular configurations.

B. TRANSIMPEDANCE RECEIVER

A block diagram of a typical receiver is shown in Fig. 16.1. The transimpedance amplifier consists of an inverting amplifier of gain, $-A$, with a feedback resistor, R. The input impedance, Zin, is R/A. The input impedance must be kept low to maximize bandwidth. The high-frequency photodiode currents will either go into the amplifier or the parasitic capacitance of the

587

HANDBOOK OF FIBER OPTIC
DATA COMMUNICATION

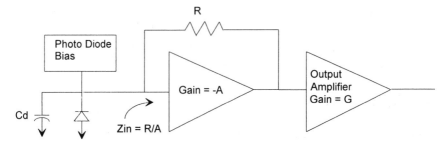

Fig. 16.1   Fiber optic preamplifier block diagram.

photodiode and wiring. Lower Zin increases the current into the amplifier. This says that R should be kept low and A kept high. There is also an output amplifier shown with a gain of G. It is used to provide additional gain and drive signals off chip at low impedance.

C.   THERMAL NOISE

The dominant thermal noise comes from the feedback resistor. It generates a noise current i_n:

$$i_n = \frac{4kT}{R} \, \text{BW}, \tag{16.1}$$

where k is Boltzmann's constant, T is temperature in degrees Kelvin, R is the feedback resistance, and BW is the noise bandwidth. This equation indicates that noise is reduced when the resistor is increased. This contradicts the previous bandwidth discussion. In addition, an increased gain, A, reduces dynamic range by reducing the maximum input. These trade-offs are summarized in Table 16.1.

Table 16.1   **Receiver Trade-offs**

Parameter to be increased	R	A
Bandwidth	Decrease	Increase
Sensitivity	Increase	Neutral
Maximum input	Decrease	Decrease

In general, the designer must find the trade-off that achieves the design objective. Moui [1] presents a good discussion of receiver design including open-loop and transimpedance amplifiers. Meyer *et al.* [2] present a concise description of transimpedance design.

D. PACKAGING

Because the photodiode lead has the smallest signals, it must be packaged with the most care. This is done to reduce parasitic capacitance and coupled noise. The packaging can fit into three categories: discrete, hybrid, and integrated.

In the discrete package, the photodiode and transimpedance amplifier are in separate packages and mounted on a common board or substrate. This is the lowest cost approach. However, it has the largest parasitics and coupled noise susceptibility. Generally, some shielding is needed to yield acceptable noise susceptibility.

The hybrid mounts the photodiode and transimpedance amplifier in die form on a common substrate. This improves parasitics and coupled noise immunity at increased cost.

The ultimate packaging is integrated. With an integrated circuit (IC) process that is compatible with a photodetector and amplifier design, one can achieve the best parasitics and coupled noise suseptability at the low cost of one IC. An example is a GaAs IC reported by Rogers [3]. The photodetector is responsive to 780-nm wavelengths and the transimpedance amplifier is implemented in metal semiconductor field-effect transistors.

E. NOISE TRADE-OFFS

Although analysis of the thermal noise contribution is important, it is quite common that coupled noise sources will dominate in the circuit's real environment. Three sources of coupled noise will be discussed.

1. Power Supply Rejection

For optical links in a data communications environment it is assumed that the receivers are imbedded in large digital systems and computers. The power supplies for those systems have noise in excess of 100 mV due to logic switching. Because digital clock rates are moving beyond 200 MHz, the noise can extend to 1 GHz. Care must be taken in the receiver design to suppress this noise and to prevent it from becoming larger than the thermal noise contribution.

An example calculation illustrates the point. If the design is intended to receive light at −20 dBm, the photodiode will have 10 μW average power or 20 μW peak-to-peak power impinging on it. A common photodiode efficiency is 0.5 mA/mW, in which case the photodiode current is 10 μA peak to peak. If the transimpedance gain and feedback resistor are −20 and 2 $K\Omega$, respectively, and the output stage has a gain of 2, we can calculate the effect of a typical power supply rejection ratio of −30 dB. The transimpedance stage will convert the 10-μA current to 20 mV peak to peak. The output stage will produce 40 mV peak to peak. One hundred millivolts of power supply noise will produce 3.2 mV of noise at the output. This is acceptable because the signal-to-noise ratio is greater than 10. Note, however, that the PSRR can be no worse than −30 dB over the entire bandwidth. This becomes more difficult at high frequencies.

2. Coupled Noise

Another often overlooked noise source is on the fiber optic transceiver itself. Most transceivers have large logic swings in close proximity to the photodiode lead. Even worse is the laser drive signal. As will be discussed, the laser signal can be as large as 25 mA peak to peak with 300 ps rise times. The typical photodiode currents at −20 dBm can be 10 μA. The laser is often physically adjacent to the photodiode. It is very easy for the laser driver current to induce signals in the receiver. The discrete package option is most susceptible, whereas the integrated is the least susceptible.

3. Electrostatic Discharge and Electromagnetic Interference Rejection

The receiver must also be immune to ESD and EMI. There are few modeling tools that the designer can use to predict ESD or EMI susceptibility. Therefore, a designer must test the design to ensure a rugged design. Again, the discrete package option is most susceptible, whereas the integrated is the least susceptible.

16.3. Laser Drivers

This section covers the critical issues involved in modulating a serial bit stream on semiconductor lasers.

A. *DESCRIPTION*

The lasers described all have the same optical power vs current characteristic as shown in Fig. 16.2. This includes longwave and shortwave lasers. The laser has a characteristic threshold current, the current at which lasing begins, and the differential quantum efficiency, the slope of the optical power vs current curve. The laser driver must convert the bit stream into the correct current levels. This means driving the correct DC bias current and AC modulation current to produce an undistorted optical output pulse as shown in the right side of Fig. 16.2.

1. Laser Topologies

Lasers come in several topologies as shown in Fig. 16.3. If the case is to be grounded, the first laser requires current source from a positive supply. The second laser requires a current sink from a negative supply. The third topology offers more latitude because the case can be grounded regardless of the driver topology. These choices will be the first step in defining a laser driver topology. All these topologies are shown with monitor diodes. These diodes are separate devices mounted in the package with the laser. They receive light from the back facet of the laser. The current from them can be used in a feedback loop to control the DC optical power. Newer devices, called vertical cavity surface-emitting lasers (VCSELs), are available. They have no back facet and therefore no monitor diode. They need to be biased with open-loop current sources.

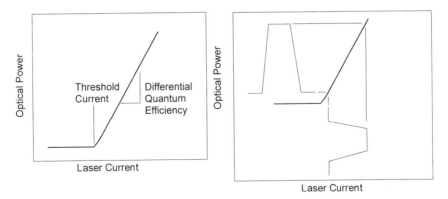

Fig. 16.2   Laser diode power vs current.

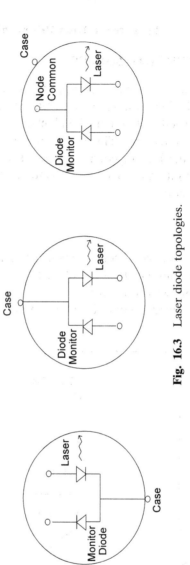

Fig. 16.3 Laser diode topologies.

B. LASER DRIVER TOPOLOGIES

An open collector laser driver is shown in Fig. 16.4. This topology makes use of high-speed NPN transistors found in many technologies. If the laser and driver are packaged with too much inductance, L1, fast rise times will cause ringing in the waveform. This is due to the high impedance of the NPN collector and the low impedance of the laser. The laser impedance can be as low as 4Ω for shortwave lasers found in CD players and as high as 50Ω for VCSEL lasers.

1. DC Bias

A basic closed-loop DC bias circuit is shown in Fig. 16.5. Current from the monitor diode flows through R_1, generating a voltage at the input of the op amp. The op amp and transistor Q_1 control the laser current so that the voltage on the negative input of the op amp equals Vref.

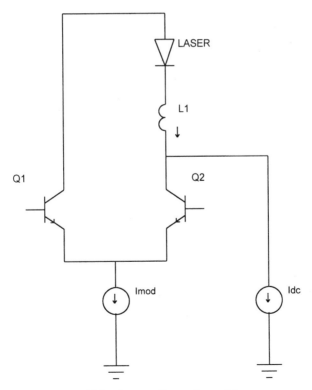

Fig. 16.4 Open collector laser driver.

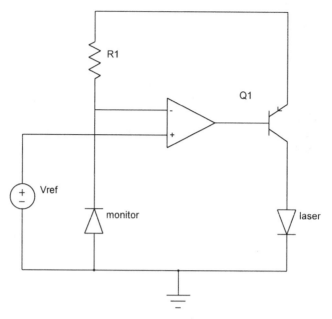

Fig. 16.5   DC laser bias circuit.

2. AC Drive

Figure 16.6 shows a 50Ω driver that reduces ringing mentioned in the first driver. It also shows the driver AC-coupled to the laser. This is needed for a common cathode laser and a positive power supply. If the laser impedance is too low a resistor may needed in series with C_1. Figure 16.7 shows an emitter follower based design. This design can provide more drive current but suffers from emitter follower ringing or oscillation that is impedance sensitive.

3. Packaging

Similar to the receiver, the packaging can fit into two categories: discrete and hybrid. Integrated is not widely used because the semiconductor requirements for lasers are incompatible with transistors.

   In the discrete package, the laser and laser driver are in separate packages and mounted on a common board or substrate. This is the lowest cost approach. However, it has the largest parasitics and electromagnetic radiation.

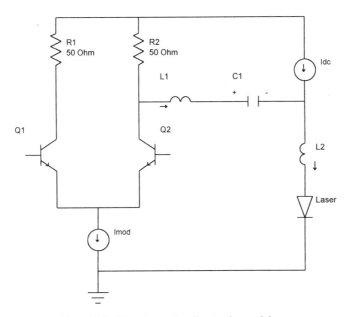

Fig. 16.6   Terminated collector laser driver.

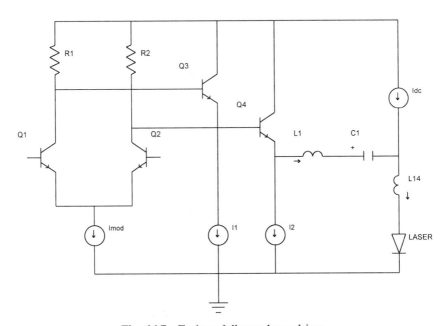

Fig. 16.7   Emitter follower laser driver.

The hybrid mounts the laser and laser driver in die form on a common substrate. This improves parasitics and electromagnetic radiation at increased cost.

4. Electromagnetic Interference

Despite the fact that fiber optic transmission inherently produces no electromagnetic radiation, laser drivers do. This is compounded by the high speeds generated by serializing a parallel data stream. For example, if there is an 8-bit parallel data bus at 100 MHz it will have rise times on the order of 2.5 ns. If the bus is transported a distance on parallel cable significant shielding will be required to bring the radiation under acceptable limits. If this bus is serialized the rise times will become eight times shorter. If a code is used the edges will be faster. This means the laser driver will be producing 300 ps edges at 10–25 mA for a typical laser. The fiber will not radiate but the wiring from the laser driver to the laser must minimize loop area and may need shielding. Figure 16.8 shows a representation of a 1 Gb/s laser driver that is in production. The laser (in a TO style package) is within 25 mm of the laser driver IC. Despite this layout the assembly needs full metal enclosure to meet class B limits.

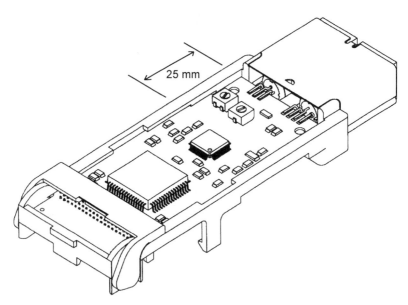

Fig. 16.8   1 Gb/s optical link card.

5. Termination and Layout Issues

Care must be used in laying out the connection between the laser driver and the laser. Parasitic capacitance and inductance of the laser and driver package must be minimized. If the path delay is a significant fraction of the rise time, transmission line techniques must be used. This may be difficult because many lasers have only a few ohms impedance. Laser drive return currents must be carefully returned to the driver. Failure to do these things will result in signal degradation and larger EMI radiation.

16.4. Clock Recovery and PLLs

A. DESCRIPTION

Two timing functions are required in a serial data link: a transmit synthesizer and a clock recovery circuit. Because these functions run at bit rates that present very demanding design challenges, the transmit synthesizer (illustrated in Fig. 16.9) must generate a bit rate clock from the byte rate or reference clock. The most common approach is to use a PLL or a delay-locked loop (DLL).

There are many good sources for PLL design [4, 5]. The particular PLL shown is based on a charge pump and loop filter [6]. This design is shown because of its popularity and ease of integration. As shown Refs. [4]–[6], the PLL can be modeled with linear small signal equations. When this is done the PLL has a simplified closed-loop transfer function of the form:

$$H(s) = K_v \frac{I_p}{N} \frac{sRC_1 + 1}{s^2 2\pi C_1 + sK_v \frac{I_p}{N} RC_1 + K_v \frac{I_p}{N}}, \quad (16.2)$$

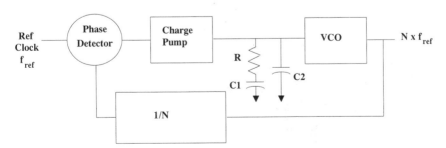

Fig. 16.9 Synthesizer block diagram.

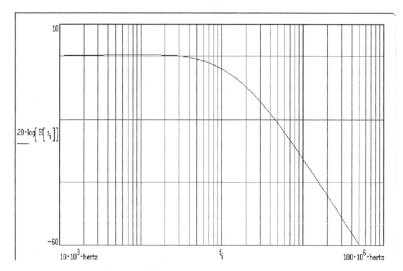

Fig. 16.10 PLL frequency response.

where K_v is the voltage-controlled oscillator (VCO) gain in Hz/V, I_p is the peak current out of the charge pump, R and C_1 are the loop filter components, and N is the divide ratio or number of VCO cycles per reference clock cycle. This function describes the low-pass transfer function of the PLL. This means that phase shifts or jitter that are slower than the loop bandwidth will be accurately transferred by the PLL. Phase shifts or jitter that are faster than the loop bandwidth will be filtered out.

It needs to be pointed out that this is a simplified transfer function. The capacitor C_2 filters out the charge pump pulses and accounts for parasitic capacitance. C_2 makes the exact transfer function third order. However, a second-order function is a good approximation because C_2 must be kept small in order to minimize excess phase shift that can harm stability. This transfer function is plotted in Fig. 16.10, where K_v is 160 MHz/V, I_p is 600 μA, R is 430 Ω, C_1 is 10 nF, and N is 10.

1.   Reference Clock Jitter

Because the reference comes from a clock on a large digital system, it will contain jitter from many digital sources in the system. The reference clock jitter has many frequency components due to system events happening at different rates. DRAM's get refreshed at a slow rate. System clocks generate noise in excess of 100 MHz. These noise sources get coupled to the reference clock through the power supply, adjacent wires, and ground bounce on logic ICs.

The effect of this jitter on the transmit synthesizer can be illustrated by examining reference clock phase shift occurring at low and high frequencies. If the reference clock shifts at low frequencies relative to the PLL bandwidth, the synthesizer will track it and propagate the jitter to the serial bit stream. If the reference clock shifts at high frequencies relative to the PLL bandwidth, the synthesizer will not be able to keep up and will filter the jitter out. At intermediate frequencies the synthesizer PLL will deliver a fraction of the jitter to the serial bit stream. The total jitter in a reference clock is a mixture of frequencies. Therefore, the synthesizer PLL will reduce but not eliminate the jitter in the reference clock. In order to predict how much jitter is propagated the designer must know the PLL bandwidth and estimate the spectral content of the reference clock jitter.

From the previous discussion, one might conclude that the synthesizer PLL bandwidth must be kept as low as possible. In the next section I will discuss why that is not always a good idea.

2. Power Supply Rejection

All practical VCOs will vary in frequency due to noise on the power supply. This can be modeled as a power supply coefficient K_v. We can illustrate the jitter this induces by again examining noise at low and high frequencies. If the supply voltage varies at a slow rate relative to the PLL bandwidth, any phase shift the VCO makes will be compared to the reference clock by the phase detector. This will generate an equal and opposite control voltage change canceling out the jitter. If the supply voltage varies at a high rate relative to the VCO frequency, not the PLL bandwidth, the VCO does not get time to phase shift before the next VCO cycle. At intermediate frequencies the PLL will generate jitter in response to the power supply noise. In contrast to the previous discussion, this indicates that the PLL bandwidth should be as large as possible to cancel out the power supply-induced jitter. In practice the bandwidth will be a compromise between these two extremes.

The power supply rejection can be modeled using the following equation:

$$P(s) = \frac{K_p}{s + \dfrac{K_v}{2\pi} \dfrac{I_p}{N} Z(s)}, \tag{16.3}$$

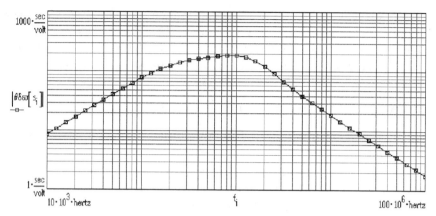

Fig. 16.11   PLL power supply rejection.

where K_p is the VCO power supply coefficient in Hz/V, K_v is the VCO gain in Hz/V, I_p is the peak current out of the charge pump, $Z(s)$ is the impedance of the loop filter, and N is the divide ratio or number of VCO cycles per reference clock cycle. This equation is plotted in Fig. 16.11. The VCO gain is 160 MHz/V. The VCO supply coefficient is 11 MHz/V, which is 5%/V of the 220 MHz VCO center frequency. N is 10, I_p is 600 μA, and the loop filter values are as follows: $R = 430\ \Omega$, $C_I = 10\ \text{nF}$, and $C_2 = 100\ \text{pF}$.

Note that the curve shows power supply-induced jitter of 200%/V at 900 kHz. This indicates that if the system has 0.1 V of power supply noise at 900 kHz, the PLL will induce 20% jitter in the serial bit stream! This is usually not acceptable. The designer must consider power supply filtering or designing a VCO with less than 5%/V power supply coefficient. This illustrates the design trade-off described previously. In this example, the designer must understand the reference clock jitter and power supply noise and design an optimal trade-off.

3.   Clock Recovery

The analysis of the transmit synthesizer reveals some basic PLL design issues. The design of the clock recovery circuit (CRC) is a more difficult task. It must derive a bit rate clock based on the incoming data stream edges. The CRC must contend with an absence of transitions due to long strings of 1's and 0's. The CRC must also tolerate jitter in the incoming bit stream caused by noise and intersymbol interference. The following sections will discuss the trade-offs between various clock recovery methods and present some examples of CRC designs. CRC designs can be separated into analog and digital architectures. The analog designs include PLLs and DLLs. The digital designs consist of oversampled architectures.

4. Analog PLL Clock Recovery

The basic analog PLL clock/data recovery block diagram is shown in Fig. 16.12.

The PLL shown in Fig. 16.12 can be analyzed with the same linearized equations discussed for the synthesizer. This PLL will generate jitter due to power supply noise, as discussed previously. It will also respond to incoming jitter in a similar fashion as the synthesizer. The incoming jitter for the synthesizer was in the reference clock. The incoming jitter to the CRC comes from the serial data stream. The data stream has jitter from many sources. Major sources of generated jitter include the synthesizer, laser noise, fiber dispersion, and receiver noise.

The CRC PLL has a variable loop gain. In the synthesizer the divider ration N was fixed. That ratio determines the number of VCO cycles between every correction from the phase detector. The larger N is, the more infrequent are the phase detector correction pulses, lowering the loop gain. In the CRC the phase detector cannot generate a correction pulse without a data transition. In the CRC N is data dependent. Therefore, loop gain is data dependent. For 101010 · · · NRZ data the gain is high and for long strings of 1's and 0's the gain is low. The CRC PLL must be designed for both extremes of gain.

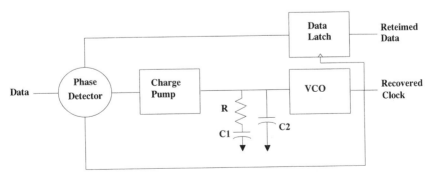

Fig. 16.12   PLL-based clock/data recovery.

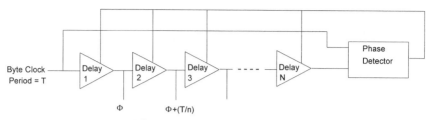

Fig. 16.13   Delay-locked loop.

5.   DLLs

A typical DLL is shown in Fig. 16.13. By making phase-shifted copies of the byte rate clock we can generate bit rate timings. If we compare this to VCO-based PLLs we see that none of the DLL circuits run at the bit rate. Therefore, a slower technology can be used. However, the DLL needs more circuits. This makes it attractive to complementary metal oxide semiconductor developers.

Several advantages of the DLL over the PLL are reported [7]. These include (i) delay is not accumulated [8], (ii) the DLL is described by a single linear equation, and (iii) the loop locks quickly. The DLL lends itself to synthesizers but is difficult to use for clock recovery because an incoming data edge can be locked to any one of the phases.

6.   VCOs

The multivibrator VCO is shown in Fig. 16.14. It has been used in a large number of designs. J. T. Wu [9] reports a VCO with a large tuning range. Its major drawback is power supply rejection. Although the design looks differential its response to power supply noise is single ended. This can be understood by understanding the switch point. If Q_1 is off, current source I_1 discharges the capacitor pulling the emitter of Q_1 low. When the emitter of Q_1 is low enough Q_1 turns on and Q_2 turns off. Any power supply noise at that time will move the base voltage of Q_1, which moves the turn-on time of Q_1.

Another major topology is the ring oscillator. It is more compatible with digital IC processing because it does not need capacitors or analog current sources [10]. It can also be made fully differential for power supply rejection [11]. The ring's main limitation is limited tuning range and difficult power supply and temperature compensation.

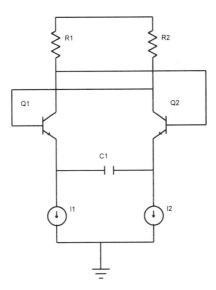

Fig. 16.14 Multivibrator VCO.

7. Phase Detectors

A two-latch synthesizer phase detector is described in Ref [12]. This design provides phase and frequency detection because for every VCO transition there is a reference transition.

The phase detector for the CRC is complicated by the absence of transitions in the serial data. The phase detector described by Deremer *et al.* [13] produces an increase and decrease pulse for every data edge. These pulses are averaged by the loop filter. A phase detector gain can be calculated for linear modeling. The design is shown in Fig. 16.15.

A single-latch phase detector is shown in Fig. 16.16. If the data rise before the rising clock edge, the clock is late. A zero is latched and Q bar is high, charging up the loop filter. If the data rise after the rising clock edge, the loop filter is discharged. Unlike the two-latch design, this method is always correcting even if there are no data transitions. The CRC phase is constantly alternating above and below phase lock. To minimize these phase excursions the loop filter must be larger than the linear design. It is impossible to use a linearized gain factor for analysis. Designers must use transient simulators to optimize this design.

The two-latch linear design provides the best trade-off between lock time and jitter because the loop filter can be smaller. However, skews in

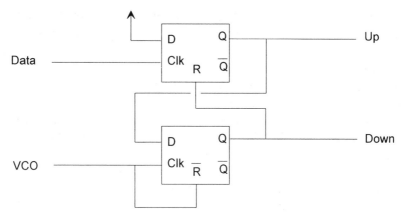

Fig. 16.15   Two-latch data phase detector.

the latches or offsets in the charge pump create a static phase error. This happens because the PLL will settle into lock when the net charge in the loop filter is zero. If there are skews or offsets, there must be a phase difference between the VCO and data in the opposite direction to compensate. Great care must be taken to reduce these skews to an acceptable level.

The single-latch design will lock slower because of the large loop filter needed to keep the phase jitter to acceptable levels. However, offsets cause no phase error because the loop is only satisfied when the clock and data edges are lined up, even with charge pump skew. The logic must only be matched up to the input of the latch. In the linear design all the logic and the charge pump must be balanced to produce equal up and down pulses when the clock is in phase with the data.

8. Digital Data Recovery

Digital data recovery is typified by oversampled techniques described in Refs. [7] and [14]. These techniques involve sampling the incoming data stream with several phases of byte rate clocks. In 3× oversampling the

Fig. 16.16   Single-latch data phase detector.

phases would be spaced at one-third bit intervals. In addition, these clocks run at the byte rate. For example, for 3× oversampling and 8-bit bytes there are 24 sampling clock phases. This generates multiple copies of the data stream. Digital circuits examine the sampled streams to find the data transitions. Once the transitions are determined, the correct data can be found by selecting data from the correct phase clock. For example in the previous example there are clock phases 0–23. If a transition is found between phases 3 and 4 the correct data can be found with phases 5, 8, 11, etc. Note also that these clock phases are synchronous with the local clock and not the remote clock that sent the data stream. Therefore, the decision regarding where to sample is constantly moving because these two clocks are within a crystal tolerance of the same frequency, typically 100 parts per million. Chen [15] describes the trade-off between jitter tolerance (defined below) and oversampling in a digital data recovery scheme.

9. Digital vs Analog Trade-Offs

For a given technology, the analog techniques can produce the highest data rates. However, the analog techniques are more difficult to integrate with adapter logic. Therefore, as technologies advance the digital techniques replace analog at a given data rate.

B. JITTER

Having discussed the jitter mechanisms in analog PLLs we can cover parameters necessary for defining the jitter in a communications system. These concepts are outlined in a Fibre Channel technical report [16].

1. Jitter Generation

Jitter generation is the jitter contained in the bit stream source. The jitter comes from the reference clock jitter and power supply noise as mentioned in earlier sections. Thermal noise and cross-talk in the transmit VCO also adds jitter.

2. Jitter Tolerance

Jitter tolerance is a measure of the amount of jitter the receiver can tolerate with no errors. It includes the jitter generation and jitter added by the media including intersymbol interference.

16.5. Conclusion

In this chapter I have summarized some of the most popular serial link circuits. Design trade-offs and real-life application concerns have been highlighted. As computer systems become faster and more widespread, the demand for high-speed serial optical connection will grow. The speed, power, and price of those connections will improve. This chapter provides designers with some information and techniques to use in those designs.

References

1. Moui, T. V. 1984 (June). Receiver design for high-speed optical-fiber systems. *J. Lightwave Tech.* LT-2(3):243–267.
2. Meyer, R. G. *et al.* 1986 (August). A wide-band low-noise monolithic transimpedance amplifier. *IEEE J. Solid-State Circ.* SC-21(4):530–533.
3. Rogers, D. L. 1986 (November). Monolithic integration of a 3 Ghz MESFET detector/preamplifier using a refactory-gate ion-implanted MESFET process. *IEEE Electron. Dev. Lett.* EDL-7(11):600–602.
4. Gardner, F. M. 1979. *Phase lock techniques,* 2nd ed.
5. Wolaver, D. H. 1991. *Phase-locked loop circuit design.* New York: Prentice Hall.
6. Gardner, F. M. 1980 (November). Charge-pump phase-lock loops. *IEEE Trans. Commun.* COM-28(11):321–332.
7. Marbot, R. *et al.* 1993. Integration of multiple point to point serial links in the gigabits per second range. Hot Links Symposium.
8. Lee, T. H., K. S. Donnelly, J. T. C. Ho, J. Zerbe, M. G. Johnson, and T. Ishikawa. 1994 (December). A 2.5 V CMOS delay-locked loop for a 10 Mbit, 500 megabyte/s DRAM. *IEEE J. Solid-State Circ.* 29(12):1491–1496.
9. Wu, J. T. 1990 (February). A bipolar 1 Ghz multi-decade monolithic variable-frequency oscillator. *ISSCC Dig. Tech. Papers,* 106–107.
10. Razavi, B. 1996 (March). A study of phase noise in CMOS oscillators. *IEEE J. Solid-State Circ.* 31(3): 331–343.
11. Walker, R. *et al.* 1992 (February). A 2-chip 1.5 Gb/s bus-oriented serial link interface. *ISSCC Dig. Tech. Papers,* 226–227.
12. Soyuer, M., and R. G. Meyer. 1990 (August). Frequency limitations of a conventional phase-frequency detector. *IEEE J. Solid-State Circ.* 25(4):1019–1022.
13. Deremer, R. L., L. W. Freitag, and D. W. Siljenberg. 1990 (September). High speed phase detector. *IBM Tech. Disclosure Bull.* 33(4):259–260.
14. Yang, C. K., and M. A. Horowitz. 1996. A 0.8 um CMOS 2.5 GB/s over sampled receiver for serial links. IEEE 1996 ISSCC Proceedings, 200–201.

15. Chen, D.-L. 1996 (August). A power and area efficient CMOS clock/data recovery circuit for high-speed serial interfaces. *IEEE J. Solid-State Circ.* 31(8):1170–1176.

16. "Fibre Channel—Jitter Working Group," draft proposed X3 Technical Report, X3 Technical Committee of ANSI.

Chapter 17 | Packaging Assembly Techniques

Glenn Raskin

Motorola, Incorporated, Chandler, Arizona 85248

17.1.   Packaging Assembly — Overview

Packaging can be thought of on many levels and serves many different functions. The basic function of packaging is the transfer of energy from one level to another. Whether that energy is electrical, thermal, or optical defines the level or type of packaging function. Traditionally, one considers the first level of packaging as that which deals with interconnection of semiconductor or optoelectronic devices to some kind of leadframe. The second level of packaging concerns itself with interconnection of the afore-mentioned leadframe to a circuit board, on which other packaged devices are also mounted.

A.   REQUIREMENTS FOR OPTOELECTRONIC PACKAGING

The requirements for the packaging drive the technologies that are used and manufacturing techniques that are incorporated. In the area of opto-electronics one is primarily concerned with the transfer of optical and electrical energy from board level to the device. For semiconductor devices the focus is typically limited to the electrical domain. Other areas of concern and interest in packaging focus on thermal transfer, mechanical strength, and electromagnetic and other considerations. The discussions in Sections 2, 3, and 4 apply to optoelectronic as well as microelectronic assemblies.

Before focusing on the specific details, it is useful to look at the overall picture. Table 17.1 attempts to capture the key considerations one must make in packaging assembly techniques. In any given application one must consider the factors in Table 17.1 and design a package that meets all of

608

HANDBOOK OF FIBER OPTIC
DATA COMMUNICATION

Table 17.1   **Considerations for Packaging of Devices**

Electrical performance
Optical performance
Thermal performance
Reliability
Cost
Manufacturability
Testability
Device protection
Size
Weight
Mechanical integrity

the requirements. It may be necessary to make compromises and trade-offs in each of the areas when making packaging decisions. For optoelectronics the requirements that drive device packaging revolve around some of the key characteristics of the system itself. If one is trying to construct a single-mode package the alignment tolerances to a single-mode fiber are much greater than those of a multimode system. These tighter alignments may require certain packaging techniques that limit the user's choice of technologies. Another key input required for optoelectronic packaging is the amount of acceptable attenuation, or coupling loss, that is allowed. If the power budget allocated for the packaging is not great, this makes efficient coupling extremely critical.

Another key consideration in optoelectronic packaging is the thermal and reliability requirements. A laser's performance and expected lifetime are much more sensitive to its operating temperature compared to classical integrated circuits. This strong dependence of lifetime and reliability on temperature has limited the applications of laser devices and driven the development of more exotic thermal packages for optoelectronics.

Finally, one of the most important considerations, and the one typically given the least ink in text, is that of cost and manufacturability. In the end, whatever technology and performance is used the product must be cost-effective to the end user. If the required technology to meet the performance needs of the product is costly, often another avenue must be taken. Because cost is a prime driver in the industry, we often must compromise in other areas of performance to meet the cost goals of the product itself.

17.2.   First-Level Interconnects

First-level interconnects are those that connect the device to the leadframe. There are a variety of interconnect technologies used in manufacturing. In this text I will review the three most common types: wire bonding, tape automated bonding (TAB), and flip chip.

A.   WIRE BOND

The most common of all first-level interconnects in the microelectronic as well as the optoelectronic industry is known as wire bonding. Wire bonding is done using a number of techniques and in a variety of packages. The flowchart in Fig. 17.1 describes a typical process flow for wire bonding of devices.

I will discuss in-depth the wire bond technologies used in the industry and also review die attach and reliability concerns. The three most common wire bonding processes are ultrasonic, thermosonic, and thermocompression.

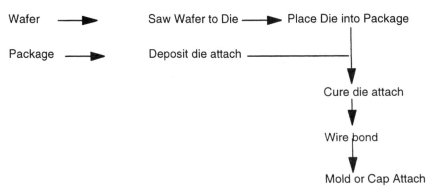

Fig. 17.1 Typical wire bond assembly flowchart.

1. Thermocompression and Thermosonic Wire Bond

Thermocompression bonding was the earliest type of wire bonding, demonstrated by AT&T in 1957 [1]. Thermocompression bonding is typically done using ultrapure (>99%) gold wire. A wire bonder is composed of a stage that holds the package and the bonding head, which feeds and places the wires. The bonding head forces the gold wire into contact with the die or substrate metallurgy that forms a bond through heating. Typical bond head temperatures between 300 and 400°C are required to form the bond.

Thermosonic bonding is very similar to thermocompression bonding except that the wedge bond is performed using ultrasonic excitation to assist in bond formation. This allows for the use of lower temperatures than in thermocompression bonding.

Thermocompression and thermosonic bonding typically form a ball bond when contacting the device and a wedge bond onto the substrate. A schematic of ball and wedge thermosonic bonds is shown in Fig. 17.2. The process flow for both thermocompression and thermosonic wire bonding is the same. In either case we begin with a capillary holding a gold wire, typically from 0.001 to 0.05 in. in diameter, near its end. The wire that extends beyond the capillary is heated using either a hydrogen flame or through capacitive discharge. This forms a ball on the end of the wire two or three times the diameter of the wire itself. This ball is then compressed and the wire is heated to form the bond on the die (the substrate is also

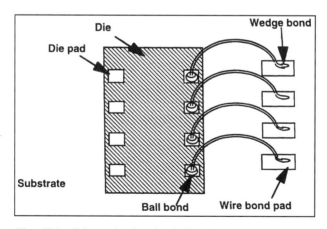

Fig. 17.2 Schematic showing ball and wedge wire bonds.

typically heated to a background temperature). The wire is then drawn through and looped over the die edge to the substrate. A wedge bond is then formed on the substrate, the wire is severed, and then the wire is fed through the capillary in preparation of the next bond [2].

2.   Ultrasonic Wire Bond

Ultrasonic bonding differs from thermocompression bonding in that it forms only wedge bonds. Industry-standard ultrasonic bonding utilizes aluminum wire and can be accomplished on very narrow pad pitches. With ultrasonic bonding the wire is typically fed through a Sono-trode or stylus. The stylus forces the wire against the bond pad and applies a burst of ultrasonic energy to the interface. This process forms a cold wedge between the wire and metal on the bond surface. The stylus then loops up above the die surface and extends to the next bond point for the next wire bond.

3.   Die Attach for Wire Bond

In order to wire bond to a device, the bottom of the device must be attached to a substrate of some kind, as shown in Fig. 17.2. The process of die attach can be accomplished utilizing a number of different materials from solders to organic adhesives to filled glass compounds [3]. Metallurgical attachment of the die to the substrate can be accomplished by metallizing the backside of the wafers. The die is then attached to the substrate using solder alloys. Solders offer low thermal and electrical resistance and are typically alloys of gold or lead.

Organic adhesives, most often epoxies, are the most common form of die attach. The adhesives are filled with precious metals to aid in thermal and electrical transfer. Epoxies provide reasonable performance and have been developed for highly manufacturable and relatively low-cost implementations. Glass adhesives, filled with silver, are a recent development that have not achieved wide acceptance in manufacturing areas.

4.   Wire Bond Reliability

Wire bonding is a common process that has achieved very high yields and reliability levels for most applications. Many of the manufacturing defects have been eliminated as wire bond technology has reached its current maturity level. The most common reliability issues encountered in wire bonding today are due to gold aluminum mixing, die attach voids, and stress-induced creep. The mixing of gold and aluminum can lead to Kirken-dahl voiding in extreme cases. This can be avoided by limiting the amount of time a unit is exposed to extreme (>400°C) temperature, by limiting the amount of impurities in the gold, and by forming the correct alloys in the bond. Stress-induced creep can occur in metallic joints due to low-melting eutectics (e.g., thallium) that weaken the grain boundaries. Once again, purity in the gold wires is key to avoiding reliability problems [4].

Reliability issues are also a concern in the die attach process. If excessive voids or bubbles are present in die attach material failure can occur with repeated thermal cycling. Failures in this case can be loss of bond line, leading to poor thermal conduction from the die to the substrate. Stresses can be so severe in die attach defects that cracking of the die itself occurs.

B. TAB

Tape automated bonding was developed in the 1960s by General Electric and is in large use today for devices that require low profiles (e.g., display line drivers) or very high lead counts. TAB differs from wire bonding in that the leadframe in this embodiment is directly attached to the device. The flowchart in Fig. 17.3 describes a typical process flow for assembling TAB devices. As can be seen in Fig. 17.3, TAB interconnection requires additional wafer processing to add a bump structure. Bump processing as well as the technology of interconnection will be discussed in the text.

1. TAB Bump Structures

Tab bump structures typically contain an adhesion layer, a barrier metal, and a plated bump. An adhesion layer helps promote adhesion between the underlying wafer metallization (typically an aluminum alloy) and the bump metal (typically gold). The composition of the layer can be copper,

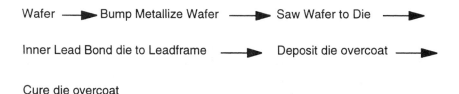

Fig. 17.3   TAB flowchart.

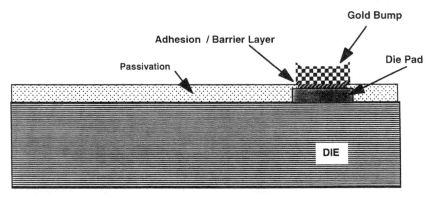

Fig. 17.4   Cross section of a device after completion of TAB bump processing.

titanium, or chromium among others. The barrier metal serves the purpose of separating the bump metallization from the underlying metal. As previously discussed, one must be concerned with gold aluminum alloys. The barrier metal prevents intermixing of the aluminum and gold metallurgy. Barrier metals are typically sputtered across the whole wafer and are composed of any number of alloys including copper, palladium, platinum, or nickel [5]. After sputtering the wafers use a mask-defined photoresist to determine where the gold bumps will be located. The bumps are plated in an electrolytic bath. Following the plating process the photoresist and unwanted adhesion/barrier metallization are striped off the wafer.

The cross-section schematic shown in Fig. 17.4 shows a device after completion of wafer processing.

2.   TAB Tape Leadframes

The TAB leadframe is typically made of copper, surface plated with either gold or tin, and comes in numerous formats (Fig. 17.5a). The leads are typically supported on a polyimide film that can be chemically etched, mechanically punched, or laser etched to provide the necessary windows where the leads need to be exposed. Schematics of a TAB tape layout and cross sections are shown in Figs 17.5b and 17.5c.

High-performance tapes have also recently been implemented that use two metal layers. The additional metal layer can be constructed to help in power/ground distribution, thereby improving the electrical and thermal performance of the leadframe.

3. TAB Bonding

Attachment of the TAB leadframe to the die is commonly referred to as the inner lead bond (ILB). The leadframe is attached to the substrate in the area known as the outer lead bond. The bond is usually a single metallurgy gold bond or eutectic gold tin bond. The ILB process is performed using either thermocompression or thermosonic single-point bonding. The most common process used today is gang thermocompression bonding. The advantage of a gang process, as opposed to wire bond or a single-point TAB attach, is in throughput. All the interconnections in a gang process are made simultaneously, in a single operation, allowing for excellent throughput. The disadvantages of gang bonding include the requirement of high levels of planarity, alignment, and uniformity across the interface. Single-point ultrasonic TAB bonding is similar to wire bonding.

Gang thermocompression bonding requires a heated tool, better known as a thermode, to join the die to the leadframe. The thermode, whether a solid block or multiple pieces, is heated to a temperature slightly above that required to make the metallurgical bond (approximately 300–400°C for tin–gold and 400–500°C for gold–gold bonding). The compression force works to guarantee good thermal contact between the thermode, leads, and the bumps. The force applied typically ranges from 10 to 50 g per lead. The dynamics of the process itself are very different when comparing a eutectic bond to a gold–gold thermocompression bond. Tin–gold eutectic bonding forms a low-melting-point liquidous alloy at the lead bump interface. This alloy allows for a lower stress, lower temperature bond compared to the single metallurgy bond.

Akin to wire bonding, thermosonic bonding techniques can also allow for lower bonding temperatures by using ultrasonic energy to aid in the attachment of the leadframe to the die. Thermosonic bonding is done in a single-point configuration and therefore does not have the throughput advantages of thermocompression gang bonding.

4. TAB Tape Reliability

TAB interconnections have shown themselves to be extremely reliable in a number of applications. As with wire bonding, care must be taken in the manufacturing process to guarantee good reliability. Key areas in ensuring reliability are in the integrity of the barrier/adhesion metals. The actual bond reliability relies on a high level of purity in the metallurgical

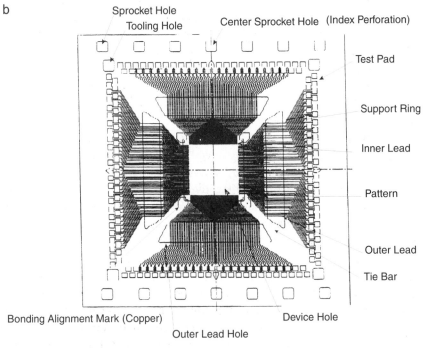

Fig. 17.5 (a) Examples of TAB tape formats. (b) TAB tape layout.

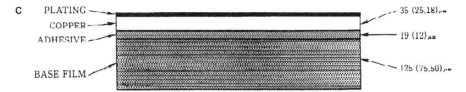

c   PLATING
      COPPER
      ADHESIVE   35 (25.18) µm
                                                            19 (12) µm

      BASE FILM   125 (75.50) µm

Fig. 17.5 *Continued* (c) Cross section of a TAB tape. (a) through (c) are courtesy of Shindo Co., Ltd. [6].

components and especially in surface quality. As with most assembly processes the time the device is exposed to elevated temperatures must be limited. This is especially true for many optoelectronic devices in which low-temperature films are often used to limit the stress and interdiffusion of the laser die.

C. FLIP CHIP

Flip-chip interconnection relies on direct attachment of the device to the substrate. Wire bond and TAB interconnects rely on peripheral attachment. In other words, the interconnection is made on the outside edges of the die. Flip chip, on the other hand, can be both a peripheral and an area array attach. By using the area array approach a much higher density of interconnections can be achieved. This is shown in Figure 17.6, in which the possible number of pads for typical wire bond pitches of 5.6 and 4.8 mils is compared to that of a typical flip-chip array pitch of 10 mils over a range of die sizes. For designs in which the number of interconnections are large, this can offer significant electrical and cost advantages. Another advantage of flip-chip attachment is that it can provide perhaps the lowest inductance interconnect between the die and the substrate. Low inductance is achieved by minimizing the length of the interconnect itself. Typical bump heights are on the order of 0.004 in. compared to 0.100 in. for wire bonds and 0.050–0.100 in. for TAB leads.

There are three major methods for flip-chip attach; C4 (controlled collapse chip connection), plated solder, and *z*-axis films. The most common embodiment is C4, which has a history closely tied with developments at IBM in the early 1960s [4]. In this embodiment an evaporated tin lead solder bump is used as the interconnect media between the die and the substrate. The plated solder solution is similar to the C4 process but relies on a plated bump rather than on an evaporative deposition. A typical

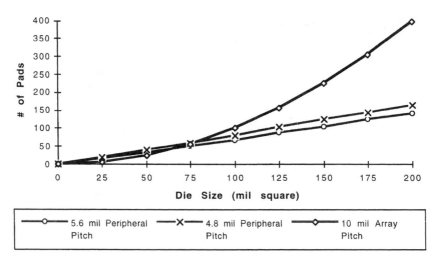

Fig. 17.6 Comparison of the number of interconnections with peripheral and array technologies.

flowchart that details the C4/plated bump processes is shown in Fig. 17.7. I will review in detail the process required to make a bump for flip-chip attach and the assembly processes required for attachment to the substrate.

The z-axis interconnects are a relatively new development that rely on adhesive films filled with conductive particles. In compression the conductive particles form a low-resistance path making contact to the substrate. Although promising, the use of conductive adhesives is currently very limited in the industry.

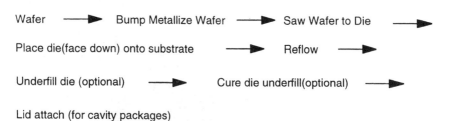

Fig. 17.7 Flow diagram for C4/plated bump processes.

1. Flip-Chip Bump Processes

The bump process metallurgy and structure can be readily compared to the bump process used in the TAB system. Whereas the TAB system is usually reliant on a gold bump system, flip chip relies on tin lead solder metallurgy.

C4 bump processing requires numerous steps to achieve the structure shown in Fig. 17.8. Wafers with vias in the top passivation over the pads are clamped to metal masks. After clamping the mask directly onto the wafer the ball-limiting metallurgy (BLM) is evaporated. Typically, this is a film of chromium, copper, and gold deposited in a single evaporator. The sandwich of metals that make up the BLM serves the purpose of adhesion and barrier layers as well as oxidation protection prior to solder deposition. The solder is then deposited typically in a separate evaporator. The metal mask is removed and the wafer is then reflowed. Reflow of the bumps is accomplished in an inert environment (usually H_2 at 350°C) that transforms the bumps into the spherical shape as shown in Fig. 17.8. The spherical shape is formed during reflow due to the surface tension of the solder bump. This surface tension is an important aspect in solder-attach processes because it provides for what is commonly known as self-alignment. Upon reflow the liquidous metal will move to the lowest energy state it can achieve. This low-energy state centers the solder ball on the BLM [7].

Solder-plated bumps are manufactured in a very similar fashion to TAB-plated bumps. The only difference in the process is that solder is plated onto the wafers rather than onto gold. Numerous solder alloys have been

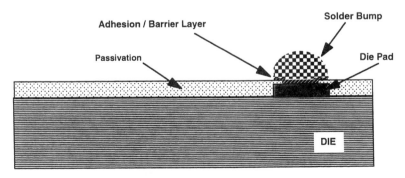

Fig. 17.8   C4 bump cross section.

considered. The most common for a solder bump is 95% lead/5% tin. This provides a high melting point (approximately 315°C) that can withstand the temperature that the package is exposed to during its attachment to the lower temperature solders (such as eutectic 63% tin/37% lead with a melting point of 183°C).

2.   Flip-Chip Substrates

Solder flip-chip attach processes are very much dependent on the substrates to which the die is being coupled. Ceramic substrates are usually metallized with copper or nickel films. Organic substrates, such as printed circuit boards, use lower melting point solders that allow for joining at lower temperatures. The footprint of the substrate is a mirror image of the die. This is shown schematically in Fig. 17.9 [8, 9].

3.   Flip-Chip Attach

Attachment of the solder-bumped die to the substrate is a fairly straightforward process. Flux is applied to the substrates to enhance attachment and the die is placed fairly accurately onto the substrate, as shown in Fig. 17.9. The assembly is then reflowed and the connection of the die to the substrate is made in one process. In this way, hundreds of interconnections can be made quite rapidly. The placement of the die onto the substrate must be good enough so that some of the bumps are on the pad, but it does not need to be precise. Upon reflow, the self-alignment of the solder helps to correct for small misalignments. After attachment the residual flux is

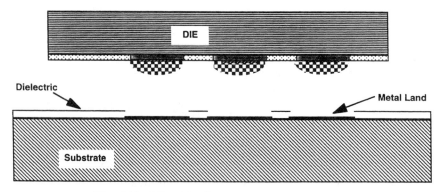

Fig. 17.9   Flip-chip die attached to a substrate.

cleaned using any of a number of solvents or water. At this point the package may be capped and assembly completed. In some applications (high bump counts, large die, and large thermal expansion mismatches between the substrate and die) an underfill is applied between the die face and the substrate. The reasons why underfill is applied will be covered in more detail in the following section. The underfill itself is an epoxy that is dispensed around the die edge. Surface tension pulls the underfill into the area between the die and substrate. Following this process an underfill cure is required.

4. Flip-Chip Reliability

Because the die are being attached directly to the substrates, the amount of thermal expansion mismatch between the two is critical. The stress caused by the mismatch is mostly absorbed in the bumps themselves. Thermal expansion mismatch can lead to excessive stress on the bumps and reliability failures if not properly managed. For alumina ceramic substrates, the coefficient of thermal expansion (CTE) is approximately $6 \times 10^{-6}/°C$, matching silicon and gallium arsenide fairly well. For organic substrates the CTE is much higher (approximately 12 to $16 \times 10^{-6}/°C$) and the mismatch stresses to the die are higher [10]. These mismatches cause a number of considerations in the application of flip-chip attachments. The first consideration is to use only materials that are well matched (such as alumina and silicon). The second consideration is that one may limit the size of the field of bumps that is allowed for flip-chip die. By limiting the bump field's size, the amount of stress on the outer bumps, where the extreme of the expansion mismatch is observed, is lessened. The third way to reduce the stress of thermal mismatch is to use an underfill epoxy between the die and the substrate. By acting as a compliant interstitial member the epoxy reduces the effective stress that is seen by the bumps themselves.

Long-term effects of solder creep and fatigue are well known and can be avoided through careful design and the matching of materials as discussed previously. If one uses these guidelines, the reliability of solder-bumped flip-chip assemblies is excellent.

17.3. Package Types

There are numerous kinds of packages and formats available in the industry. The form factor of a package and its material makeup are driven by the application in which it will be used. Packages provide mechanical and environmental protection to the device while providing the electrical inter-connections between the device and the circuit board. Packages in the optoelectronic industry can be divided into three categories; metal, plastic, and ceramic.

A. *METAL PACKAGES*

Metal packages were the first packages used for the earliest transistors. As the microelectronics industry moved toward higher levels of integration, other packaging technologies were developed. The ceramic and plastic technologies were driven by the higher integration levels and the need for greater interconnection density between the device and the package. Metal packages in the form of transistor outline (better known as TO cans) are used widely in the optoelectronic industry today as the basis for numerous devices. These packages provide a cost-efficient method to package devices within an optoelectronic package. Schematic drawings of a TO can are shown in Figure 17.10 [11].

TO cans utilize wire bonds (as discussed in Section 2.A) to interconnect the device to the package. As can be seen from Fig. 17.10b, the interconnection is made to the base of the package and to a post connected to the leads. TO cans are often integrated into optoelectronic modules in which

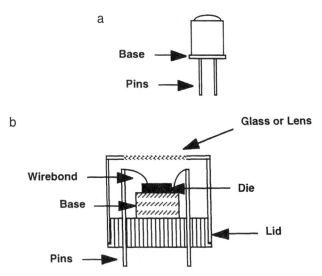

Fig. 17.10   (a) TO can outline. (b) TO cross section with interconnection.

other components (integrated circuits, discretes, etc.) are present. This provides the end user with a compact subassembly that serves a number of uses in optoelectronic products.

B.   PLASTIC PACKAGES

In a plastic package the die is typically attached to a leadframe with any of the interconnection technologies discussed in Section 2. The leadframe and device are then encapsulated using a molding process. The molded plastic then serves as the package of the device. This embodiment is very popular in the microelectronics industry because it provides for a low-cost, highly automated assembly. In optoelectronics, molded packages are less common due to the thermal and performance limitations of the devices.

The first plastic packages used for integrated circuits were introduced in the 1960s and are known as flat pacs. These packages were used to replace metal cans in which higher lead counts and surface mounting was desired. Closely following the flat pack was the introduction of the dual in-line package (DIP). Both the flat pack and the DIP are cavity packages that are wire bonded and then the plastic is molded around the body and leads of the package. Following the introduction of the flat pac and the DIP packages, a number of other outlines were developed. Examples of flat pac and the DIP outlines are shown in Fig. 17.11a and 17.11b [12].

The key elements in a plastic package are the leadframe and the mold compound. Lead frames are typically made of copper with small amounts of iron, tin, phosphorous, or other elements to form an acceptable alloy. The design of the leadframe is optimized for manufacturing. The leadframe must accommodate the flag (which is where the die is attached), the fingers that radiate around the die (where wire bonds will be attached), and the exterior pins that are attached to the circuit board. Lead frames are typically mechanically stamped or chemically etched from copper sheets (though other alloys such as alloy 42, a mix of nickel and iron, are also widely used). After formation the flag and bond fingers are plated with silver or gold, often utilizing a nickel barrier layer. The leadframes are formed in strips that allow for automated equipment to handle a number of packages simultaneously.

The mold materials are usually thermoset compounds. The most common of these compounds are epoxy–Novolac-based molding materials. Compounds themselves are based on many components such as the resin, curing agent, mold release, accelerator, flame retarder, filler, and colorant. The

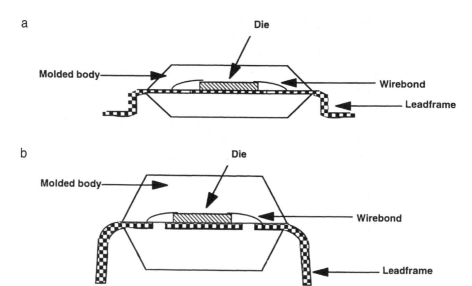

a

b

Fig. 17.11   (a) Cross section of a flat pac. (b) Cross section of a DIP.

mold compound must adhere well to both the leadframe and the surface of the device. Transfer molding typically preheats the resin pellet and then forces the compound into the mold cavity that holds the leadframe. The material actually flows as a viscous liquid when the viscosity drops enough due to the heat and force applied. As the resin flows through the small gates of the mold cavity it is heated through friction. This heating allows the material to cure and compact in a relatively short time. Curing of the plastic compound occurs in the mold, and in some packages it also occurs in a separate process known as postmold cure.

Some of the key concerns with molding around a device include how the mold compound affects the wire bonds that are present. Wire sweep is one of the major causes of defects in molded plastic packages. While the mold compound is filling the cavity the flow front of the material can displace the wires so that shorting or even wire fracture can occur. Careful design of the leadframe, cavity, and wire bond placement can minimize this effect. Other concerns with plastic package manufacture center around residual stresses of the molded body, voids in the mold compound, mold/ device interface or mold/flag interface, and moisture-related issues. The susceptibility of plastic packages to moisture absorption and adsorption is one of the major limiters to its applications.

C. CERAMIC PACKAGES

Ceramic packages are used extensively in the microelectronic and optoelectronic industries. They come in a variety of forms, such as DIPs, flat packs, chip carriers, and pin-grid arrays (PGAs). Many of the formats discussed in the arena of plastic packages are offered in ceramic versions and vice versa. Ceramics cover a broad area of materials that offer a variety of material characteristics. This allows the flexibility of choosing a ceramic material that can best meet one's application. Ceramics are available that have thermal expansion coefficients that match a spectra of materials, from semiconductors ($3 \times 10^{-6}/°C$) to metals ($17 \times 10^{-6}/°C$). Ceramics thermal conductivities range from the insulator to conductor regime (up to $220 \text{ W/m} \cdot \text{K}$) and dielectric constants from 4 to 10,000 [13]. The other major advantages of ceramics are its dimensional stability and its resistance to high-temperature processes. These characteristics allow ceramics to be used for highly integrated, densely routed packages and in ones in which hermetic sealing is a requirement. In these respects, ceramic packaging services an area of applications not covered by either of the materials discussed previously. Ceramic packages are made from a slurry of alumina, glass, and an organic binder or by utilizing dry processes and sintering of preforms.

For higher density of routings, ceramic packages utilize films of dielectrics and conductors on top of the alumina base layers. The two major technologies used in this area are thin film and thick film. Thin-film technology allows for higher routing density by utilizing lithographic techniques to define the conductors. Thick-film ceramic substrates are often used where seperate power and ground planes are desired. Whether using thick or thin film, the ability to have multilayer packages is a significant advantage in the areas of not only routing density but also thermal and electrical distribution.

Ceramic packages are often used for "high-reliability" applications. This is based on the ability of ceramic packages to be hermetic. By hermetically sealing the lids of the package and controlling the environment of the die cavity, ceramic packages can be made to be more resistant to caustic environments than can plastic packages. There are specific considerations that one must give to the reliability of ceramic packages. The pitfalls that arise with these package types are usually related to corrosion problems and mechanical cracking in the packages. Corrosion can be avoided by careful avoidance or removal of ionic contamination. Lid sealing of the package is also critical and must be monitored continuously in a manufacturing environment. Cracking must be managed through good manufacturing handling and careful treatment of the different materials used in the package, so as not to create a high-stress situation.

17.4.   Package to Board Attach

Attachment of the packaged device to the board is also a key consideration in the manufacturing environment. This second level of interconnection is often done to cards, boards, flexible substrates, and back planes. The key considerations at this level of attachment are its ability to be reworked, its density, and, of course, its performance. The two major types of second-level interconnection can be classified as pinned and surface-mount technologies (SMT).

A.   PINNED PACKAGES

Pinned packages are a common interconnection on nearly all boards in the industry today. The most common pinned packages are the DIP and PGA formats discussed earlier. Density of a pinned package is strongly dependent on the format of the package as shown in Figure 17.12. The density relationship between peripheral and array layouts shown in Fig. 17.12 indicates that for large pin counts area arrays allow for higher packing densities.

Pins on a ceramic package are attached using a number of different methods. The two most common methods require that the pins be inserted

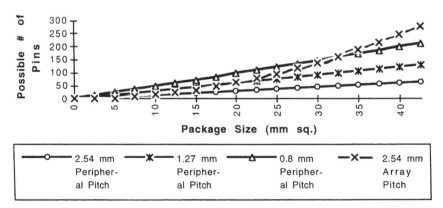

Fig. 17.12   Graph of the possible numbers of pins vs package size for a variety of pin pitches.

through the package or brazed onto the surface. Inserted pins travel through the body of the package and therefore restrict trace routing and the placement of the device to where there are no pins. In this way we have a peripheral pin layout. Brazed-on pins allow the package designer more flexibility because the device or routing metals can be directly above the pins. Therefore, brazed-on pins are often used for full-array pinned packages. Of course, in the industry through-hole pins are used in partial array configurations where a few rows of peripheral pins are used and the active routing is placed either between the pins or in areas where there are no pins present. Schematics of through-hole and brazed-on pins are illustrated in Figs. 17.13a and 17.13b.

A pinned package is attached to the board through a socket (which allows for easy replacement of the component) or, more permanently, solder attached to the board. The pins on the package are attached to the board, which has a grid of plated-through holes filled with solder. The package is then wave-soldered into place. Wave soldering allows the solder deposition and joint formation to occur in one continuous process. In wave soldering the board and the components are exposed to flux, molten solder (typically eutectic lead tin at 230–260°C) for 3–10 s and then cleaned. In this fashion multiple components can be attached to a board at one time.

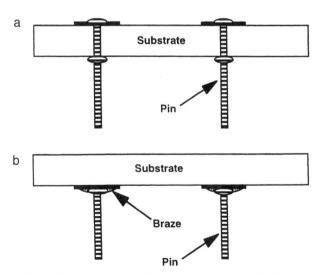

Fig. 17.13 (a) Schematic of a pin through a substrate. (b) Schematic of a pin brazed onto the surface of a substrate.

One of the major advantages of the wave-solder process is that the packages are exposed to relatively low temperatures (120–160°C) for single-sided boards. The reason for this is that the exposure of the board to the wave-solder temperature occurs on the underside of the board. This limits the amount of thermal stress that is applied to the package.

B.   SURFACE-MOUNT PACKAGES

Surface-mount packages cover a broad number of formats as discussed previously (such as PLCC, SOJ, QFPs, TAB, and BGAs) and can provide a number of formats for the leads (gull-wing, J-lead, leadless, and butt joints). The use of SMT in the industry is growing due to reduced board space, cost, and the promise of higher performance solutions. Interconnection of SMT packages is considered to be more difficult and has been the focus of a great deal of industry reseach during the past decade. The focus of this development has been in allowing for smaller pitches (and sometimes blind mating) and in making the surface-mount joints less susceptible to thermal-cycle fatigue failures. The problem of thermal-cycle fatigue is analogous to the problems discussed in Section 2.C.4 because these joints are typically composed of solder.

Joining a SMT to a board can be accomplished through the same wave-soldering process described in Section 4.A. This process is in common use for SMT packages with lead pitches above 1.27 mm but is not considered adequate for packages with finer pitches due to solder-shorting issues. The most common processes used for SMT attach are known as vapor phase reflow (VPR) and infrared reflow (IR). These processes are often preceded by an application of solder paste to the mother board and treatment of the paste prior to the introduction of the packaged components. For VPR, the board and mounted components are loaded into a chamber containing vapor and boiling liquidous perfluorocarbon compounds. Perfluorocarbons are used because they have narrow and predictable boiling temperature ranges and are fairly inert. The assembly is rapidly brought to the desired temperature (175–250°C) and then reflowed [14].

One of the major advantages in VPR is that the processing time needed is very rapid, typically less than 5 min. This advantage is also the source of some of the significant problems with VPR. The rapid temperature rise can cause thermal shock of the components, problems with solder uniformity during reflow, and component movement due to solder wetting.

IR utilizes IR lamps providing radioactive energy to the surface components. The wavelength used by the IR lamps is usually between 1 and

5 μm. This technique is good for surface heating but causes problems with differential heating of the board surface. Differential heating is caused by different absorption of the IR energy by the components on the board. Often, the bodies of the package are heated to higher temperatures more readily than the solder lands. This problem can exasperate the temperature shock problems and lead to substantial temperature gradients across an assembled board. To overcome these shortcomings, IR is typically coupled with connective heat transfer to provide for a more stable process [15].

Other second-level interconnection technologies are also being investigated. These technologies include laser reflow, pressure interconnections, and conductive adhesives. These techniques have not found wide acceptance to date but continue to be investigated.

17.5. Optical Interconnect

The previous discussions incorporated manufacturing techniques that are common to both the microelectronics and optoelectronics industries. In this section I will focus on some of the manufacturing technologies specific to optoelectronic devices. The specific challenges encountered with packaging optoelectronic devices revolve around the transfer of optical power between the device and the outside world.

A. COUPLING

In the regime of assembly techniques, we are concerned with how efficiently we can couple light into (for detectors) or out of our devices [for light-emitting diodes (LEDs) and lasers]. This coupling can be in the free space regime (e.g., CD ROMs, bar code scanners, or imaging) or be concerned with the transfer of data over a fiber of any of a number of compositions. The assembly technologies used are very dependent on the application and what level of coupling is required. For coupling of devices to a single-mode glass fiber (in which the core diameter is on the order of 6–8 μm) alignment tolerances are quite tight. Coupling of devices to multimode glass fibers (in which the core diameter is on the order of 50–150 μm) is less critical, and coupling to large-diameter plastic fibers (>1 mm) requires even less attention to coupling accuracy.

The manufacturing techniques that are required to meet the application's coupling needs are quite varied. The simplest method of coupling a device

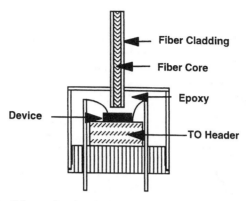

Fig. 17.14   Schematic of a device that is directly coupled to a fiber.

to a fiber is known as direct or "butt" coupling. Fig. 17.14 shows a typical direct-coupled configuration [16]. This direct coupling is often used for plastic fiber and some multimode applications but is not adequate for many multimode and nearly all single-mode applications.

In addition to butt coupling, light interconnection to fibers can also be accomplished by using waveguides, lenses, and prisms. This is especially useful when manufacturing tolerances need to be enhanced, as in the case of the smaller core fibers.

B. WAVEGUIDES

Waveguides come in a number of configurations. These configurations include those made of polymers, ceramics, glass, and semiconductors; however, a waveguide can be formed from essentially any family of materials whose refractive indexes are different. In some cases, a waveguide allows for passive coupling of light to or from the device to the fiber. In other cases, waveguides are used to transfer light or to alleviate difficult tolerances.

The most commonly used waveguide is the silica fiber itself. Many of the processing techniques used for waveguide manufacture marry polymer or wafer technologies to those used in the manufacture of fiber and connectors. Attachment of the waveguide to the device can be directly coupled and often acts as an intermediate between the device itself and the fiber.

C. LENSES

Lenses are commonly used in the optoelectronic industry. Lenses serve to focus and shape the light emitted from a laser or LED into a fiber. Nearly all single-mode fiber optic products use some kind of lens to help focus

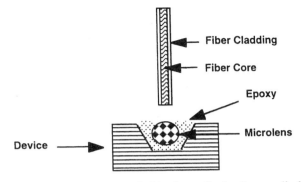

Fig. 17.15   Microlens epoxied in the well of a Burrus diode.

and shape the beam profile for coupling to a fiber. In single-mode fiber applications one must contend with small core diameters (6–8 μm) and relatively low acceptance angles (numerical aperture = 0.15, α = 8.6°). The lenses and fiber must be placed at the focal distance from the light source to achieve the smallest spot sizes and most efficient coupling. In many cases this involves accurate placement of a semispherical lens on the surface of the optoelectronic device itself. This lens is secured through the addition of an adhesive. The fiber itself can also be secured at the focal distance from the lens by the same adhesive material. Figure 17.15 shows the use of a microlens in the well of a Burrus diode. This is often used in concert with active alignment to achieve optimal coupling. Although rigorous, this technique has provided the best solution for single-mode optical coupling in the industry.

D.   PRISMS

Prisms are also used in the interconnection of optical energy. In the case of prisms, one can bend light as well as selectively filter out unwanted energies. Prisms offer interesting alternatives in the selection of which radiation energies are desired. Bending light can be quite useful in coupling from fibers to devices when the end of the fiber is not in the same plane as that of the active devices.

References

1. Anderson, O. L. 1975 (November). Bell Laboratories Record.
2. Steidel, C. A. 1982. Assembly techniques and packaging. In *VLSI technology*, ed. S. M. Sze, 551. New York: McGraw-Hill.

3. Shukla, R., and N. Mencinger. 1985 (July). A critical review of VLSI die-attachment in high reliability application. *Solid State Tech.,* 67.
4. Koopman, N., T. Reiley, and P. Totta. 1989. Chip-to-package interconnections. In *Microelectronics packaging handbook,* eds. R. Tummala and E. Rymaszewski. New York: Van Nostrand Reinhold.
5. Kawanobe, T., K. Miyamoto, and M. Hirano. 1983 (May). Tape automated bonding process for high lead count LSI. *33rd Electron. Comp. Conf. Proc.,* 221–226.
6. Shindo Company, LTD. Shindo TAB brochure.
7. Miller, L. F. 1969 (May). Controlled collapse reflow chip joining. *IBM J. Res. Dev.* 13:239–250.
8. Fired, L. J., J. Havas, J. S. Lechaton, J. S. Logan, G. Paal, and P. Totta. 1982 (May). A VLSI bipolar metallization design with three-level wiring and area array solder connections. *IBM J. Res. Dev.* 26:362–371.
9. Ohshima, M., A. Kenmotsu, and I. Ishi. 1982 (November). Optimization of micro solder reflow bonding for the LSI flip chip. Second Internation Electronics Packaging Conference.
10. Greer, S. E. 1978 (May). Low expansivity organic substrate for flip-chip bonding. *28th Electron. Comp. Conf. Proc.,* 166–171.
11. Motorola Inc. 1983. Optoelectronics Device Data.
12. Robock, P., and L. Nguyen. 1989. Plastic packaging. In *Microelectronics packaging handbook,* eds. R. Tummala and E. Rymaszewski. New York: Van Nostrand Reinhold.
13. Tummala, R. 1989. Ceramic packaging. In *Microelectronics packaging handbook,* eds. R. Tummala and E. Rymaszewski. New York: Van Nostrand Reinhold.
14. Hall, W. J. 1982. Vapor phase processing considerations for surface mounted devices. *Proc. Int. Electron. Pack. Soc. Conf.,* 41–46.
15. Dow, S. 1986 (October). Infrared reflow soldering: Another approach. *Surface Mount Tech.,* 28–29.
16. Midwinter, J. 1977. *Optical fibers for transmission,* 81. New York: Wiley.

Chapter 18 | Alignment Metrology and Techniques

Darrin P. Clement

Lasertron, Incorporated, Bedford, Massachusetts 01730

Ronald C. Lasky

Cookson America, Incorporated, Providence, Rhode Island 02903

Daniel Baldwin

Manufacturing Research Center, School of Mechanical Engineering,
Georgia Institute of Technology, Atlanta, Georgia 30332

18.1. Introduction

This chapter will provide the reader with some benchmarks to use when determining characteristics of coupling in actual components. The theory behind such data will only be presented in the references. Alignment will be discussed in both optical and mechanical contexts. We will assume that the reader has a familiarity with the concepts herein and that this chapter will be used primarily as a reference.

We will then discuss the current techniques for achieving reliable mating between optical components for data communication. Active manual, active automated, and passive automated processes will be explored with an emphasis on expected performance and relative costs. It should be noted that throughout the chapter, we will not discuss telecommunications applications except as they may apply to data communication in the present or near future.

18.2. Interface Definition and Importance

An optical "interface" includes several interconnection elements. The cases considered here involve the mating of an optical source to an optical fiber and/or the coupling between a fiber and a photodetector.

HANDBOOK OF FIBER OPTIC
DATA COMMUNICATION

A "source" may be either a laser diode (LD), a light-emitting diode (LED), or a vertical cavity surface-emitting laser (VCSEL). A lens is typically situated between the source and the fiber waveguide to increase coupling efficiency.

A packaging structure for datacom contains all of these elements and provides the required degree of precise optomechanical alignment among all elements. This structure is an optical subassembly (OSA), and we distinguish those for transmitter and receiver ends of the link as TOSAs and ROSAs, respectively. With the optoelectronic elements packaged in this way, the problem becomes an optomechanical one of aligning the OSA to fibers.

When using an LED, the fiber into which light is coupled would be multimode, whereas single-mode fibers require LD sources. In either case, the fibers used must be ferruled in order to provide durability and accuracy in optomechanical alignment. These cylindrically ferruled fibers are joined to the OSA module by "plugging" them into a bore (an alignment sleeve) located in the OSA. The ability for the ferrule to stay within the bore is characterized through spring and retention forces. Typically accepted values are 7.8–11.8 N for the ferrule spring force and 2.9–5.9 N for the bore retention force [1, 2].

A popular design scheme for datacom is the "connectorized" module. Connectorized in this context is the provision of an optomechanical construction at the module that can provide secure alignment and retention of the fiber while also allowing simple manual detachment and reconnection.

Generally, the trend in datacom hardware is to employ "duplex" connectors/modules, meaning that the connector hardware accommodates a pair of fibers, one each for the sending and receiving paths. Thus, at the transceiver module, the TOSA and ROSA are situated alongside each other and both fibers can be connected or detached in the same manual operation. The fiber distributed data interface, Fibre Channel Standard (a derivative of SC) and the Enterprise System Connection series of duplex connectors/modules are examples that embody this design [3]. Ribbon connectors are used for array devices whereby multiple fibers (usually multimode) are coupled into a linear (one-dimensional) array of lasers or detectors. These are all discussed in Chapter 1. To date, several new connector types are under development for use with VCSEL array transmitters.

18.3. Light Coupling

18.3.1. MIN/MAX COUPLED POWER LIMITATIONS

Because we are concerned with the pluggability of fiber optic connectors into OSAs, both mechanical and optical characteristics must be considered when determining the amount of coupling deviation acceptable. In order to achieve successful data transmission, criteria must be established that put a lower limit on what is considered "enough" coupled power. The coupled power range (CPR) is a measure of how much variation in light transfer is allowed when mating any given fiber with any given transmitter optical subassembly to ensure adequate system performance and yet not violate laser safety requirements.

Definition.   Coupled power range is the allowable difference between the minimum and maximum allowed power.

In order to account for all the losses along a system's length, a concept referred to as the "link budget" must be understood and has been explained in Chapter 7.

Whenever a receiver is said to have a sensitivity value, it must also be understood that this is with respect to an attainable bit error rate (BER). Data communication often requires a BER of 10^{-15}, which means that only 1 in 10^{15} bits can be erroneous. Telecommunication requirements are more relaxed and can be as low as BER $= 10^{-9}$. Some data links have specifications midway between the two values (e.g., 10^{-12}). See Ref. [4] for a trade-off analysis.

For example, a series of components (fiber connectors, fiber length, couplers, etc.) could result in approximately 13-dB total link loss. The minimum light that a typical photodetector can receive for a successful BER of 10^{-15} is approximately -23 dBm. This leaves a minimum launching power of -10 dBm. Thus, the TOSA and connector must achieve at least -10 dBm of coupling every time (e.g., within a 3 σ limit, or less than 27 faulty connections per 10,000). These numbers will vary depending on particular system implementations.

Maximum coupled power is a strong concern in laser systems (compared to LED systems) simply because of the higher available intensities. Some local area networks are user accessible, meaning that nontechnicians are able to connect/disconnect optical fiber cables between computers. This access provides robustness but also presents potential safety hazards to the user. Additionally, there is concern that technicians may be exposed to focused/amplified light when inspecting an optical link for problems. International laser safety standards place a limit of less than 0.600 mW (-2.2 dBm) of optical power in the link [5], whereas the American National Standards Institute standard requires less than 100 s of exposure to 1 mW of light at $\lambda = 1.3$ or 1.55 μm [6]. See Chapters 7 and 8 for more information regarding planning a link and safety considerations.

This situation is unique to datacom because the telecom industry typically has user-inaccessible optical fibers. Thus, if the telecommunication power received at a detector is too weak, the laser output can simply be increased with much less concern for safety. Their limit of maximum coupled power is more likely to be dictated by the characteristics of the components. Recent developments have provided a different perspective on laser safety concerns. The alignment of source to fiber need not be controlled so as to keep the launched power below a certain value. Automatic power controls and open fiber controls, which electronically adjust the power while monitoring the transmission, are permissible under International Electrotechnical Commission (IEC) rules.

Another reason to have limits on output power, even with LED sources, is receiver saturation. Receiver sensitivity can be very high, approaching -40 dBm or better, but the dynamic range of the receiver may be limited. If the power at the detector is too high and saturation occurs, response speed can be degraded to a point at which link performance may be compromised.

18.3.2. MULTIMODE VS SINGLE-MODE CONNECTIONS

We should at this point contrast the cases of single- and multimode. In doing so, we will necessarily have to compare the situation of laser sources with that of LED sources. Although a laser can be used with either fiber type, LEDs are for all practical purposes suitable only for multimode fiber.

Let us assume two hardware sets, one for single mode (SM) and one for multimode (MM), that have the same absolute tolerances on dimensions, angles, and distances. For example, a loss of 1 dB in a SM system could occur with a transverse displacement of the fiber core by approximately 1 μm. It takes approximately 15 μm of similar multimode displacement (50-μm diameter core) to give rise to the same fractional loss of power [7]. The analysis of coupling geometry and efficiency is more difficult for the multimode case; however, connector hardware parts that fail optomechanical tolerances for single mode are often serviceable for multimode applications. See Refs. [8–14] for details.

Multimode fiber used in conjunction with LED sources can be very low cost. Currently, there are a number of commercial offerings of fiber optic transceivers designed for use with multimode fiber that provide distance–bandwidth products in the hundreds of Mb/km. Requirements for precision in the critical dimensions of the multimode connector interface are readily achievable with good- to high-quality machining. In certain applications, even a plastic fiber with core diameter of 100 μm may suffice.

We must bear in mind that the discussion has emphasized the need for short-distance data transmission at high data rates. For much longer distances, single-mode fiber should be used to ensure high bit rates in excess of 1 Gb/s. For this reason, the long-term strategy may be to use single-mode fiber "to the curb" supplanted by multimode fiber "to the home or office." Although MM fiber is no cheaper than SM fiber, it has been possible to reduce component cost with multimode solutions.

However, many companies are now marketing an "array" type of transmitter/receiver set. In these parallel transmission systems, an array of VCSELs are coupled into a ribbon fiber. The highly directional VCSELs coupled into multimode fibers provide relaxed mechanical tolerances in order to achieve consistent coupling.

18.4.   Elements of Coupled Power

The practical coupling considerations for datacom connections are outlined in this section. Coupling to the TOSA and ROSA requires different considerations. The large active area of a typical photodetector and the larger MM fiber core diameter result in the fiber/ROSA interface having much less stringent optomechanical constraints than the TOSA/fiber interface. Thus, the TOSA/fiber interface will be emphasized in this chapter.

Poor control of the many tolerances involved can cause two basic problems in connector/module mating. The first problem may be observed when one plugs the same connector into the same module and obtains extensive variations in the launched power. This variation is referred to as plug repeatability (PR). Two decibels is typically the most extensive acceptable PR variation.

Definition.   Plug repeatability is the variation in coupled power for multiple connections between the same components.

The second variation is referred to as cross-plug range (XPR) and is measured by subtracting the lowest power reading from the highest power reading in a large sample of different connectors mated to the same module. A large cross-plug range is of concern because it is a significant source of the variation of the power launched to any given link. In many cases, improving the "peak" performance is less important than ensuring a small change in performance when interchanging connections.

Definition.   Cross-plug range is the difference between the measured lowest power and highest power for matings between multiple connectors and the same transceiver.

In order to reduce XPR between components, we must establish the sources of coupling variations. Once these are defined, it becomes an engineering problem to optimize component performance.

18.4.1. OPTICAL ALIGNMENT

An optical parameter that plays an important role in coupling is the beam itself. Most beams can be characterized as approximately Gaussian: for example, a SM fiber's fundamental mode, an LD's far-field (although elliptical), the distribution of modes in a MM fiber, and the beam from an LED. Values for each of these parameters vary widely and often must be determined through direct measurement [15]. Rather than use separate terms for each (e.g., mode field diameter, spot size, numerical aperture, etc.), we will simply use source beam spot (SBS) and fiber beam spot (FBS), although the reader should be aware that different definitions of "width" are used: $1/e$, $1/e^2$, full width at half maximum, etc. When you are given a value for a beam spot, make sure you know which definition has been used. Even when all mechanical alignments are perfect, a difference in beam spots between components will reduce the coupling efficiency. Although the FBS is well measured and controlled, the SBS is usually poorly known or controlled.

18.4.2. PLANAR ALIGNMENT

Physical misalignments are the most obvious source of coupling variation when mating two optical components such as a ferrule and an OSA bore. We can define the interface plane as the plane normal to the optical axis.

Longitudinal displacements occur along the (optical) z-axis and are not significant [11, 16]. We do consider some lateral misalignment in the x and y directions, or r (where $r^2 = x^2 + y^2$).

Ferruled fibers are inserted into bores in the OSAs that house the transmitter or receiver. Two style of bores are available: solid and split sleeve. A solid bore has complete rotational symmetry in the sense that it is a "perfect" cylinder as shown in Fig. 18.1a. It is typically constructed of a rigid, ceramic material. The split-sleeve bore, in contrast, is separated lengthwise along the cylinder wall, as shown in Fig. 18.1b. Its composition is of a more flexible material. The split-sleeve bore is designed so that its effective diameter will enlarge slightly to accommodate a larger ferrule.

The solid bore inner diameter (BD) and ferrule diameter (FD) also determine coupling performance. The difference in these two diameters creates an offset in solid bore TOSAs that can vary from insertion to insertion in the same module/connector pair. Hence, these two elements are critical in plug repeatability for solid bore TOSAs. Typical values for the BD are 2.5010–2.5025 mm. For the ferrule dimensions, some specifications require a diameter of 2.4992–2.5000 mm, within a 3-standard deviation criterion. See Fig. 18.2 as a reference. These parameters give the largest ferrule-to-bore difference possible of 3.3 μm. The minimum difference is 1 μm, and assuming normal distributions the average would be 2.15 μm.

Beam centrality (BC) is the position of the laser beam axis with respect to the mechanical axis of the bore. It is desirable that the two be coaxial (making BC = 0) so that the laser beam center is exactly in the center of

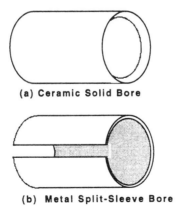

(a) **Ceramic Solid Bore**

(b) **Metal Split-Sleeve Bore**

Fig. 18.1   Solid and split-sleeve bores [17].

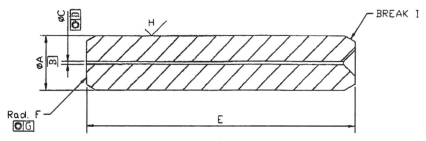

Fig. 18.2 Example schematic of a ferrule.

the bore. The ferrule/core eccentricity (FCE) is the corresponding centrality misalignment element in the connector. The FCE is typically controlled to <1.6 μm. It is non-zero when an imperfect ferruling process results in a fiber core that is nonconcentric with the ferrule.

These parameters are shown in Fig. 18.3. Note that to ease interpretation, it is assumed for now that the ferrule fits "perfectly" within the bore so that the ferrule and bore centers coincide. (The degree of "coincidence" is explained below.) Therefore, we can define an angle, ψ, between the directions of BC and FCE with respect to some fixed coordinate system. (ψ itself does not constitute a new "element.") For the BC and FCE given, the ψ value will be randomly distributed throughout 360°C. For $\psi = 0°$,

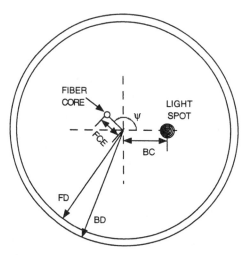

Fig. 18.3 BC and FCE (exaggerated scale) [17]. *Note that the ferrule and bore centers coincide.

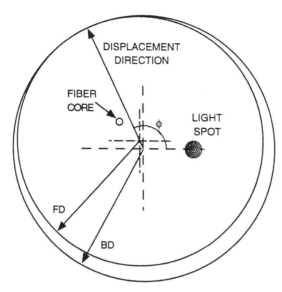

Fig. 18.4   BD and FD (exaggerated scale) [17].

the misalignments will have the least effect, whereas $\psi = 180°$ will place the fiber core the farthest from the light spot. The scales of the BC and FCE have been grossly enlarged in Fig. 18.3 for clarity.

These same elements are shown in Fig. 18.4 for an arbitrary direction, ϕ, of displacement. If FD $-$ BD $= 0$, then ϕ is meaningless and we revert back to Fig. 18.3. Depending on the ψ orientation between BC and FCE, ϕ can either improve coupling or degrade it. Because ψ and ϕ are random variables, the best we could hope for is to establish a range of possible distances between the fiber core and the light spot.

The random fluctuation of all these dimensions can be reduced during manufacturing through a process known as tuning, whereby the eccentricity is intentionally aligned in a preferred direction [1].

It should be observed, however, that split-sleeve TOSAs will not be sensitive to these elements; by design, the split-sleeve walls flex to achieve a snug fit for a wider range of FD.

18.4.3. ANGULAR ALIGNMENT

The ferrule/bore diameter difference can lead to a source of misalignment known as the angular offset, or tilt. However, due to the small lateral displacement of BD $-$ FD $= 3.3$ μm, and the long length of the bore of approximately 4 mm, this tilt will be small. The largest tilt due to the ferrule/bore difference would be 0.05°. This can be shown to contribute insignificantly to the CPR [16].

There are, however, other possibly significant sources of tilt. One potential for tilted beams can be found in the TOSA. Even if the beam center spot can be successfully imaged to the bore center (i.e., BC = 0), the laser may possess some tilt as shown in Fig. 18.5. This tilt is known as the pointing angle (PA). As explained previously, before the laser and bore are welded, the laser is repositioned for maximum coupling. If there is a significant PA to the laser light, the weld will still be optimum for a small BC, as shown, and the engineer will be oblivious to the fact that any tilt is present!

The pointing angle can be as large as 3° and must be taken as an unknown. It is hoped that as datacom applications become more common, the need for well-controlled tolerances will drive manufacturers to produce modules with much smaller laser tilts. A necessary precursor to this optimization is the development of an accurate measurement technique.

The fiber end-face angle (EFA) is due to the polishing process of the ferruled fiber. It is common to intentionally polish fibers at angles ranging from 5 to 10° in order to reduce back reflections from the fiber surface. The de facto industry standard is 8°.

However, even supposedly "flat" fibers may have a non-zero EFA. Fibers are typically arc-polished so that the hemispherical tip deforms slightly when two ferruled fibers are butt-coupled. This process, known as the physical contact polish, assumes that the polished apex coincides with the fiber core center. If the polishing tool is not precisely aligned, the end-face normal of the fiber will not be parallel to the fiber/ferrule axis. A non-zero fiber FCE further complicates this situation.

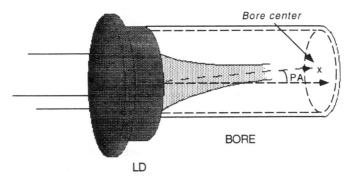

Fig. 18.5   Laser PA [17].

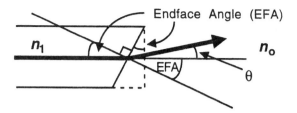

Fig. 18.6 Fiber EFA.

Refraction of the entering/emerging light will cause the effective light axis to be nonparallel, or tilted, with respect to the alignment axis. As seen in Fig. 18.6, a beam emerging from an obliquely polished fiber is refracted according to Snell's Law:

$$n_1 \sin(EFA) = n_0 \sin(EFA + \theta). \tag{18.1}$$

It can easily be seen that when a fiber end face is polished at an angle EFA, the beam emerges at an angle θ with respect to the fiber axis (which is an angle EFA + θ with respect to the end-face normal). Using the small angle approximation, θ can be found:

$$n_1 (EFA) \approx n_0 (EFA + \theta)$$

or

$$\theta \approx \left(\frac{n_1}{n_0} - 1\right) EFA. \tag{18.2}$$

Using 1.47 for the fiber's index of refraction, and a value of 1 for air, the relation between θ and EFA is $\theta \approx (0.47)$ EFA. Note that for analytical purposes, θ can be seen as the fiber analog to the TOSA pointing angle.

Random sampling of nonrejected spherically polished fibers shows up to 0.3° of effective tilt, θ. The measurement of the end-face angle can typically be made using interferometric techniques so that samples with, for example, >0.1° EFA can be rejected, but this will produce large numbers of unused cables, increasing manufacturing costs.

18.4.4. MODULE ALIGNMENT

The module itself can also contribute to coupling variations through the subassembly misalignment (SAM) and connector ferrule float (FF). SAM is the axial misalignment of the entire OSA/bore assembly from its desired true position within the transceiver module. Remember that we are discussing duplex interconnections so that both the transmitting and receiving bore/ferrule matings must be aligned (Fig. 18.7). If one of the couples cannot be mated, then by design the other pair would be prevented from achieving successful plugging.

FF is the amount of lateral movement or float possessed by the connector ferrules. Too rigid a setting would impede ease of use and/or would make component damage more likely. Thus, spring forces built into the design allow a little "give." FF could compensate for a slight SAM. Of course, too large a FF could also prevent proper mating because of the "sloppiness" of the ferrule position. In split-sleeve OSAs, excessive SAM may cause a large off-axis force on the split sleeves that can cause plug repeatability problems or damage the split sleeve. It is important to note, however, that these two elements do not strongly affect the plug repeatability or cross-

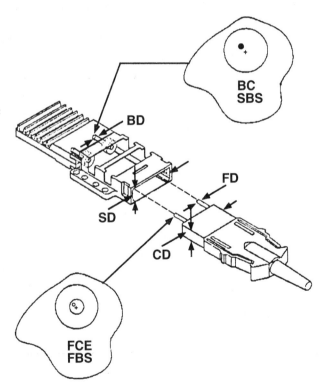

Fig. 18.7 Connector/module mating elements.

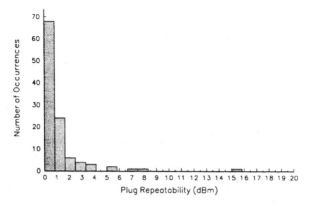

Fig. 18.8   Split-sleeve bore PR [17].

plug range for solid bore systems because the bore stiffness overwhelms them. A detailed account of the forces involving FF and SAM is given in Ref. [2].

Other characteristics are the module shroud dimension (SD) and the connector body dimension (CD). The clearance between the two is the variable that affects plug repeatability and cross-plug range. Excessive clearance can create several mechanically stable plug modes that can affect plug repeatability in the same module/connector pair. Data indicate that this clearance has a much stronger effect on split-sleeve OSAs than on solid bores (Figs. 18.8 and 18.9).

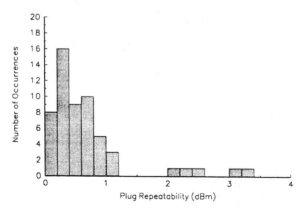

Fig. 18.9   Sold bore PR [17].

18.4.5. PUTTING IT ALL TOGETHER

For reference, Table 18.1 provides typical values for the 12 elements.

A typical CPR might be from -2.2 to -10 dBm—a 7.8-dBm range. All of the variables that affect the CPR are usually analyzed with a Monte Carlo technique to ensure that the CPR can be achieved. The CPR can be viewed as the sum of variations due to aging effects on the electrical components and to changes in coupling from mechanical/optical variations in the connectors and TOSAs.

In general, all 12 elements affect XPR. However, if we limit ourselves to solid bore TOSAs, only 8 variables are operable. Four variables are related to the connector and 4 to the module. These variables are

Module	*Connector*
BD	FD
BC	FCE
PA	EFA
SBS	FBS

Table 18.1   **Typical Values of the 12 Elements**

		CPR influence	
Element	*Typical value*	**Solid bore**	**Split sleeve**
SBS	5–20 μm[a]	√√	√√
FBS	9.5 $\pm$ 0.5 μm for SM[a]	√	√
BC	3.3 $\pm$ 2.0 μm	√√	√√
FCE	<1.6 μm (0.7 μm)	√	√
BD	2.5010–2.5025 mm	√	
FD	2.4992–2.5000 mm	√	
PA	<3°	√√	√√
EFA	0.1–0.2° (barring rejection)	√	√
SAM	Spec. $\pm$ 0.34 mm		√
FF	Spec. $\pm$ 0.34 mm		√
SD	Spec. $\pm$ 0.125 mm		√
CD	Spec. $\pm$ 0.125 mm		√

Note. From Ref. [17]. √, determines the influence of a parameter's variability. If a parameter either has little variation or has little effect on the coupling efficiency to begin with, it will be ranked with fewer √. Given these measurements, it is desirable to predict how much variation to expect from a given product that could include any combination of values.
[a] Using $1/e^2$ width definition.

Each of these factors will produce variability in both PR and XPR.

The actual distance between the fiber core and the laser spot is essentially determined by BC, FCE, BD, and FD as shown in Fig. 18.10 (compare with Figs. 18.3 and 18.4). Knowing these quantities along with the directions of displacement ψ and ϕ, we can develop a relation for the distance R:

$$R^2 = \left[BC - FCE \times \cos \psi - \frac{1}{2}(BD - FD) \times \cos \phi \right]^2$$
$$+ \left[FCE \times \sin \psi + \frac{1}{2}(BD - FD) \times \sin \phi \right]^2. \tag{18.3}$$

It is clear that the maximum value for R occurs when ψ and ϕ are 180° and the minimum occurs for both directions equal to zero. Thus,

$$R_{max} = BC + FCE + \frac{1}{2}(BD - FD)$$

and

$$R_{min} = BC - FCE - \frac{1}{2}(BD - FD). \tag{18.4}$$

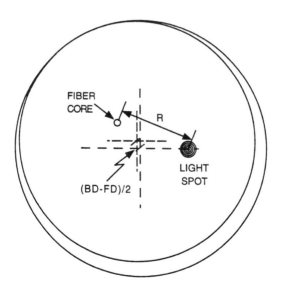

Fig. 18.10 Fiber core and laser spot separation [17].

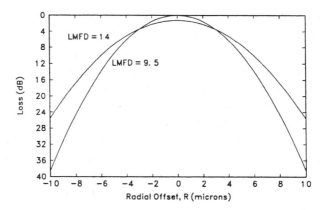

Fig. 18.11   Coupling loss as a function of radial offset [17].

Clearly, BC, FCE, and DB − FD can give rise to an R_{min} of zero. Figure 18.11 shows the theoretical results for coupling from LDs with SBS = 9.5 and 14 μm to a standard fiber with FBS = 9.5 μm for purely radial offset R. Actual coupling losses can be up to 3 dB greater than those shown in Fig. 18.11, but the shapes of the experimental curves are surprisingly close to those in Fig. 18.11. It is believed that the difference is due to lens aberrations in the TOSA [7, 18].

Different modules will necessarily have different PAs, even if BC is well controlled. Figure 18.12 shows the effect on coupling loss that these various

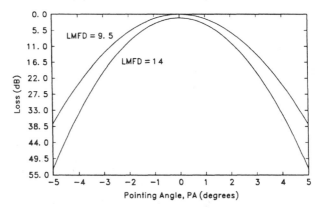

Fig. 18.12   Coupling loss as a function of laser pointing angle [17].

pointing angles will create. Generally speaking, this variation can be responsible for up to 10-dB differences in loss between matings of different modules with a given connector. However, plug repeatability will not likely be affected because the PA for a given module can be assumed constant.

The fiber EFA has a much less severe influence on coupling efficiency when it is assumed that other misalignments are small. In fact, because it is common that EFA $< 0.2°$, one can use Fig. 18.12 to estimate its effect on overall coupling loss, provided one sets PA $= \theta = 0.47$ (EFA) from Eq. (18.2). The variability comes from different insertions of one connector/module pair. With 360° of possible rotation within the bore, a fiber with an EFA $= 0.2°$ will produce minimal coupling variation.

It should be noted, however, that many applications call for intentionally angled end faces to reduce back reflection into the source. The industry standard for the EFA in this case is 8°. Here, for the same parameter values as in the previous paragraph, different insertion orientations have dramatic effects, as shown in Fig. 18.13 [8].

It is well worth noting that in our discussions, we have primarily considered isolated misalignments: radial, axial, and angular. When all these are present, they can each become more significant to coupling. When the radial offset is significant, the PA has a more dramatic effect and vice versa. Nemoto and Makimoto [16] give analytic solutions when the three misalignment "categories" can be assumed to occur in the same plane. Out-of-plane misalignments are explored in Ref. [8].

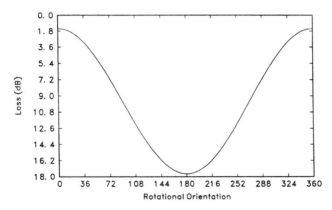

Fig. 18.13 Change in coupling loss during fiber rotation for a large end-face angle [17].

18.5. Alignment Techniques

Alignment technology is critical to the production of low-cost optodatacom components. It is considered by some to be the critical issue that may delay the proliferation of optical data communication. Currently, it can cost up to $100 to align and secure a laser diode in a SM optical subassembly. Developing technologies to reduce this cost by one or two orders of magnitude is paramount to the proliferation of optodatacom into the consumer markets.

18.5.1. ALIGNMENT REQUIREMENTS

Although SM alignment requirements are often stated as submicrometer, experience has shown that 1–3 μm control of the center of the laser spot to center of the SM fiber is usually adequate. MM technology is relaxed to approximately five times that of SM technology. There are active (light source turned on), manual alignment technologies for manufacturing available in the late 1990s to perform both SM and MM assembly. Unfortunately, these techniques are slow, serial, and hence expensive. Passive and active automated alignment techniques are on the horizon that may address this need.

18.5.2. CURRENT ALIGNMENT AND
SECURING TECHNOLOGY

As stated in previous chapters a transceiver module generally consists of a ROSA and a TOSA assembled on a substrate with associated drive and mux/demux circuitry. The critical alignments that we are discussing are needed to assemble the TOSA. The alignment tolerances in assembling the ROSA are considerably more relaxed as discussed previously. To review the assembly process and structure of the TOSA see Chapter 5.

In a typical TOSA the optical axis of the laser (or other source) is first aligned to that of a lens (Fig. 18.14). This alignment is currently performed with the light source active, using some type of robot. When the alignment is optimized the robot measures a maximum in light output. At this point the securing of the lens and laser positions is performed, usually using some type of welding or occasionally adhesive. The welding process, although fast and strong, can also result in a shifting as the weld cools and shrinks. The assembled laser and lens are then aligned to the center of the bore of

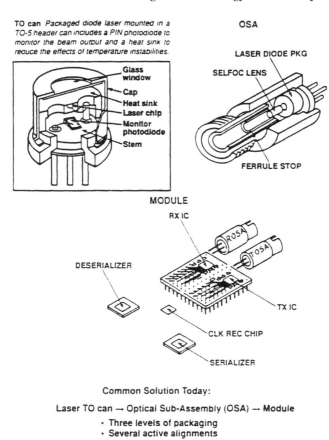

TO can *Packaged diode laser mounted in a TO-5 header can includes a PIN photodiode to monitor the beam output and a heat sink to reduce the effects of temperature instabilities.*

OSA

Glass window
Cap
Heat sink
Laser chip
Monitor photodiode
Stem

LASER DIODE PKG
SELFOC LENS
FERRULE STOP

MODULE

RX IC
DESERIALIZER
ROSA
TOSA
TX IC
CLK REC CHIP
SERIALIZER

Common Solution Today:

Laser TO can → Optical Sub-Assembly (OSA) → Module
  • Three levels of packaging
  • Several active alignments

Fig. 18.14 A typical OSA.

the TOSA. This process is also performed actively, aligning the light from the laser/lens combination with the bore center by measuring light coupled into a ferruled fiber in the bore. Using a ferruled fiber in the bore contributes a 1- or 2-μm offset immediately because there is clearance between the ferrule and bore and the fiber is never perfectly centered in the ferrule. These errors were discussed previously in this chapter: FCE and BD − FD. After the laser/lens combination is aligned by the robot to the bore/ ferrule, another weld is formed to secure these two components. Unfortunately, another "weld shift" is usually experienced. Hence, with the weld shift and bore to ferrule error, it is typically difficult to obtain better than

3-μm alignment in a welded TOSA. These steps are time consuming and hence expensive, especially in SM technology.

18.5.3. ACTIVE AUTOMATED ALIGNMENT

Active automated alignment involves aligning a source to an optical detector (i.e., axis) while both source and detector are active. A primary advantage of active alignment techniques is that they produce known good assemblies with optimum or near-optimum alignment. This is in contrast to passive alignment techniques, which must be tested postmanufacture. Cost is believed to be the primary disadvantage with this alignment technique. The prevailing activity in optoelectronics packaging is to develop passive alignment technology for low-cost assembly. The most common method for active automated alignment involves mechanical alignment techniques. Recent developments in selective silicon etching and silicon micromachining have enabled development of combined active and passive alignment techniques, simplifying the assembly process.

There are numerous automated alignment techniques reported in the literature, with each having unique advantage. A representative set can be found in Refs. [19–23]. A typical assembly process involves first the temporary insertion of an optical fiber into a large area photodetector to determine the maximum optical signal strength. This signal strength is used to determine the maximum relative signal strength during active alignment. Then the optoelectronic package is assembled and the source and detector are activated. Next, after fiber is removed from the photodetector, it is inserted into the optoelectronic package and brought into the vicinity of the photodetector or source depending on the function of the package. Micromanipulation is then used to align the fiber to the photodetector or source. The end of the fiber is manipulated in five degrees of freedom (i.e., x, y, z, θ_x, and θ_y) while the signal strength on the detector is monitored. The photodetector signal strength is maximized to find the optimum alignment. At this point, the fiber is permanently affixed in the package to secure the aligned position. Typical fixturing processes include soldering, adhesive attach, and welding.

Active automated alignment in optoelectronic assemblies is typically accomplished using one of three techniques for maximizing the detector signal as the fiber is precision aligned relative to the detector. Near-optimum alignment is achieved using simple linear five-axis optimization in which the fiber is moved independently along each of five axes (i.e., x, y, z, θ_x, and θ_y) to identify a local maximum signal strength [22]. Each axis is traversed independently, whereas previously analyzed axes are fixed in their local maxima positions. The primary drawback with this technique is that it rarely produces optimum alignment and is prone to locating local maxima having signal strength considerably lower than the maximum. A second active alignment technique utilizes minimum path search techniques to determine the optimum alignment (Fig. 18.15). These search techniques are derived from fundamental research performed in robotic motion and automated assembly. Minimum path searches involve incrementally moving the fiber along each degree of freedom (i.e., δx, δy, δz, $\delta\theta_x$, and $\delta\theta_y$) and measuring the signal strength. The path motion space is then analyzed to identify the incremental motion yielding the maximum positive gradient

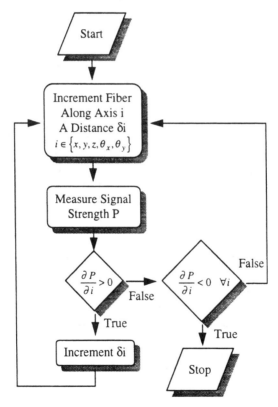

Fig. 18.15 Typical process flow for minimum path search technique for active automated alignment.

of the signal strength. The path variables are then incremented based on the local optima. The fiber is again incrementally moved along each degree of freedom and the path of maximum response is again identified. This process is repeated until all combinations of incremental moves generate negative signal strength gradients corresponding to the optimum alignment. Identification of localized maxima is a potential problem with this technique, although various techniques have been developed to minimize the problem. With this technique, the fiber is incrementally moved along a path of increasing signal strength until the maximum is reached.

The third technique for active automated alignment is direct feedback control of the micropositioning system during active alignment [24]. One technique used for feedback control utilizes a phase-sensitive detector in the optoelectronic assembly and drives the fiber alignment axis using a sinusoidal dithering motion superimposed on a monotonic translation. The output of the phase-sensitive detector is directly proportional to the spatial offset error (used for feedback control) of the optical fiber from the optimum alignment position (Fig. 18.16). Moreover, the polarity of the phase-sensitive detector output corresponds to the positive and negative approach directions for the optimum alignment position. For zero-offset positioning

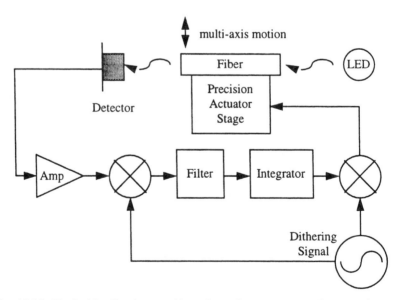

Fig. 18.16 Typical feedback control loop for active automated alignment (adapted from Ref. [24]).

of the fiber from optimum (within 0.1–1%), an integral feedback controller can be used. Additional options include proportional–integral and proportional–integral–derivative as feedback control systems. The primary advantage of this technique is its rapid response and settling time for locating an optimum alignment. For multiaxis alignment, this technique has a significant speed advantage in that it allows optimization of the various axes simultaneously. This is done by having the micropositioning system actuators for the different axes dithered at different frequencies. Different phase-sensitive detectors are then used to extract the spatial offset information pertaining only to the appropriate axis. The spatial offset errors from each phase-sensitive detector are then used as the feedback control error signal for actuating the independent axis transducers controlling the fiber position.

Assembly of optical fiber is typically achieved using a microactuator to mechanically align the fiber relative to a detector or source. The manual alignment technique is the most common method. Here the fiber is grasped with a microgripper attached to a precision multiaxis stage and translated/rotated using precision lead screw actuators. To minimize vibration and backlash, the stages typically ride on air bearings and are mounted to large granite slabs. Actuator drive systems incorporate submicrometer encoders and closed-loop position control to ensure precise alignment.

18.5.4. *HYBRID ACTIVE/PASSIVE AUTOMATED ALIGNMENT*

Due to the excessive cost associated with five degrees of freedom, precision actuators, and the relatively low-volume utilization of the precision assembly systems, recent efforts have focused on combined active and passive alignment assembly. Numerous techniques have been reported using various degrees of passive alignment [25–31]. Tewksbury *et al.* [32] also present a comprehensive review of optical interconnect technology. The primary advantage of combined active and passive assembly technology is significantly reduced cost. This is because combined active and passive alignment requires only one or two precision actuators compared with five for a full degree of freedom assembly. Numerous combined active and passive alignment techniques have been developed. We will focus on a few representative techniques here.

Due to the relatively high cost of optoelectronic component assembly and optical fiber alignment, there has been a heightened interest in fiber array assembly technology. Jackson *et al.* [33] present several approaches for aligning linear arrays of optical fibers (similar to ribbon cables used in electronics) to multichannel GaAs lasers and detector arrays. The assembly process is simplified using anisotropically etched precision silicon V grooves etched in a rectangular silicon block. The fibers are secured in the V grooves using high-temperature adhesives or solders forming a passively aligned subassembly. In the case of solder attach, the fibers must be metallized with a solder-wettable metallization system to ensure reliable adhesion and robust performance. This is particularly important for subsequent processing steps that can subject the assembly to numerous high-temperature thermal excursions. The fiber array and V-groove assembly significantly reduces assembly complexity by transforming the multifiber array into a single-fiber subassembly that can be aligned and assembled as a single unit. This enables multiple interconnects to be formed during a single assembly process and significantly reduces the cost per interconnect. To further simplify assembly of the fiber array to the vertical GaAs laser array or the detector array, the fiber V-groove subassembly is beveled on one end (typically at a 35 or 45° angle relative to the horizontal). The beveled end acts as a mirror reflecting the vertically transmitted light from the lasers into the fibers or from the fibers into the detectors. In the case of the 45° bevel, the fiber end must be metallized to enhance reflectivity and reduce coupling losses; the 35° bevel totally reflects the transmitted light and does not require metallization. The alignment of the fiber array to the laser array and/or detector array is achieved using standard precision actuator systems. To simplify the alignment process, photolithographically defined ridges and contacts are produced on the laser array and/or detector. These act as alignment faducials for precision alignment of the fiber array. This assembly can have an angular registration better than $\pm 1°$ and positioning alignment within $\pm 3\mu m$ [33] (Fig. 18.17). As one would expect, these tolerances are well within the requirements for multimode assembly.

A second representative assembly processing using combined active and passive alignment is presented by MacDonald *et al.* [34]. The particular assembly integrates a laser, a back-facet monitor diode [positive–intrinsic–negative (PIN)], and coupling optics and is based on a silicon subassembly. As shown in Fig. 18.18, the silicon source submount houses the laser. Registration of the active area of horizontal (side) emitting laser to the source submount is achieved with a solder column bond site as shown in Fig. 18.19. A large pad of solder is sputtered onto the source submount bond site, and textured compression bond sites are formed on stand-off columns of the source submount using selective silicon etching. The laser is then tacked, junction side down (i.e., the metallized side down), to the

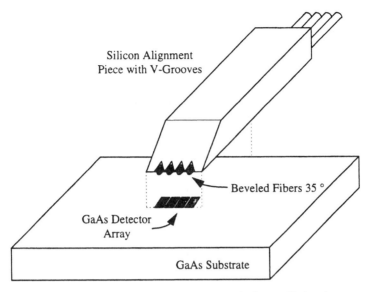

Fig. 18.17 Beveled fiber array coupling approach for a GaAs detector array (adapted from Ref. [33]).

texture bond site. Mechanical alignment of the laser is achieved using a precision actuator system capable of compression bonding. Permanent bonding of the laser is achieved by reflowing the solder, which balls due to surface tension effects, forming a permanent electrical and mechanical bond. The process typically results in a ± 2-μm registration of the laser active area to the source submount. The source and detector submounts also have selectively etched cavities and V grooves. One such element is the turning mirror in the source submount that reflects back transmitted light from the laser onto the surface-illuminated PIN on the detector submount. Key elements to the assembly are the front and back facet ball lenses, which are placed in selectivity etched lens holders (V grooves intersecting at 90°) that precisely establish the position of the spheres relative to the surface of the source and detector submounts. The position of the two lenses is controlled by photolithography and etching to within ± 1 μm. When the two submounts are assembled, the surfaces of the two submounts are brought into contact, aligning the laser and lenses in the direction normal to the submount surface (the z direction). The axial degree of freedom along the optical signal path is establish by the internal references (V grooves and columns) defined by photolithography and selective etching.

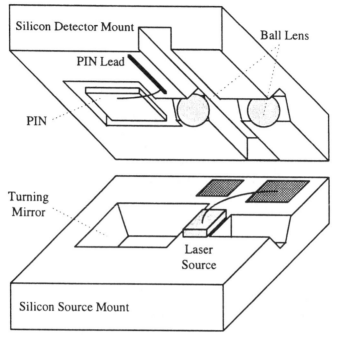

Fig. 18.18 Schematic drawing of laser source and back facet monitor subassembly (adapted from Ref. [34]).

The lateral degree of freedom is established by sliding the two submounts relative to one another. Alignment is optimized along a single axis by actively monitoring the near-field intensity at the output of the front facet lens. Therefore, in this assembly technique, active alignment is used on only a single axis. The remaining axes are aligned passively.

A third combined active and passive alignment assembly process is presented by Solgaard *et al.* [35]. Their innovative optoelectronic packaging technology uses silicon surface micromachined alignment mirrors for active alignment of an optical fiber to a laser or detector. Their alignment technique is based on silicon optical bench technology in which active and passive optoelectronic components are assembled on a silicon chip. Typically, accurate passive positioning of the components is achieved by solder bumps and mechanical stops that are defined using conventional photolithography. Although this technology has found wide application, it does not provide adequate positioning accuracy for high-performance applications such as CATV laser modules. The technology Solgaard *et al.* have

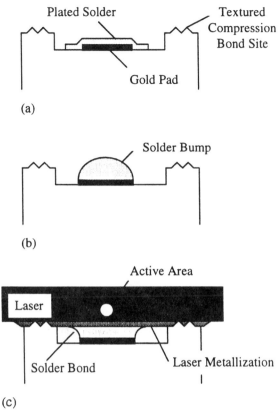

Fig. 18.19 Solder column bond site featuring textured compression tacking and bump bonding (adapted from Ref. [34]). (a) As plated. (b) After solder reflow. (c) Laser tacked into textured surface and solder bump bonded.

developed incorporate movable, micromechanical structures that are fabricated directly on the silicon chip by polysilicon surface micromachining [34]. The independent optoelectronic components are assembled to the package using standard optical-bench technology. The inherent inaccuracy of the passive alignment of the optoelectronic components is corrected by movable aligning microstructures as shown in Fig. 18.20.

Two schemes are used for alignment. The first uses an external servo feedback control system to automatically align the micromechanical structure. After alignment, the servo is disconnected and the micromechanical structure is left in a permanent position. The second technique is to integrate the servo feedback control, detector, and source in the package so that accurate alignment is guaranteed throughout the life of the package.

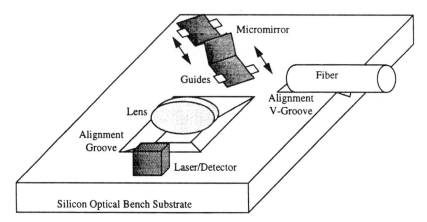

Fig. 18.20 Fiber-coupled, semiconductor laser module with on-chip alignment mirror.

The alignment mirrors are micromachined polysilicon plated with gold to provide high reflectivity. As shown in Fig. 18.21, the lower portion of the micromachined alignment mirror is connected by polysilicon hinges to another polysilicon plate that can translate linearly (i.e., a slider plate) along polysilicon guides etched in the silicon carrier. The top of the mirror is also hinged to a back support, which is hinged to a second slider. Moving the sliders simultaneously translates the mirror linearly. Translating one slider relative to the other tilts the mirror to a controlled angle. The two degrees of freedom (translation and tilt) provides the necessary beam alignment to compensate for the transversal offset of the lens and laser assembled using the conventional silicon optical-bench technology.

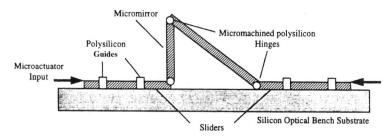

Fig. 18.21 Schematic of micromachined micromirror structure.

Potential concerns with this alignment technology include possible flex-ing of the structure during motion, dimensional tolerance loss due to loose hinges, and the possibility of slider motion during shock and vibration. However, due to the extremely light masses, inertial shock forces and vibration acceleration do not translate the micromirrors. This was demon-strated for shock tests up to 500 G. The micromachined structure is held in place by the friction forces in the slides and hinges.

18.5.5. PASSIVE AUTOMATED ALIGNMENT

Passive automated alignment involves aligning a light source to an optic axis automatically without turning the light source on. This approach has a disadvantage in that it is not possible at the time of manufacture to determine if the alignment is successful because the light source is not turned on. However, with state-of-the-art manufacturing techniques any manufacturing loss due to this phenomenon could be minimized.

One of the most promising technologies for passive automated align-ment involves using the surface tension of solder. This technique is shown in Fig. 18.22. This self-alignment phenomenon was first observed at IBM [36]. Because their integrated circuit technology uses solder balls to inter-connect, it was observed that if one did not precisely place the integrated circuit (IC) on the package, then the surface tension of the melted solder balls would ensure the centering of the IC. To use this technique to assemble the light source to a lens or fiber, the following process could be used. The light source is placed with an accuracy of approximately 25 μm by a standard IC die placement machine. Upon reflow of the solder, the surface tension moves the light source into alignment with the pads (Fig. 18.23). Hence, the light source and fiber are aligned with approximately the same accuracy as the pads were to the optical axes. Lee *et al.* and Lin *et al.* [36–39] have shown that this technique is capable of accuracies down to micrometer levels.

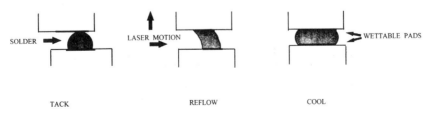

Fig. 18.22 Solder ball surface tension.

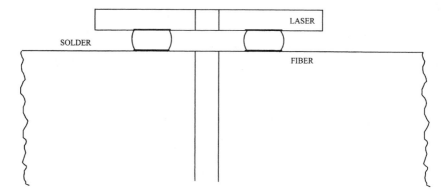

Fig. 18.23   Solder reflow alignment.

With the advent of VCSELs it is likely that a light source may be passively mounted directly to a fiber. A possible process to perform this assembly might be as follows. Photolithographic processes are used to form the pads (likely copper) on both the VCSEL and the fiber (Fig. 18.24). Current art

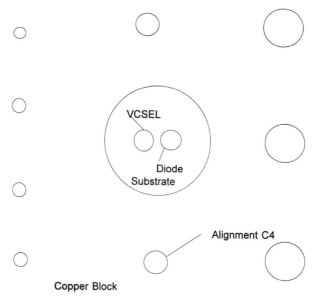

Fig. 18.24   Solder pad layout (top view).

enables better than micrometer location of the pads with respect to the VCSEL optic axis, in the VCSEL process. Unfortunately, a new tool would likely need to be invented to form the pads on the fiber to within micrometer accuracy with respect to the optic axis of the fiber. This tool is shown in Fig. 18.25. To form the pads accurately with respect to the optic axis of the fiber, the tool must focus on light emitted from the fiber and use this point as a reference to photolithographically image the pads in the correct spot. The lasers shown on this tool would image the pads in the photoresist on the fiber stub. To perform this operation the lasers would have to be

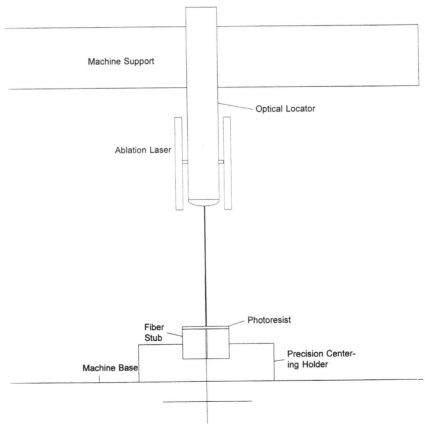

Fig. 18.25 Proposed tool for formation of solder pads on fiber stub.

moved to the proper location. This movement would require state-of-the-art motion systems and good vibration damping to achieve the desired micrometer tolerance control. An additive or subtractive could be used to form the metal pads.

Because the SBS of the VCSEL can be controlled by design, it is possible to select it to match that of a SM fiber. In this case a lens is not needed to achieve reasonable coupling. Hence, a structure such as that in Fig. 18.26 would be possible. The fiber is tilted to minimize reflected light into the laser. It is evident that the solder balls require a gradation in size. This size gradation is possible with solder jetting technology [40].

The amount of light coupled will be critical to the successful functioning of the link. The theoretical coupling of light as a function of the distance

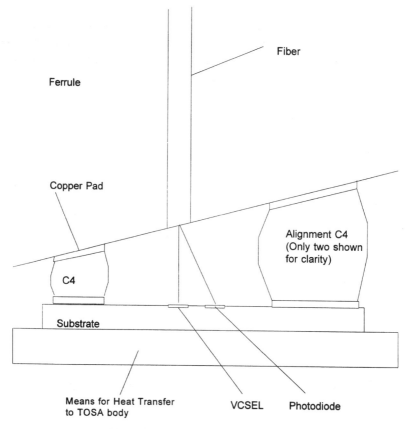

Fig. 18.26   Passive alignment with solder balls.

of the VCSEL from the fiber and its transverse offset is shown in Fig. 18.27. A mode diameter of 10 μm has been chosen for both the fiber and the VCSEL in this calculation. As seen in Fig. 18.28, the coupling effects from longitudinal (z) offset are far less dramatic. The coupling model of Nemoto and Makimoto [16] has been used. Our experience has shown that this model provides excellent relative results; however, absolute amounts of light coupled are typically 30–50% less. This discrepancy is likely due to abberration in the light source.

Although somewhat more cumbersome, it is still possible to use this approach to align edge emitting light sources to lenses or fibers, albeit with less coupling due to the typical mismatch between the SBS and the FBS.

To date, this process has still not been implemented into practice. Our belief is that the process could be implemented with initial costs of tens of dollars per SM alignment for process volumes of less the one hundred thousand. As volumes approach a million, cost per alignment should approach several dollars or less. These kinds of price reductions will be needed to enable optoelectronic communication to become a consumer product.

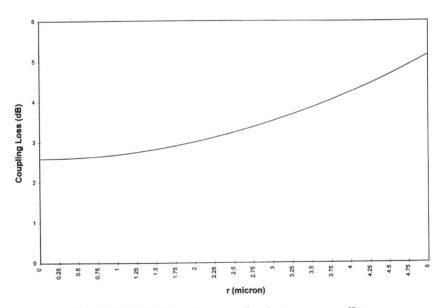

Fig. 18.27   VCSEL-to-fiber coupling for transverse offset.

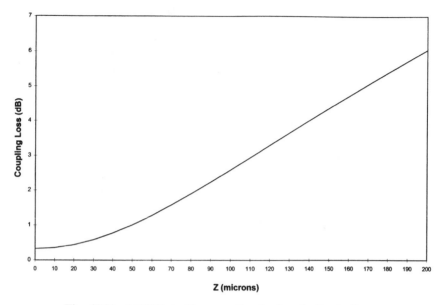

Fig. 18.28   VCSEL-to-fiber coupling for longitudinal offset.

18.6.   Conclusion

Analyzing coupling efficiencies and the effects of alignment technology is an inherently complex process. This chapter has outlined the relevant parameters and the principles needed. However, for detailed calculations, many references have been presented that explore theoretical and experimental techniques. Currently, all models show a slight deviation in absolute coupling efficiencies but the relative results are well established.

Various alignment methods have been presented but it is understood that this is an area that is still under development for optodatacom. Passive automated alignment of components is the ultimate goal of optodatacom manufacturing. Parallel components and VCSEL technology in particular present new opportunities to explore new manufacturing techniques and to reduce the cost of components.

References

1. Vukovic, M. 1996 (July). Analyze connector parameters to realize high performance. *Lightwave*, 50–56.

2. Paz, O., H. B. Schwartz, D. E. Smith, and E. B. Flint. 1992 (December). Measurements and modeling of fiber optic connector pluggability. *IEEE Trans. Comp. Hybrids Manuf. Tech.* 15(6):983–991.

3. Aulet, N. R., D. W. Boerstler, G. DeMario, F. D. Ferraiolo, C. E. Hayward, C. D. Heath, A. L. Huffman, W. R. Kelly, G. W. Peterson, and D. J. Stigliani, Jr. 1992 (July). IBM enterprise systems multimode fiber optic technology. *IBM J. Res. Dev.* 36(4):553–576.

4. Saleh, B. E. A., and M. C. Teich. 1991. *Fundamentals of photonics.* New York: Wiley.

5. International Electrotechnical Commission (IEC). International standard 825-1 and 825-2, safety of laser products. Central Bureau of the IEC, 3 rue de Varembe, Geneva, Switzerland.

6. American National Standards Institute. 1988. American National Standard for the safe use of optical fiber communication systems utilizing laser diode and LED sources. American National Standards Institute No. Z136.2.

7. Kawano, K. and O. Mitomi. 1986 (January). Coupling characteristics of laser diode to multimode fiber using separate lens methods. *Appl. Opt.* 15(1): 136–141.

8. Clement, D. P., and U. Österberg. 1995. Laser diode to single-mode coupling using an "out-of-plane misalignment" model. *Opt. Eng.* 34(1):63–74.

9. Hudson, M. C. 1974 (May). Calculation of the maximum optical coupling efficiency in multimode optical waveguides. *Appl. Opt.* 13(5):1029–1033.

10. Kogelnik, H. 1964. Coupling and conversion coefficients for optical modes. Proceedings of the Symposium on Quasi-Optics, June 8–10. In *Microwave research institute symposia series,* Vol. XIV. Brooklyn, NY: Polytechnic Press of the Polytechnic Institute of Brooklyn.

11. Marcuse, D. 1977 (May/June). Loss analysis of single-mode fiber splices. *Bell Syst. Tech. J.* 56(5):703–718.

12. Miller, C. M. 1986. *Optical fiber splices and connectors.* New York: Dekker.

13. Neumann, E. G. 1988. *Single-mode cables.* Heidelberg: Springer-Verlag.

14. Snyder, A. W., and J. D. Love. 1983. *Optical waveguide theory,* Part 1. London: Chapman & Hall.

15. Anderson, W. T., and D. L. Philen. 1983 (March). Spot size measurements for single-mode cables—A comparison of four techniques. *J. Lightwave Tech.* LT-1(1):20–26.

16. Nemoto, S., and T. Makimoto. 1974. Analysis of splice loss in single-mode fibres using a Gaussian field approximation. *Opt. Quantum Electron.* 11:447–457.

17. Lasky, R. C., U. Österberg, and D. J. Stigliani, eds. 1995. *Optoelectronics for datacommunication.* New York: Academic Press.

18. Karstensen, H., and K. Drögemuller. 1996 (May). Loss analysis of single-mode fiber couples with glass spheres or silicon plano-convex lenses. *J. Lightwave Tech.* 8(5):739–747.

19. Bargar, D. S. 1988. An automated fiber alignment, fixing, and hermetic sealing system. *SPIE* 994:11–17.

20. Bristow, J., Y. Liu, T. Marta, K. Johnson, B. Hanzal, A. Peczalski, S. Bounnak, Y. S. Liu, and H. Cole. 1996. MCM board level optical interconnects using passive polymer waveguides with hybrid optical and electrical multichip module packaging. *SPIE* 2691:18–24.

21. Gabler, C., L. Li, S. Hackwood, and G. Beni. 1986. An optical alignment robot system. *SPIE* 703:8–28.

22. Goldman, L. S. 1972 (May 15–17). Proceedings of the 22nd Electronic Components Conference, 332–339, Washington, DC.

23. Karioja, P., K. Tukkiniemi, V. Hikkinen, and I. Kaisto. 1993. Inexpensive packaging techniques of fiber pigtailed laser diodes. *SPIE* 1851:48–53.

24. Goodwin, J. 1986. Dynamic alignment of small optical components. *SPIE* 703:2–7.

25. Han, H., J. E. Schramm, J. Mathews, and R. A. Boudreau. *SPIE* 2691:118–123.

26. Kalman, R. F., E. R. Silva, and D. F. Knapp. 1996. Single-mode array optoelectronic packaging based on actively aligned optical waveguides. *SPIE* 2691:124–129.

27. Kilcolyne, M. K., J. W. Scott, and J. Plombon. 1993. Packaging of optical interconnect arrays for optical signal processing and computing. *SPIE* 1851:80–88.

28. Schmid, P., and H. Melchior. 1984. Coplaner flip-chip mounting technique for picosecond devices. *Rev. Sci. Instrum.* 55:1854–1858.

29. Spitzer, M. B., D. P. Vu, and R. P. Gale. 1996. Applications of circuit transfer technology to displays and optoelectronic devices. *SPIE* 2691:34–42.

30. Weber, R., F. Fidorra, M. Hamacher, H. Heidrich, and G. Jacumelt. 1993. Multi fiber/chip coupling and optoelectronic integrated circuit packaging based on flip chip techniques. *SPIE* 1894:149–152.

31. Yap, D., W. W. Ng, D. M. Bohmeyer, H. P. Hsu, H. W. Yen, M. J. Tabasky, A. J. Negri, J. Mehr, C. A. Armiento, and P. O. Haugsjaa. 1996. RF optoelectronic transmitter and receiver arrays on silicon waferboards. *SPIE* 2691: 110–117.

32. Tewksbury, S. K., L. A. Hornak, H. E. Nariman, and S. M. Lansjoen. 1993. Opportunities and issues for optical interconnects in microelectronic systems. In *Optoelectronics: Technologies and applications,* eds. A. Selvarajan, K. Shenai, and V. K. Tripathi. Bellingham, WA: SPIE.

33. Jackson, P., A. J. Moll, E. B. Flint, and M. F. Cina. 1988. Optical fiber coupling approaches for multi-channel laser and detector arrays. *SPIE* 994:40–47.

34. MacDonald, W. M., R. E. Fanucci, and G. E. Blonder. 1993. Si-based laser sub-assembly for telecommunications. *SPIE* 1851:42–47.

35. Solgaard, O., M. Daneman, N. C. Tien, A. Friedberger, R. S. Muller, and K. Y. Lau. 1995. Optoelectronic packaging using silicon surface-micromachined alignment mirrors. *IEEE Photon. Tech. Lett.* 7:41–43.

36. Lin, W., S. K. Patra, and Y. C. Lee. 1995. Design of solder joints for self-aligned optoelectronic assemblies. *IEEE Trans. Comp. Packaging Manuf. Tech. Part B,* 18:543–551.
37. McCroarty, J., B. Yost, P. Borgesen, and C. Y. Li. *Mater. Res. Soc. Symp. Proc.* 264:423–435.
38. Patra, S. K., J. Ma, V. Ozguz, and S. H. Lee. 1994. *SPIE* 2153:118–131.
39. Patra, S. K., and Y. C. Lee. 1991. *Electron. Packaging* 113:337–342.
40. Snyder, M. D., R. Lasky. 1995. Proceedings of the MRS, Spring 1995 meeting. San Francisco.

Part 5 | The Future

Chapter 19 | Market Analysis and Business Planning

Yann Y. Morvan

MPM Corporation, Medway, Massachusetts 02035

Ronald C. Lasky

Cookson America, Inc., Providence, Rhode Island 02903

19.1. Introduction

Performing an effective market survey and business plan may be the most important aspect of any successful technology. However, because most engineers have little experience with this topic, one is seldom performed until late in a program and when done is usually poorly performed. This weakness has been especially true in optoelectronics. There are numerous examples of sound technology that did not get implemented because of the lack of market analysis or a sound business plan. It is hoped that this chapter will help the reader to avoid pitfalls.

19.2. The Need for Applications

When Apple shipped its first MacIntosh in 1984 there were no customers crying for a product with its features. However, it was phenomenally successful. This success argues strongly against the philosophy of providing what the customer wants. Masseurs Job *et al.* anticipated what the customer would want and delivered a complete workable system. In today's high-technology world, this approach is the true winner. The lackluster companies will be content to provide only what the customer wants. On the other end of the spectrum is a technology that is impressive and anticipates a need but is not complete. A good, but perhaps facetious example would

HANDBOOK OF FIBER OPTIC
DATA COMMUNICATION

be the delivery of a Pentium integrated circuit (IC) in 1985. It is truly a great accomplishment, but there is no infrastructure to support it or make it useful. No circuit boards or buses exist that can support the high clock speeds, and there are no applications that require the fast speeds or supporting software. This example stresses the need to have technology support a complete system that performs a useful function.

Thus, when delivering a technology component there must be a need that it fills in the technology infrastructure for an existing or future application. An optoelectronics example would be a very low-cost fiber distributed data interface (FDDI) transceiver module. The FDDI standard, although its use has not emerged as rapidly as hoped, now has applications and a complete technology infrastructure. Hence, someone producing this transceiver can find customers that have a use for it. In 1992 a major company developed a single-mode (SM) Gb/s transceiver module. It was initially developed for internal use but was eventually targeted for the external market. There was much excitement for it because it seemed to support the information super-highway, but in 1992 no applications existed. Unfortunately, it did not attract much serious interest because the needed infrastructure did not exist.

19.3. Supporting Technology Infrastructure

The issue of supporting technology infrastructure needs some clarification. The previous example of the FDDI transceiver is used. A transceiver such as this one requires much technology to support it. Printed circuit boards (PCBs) that can support Gb/s data rates typically need to handle 20 channels of 50 Mb/s. In 1992 PCBs of this complexity were not common. The trans-ceiver did have a built-in serialize/deserialize (similar to mux/demux) func-tion. Many transceivers do not, hence the availability of low cost ICs to perform this function is crucial. These ICs are often available. Other inter-face hardware to the system of interest is also needed. At leading-edge data rates it will typically not be available unless there is a driving application. Assuming that the hardware issues are settled, what about software? To be useful, a transceiver usually has to talk to systems with different types of protocols. An example might be IBM microchannel architecture to AppleTalk. This software did not exist.

All these issues arose in late 1992 when a team of electrical engineering students was commissioned to use two of the Gb/s transceiver modules to enable an IBM RISC workstation to communicate with a DEC workstation. This application was extremely simple compared to likely use for this product, in something like a complex switch for the Fibre Channel Standard. The team initially felt that the assignment would be easy. However, after 8 months it could only produce a report on how it would solve the problem if it had more time and money. The following were their conclusions:

1. The data rate is so high that the team had to buffer the data between computers.

2. No software drivers existed that could handle the two different types of systems: IBM vs DEC. It would be a 10-man year effort to develop them for this simple application.

3. The team could find no PCBs or ICs to use to provide the hardware interface support needed to connect the module to the two computers.

It should not be surprising that the developers of this transceiver could not sell it. There were few applications requiring this data rate and no supporting hardware and software that would be needed to implement an entire, sellable function.

The technology that one is supplying must support a complete, sellable function for which there is a current or future application. If the technology predates the necessary hardware and software that are needed to support it, it may fail. One takes a tremendous risk if one produces a technology element for which the supporting elements are undeveloped.

19.4. Implementing a Market Survey

Performing a marketing survey should be one of the first tasks in a technology program. Many technologists take pride in their understanding of the market, but this pride is often misplaced because a thorough knowledge of a technology market is a full-time job.

Marion Harper Jr., a marketing executive, once said: "To manage a business well is to manage its future: and to manage the future is to manage information" [11, p. 103]. A market analysis is the prelude of planning, implementation, and control of a marketing strategy. The launch of any new product must go through this process in order to be successful. Knowing your market is not an option, it is a prerequisite for anyone involved in the product's design. The intent of this section is to offer a basic understanding of the major steps to follow when conducting a market analysis.

A. FIRST STEP: INDUSTRY DESCRIPTION OUTLOOK

The Standard Industrial Classification (SIC) manual is the first source to consult when one conducts market research. The researcher must give a detailed description of the primary industry in which the product is manufactured: size (present and over time), characteristics and trends, main customers (industrial, individuals, or government?), etc.

B. SECOND STEP: SEGMENT, TARGET, AND POSITION

Segmenting the market helps to obtain a clear view of one's customer. A segment consists of a group of potential customers (individual or organization) who are similar in the way that they value the product, in their patterns of buying, and in the way they use the product.

In order to be effective, the segmentation must follow Kotler's rule: It must be measurable, accessible, substantial, and actionable [11, p. 252]. In every segment, all the following items should be checked:

Price levels

Purchasing behavior

Helpful sources of information

Best media

Demographic information

Competitive products

Distribution channels

For a clearer view of the target market, it is interesting to establish a segmentation grid that summarizes all the characteristics of each segment (Table 19.1).

Once all the segments of the market have been identified according to different variables (demographic, geographic, psychographic, or behavioral), it is pertinent to select the segment that represents the most suitable opportunities to the company. To do so, the researcher must assess its size and anticipated growth as well as the company objectives and resources. In a target segment, not only strengths are required but also superiority over competition.

Anticipating the market penetration can be achieved after the interpretation of the data collected during the market research. Thus, the company can estimate its market share, the number of its customers, and its geographic coverage. Any trend or expected change should always be kept in mind

Table 19.1   **Segmentation Grid**

	Usage rate		
	Segment No.		
	1	**2**	**3**
Products purchased	. . . . .	. . . . .	. . . . . .
Product benefits and attributes	. . . . .	. . . . .	. . . . . .
Key information sources and influences	. . . . . .	. . . . .	. . . . . .
Best media	. . . . . .	. . . . .	. . . . . .
Price limit	. . . . . .	. . . . .	. . . . . .
Best distribution channels	. . . . .	. . . . .	. . . . . .
Demographic information	. . . . .	. . . . .	. . . . . .
Segment size (example)	35%	50%	15%
Market size = 100%			

Note. From Harmer, R., *New Product Development Project Workbook.* Boston: Boston University School of Management, 1994, pp. 3–4. Comment: The highest percentage for segment 2 does not mean that it is the most enticing but simply that it represents 50% of the entire market. The biggest segment of a market is not always the best suited to the company's resources.

when making a decision. The environmental analysis shall never be forgotten. Its main parameters are economic forecasts, governmental regulations, demographic and social trends, and technology breakthroughs.

Even if the company greatly focuses on its target segment, it is wise to collect precious information, such as needs, demographics, and significant future trends, on the secondary target segments because they may become attractive in the future.

The last assignment is the position step. This final stage can be achieved with the analysis of one's competition. Identifying competitors and listing their strengths and weaknesses is a good way to refine one's positioning strategy. The company must place its product, in the customer's mind, at a higher value than any other competing products. It should emphasize the appropriate competitive advantages that the product may offer and base its communication on them.

C. THIRD STAGE: DEFINING THE MARKETING STRATEGY

The third stage consists in defining the marketing strategy based on all the information collected throughout the market analysis (Fig. 19.1). It is clear that the marketing strategy is not a static formula but should always be

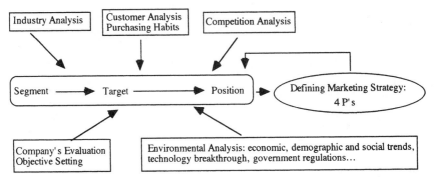

Fig. 19.1   The market analysis diagram. From Morvan, Y., Université Jean Moulin, Lyon, France. Copyright 1996.

modified according to the market evolution and the life cycle of the product. The "four P's," in marketing terminology, represent the four pillars of the strategy: product, price, promotion, and place (i.e., distribution). A market analysis on the transmitter optical subassembly (TOSA), presented in the Appendix, gives an illustration of most of the points discussed previously.

D.   NOTE ON DATA COLLECTION

A market analysis consists of collecting data that come in two forms: primary and secondary. Primary data are collected for a specific goal through methods such as personal interviews, focus groups, questionnaires, experiments, or observation. On the contrary, secondary data consist of information available at some source without a specific goal. Secondary data (Table 19.2) include the company's internal information, government publications, periodicals and books, commercial data, and international data. Most secondary data can be obtained from public libraries or university libraries specialized in business. Let us point out that, nowadays, the Internet represent a wonderful source of information, often at no charge.

19.5.   Business Planning

Jeffry Timmons wrote in his book, *New Venture Creation Entrepreneurship in the 1990's,* "Planning is a way of thinking about the future of a venture; that is, of deciding where a firm needs to go and how fast, how to get there,

Table 19.2   **Sources of Secondary Information**

Industry analysis

    Directories: SIC manual, *U.S. Industrial Outlook* (USDOC), *Standard & Poor's Industry Surveys*, *Census of Manufacturers* (USDOC), *Industry Norms and Ratios* (Dun & Bradseet), *Market Share Reporter*

    Databases: Nexis, Investext, ABI/Inform, F&S Index: United States and International

    Others: Books, periodicals, Internet

Competition analysis

    Directories: *Ward's Business Directory, Million Dollar Directory, Standard & Poor's Register of Corporations, Moody's Manual, Principal International Business, International Directory of Corporate Affiliations*

    Databases: Nexis, Dialog databases, Dow Jones News/Retrieval, Global Vantage PC Plus, American Business Disk

    Others: Company brochures, books, periodicals, Internet

Environmental analysis

    Government publications: *Statistical Abstract of the U.S., County and City Data Book*

    Periodicals and books: General business and economic magazines and newspapers such as *Business Week* or *The Wall Street Journal*, specialized newsletter and periodicals

Company's evaluation

    Internal source: Income statement, balance sheets, ratio analysis, prior research reports

---

Note. From Morvan, Y., Université Jean Moulin, Lyon, France. Copyright 1996.

and what to do along the way to reduce the uncertainty and to manage risk and change."

    Business planning aids in developing the marketing strategy to be carry out based on the market analysis described previously and on the resources and objectives of the company. It is commonly included in a document called "business plan." Let us now evaluate the purpose of such a document and study its content.

A. THE PURPOSE OF A BUSINESS PLAN

Figure 19.2 describes the main advantages of having a business plan. It sets the objectives of the company based on the knowledge of its resources (physical, financial, and human), its market, and its environment. It offers

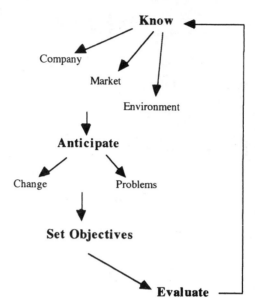

Fig. 19.2 Why a business plan? From Morvan, Y., Université Jean Moulin, Lyon, France. Copyright 1996.

a long-term view for the manager, which helps to anticipate future trends. The business plan is a document that anyone inside the company can access to evaluate performance. It can also be utilized externally by the firm to attract investors or to communicate with outside players such as suppliers.

The purpose of a business plan is to

1. Evaluate capacity and potential
2. Set objectives
3. Evaluate performance
4. Communicate

B. THE CONTENTS OF A BUSINESS PLAN

A traditional business plan includes the following topics:

Executive summary
Company description
Product description

Market analysis and marketing strategy

Financial analysis

A business plan must be clear, comprehensive, and concise. Always keep in mind that your reader does not necessarily know the nature of your business; therefore, it is useless to overwhelm him or her with too many technical terms.

The following outline can help to understand what a business plan should include. It is only a suggestion and should be regarded as such. Not all business plans are alike, simply because every company is unique:

Business plan outline

1. Executive summary
   a. Purpose of the plan
   b. Market analysis
   c. Company and product descriptions
   d. Financial data
2. Company description
   a. Nature of business
   b. Ownership
   c. History
   d. Location and facilities
3. Market analysis
   a. Industry
   b. Market segmentation and target segments
   c. Competition analysis and positioning
   d. Environmental trends affecting the market
4. Product description
   a. Benefits and competitive advantages
   b. Product life cycle
   c. Copyrights and patents
5. Marketing strategy
   a. Market penetration strategy
   b. Pricing strategy
   c. Distribution strategy
   d. Promotion strategy
6. Management and ownership
   a. Legal structure and organization chart
   b. Brief description of key managers (compensation, skills, etc.)

   c. Board of directors
   d. Shareholders
 7. Financial data
   a. Historical documents
        Income Statement
        Balance sheet
        Cash flows
        Key ratios analysis
        Break-even analysis
   b. Prospective documents
        Income Statement
        Balance sheet
        Cash flows
        Key ratios analysis
        Break-even analysis
        Capital budgets
 8. Appendices
   a. Resumes of key managers
   b. Pictures of products
   c. Professional references
   d. Market studies
   e. Pertinent published information
   f. Patents
   g. Significant contracts

Writing an effective business plan is not an easy task but you can obtain precious help from college business schools, another business owner, the local chamber of commerce, library/bookstore, or business association. Many software programs are also available to assist you in this work, especially with regard to the financial aspects.

VI. Summary

The key steps in market analysis are market segmentation, market targeting, and market positioning. The segment, target, and position process is the prerequisite for a successful commercialization. The market analysis is often included in the business plan. Business planning gives an evaluation of the capacity and potential of the company; it also sets objectives, evaluates performance, and represents an internal and external medium of communication.

Appendix: Market Analysis on a Transmitter Optical Subassembly

Introduction

This summary presents information obtained through literature research concerning the market analysis on a TOSA.

The product has the following characteristics:

- Data rates above 1 Gb/s; distance without repeaters >20 km
- Single-mode light 1.3 or 1.55 μm in wavelength; rise and fall times under 0.3 ns; spectral width under 1 nm
- Most likely applications will be asynchronous transfer mode (ATM) application at Gb/s speeds and Fibre Channel Standard applications at Gb/s speeds

The purpose of this work is to assess the overall industry of fiber optic components, differentiate the segments of this market, define the target segment and its trends, and initiate a study of the competition.

Industry Description and Outlook

DESCRIPTION

The optical fiber, characterized by low transmission loss and immunity to electromagnetic interferences, provides the transmission speeds and high-volume data handling that are currently driven by the consumer needs.

The U.S. fiber optics industry comprise cables, optoelectronic components (transmitters, receivers, and fiber amplifiers), connectors, and passive optical devices. Telecommunications, data networks, and cable television are the main users of fiber optic equipment.

The transmission of all types of signal, including voice, and video data, will be predominantly digital beyond the year 2000. In 1993, 50% of the fiber optic networks carried only data, whereas 21% carried voice, data, and video at the same time [1] (Fig. A.1).

INDUSTRY CHARACTERISTICS AND TRENDS

More than 50% of U.S. fiber optic networks have been installed since 1991. The change from copper to fiber indicates that a greater capacity, or bandwidth, is required to deliver the broad array of services proposed for the superhighway. The need for greater bandwidth is evidenced in part by the fact that transmission of data over telephone is increasing by approximtely 20% a year [2].

This industry is expected to enjoy continued growth throughout the 1990s. In 1993, the North American consumption of fiber optic components was worth $1.93 billion and is estimated to reach $9.14 billion in 2003. Therefore, the consumption of optoelectronics in North America is expected to grow at an average annual rate of 18.75% between 1993 and 2003 [3] (Table A.1). Although telecommunications still leads demand (Table A.2), cable TV (CATV) and data communications are growing more rapidly (Table A.3), and have become significant markets with different needs because they consume more optoelectronic interfaces than does the telecommunication industry. The fastest fiber optic optoelectronics growth over the next decade should be in enterprise or premises data communications network applications. In this market, fiber optics must compete against the cost of unshielded twisted-pair copper wire, coaxial cable, and electronic terminations and connectors. The total North American consumption of optoelectronics components in enterprise networks should expand by 32% per year between 1993 and 2003. Over the next 10 years, the most important

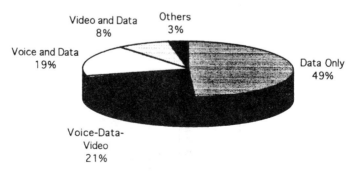

Fig. A.1 Fiber optic networks. Source: Kessler Marketing Intelligence, 1993.

Table A.1 **North American Consumption of Fiber Optic Components**

Component type	1993		1998		2003		Average annual growth rate (%/year)	
	$ (Billion)	%	$ (Billion)	%	$ (Billion)	%	1993–1998	1998–2003
Cables	1.19	61.6	2.28	55	4.9	53.7	13.9	16.5
Optoelectronics	0.57	29.5	1.35	32.6	3.18	34.8	18.8	18.7
Connectors	0.12	6.4	0.31	7.3	0.61	6.6	20.8	15.1
Passive optical devices	0.05	2.5	0.21	5.1	0.45	4.9	33.2	16.5
Total production	1.93	100	4.15	100	9.14	100	16.5	17.1

Table A.2   **North American Fiber Optic Installation Apparatus**

Telecommunications	75.0%
Premises data networks	18.0%
Cable TV	7.0%

Note. Distribution of fiber market apparatus installations is shown in percentage. Total market in 1992 was $384 million. Source: Lightwave, 1993 (August), p. 24, from Electronicast Corp.

growth of optoelectronics-electronic demand in enterprise networks should be in premises ATM switches (Table A.4). Light-emitting diodes and laser diode-based transmitter/receiver unit prices are expected to drop over the next decade. The laser diode-based transmitter/receiver share should drop to 56% by 1998 and then rebound to 59%, or $1.67 billion, in 2003.

New equipment used for multimedia applications (full-motion video and supercomputer visualization), digital high-definition TV, video conferencing, distributed high-speed supercomputer networks, and other wide-bandwidth applications will enter the market within 5 years and drive an explosive expansion of bandwidth demand. Telecommuting and distance learning are also promising applications.

World Fiber Optics Industry

Currently, the world market of optoelectronics is worth $5 billion and is predicted to reach $10 billion by the end of the century. The U.S. Department of Commerce notes that Asian-Pacific countries, with more than two-thirds of the world population, represent one of the fastest growing markets for fiber optic equipment. Emerging countries in the Caribbean, Latin America, and Eastern Europe will also provide opportunities as these countries upgrade antiquated networks (Brazil, Mexico, etc.). The Eastern European market amounted to $42 million in 1991 and is forecasted to be $1 billion by the end of the century. The Asian-Pacific countries represented a $1 billion market in 1993, a market should double by the Year 2000. The world's largest underdeveloped telephone market is China. The Chinese government wants to raise the ratio of 2 lines/100 people to 10 lines/100 people by the Year 2000 and 40 lines by the Year 2020 [4].

Table A.3 North American Consumption of Fiber Optic Optoelectronics (by Application)

Application	1993		1998		2003		Growth rate (%)	
	$ (Million)	%	$ (Million)	%	$ (Million)	%	1993–1998	1998–2003
Telecommunications	789	75	1148	56	3180	48	8	23
Enterprise (premises) data	166	16	627	31	2853	43	30	35
Cable TV	38	4	107	5	260	4	23	19
Military/aerospace	45	4	102	5	212	3	18	16
Specialty applications	18	2	49	3	158	2	22	26
Total consumption	1056	101	2033	100	6663	100	14	27

Source: ElectroniCast Corp.

Table A.4 Enterprise Network Consumption of Fiber Optic Optoelectronics (by Network Product Category)

Network product	1993		1998		2003		Growth rate (%)	
	$ (Million)	%	$ (Million)	%	$ (Million)	%	1993–1998	1998–2003
Premise ATM	2.3	1	191	30	1704	60	142	55
FDDI	36	22	83	13	40	1	18	–14
Fast Ethernet	0	0	96	15	364	13		31
Fiber Channel	0	0	62	10	441	15		48
Conventional data Communications	128	77	194	32	303	11	9	
Total consumption	166.3	100	626	100	2852	100	30	9

Source: ElectroniCast Corp.

U.S. producers and consumers of fiber optic equipment appear to be increasingly optimistic about the potential demand for the expanded service offerings possible through fiber optic technology.

Target Markets

The optical communication market includes applications in telecommunication, computing cable television, and automobiles. Currently amounting to approximately \$3 billion, this market is supposed to grow to more than \$30 billion by the Year 2003, which represents an annual growth of 26% between 1993 and 2003 (Fig. A.2). Over the next 20 Years, the technology used in both telecommunications and data communications is going to evolve. (Fig. A.3) presents the optical communications product road map.

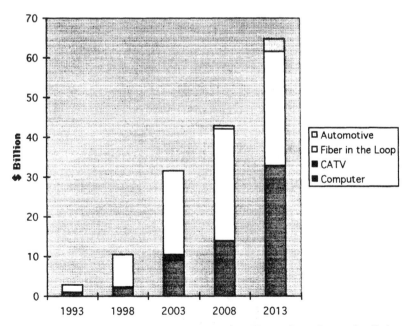

Fig. A.2   Optical communications market size. From Optoelectronics Industry Development Association [5].

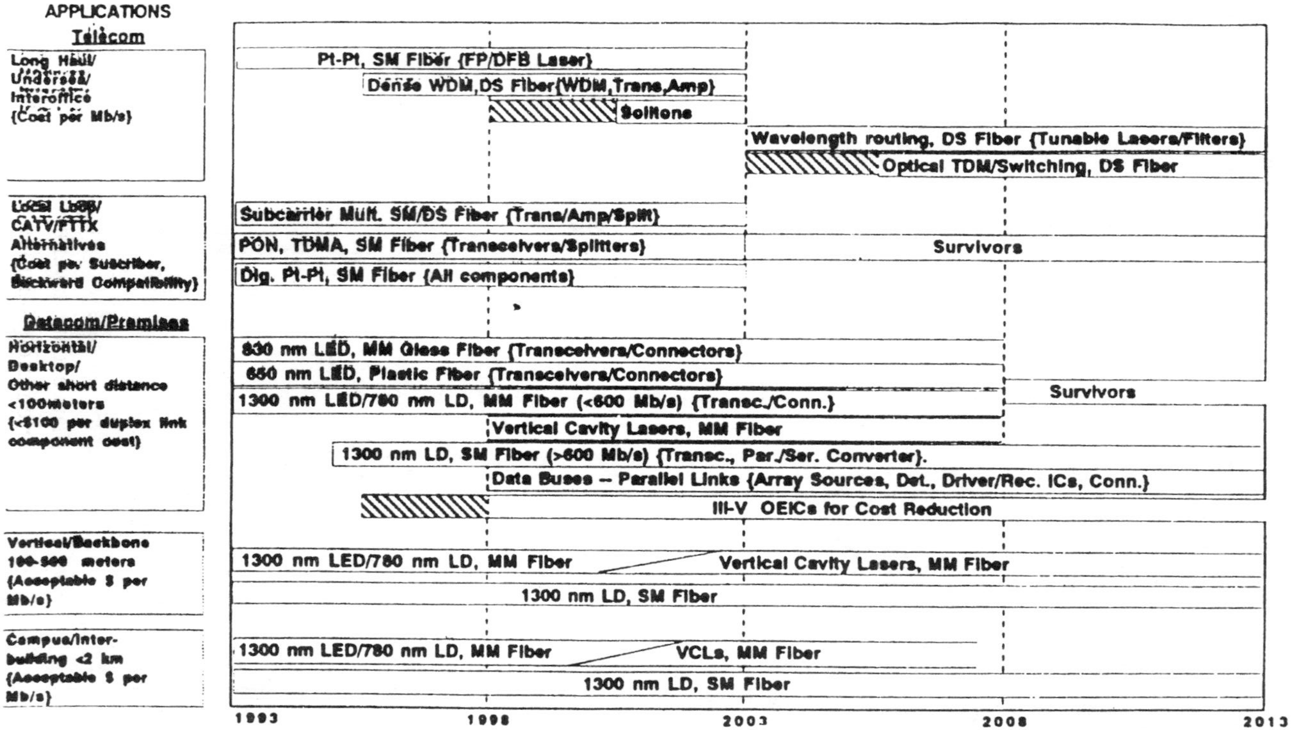

Fig. A.3 Optical communications product road map.

The data communications market for fiber optic equipment, which was just $0.8 billion in 1993, is forecast to grow to $2.6 billion in 1998 and to $6.3 billion in 2003 (i.e., annual growth of 23% from 1993 to 2003). The Optoelectronics Industry Development Association (OIDA) divides the optical communication equipment market in four segments:

1. Long-distance telecommunications (trunk and municipal area networks): This is the high-performance segment of optical communications, characterized by repeaterless spans of 5–100 km and high speeds (0.6–2.5 Gb/s). A key requirement is high-performance lasers.

2. Shorter distance telecommunications (such as fiber in the loop): This segment is characterized by spans of 1–10 km and speeds of 50–622 Mb/s. This market segment is growing and is highly cost sensitive. U.S. industry lags behind Japan in this segment and needs improvement, primarily in the form of lower cost manufacturing, packaging and alignment technologies, and better epitaxial growth and processing of lasers.

3. High-performance data communications: This segment is characterized by long distances (for data communications) of 300–2000 m or relatively high speeds (200–1000 Mb/s). It is cost sensitive. U.S. industry is competitive, but improvements in the form of cost reductions are needed.

4. Low-cost data communications: This segment is characterized by short distances (<300 m) or low speed (<200 Mb/s). This segment is very cost sensitive. Component costs need to be competitive with wire or small relative to installation costs, and currently are in many cases. Individual U.S. suppliers are strong, but users would like more domestic suppliers. The main barrier of this segment is the lack of familiarity with optical technology on the part of installers and users.

Much of the market growth is expected to be in segments 2–4, which are all characterized by higher volumes and higher cost sensitivity than the first segment, which has been the largest segment to date [5].

The segment, high-performance data communications, which we are targeting, is expected to experience considerable growth during the next decade.

HIGH-PERFORMANCE DATA COMMUNICATION SEGMENT

Characteristics

The main drawback that the fiber optic industry is facing is the high cost of optoelectronic interfaces. The cost of optoelectronic transmitters, in particular, can add significantly to the cost of hardware such as hubs, concentrators, and network cards. In 1992, according to Frost & Sullivan, the average price of any kind of fiber optic transmitter was $600, whereas the other components cost much less: an eight-fiber cable cost approximately 59¢/ft, an average of $100 for a receiver, an average of $20 for a connector, and an average of $170 for a coupler [6].

Optical communication technologies are being used increasingly for shorter distance communication, where much of the growth is expected and where lower cost components are required. Therefore, the reduction of the transmitter cost is a crucial factor for the growth of the high-performance data communication segment.

A faster local area network (LAN) is of concern to not only large companies but also small companies. According to Fig. A.4, companies earning between $50 and $500 million per year are also considering high-speed LAN solutions, in some cases more aggressively than larger firms.

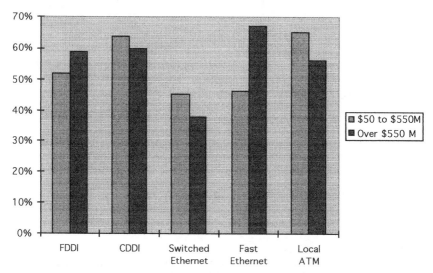

Fig. A.4 Percentage of larger and smaller companies that expect to be evaluating higher speed transmission. Source: Sage Network Research & RFTC, Inc., 1993.

Fiber and optoelectronic component manufacturers are also expecting a local loop fiber explosion (Table A.5). Deepak N. Swamy, a senior analyst for KMI Corp., predicts that,

By the Year 2002, the U.S. market for fiber optic cable and equipment for loop applications will increase to more than $5 billion. By then, local exchange carriers will have installed 7.9 million local fiber access lines that may each connect hundreds of customers. Meanwhile, cable television companies will connect more than 40 million subscribers to small-radius nodes. [6]

In short, the critical need for this segment is low-cost optoelectronic components able to sustain the increasing transmission of data. The TOSA would fully satisfy this need.

High-speed telecommunications standards are synchronous optical net works (SONET) and synchronous digital hierarchy (SDH). The standard of the data communication segment is Fibre Channel Standard (FCS).

Table A.5   **The Future of Fiber Applications**

	%	
Application	**1993**	**1999**
Long-haul terrestrial	52.3	38.2
Under sea	22.6	7.5
Subscriber loop	14.7	35.3
MAN	4.3	6.8
LAN	6.1	12.2

Note. According to "World Fiber Optic Communication Markets," a report released by Market Intelligence Research Corp., although long-haul fiber will continue to present the majority of total worldwide fiber cable revenues, it is expected to lose market share. Long-haul applications include terrestrial and undersea cabling. The strongest foreseen growth application for fiber optics is the subscriber loop. Its revenues are expected to jump to well over double its present market share by the end of the decade. Other gainers are expected to be LANs and metropolitan area networks (MANs). MANs are already enjoying popularity as high-bandwidth communication links and are expected to continue to do so. Fiber deployment in the LANs is expected to continue increasing to address the growing needs of private network users. Source: Telecommunications, 1993 (September), p. 12.

SONET is a North American telecommunication standard for a high-capacity fiber optic transmission system. The transport system is based on the principles of synchronous multiplexing, which offers compatibility with today's data rate and tomorrow's capacity. The traditional speed rates are 51.84, 155.52, and 622 Mb/s, and 2.48832 and 9.95328 Gb/s.

SDH is an international standard similar to SONET, but it uses the terminology transport module to refer to an optical transmission rate.

Fibre Channel is an example of a MAN that is optimized for transfers of large amounts of data between high-performance processors, disk and tape storage systems, and output devices, such as laser printers and graphic terminals. It converts bytes of data into a serial transmission stream at signaling rates of 132.813 and 256.625 Mbauds and 1.0625 Gbauds.

Trends

Although telecommunications remains the main customer of optoelectronic components much of the growth is expected to occur in enterprise data networks (Fig. A.5).

On April 19, 1995, the *Wall Street Journal* stated,

Institutions are rushing to build computer networks to use as competitive tools. . . . . Much of the buoyancy in the industry's fortunes also comes from

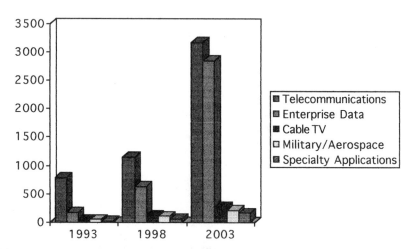

Fig. A.5 North American Consumption of Fiber Optic Optoelectronics. Source: Electronicast Corp., 1993, *Lightwave*.

changes in the way corporations use computers. The trend toward hooking computers into vast networks is accelerating, boosting sales of the "server" computers—like
Sun's and H. P.'s—that tie networks together as well as the software—such as
Novell Inc.'s—that runs on networks. Sun attributed its earning surprise to strong
sales in networking and Internet markets.

This trend is confirmed in Table A.6, in which the feeder/local telecom
sector is expected to grow by 31.5% every year until 1998.

The following chart shows the forecast revenues for sending high speed
data communication over fiber optic cable:

	$(Billion)
1993	0.8
1998	2.6 (Annual growth rate 22.9%)
2003	6.3

Source: Data over fiber, OIDA, 1994, Computer Reseller
News, p. 61. Copyright © 1994 by CMP Publications, Inc.,
600 Community Drive, Manhasset, NY 11030. Reprinted
from Computer Reseller News with permission.

This chart demonstrates that high-speed data communication has a bright
future because its revenues are expected to grow at an annual compound
rate of 22.9%.

The future of high-speed rate lasers or transmitters is in the short data
transfer of the high speed computer interconnects. Distances are on the
order of several thousands feet compared with telecommunication's thousands of kilometers. The speed rate determined by the High Speed Com-

Table A.6 **Forecast Market for Fiber Optics (by Application Sector)**

	Year		
	1992	**1998**	**Annual growth rate (%)**
Long-haul telecom	1668	1,715	0.5
Feeder/local telecom	1364	7,048	31.5
Multimode	1807	2,053	16.8
Cable TV	198	800	26.2
Other	120	560	29.3
Total	4157	12,176	19.6

Source: KWI Corp., 1993 (December 1), Lightwave, p. 9. Copyright 1993 by Lightwave, PennWell
Publishing Co., Nashua, NH, USA.

mittee of the American National Standards Institute (ANSI) will be 500 Mb/s to 1 Gb/s. Video is another promising application for high-speed transmitters, especially CATV.

The use of optical fibers and, therefore, the use of transmitters in subscriber loop systems will continuously increase. Indeed, the optical data communications business should experience a tremendous development during the next decade as fiber is extended from the central office further toward business and residential customers and as the use of higher capacity local area networks increases. The different configurations of subscriber loops are the following:

Fiber-to-the-curve (FTTC)

Fiber-to-the-office (FTTO)

Fiber-to-the-home (FTTH)

FTTC and FTTO use metallic cables near subscribers, whereas FTTH brings the optical fiber directly to the home of the subscribers. Residences and small businesses would be the beneficiaries. FTTH is considered to be the final step. However, FTTH is difficult to realize without lowering the cost of optical receivers and transmitters.

Among the optical subscriber loop networks using single-mode fibers, the following different methods of multiplexing are used:

Space-division multiplexing (SDM)

Wavelength-division multiplexing (WDM)

Directional-division multiplexing (DDM)

Time-compression multiplexing (TCM)

SDM is the most expensive because two fibers are needed; one for the upstream transmission and one for the downstream transmission. In WDM, two different wavelengths are performed over a single fiber (usually 1.3 and 1.5 nm) that require two lasers, increasing the cost. In DDM, the laser diode and photodiode are combined by using a coupler. A TCM system is close to DDM, when a diode is in the transmission mode and the other is in the receiving mode ("ping-pong transmission"). The maximum information bit rate in the TCM system is 1.5 Mb/s.

SURVEY

In order to present a clear evaluation of the high-speed data communication segment, a survey has been designed (Fig. A.6). Thanks to the results, we will be able to acquire precious information on the following:

SURVEY

Hello,

Our company is interested in developing an optical transmitting subassembly for use in high-speed fiber datacommunication. We would like to develop a product with the following characteristics:

- Data rates > 1 gigabit/sec, distances without repeaters >20 km.

- Single-mode light 1.3 or 1.55 microns in wavelength, rise and fall times < 0.3 nanoseconds, spectral width < 1 nanometer.

- Most likely applications will be:
  Asynchronous Transfer Mode (ATM) applications at Gb/s speeds
  Fiber Channel standard applications at Gb/s speeds

Would you please take the time to respond to the following questions? Thank you.

I) How would you describe your organization?
a) Telecom Industry
b) Enterprise Datacom Industry
c) Cable TV
d) College/University
e) Government
f) Other_____

II) How many 1 Gbit/s optical subassemblies does your organization currently purchase annually?
a) 0
b) 1-99
c)100-1000
d) 1000-10,000
e) >10,000

III) How many Gbit optical subassemblies does your organization foresee purchasing in 1996?
a) 0
b) 1-99
c)100-1000
d) 1000-10,000
e) >10,000

IV) How much would you expect to pay for an optical subassembly as described above?
a) >$5000
b) $3000-$4999
c) $1500-$2999
d) $500-$1499
e) <$500

V) Do you believe that low price implies low quality?
a) Yes
b) No

Fig. A.6 Survey.

VI) Which characteristic of current commercial optical subassemblies are you most
dissatisfied with?
a) Speed
b) Price
c) Other_____

VII) Which of the following applications does your organization presently use?
a) FDDI
b) ATM
c) Fiber Channel
d) ESCON
e) Hybrid (explain) _____
f)Other_____

VIII) Which type of fiber conection would you prefer?
a) FC
b)ST
c)SC
d)ESCON
e) Other_____

IX) What wavelength will your organization be using in 1996?
a) 1.55μ
b)1.3μ
c) 700-900nμ

X) When do you envision such a transceiver would be desireable for your network?
a) Immediately
b) 1-2 years
c) 3-5 years
d) over 5 years
e) never

XI) From which of the following sources do you receive information on optical transceivers?
a) Professional Journals
b) Personal references
c) Trade Shows
d) Trade Magazines
        d1) Laser Focus World
        d2) Lightwave
        d3) Lasers and Optronics
        d4) Photonics Spectra
        d5) Fiberoptic Product news
        d6) Other_____
e) Other _____

XII) How would you describe your position within your organization?
a) Product Development
b) Purchasing
c) Sales
d) Management
e) Research
f) Other_____

Fig. A.6   *Continued*

Size of the target market

Critical needs

Price level

Weight of each application (FDDI, ATM, Fiber Channel, ESCON, etc.)

Media to use to reach this segment

Measure of the desirability of the TOSA

Forecasted sales

Potential customers

Competition

COMPANIES CONTACTED

Approximately 25 companies in the United States were identified as manufacturers of digital transmitters in the *Thomas Register* and the buyer's guide 1995 from *Laser Focus World* magazine. Receiving catalogs and price lists of their products will enable a clear identification of the competition by product line. The following is a list of the companies contacted:

Name of Company	*Phone No.*
Advanced Fiber Optic Technologies (CA)	818-357-0159
Analog Modules Inc. (FL)	407-339-4355
AT&T Microelectronics (PA)	610-712-5133
Broadband Communications Products Inc. (FL)	407-728-0487
Fiber Optic Center (MA)	800-473-4237
Fiber Options Inc. (NY)	516-567-8320
Force Inc. (VA)	703-382-0462
Hewlett-Packard Co. (DE)	800-545-4306
Laser Diode Inc. (NJ)	908-549-9001
Lasertron Inc. (MA)	617-272-6462
Litton Poly-Scientific (VA)	703-953-4751
Math Associates (NY)	516-226-8950
MRV Technologies (CA)	818-773-9044
NEC Electronics Inc. (CA)	415-960-6000
Optical Communication Products Inc. (CA)	818-701-0164
SI Tech (IL)	708-232-8640
United Technologies Photonics (CT)	203-769-3000

Name of Company	Phone No.
Ross Engineering (CA)	800-654-3205
Philips Broadband Networks (NY)	800-448-5171
Telecommunication Techniques Corp. (MD)	800-638-2049
Telco Systems (MA)	617-551-0300
LNR Communication, Inc. (NY)	516-273-7111
FiberCom, Inc. (VA)	800-537-6801
Optek Technology Inc. (TX)	214-323-2301

NEW PRODUCTS

In this very high-technology market, it is important to be aware of the new projects undertaken by other companies that could create either direct or indirect competition.

Scientists at Bell-Northern Research (BNR), the R&D arm of Northern Telecom, have developed a low-cost optoelectronic device. Known in scientific circles as a Mach–Zehnder (MZ) optical modulator, made of III–V semiconductor, it has a speed rate of 10 Gb/s. In a BNR test, scientists used the semiconductor MZ prototype to successfully maintain 10 Gb/s transmission over conventional glass fiber at distances exceeding 100 km with minimal distortion or signal loss. The BNR model is less than 1% of the size of existing units. Its semiconductor design requires less power and is more resistant to temperature and vibration than other units. BNR's MZ also has the potential to be economically mass produced [7].

At Supercomputing '94, on November 15, the IBM Research Division demonstrated its new Rainbow 2 all-optical network. The network prototype can support 32 nodes, each running at 1 Gb/s. IBM said that,

They are developing low-cost high-speed networks to answer increasing demands in scientific, medical, university and other research environments for networks with one Gbps per node throughout rates. . . . Multimedia applications, in particular, will take advantage of this speed in areas such as full-motion video and supercomputer visualization. . . . University and other research environments already need local area networks and metropolitan area networks that will provide supercomputer support of high-performance workstations with live-motion color graphics and supercomputer visualizations. Medical imaging is another important applications [8].

Position

The TOSA uses vertical cavity surface-emitting laser (VCSEL) technology, which is predicted to become, according to the OIDA, a noticeable technology in data communication by the Year 1998:

The Fibre Channel Systems Initiative (FCSI) and the Fibre Channel Association (FCA) announced on March 14, 1995 a major leap in possible data communication speeds with the ANSI Committee adoption of Fibre Channel standards for 2 and 4 Gbps data rates. This effectively quadruples the previous Fibre Channel ANSI standard of up to 1 Gbps and provides the standards for the fastest data communication speed possible to date. This major advancement in data communication speed is made possible through Fibre Channel's enabling implementation of Vertical Cavity Surface Emitting Laser (VCSEL) technology, a new technology that provides a practical, cost-effective means of using lasers to transmit data at ultra-high speeds. The ultra-high data transfer speed is ideal for such applications as motion picture and video production where large amounts of audio and other digitized information is manipulated [9].

VCSELs offer several important advantages over conventional edge-emitting visible diodes, including surface-normal output, ease of fabrication into two-dimensional arrays, a circular beam with little angular divergence, wafer-level testing and additional control over the lasing wavelength. Visible VCSELs may be an enabling technology for many advanced applications such as plastic-fiber-based communications for LANs, 2-D visible arrays for displays (including laser-projection), printing applications and optical memory. [10]

According to an article published in *Photonics Spectra* in February 1995, VCSELs appear to be the "key technology for future data communications." It was first introduced on the market in 1992 by Photonics Research Inc. and Bandgap Technology Corp., both of which later became a single company called Vixel Corp.

VCSELs offer many advantages:

Easy to manufacture

Very low cost

High performance

Multichannel lightwave transmitters

Compact

Low-divergence circular beams

High wall-plug efficiency

Low drive current of a few milliamps

Low drive voltage of a few volts

The primary applications of VCSELs are the following:

Fiberoptic communication

Optical storage

Laser printing

Laser scanning

Optical sensing

Applications of VCSELs in data communications include:

Massive parallel processing

Interconnections of workstations and high-performance PCs and video file servers

Premises switching

Optical LANs

Inter- and intracabinet switching and Fiber Channel switching

VCSELs lower the cost and increase the performance of optoelectronic devices. Today, their wavelength ranges from 630 to 1050 nm and is expected to be from 480 to 1300 nm in the Year 2000. The primary function of VCSELs is the transport of information at high speed over single or multiple channels in which the use of this technology will expand. They fulfill the needs created by the high-performance data communication segment.

Conclusion

This analysis shows that the fiber optic industry will experience large growth during the next decade. The North American consumption of fiber optic components is expected to increase at an annual rate of 20% in the 1993–2003 period. The global market provides a similar perspective with countries such as China, Brazil, Mexico, or the Asian Pacific countries. Although telecommunications remains the main customer of this industry, the most rapid growth will occur in data communications with a rate of 30% compounded annually during the next 10 years. By segmenting the overall market, we realize that the high-performance data communication segment shows a critical need: a reduction in the cost of optoelectronic interfaces. Indeed, the current trend is toward the local networks that require lower cost in order to be profitable. This need could be satisfied by the low-cost TOSA.

The market analysis is certainly not finished. The results of the survey will enable us to measure the desirability of this product and give us valuable information on this particular market segment. However, personal interviews and focus groups will need to be conducted. Personal interviews will include a discussion with professionals in the field or with potential customers, whereas a focus group will give many reactions to the presentation of the concept. A study of the competition and an evaluation of the regulatory restrictions are other aspects of the market analysis that should be completed.

References

1. *Lightwave.* 1994 (January). p. 15. Copyright 1994 by Lightwave, PennWell Publishing Co., Nashua, NH, USA.
2. *Standard & Poor's.* 1994 (October). p. T43. Reprinted by permission of Standard & Poor's, a division of The McGraw-Hill Companies.
3. *Lightwave.* 1993 (December). p. 42. Copyright 1993 by Lightwave, PennWell Publishing Co., Nashua, NH, USA.
4. *U.S. Industrial Outlook.* 1994.
5. Optoelectronics Industry Development Association. 1994. *Optoelectronic technology road map.*
6. *Photonics Spectra.* 1994 (February).
7. *Canada Newswire.* 1994 (January 13).
8. *Business Wire.* 1994 (November 15).
9. *OIDA.* Edge Publishing. 1995 (March 20). No. 347.
10. *Photonics Spectra.* 1993 (May).
11. Kotler, P., and Armstrong, G. Principles of marketing, 6th edition, Prentice Hall, Englewood Cliffs, NJ.

Chapter 20 | New Devices for Optoelectronics: Smart Pixels

Barry L. Shoop
Andre H. Sayles
Daniel M. Litynski

Photonics Research Center, Department of Electrical Engineering and Computer Science, United States Military Academy, West Point, New York 10996

20.1. Historical Perspective

Optics has long held the promise of high-speed, high-throughput, parallel information processing and distribution. The focus of its early applications was on analog signal processing techniques such as the optical Fourier transform, matrix–matrix and matrix–vector processors, and correlators. During this period, optics was used almost exclusively for front-end, preprocessing of wide-bandwidth, high-speed analog signals that were subsequently processed using digital electronic techniques. Although work continues on special-purpose analog applications, much of the recent optics work has focused on semiconductor devices for logic, switching, and interconnection applications.

Prior to 1980, optics research could generally be categorized as foundational with numerous basic device and proof-of-concept demonstrations that would potentially provide the framework for future optical system demonstrations. In the early 1980s, the focus of the community tended toward all-optical techniques. During this period, the community also began to collectively work on system-level demonstrations in an effort to gain acceptance and recognition in a predominantly electronics-based society. In 1984, B. K. Jenkins and colleagues [1] at the University of Southern California demonstrated an optical master/slave flip flop using holograms and imaging techniques. In 1985, D. Psaltis and Abu-Mostafa [2] at the California Institute of Technology demonstrated an optical implementation of an artificial neutral

704

HANDBOOK OF FIBER OPTIC
DATA COMMUNICATION

ISBN: 0-12-437162-0

network. In the late 1980s, the first digital optical computer, demonstrated by A. Huang [3] at AT&T Bell Laboratories, showed the potential for an optical architecture to implement the functionality necessary for computing functions. However, this was far from a fully programmable optical computer. Before long, it was recognized that all-optical processing was synonymous with special-purpose processing and therefore was limited in its applications. By the mid-1980s, the term photonics had been coined, reflecting the growing ties between electronics and optics. Continued advances in materials science and semiconductor growth techniques led to novel device structures that could be engineered at the atomic level. These new quantum structures provided optical-to-electronic conversion at the quantum level that improved device efficiency and provided the potential for direct integration with classical semiconductor circuits. Since then, the field has grown considerably, primarily as the result of a number of successful demonstrations of high-speed, high-throughput systems along with the commercialization of several other systems.

Recently, the area of optical information processing has progressed to yet another level. Systems that incorporate smart pixel-based processing using hybrid technologies [4] are being employed to take advantage of the parallelism and throughput of optics and the general-purpose processing ability of very large-scale integration (VLSI) techniques. From about 1988 to the present, AT&T's Bell Laboratories has demonstrated several generations of optical switching fabrics. System$_6$ demonstrated a switching fabric capable of supporting 4096 optical inputs and 256 optical outputs and operating at a maximum rate of 450 Mb/s [5]. AT&T's most recent system demonstration, System$_7$, produced an interconnection fabric with 512 optical inputs and 512 optical outputs with individual channels tested above 600 Mb/s [6]. This most recent switching fabric demonstrated channel data rates on the order of a gigabit per second (10^9 b/s) with nearly 1000 total system connections, yielding an aggregate throughput approaching 10^{12} b/s. This smart pixel technology is also being successfully applied to applications such as optical neural networks [7], image processing techniques [8], and telecommunications applications [9–11]. The fundamental question at this point is, what are the prospects for further breakthroughs in this technology and what role will smart pixel technology play in future optoelectronic and data communication applications?

To address this question, we will begin by considering some of the early semiconductor device developments that led to and continue to dominate a significant part of smart pixel technology.

20.2.   Multiple Quantum Well Devices

Nonlinear operations are fundamental to any processing, switching, or logic operations. Much of the early work in this area was motivated by the idea that perhaps optical devices could avoid some of the inherent speed limitations exhibited by electronic devices. During the early 1980s, this search for nonlinear optical devices coincided with a maturing of sophisticated compound semiconductor growth techniques and advances in material sciences. With improvements in growth techniques, researchers were able to engineer semiconductor devices at the atomic level, one atomic layer at a time, and were able to create novel optoelectronic devices such as the multiple quantum well device.

Multiple quantum well (MQW) modulators consist of alternating thin layers of two semiconductor materials, the most studied to date being gallium arsenide (GaAs) and aluminum gallium arsenide (AlGaAs). The thin crystal layers are typically grown using advanced growth techniques such as molecular beam epitaxy (MBE) or metal-organic chemical vapor deposition (MOCVD), which have the ability to grow these layers with atomic precision. The operation of these devices can best be understood by considering the effect of the layered semiconductor structure on the electrons and holes within the material. Using GaAs and AlGaAs as the material example, the bandgap energy of GaAs (1.424 eV) is lower than that of AlGaAs (1.773 eV) and, as a result, the electrons and holes in the semiconductor material see a minimum energy in the GaAs "well" material and therefore the AlGaAs material on either side acts as a "barrier." The semiconductor layers are so thin that the electron–hole pairs behave as "particles-in-a-box," often described in elementary quantum mechanics. The resulting quantum confinement causes discrete energy levels of the electron–hole pairs. One important consequence of this energy discretization is that very strong exciton absorption peaks appear at the edges of these steps, even at room temperature.

20.2.1.   THE QUANTUM-CONFINED STARK EFFECT

When an electric field is applied perpendicular to the layers, the electrons and holes move to lower energies, and the optical transition energy decreases, resulting in a shift in the wavelength of the absorption peak. This can be understood with the aid of the energy band diagrams shown in Fig. 20.1.

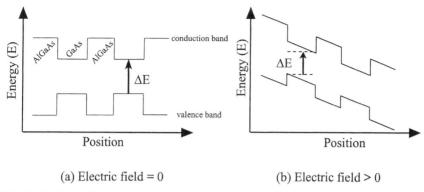

(a) Electric field = 0   (b) Electric field > 0

Fig. 20.1   Band diagram for GaAs/AlGaAs MQW structure. (a) No applied electric field; (b) applied electric field.

Figure 20.1a shows the energy band diagram under conditions of no applied electric field. Here, the optical transition energy is represented by the length of the arrow. When an electric field is applied, the band diagram tilts as shown in Fig. 20.1b, both the electrons and holes move to lower energies, and the optical transition energy decreases. Because photon energy is inversely proportional to wavelength, this decrease in optical energy corresponds to an increase or red-shift in the wavelength of the absorption peaks. This shifting of the absorption peaks with applied field is the underlying principle of the quantum-confined Stark effect (QCSE) [12, 13]. This is shown in the responsivity characteristics of Fig. 20.2.

20.2.2.   THE SELF-ELECTROOPTIC EFFECT DEVICE

Because a change in optical absorption at a specific wavelength can be affected by a change in the applied electric field, QCSE or electroabsorptive modulators can be produced. Here, optical intensity modulation is achieved using an external electric field that is applied to the modulator. In most instances to date, the structure used in these modulators is a positive–intrinsic–negative (*PIN*) diode in which the quantum well layers are placed within the intrinsic region of the diode. The structure for a specific reflection-type MQW modulator, called the self-electrooptic effect device (SEED) [14], is shown in Fig. 20.3. Here, P_{inc} is the incident optical power, P_{ref} is the reflected or output optical power, I_p is the series current in the external electric circuit, which by Kirchoff's current law is the photocurrent generated by the MQW modulator, and V is the applied electric field across the

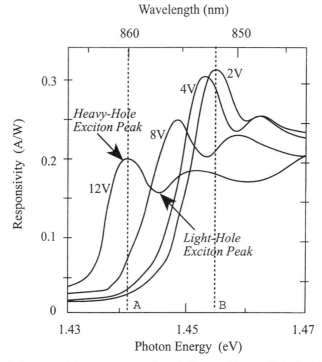

Fig. 20.2   Responsivity versus wavelength for specific applied electric fields.

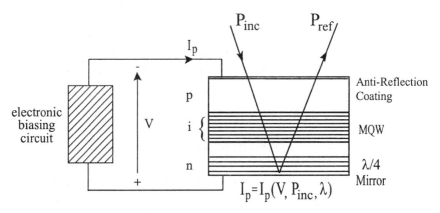

Fig. 20.3   Schematic diagram of a reflection-type MQW modulator.

MQW structure. A quarter-wave dielectric stack provides a high-reflectivity mirror at the rear of the device. An important advantage of this type of structure is that the PIN diode is operated in a reverse-bias configuration and therefore very low operating energies can be achieved while maintaining large electric fields across the quantum wells.

This type of electroabsorption modulator has enjoyed a great deal of success and widespread use. Switching speeds of 33 ps have been experimentally demonstrated [15], although this does not appear to represent any fundamental limit. Input optical energies on the order of 1 pJ for a 10 × 10-μm device are typical of current device technology. Arrays as large as 64 × 32 have been used for optical logic and memory operations [16]. There are, however, some less than desirable characteristics of these types of electroabsorption modulators. Contrast ratios for SEED-type modulators are typically low, on the order of 4 : 1. Also, because the operation of these modulators relies on the shift of narrow exciton absorption peaks, optimum performance requires an accurate and stable monochromatic source.

20.2.3. OTHER ELECTROABSORPTION MODULATORS

Another important class of electroabsorption modulator is the asymmetric Fabry–Perot modulator. Here, a partially reflecting mirror is grown on the front of the reflection-type electroabsorption modulator creating a Fabry–Perot étalon around the MQW structure. The cavity produces an increase in the field strength at the Fabry–Perot wavelength that produces larger shifts of the exciton absorption peaks than in a modulator without an étalon. This particular modulator structure has gained considerable attention because of its lower insertion loss, higher contrast ratios, and lower operating voltages [17–19]. Although contrast ratios in excess of 100 : 1 have been reported [20], the wavelength sensitivity in these electroabsorption modulators is even more pronounced than with SEED-type modulators.

20.2.4. FUNCTIONALITY

To first order, the MQW modulator can be modeled as a parallel plate capacitor, while an analysis of the circuit results in a first-order differential equation. For a stable solution to exist, it can be shown that the change in responsivity with applied field must be positive, resulting in a mode of operation in which the absorption of the MQW modulator is directly proportional to the current in the electronic bias circuit. If a constant current source is used as the external bias circuit, optical level shifting [14] or noninterferometric optical subtraction [21] can be realized. For this mode of operation, the modulator is operated at a wavelength at which the responsivity of the modulator increases with increasing applied voltage, shown by the dashed line at point A in Fig. 20.2. If, however, the modulator is operated at a wavelength such that the responsivity decreases with increasing applied voltage, such as point B in Fig. 20.2, the solution to the differential equation is unstable and the resulting functionality is bistability.

To date, the majority of the applications using electroabsorption modulators have used the bistable switching functionality of the modulator. Applications such as optical logic [22], optical interconnection [23], and analog-to-digital conversion [24] have been demonstrated with this functionality. By integrating two MQW modulators in series, a symmetric SEED (S-SEED) [22] is formed. This device provides improved bistable characteristics, freedom from critical biasing issues characteristic of bistable devices, and additional features such as time-sequential gain [16]. A smaller number of researchers have investigated applications of the analog functionality including image processing [25] and laser power stabilization [26]. A good review of the analog functionality of SEEDs can be found in Ref. [27].

20.3. Smart Pixel Technology

Although the advent of the MQW modulator provided an important component for optical switching and information processing systems, there remained a growing need for programmability and increased functionality. Smart pixels were the next step toward finding this flexibility.

Until this point, individual optical devices provided enhanced capabilities but generally only with regard to a specific point process. Cascading optical devices to provide increased complexity and functionality was difficult and impractical due to fan-out and losses in the optical path. The concept of the smart pixel was to integrate both electronic processing and individual optical devices on a common chip to take advantage of the complexity of electronic processing circuits and the speed of optical devices. Arrays of these smart pixels would then bring with them the advantage of parallelism that optics could provide.

20.3.1 APPROACHES TO SMART PIXELS

There are a number of different approaches to smart pixels that generally differ in the way in which the electronic and optical devices are integrated. Monolithic integration, direct epitaxy, and hybrid integration are the three most common approaches in use today.

20.3.1.1. Monolithic Integration

Monolithic integration is a technique that allows both the electronics and the optical devices to be integrated in a common semiconductor material in a single growth process or by utilizing a regrowth technique. The material of choice in most cases is a compound semiconductor material such as GaAs, indium gallium arsenide (InGaAs), or indium phosphide (InP). Potentially, this approach would produce higher speed smart pixels. However, the simultaneous fabrication of both the electronic and optical circuits on the same substrate generally results in an overall system with less than optimum performance.

There are generally two types of monolithically integrated smart pixels: ones that incorporate passive optical modulators and ones that incorporate active optical emitters.

20.3.1.1.1. Modulators

Probably the most studied monolithically integrated smart pixels to date are those that incorporate MQW-type optical modulators. Figure 20.4 shows a schematic of a SEED that has been monolithically integrated with a field-

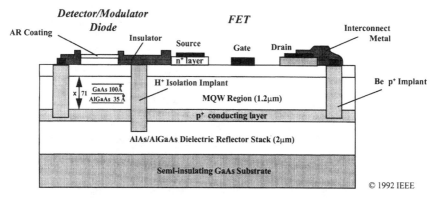

Fig. 20.4 Monolithic integration of FET with a SEED [28].

effect transistor (FET) [28]. This particular device was one of the first attempts at smart pixel development using MQW modulators. In the experimental demonstration, a PIN photodetector and load resistor, a depletion-mode GaAs–AlGaAs heterostructure FET (HFET) and self-biased HFET load, together with an output AlGaAs MQW optical modulator, were monolithically integrated within a 50×50-μm area. Here, the FET provides electronic gain and also electronic control of the SEED modulator.

Since this first demonstration, FET–SEED-type experiments have demonstrated 32×16 switching fabrics in which each pixel contained 25 FETs and 17 PIN diodes operating at switching speeds as high as 155 Mb/s [5].

20.3.1.1.2. Emitters

The second type of monolithically integrated smart pixel integrates active emitters such as light-emitting diodes (LEDs) or laser diodes with electronic circuitry. One such project integrates LEDs, optical FET (OPFET) photodetectors, metal–semiconductor–metal (MSM) photodetectors, and GaAs MESFET electronics in a common 0.6-μm GaAs process [29, 30]. Figure 20.5 shows the three stages of integrating GaAs LEDs with GaAs VLSI circuitry using a regrowth technique. Here, GaAs-based heterostructures are epitaxially grown on fully metalized commercial VLSI GaAs MESFET integrated circuits.

LED-based structures are generally power-inefficient devices requiring high drive current and producing low optical output power. The broad angular distribution of the emitted light also introduces optical cross-talk in densely packed arrays of devices. However, the advantage of using LED structures is that they are generally simpler to fabricate and are less susceptible to thermal variations than laser diodes.

Surface-emitting lasers have also been monolithically integrated to form smart pixels [31]. Most of this work to date has focused on a specific type of surface-emitting laser called the vertical cavity surface-emitting laser (VCSEL). Figure 20.6 shows one specific demonstration of the monolithic integration of a VCSEL with a MSM photodetector and several metal semiconductor field-effect transistors (MESFETs) [32]. In this demonstration, the MSM photodetector and MESFET layers were deposited using MBE, whereas the highly doped InGaP etch-stop layer, the GaAs buffer layer, and the VCSEL layers were deposited using MOCVD. The VCSEL layers contain a C-doped top-distributed Bragg reflector, a Si-doped bottom DBR, and an active layer consisting of quantum wells. The total thickness

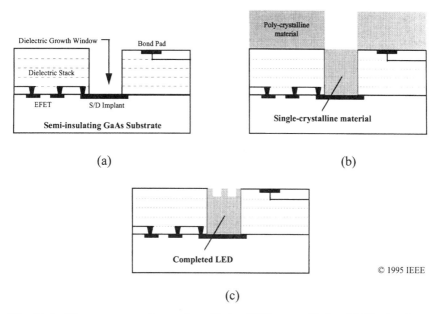

(a)   (b)

(c)

Fig. 20.5 Three stages of growing GaAs LEDs on GaAs VLSI circuitry. (a) Custom-designed GaAs VLSI chip with dielectric growth window and source/drain implant; (b) single-crystal material grows in the dielectric window and poly-crystalline material deposits on the overglass and bond pads; and (c) planarizing polycrystalline etch, current confining mesa etch, Si_3N_4 chemical vapor deposition, and AuZn–Au-ohmic contact evaporation complete the process [30].

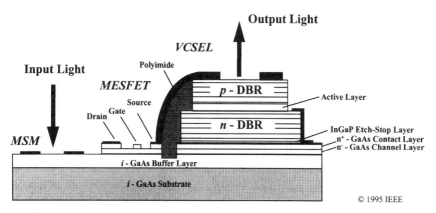

Fig. 20.6 Schematic cross section showing the monolithic integration of a VCSEL, an MSM photodetector, and a MESFET [32].

of the VCSEL layer is approximately 8 μm. Here, input light is detected by the MSM and amplified by the MESFET, which then controls the current through the VCSEL, thus modulating the output light. Functionality of both NOR- and OR-type operations with optical gain and a 3-dB bandwidth of 220 MHz was demonstrated. The operating wavelength in this particular demonstration was 860 nm.

Some of the advantages of VCSELs are inherent single longitudinal-mode operation and small divergence angle. However, historically, VCSELs have suffered from high series resistance and threshold voltage and nonuniformity in the VCSEL characteristics across arrays of devices. The high series resistance limits the optical output power, the temperature range for continuous wave operation, and the overall power efficiency of the laser. VCSELs with output optical power of 1 mW, threshold current of 2 mA, series resistance of 22 Ω, and power efficiency of 5% have been demonstrated [33]. The nonuniformity in the VCSEL characteristics is typically manifest in variations in output power and threshold and currently remains a significant problem for smart pixel arrays.

20.3.1.1.3. Observations

Although this integration approach promises improved performance, there are some limitations to its use. These limitations stem from two distinct issues: simultaneous optimization of both the optical devices and the electronic circuitry and a general lack of maturity of compound semiconductor technology. The optimization techniques used for the electronic circuitry that provides the increased functionality to the smart pixel are, in most cases, not the same as the optimization techniques required to produce high-quality optical devices. As a result, trade-offs must be made that ultimately reduce some performance metric associated with the overall smart pixel system. If a regrowth-type process is used, then additional processing complexity is also introduced. The second issue, technology maturity, is associated with the differences in the technological maturity of silicon processes compared to those of compound semiconductors. Mainstream silicon has enjoyed considerable success during the past several decades and, as a result, has benefited tremendously in terms of technology advancement and development. Compound semiconductors, however, have been used primarily for special purpose, typically high-frequency applications, and therefore the entire technology infrastructure is not as well developed. As a result, the tools necessary for designing, modeling, and fabricating compound semiconductor devices are not as well developed as those used for silicon devices. Taken together, these two issues severely limit the usefulness of monolithic integration as a viable integration approach for smart pixel technology.

20.3.1.2. Direct Epitaxy

Direct epitaxy of compound semiconductors onto silicon is another approach to the integration of smart pixels. Here, the compound semiconductor devices are directly grown onto the silicon crystal substrate. Because a significant amount of recent research has focused on the use of optical interconnects to alleviate some of the input and output bottlenecks associated with high-performance silicon integrated circuits, this integration technique appears to be a natural match. The principal limitation of this integration approach is a result of the large lattice mismatch between the silicon and compound semiconductor materials: approximately 4% for GaAs grown on silicon. In order to overcome this mismatch, a high-quality buffer layer is typically grown between the compound semiconductor and the silicon substrate. This buffer layer usually takes the form of a strained-layer superlattice, which provides an intermediate lattice constant between that of the compound semiconductor and the silicon substrate for crystalline bonding.

Figure 20.7 shows the structure of a GaAs–AlGaAs multiple quantum well reflection modulator grown directly on a silicon substrate using direct epitaxy techniques [34]. Here, a 1000 Å n^+ GaAs buffer layer is grown atop an n^+ silicon substrate. After the entire growth process, etching is used to fabricate the optical devices. In this particular demonstration, the etching process produced cracks in the strained-layer superlattice, which is common in this approach. Although these cracks did not limit device performance, they did limit the overall usable device area.

To date, integration of optoelectronic devices with electronic circuitry using direct epitaxy has focused on growing passive optical devices such as MQW modulators on silicon. The difficulty associated with active emitter integration using this approach can be attributed to differences in the thermal expansion coefficients of the two dissimilar materials. For GaAs and silicon, this difference in thermal expansion coefficients is 50%. Active optical emitters that generate large amounts of heat are therefore problematic for this integation approach. Even with passive optical devices, the large differences in the thermal expansion coefficients can result in a potentially unstable bond between the two dissimilar materials. Catastrophic failure

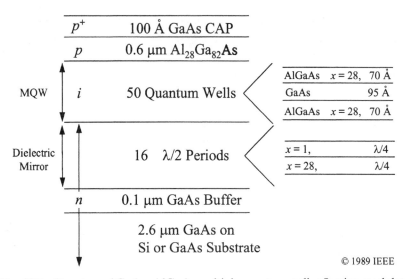

Fig. 20.7 Structure of GaAs–AlGaAs multiple quantum well reflection modulator grown on silicon substrate using direct epitaxy techniques [34].

of these devices typically occurs at this interface as a result of stress, strain, and sheer of the crystalline material. Reduction of on-chip heat generation is therefore an important consideration in the design of smart pixel architectures using this integration technique.

Another problem associated with direct epitaxy is that the integrated circuit must be subjected to temperatures as high as 850°C during the compound semiconductor growth. This exposure to high temperatures can cause degradation to some or all of the transistors in the silicon circuitry. Also, the metallization of the integrated circuit must be performed after the growth of the compound semiconductor because aluminum, which is the standard metal used in silicon VLSI circuitry, degrades at such high temperatures.

20.3.1.3. Hybrid Integration

Hybrid integration is the third approach to developing smart pixels. Here, the optical devices are grown separately from the silicon electronic circuitry. In a subsequent processing step, the optical devices are bonded to the silicon circuitry using a variety of bonding techniques. These include flip-chip bonding, epitaxial liftoff and subsequent contact bonding, and the creation of a physical cavity atop the silicon circuitry and flowing-in optical material such as liquid crystal material. Using this third approach, both the optical devices and the electronic circuits can be independently optimized, resulting in an overall optimization of the smart pixel system. In contrast to the direct epitaxy approach, hybridization allows the silicon circuitry to be fully fabricated before attaching the optical devices and therefore no high-temperature exposure of the silicon circuitry is incurred in the process.

In the following sections, we will consider the three most popular hybrid integration techniques: flip-chip bonding, epitaxial liftoff, and liquid crystal on silicon.

20.3.1.3.1. Flip-Chip Bonding

In this integration approach, bonding pads are incorporated into both the compound semiconductor optical devices and the silicon circuitry and then, in a subsequent processing step, associated pads are brought into contact and a mechanical solder bond is effected. In some of the earlier approaches [35–38] to flip-chip bonding, gold–gold or gold–semiconductor bonds were employed. These types of bonds require a high-temperature anneal, which again can be detrimental to the silicon circuitry. In recent flip-chip approaches [39, 40], both the p and n contacts are attached to the silicon, eliminating the need for the high-temperature anneal. Because the optical devices are inverted during the flip-chip bonding process, the substrate must be either transparent to the optical radiation or removed during a subsequent substrate removal step allowing access to the optical devices. As with monolithic integration, both passive modulators and active emitters can be integrated using this approach.

Modulators.   Figure 20.8 shows one particular example of hybrid integration in which a SEED MQW modulator is flip-chip bonded to complementary metal oxide semiconductor (CMOS) silicon circuitry [39]. In this approach, a three-step hybridization process is employed. First, the MQW modulator is fabricated and gold and gold–germanium–gold contacts are deposited. Subsequently, indium contacts are deposited on the existing contacts, completing the MQW contact. The silicon circuitry is then patterned with aluminum contacts of the same size and spacing as those on the MQW modulators, and finally indium is deposited on the aluminum contacts. The modulator and silicon contacts are then aligned and heated to 200°C, causing the indium from each contact to melt and fuse. In the second step, epoxy is flowed between the samples, providing additional mechanical support. Finally, the GaAs substrate is removed using a chemi-

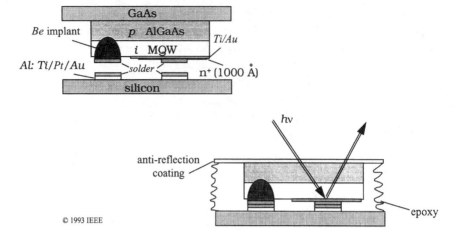

Fig. 20.8   Hybrid integration: CMOS SEED [39].

cal etch and an antireflection coating is applied to complete the hybridization.

CMOS SEED technology has demonstrated tremendous potential for use in smart pixel applications. Early devices demonstrated superb heat sinking capability and ohmic contact, cycling from 30 to 100°C with no degradation in device performance [39]. Later, research focused on device yield and demonstrated 32 × 32 arrays of MQW modulators flip-chip bonded to silicon CMOS circuitry with greater than 95% device yield [40]. In the first demonstration of this technology to smart pixels, an 8 × 8 array of CMOS–SEED switching nodes demonstrated 250 Mb/s operation [41]. Recently, circuits with 4096 optical detectors, 256 optical modulators, and 140,000 FETs operating beyond 400 Mb/s have been demonstrated [42].

The majority of the previously mentioned smart pixels operate at 850 nm. Other similar hybrid smart pixel arrays have also been demonstrated operating at 1.6 μm [43]. We will discuss wavelength selection as a design consideration in Section 20.4. Another important point is that this technique is not limited exclusively to silicon CMOS but could be used just as effectively with other silicon circuit families such as emitter-coupled logic, bipolar, and bipolar CMOS.

Emitters.   Active emitters are the alternative to passive modulators when considering hybrid integration of smart pixels. These emitters come in the form of either LEDs or laser diodes. The majority of the work on active emitters has focused on VCSEL emitters for parallel processing and inter-connect applications. Some of the advantages attributable to VCSELs are inherent single longitudinal-mode operation, small divergence angle, low threshold current, and the capability of high bit-rate modulation. The out-of-plane geometry provided by these devices is particularly advantageous for smart pixel applications. One-dimensional arrays of 64×1 individually addressed VCSELs [44] and two-dimensional arrays of 7×20 VCSELs [45] have been demonstrated. In Ref. [46], an eight-element section of an 8×8 array of VCSELs has experimentally demonstrated 622 Mb/s operation, showing the potential for high-speed operation. VCSELs are finding applications in conventional duplex transceivers and single devices and small arrays are being used for intramachine bus applications. However, heat dissipation and nonuniformity of the output characteristics remain a concern with the application of these emitters to smart pixel applications. Also, because the majority of VCSEL integration demonstrations to date have integrated contiguous arrays that include the substrate, the silicon circuitry underlying the VCSEL array is completely covered and not available for detector integration. As a result, detectors must be integrated with the VCSELs if detectors are to be included in the smart pixel architecture. Until these two issues are resolved, the usefulness of these devices for smart pixel applications remains limited.

20.3.1.3.2. Epitaxial Liftoff

A slightly different approach to hybrid integration of smart pixels is that of epitaxial liftoff [47, 48]. Here, thin compound semiconductor optical devices are grown separately from the silicon circuitry, removed from a sacrificial growth substrate using a technique called epitaxial liftoff (ELO), and then bonded to the CMOS silicon circuitry using either Van der Waals (contact) bonding or adhesion layers between the host and ELO epitaxial layer. A recent approach enables the alignment and selective deposition of both individual and arrays of devices onto a host substrate and also allows the devices to be processed on both the top and the bottom of the epitaxial sample [49].

This improved approach to the original ELO process, which uses a transparent polyimide transfer diaphragm, is illustrated in the seven-step process shown in Fig. 20.9 [50]. The as-grown material, with the epitaxial layers of interest atop the sacrificial etch layer, is shown as step 1. First, any contacts or coatings that are desired are applied to the as-grown material

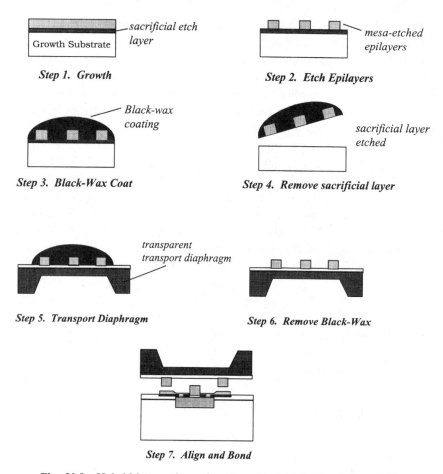

Step 1. Growth

Step 2. Etch Epilayers

Step 3. Black-Wax Coat

Step 4. Remove sacrificial layer

Step 5. Transport Diaphragm

Step 6. Remove Black-Wax

Step 7. Align and Bond

Fig. 20.9 Hybrid integration using the epitaxial liftoff technique [50].

and individual devices are defined through standard mesa-etching techniques. Next, the sample is coated with a black-wax handling layer by either melting or spray coating the wax onto the material. The sample is then immersed into a selective etch solution, which etches away the sacrificial layer and separates the epitaxial layers from the growth substrate. In steps 5 and 6, the thin-film devices, embedded in the back wax, are contact bonded to a transparent diaphragm and the black wax is removed to reveal the thin-film devices. To transfer and bond these thin-film devices to the host substrate, the transparent diaphragm is inverted so that the devices face the host substrate. The thin-film devices are visually aligned and then a pressure probe is applied to transfer individual, subarray, or the entire array of devices to the host substrate. High-quality thin-film devices have been successfully fabricated using this process. Thin-film devices as large as 2×4 cm [51] and film thicknesses ranging from 200 Å [51] to 4.5 μm [52] have been reported. To date, thin-film devices made of AlGaAs, InGaAsP, CdTe, and CdS have been integrated with host substrates, which include silicon, lithium niobate, and various types of polymers.

One of the advantages of this type of hybrid integration technique is that because the thin-film devices are inverted from the growth structure, both the top and the bottom of the epitaxial sample are available for processing. This is an important feature when considering three-dimensional interconnect applications such as in Ref. [52]. One of the limitations of this approach is that, to date, the process of transferring the thin-film devices to the host substrate has been labor intensive. However, recently, thin-film detector arrays as large as 64×64 have been demonstrated [53], opening the possibility of integrating large, homogeneous arrays of thin-film devices into smart pixel applications.

20.3.1.3.3. Liquid Crystal on Silicon

Techniques other than compound semiconductor on silicon are also being explored for smart pixel applications. Liquid crystal on silicon (LCOS) is another hybrid integration technique that has demonstrated tremendous potential for smart pixel applications. In this approach, a cover glass that has been coated with a transparent conducting material is placed on top of the silicon circuitry separated either by spacers or the natural contour of the processed circuitry to create a physical cavity. The liquid crystal material is then flowed into the cavity and placed in direct contact with the silicon circuitry. An electric field is created between the glass electrode and metal pads that are fabricated on the silicon chip, defining individual pixels at the locations of the metal pads. The metal pads also provide a high-reflectivity mirror from which the optical signal is reflected. Modulation of the electric field results in a modulation of the polarization, the intensity, or the phase of the illumination incident on the pixel.

Liquid crystal materials have previously found numerous applications in the area of projection display systems and are particularly attractive for smart pixels because of their large birefringence, low voltage operation, and their ease of fabrication for large arrays.

Liquid crystals most commonly used for optoelectronic applications have elongated molecules that are shaped like tiny rods. The directional alignment of these rods varies from completely random in the liquid phase to almost completely ordered in the solid phase. When a liquid crystal is heated, usually above 100°C, the molecules are randomly oriented and the material is isotropic. As the material cools, the molecules orient themselves, predominately in one direction. There are three types of liquid crystals: nematic, smectic, and cholesteric. In nematic liquid crystals, the molecules tend to be oriented parallel but their positions are random. In smectic liquid crystals, the molecules are again parallel, but their centers are stacked in parallel layers within which they have random positions. Cholesteric liquid crystals are a distorted form of the nematic liquid crystal in which the molecular orientation undergoes a helical rotation about a central axis.

In the early 1970s, Meyer *et al.* [54] recognized that smectic liquid crystals exhibited ferroelectricity, a permanent dipole moment that can be switched by an externally applied electric field. As a result, the polarization vector of light propagating through a ferroelectric liquid crystal material can be altered by applying an electric field to the liquid crystal. Optoelectronic applications using these types of liquid crystals take advantage of the material's large birefringence, Δn. Typical values for Δn range between 0.1 and 0.2. With this large birefringence, a controllable half-wave plate can be fabricated. A π phase shift between the ordinary and extraordinary waves of light propagating through the liquid crystal material results if the thickness of the material d is set equal to $d = \lambda_0/(2\Delta n)$, where λ_0 is the wavelength of light in a vacuum. For $\lambda_0 = 0.632 \ \mu$m and $\Delta n = 0.15$, $d = 2 \ \mu$m requiring a very thin layer of liquid crystal material to produce a half-wave plate. Binary phase modulation is accomplished by using an input polarizer oriented along the bisector of the two maximum switching states and an output polarizer oriented perpendicular to the input polarizer. Analog phase modulators can be made directly using nematic liquid crystals because the birefringence of these devices is directly voltage controlled. Currently, ferroelectric liquid crystals are used almost exclusively for LCOS smart pixels.

Figure 20.10 shows a cross-sectional schematic of the construction of one type of LCOS smart pixel. Here, an optically flat glass plate is coated with a thin transparent conducting electrode such as indium tin oxide (ITO) and is then placed on top of a conventional silicon integrated circuit. A cavity is then created between the glass and the processed silicon circuitry. Either polyball spacers are epoxy-bonded to the processed silicon CMOS circuitry or the natural contour produced by the multilayered silicon growth process forms the cavity. An alignment layer of obliquely evaporated silicon monoxide or rubbed polyvinyl alcohol is deposited onto the ITO. This

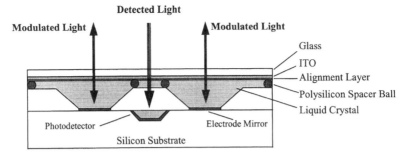

Fig. 20.10   Cross-sectional diagram of liquid crystal on silicon structure.

alignment layer makes contact with the liquid crystal material and induces spatial alignment of the liquid crystal molecules.

An important issue that must be considered when designing LCOS smart pixels is DC balancing of the external applied electric field. Ionic impurities in the liquid crystal material can cause a performance degradation if measures are not taken. Under the influence of an electric field, these ions drift and accumulate at the interface between the alignment layer and the liquid crystal material, causing a reduction of the total effective field. If the field is subsequently removed, these ions generate an electric field of the opposite polarity that could cause the liquid crystal to switch to the opposite state. DC balancing is an active design measure in which this reverse field buildup is eliminated by driving the device with an AC signal, actively cycling the device between states to ensure the net field equals zero. This DC balancing requirement necessitates serious consideration during the design phase of any system application of LCOS smart pixels.

Smart pixel arrays that use LCOS integration have found applications in many different areas including optoelectronic neural networks [55], image processing [56], interconnects [57], and competitive learning and adaptive holographic interconnections [58]. An excellent review of LCOS technology can be found in Ref. [59].

20.3.2. PERFORMANCE METRICS

In order to characterize and compare the performance of different types of smart pixels, standard performance metrics must be developed. Numerous methodologies exist by which the performance of smart pixel technology can be characterized and evaluated. In this section, we consider two performance metrics that have been used by others [60] in the community: aggregate capacity of the smart pixel array and complexity of the individual smart pixel.

20.3.2.1. Aggregate Capacity

Aggregate capacity is a measure of the total information handling capability of the smart pixel array. It combines the individual channel data rate with the total number of channels in the array to produce an aggregate information carrying capacity. Figure 20.11 shows a number of smart pixel demonstrations plotted as a function of channel data rate and array connectivity. This listing of demonstrations is not meant to be all-inclusive but instead attempts to represent the trend in smart pixel aggregate capacity. Notice also in Fig. 20.11 the terabit aggregate capacity regime — aggregate capacity in excess of 10^{12} b/s.

20.3.2.2. Complexity

The underlying concept of smart pixels is to provide increased functionality and programmability to arrays of optical devices. One measure of this level of functionality and programmability deals with the number of transistors

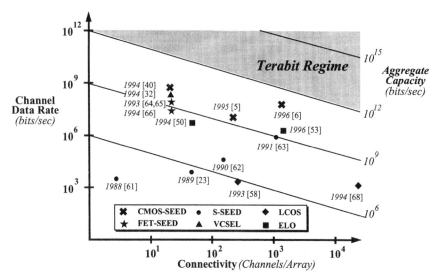

Fig. 20.11 Aggregate capacity of several representative smart pixel demonstrations. The year of the demonstration and the reference are also shown. Channel data rates are maximum channel data rates experimentally demonstrated.

associated with each individual smart pixel. Higher transistor counts indicate more processing capability per smart pixel and therefore more functionality. Figure 20.12 shows a number of smart pixel demonstrations plotted as a function of individual channel data rate and number of transistors per smart pixel. Conspicuously absent from this plot are the S-SEED demonstrations previously plotted in Fig. 20.11. The reason for this purposeful omission is that the S-SEEDs do not meet the standard definition of smart pixels in that they have no additional transistors or drive electronics to provide increased flexibility, functionality, or programmability.

One can deduce the current level of smart pixel complexity as well as identify the smart pixel technology capable of supporting a specific application by first considering the number of transistors necessary for some simple operations. A simple CMOS inverter requires 2 transistors, and two-input CMOS NOR and NAND gates each require 4 transistors to provide this functionality. A two-input exclusive-OR gate requires 8 transistors and an and-or-invert full-adder requires 24 transistors. A wide-range transconductance amplifier that produces a smooth sigmoidal function for thresholding requires 12 transistors.

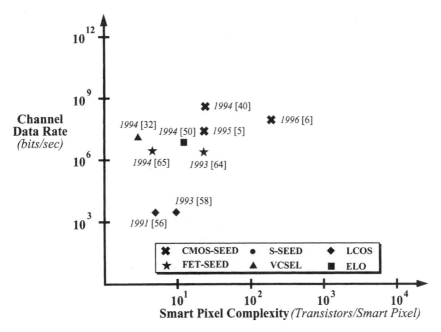

Fig. 20.12 Channel data rate versus complexity for several representative smart pixel demonstrations.

20.4.   Design Considerations

Smart pixel design considerations depend largely on the specific application
and fabrication technology. Because smart pixels cover a wide range of
applications and approaches, this section will not attempt to present an all-
inclusive set of design considerations. Instead, some of the more universal
constraints that apply to current technologies will be discussed.

20.4.1.   SILICON TECHNOLOGY ROAD MAP

Because the majority of smart pixel technology relies heavily on silicon
electronics, it is essential to consider projections for mainstream silicon
technology when considering future capabilities for smart pixels. Selected
excerpts from the 1994 Semiconductor Industry Association *National Technology Roadmap for Semiconductors* [69] are provided in Table 20.1. The
reductions in feature size and improvements in clock speed and chip size
will have a profound impact on the information processing capability of
silicon-based smart pixels. If we consider a pixel of constant physical area,
as the process feature size decreases, the number of devices per pixel will
nearly double every 3 years. This translates directly to increased smart
pixel processing and functionality.

20.4.2.   WAVELENGTH SELECTION

Wavelength selection is an important consideration in any optical system
design. Applications often dictate a specific wavelength such as in long-
haul optical fiber systems, in which dispersion-shifted fibers are used and the
preferred wavelength is 1.55 μm. Similarly, applications using mainstream
compound semiconductors, such as GaAs and AlGaAs, operate at wavelengths near 850 nm. Optical spot size is a key consideration in smart pixel
system design. Because spot size is proportional to wavelength, shorter
wavelength light will allow for smaller detectors and higher density smart
pixels. As a result, shorter wavelengths may be more attractive for near-
term smart pixel applications that are based on high-density arrays and
free-space optical inputs. Pixel area can be decreased even further by
utilizing GaN devices that operate at approximately 400 nm, potentially
reducing the diffraction-limited spot size by nearly a factor of four in
comparison to 1.55-μm sources. In Ref. [70], optical-to-optical modulation using a MQW modulator and a PIN photodiode was experimentally
demonstrated. Here, digital optical information modulated on a carrier

Table 20.1 Silicon CMOS Road Map Technology Characteristics

Year of first DRAM shipment	1995	1998	2001	2004	2007	2010
Minimum feature size (μm)	0.35	0.25	0.18	0.13	0.10	0.07
Number of chip I/Os						
Chip-to-package (pads), high performance	900	1350	2000	2600	3600	4800
Number of package pins/balls						
Microprocessor/controller	512	512	512	512	800	1024
ASIC (high performance)	750	1100	1700	2200	3000	4000
Package cost (cents/pin)	1.4	1.3	1.1	1.0	0.9	0.8
Chip frequency (MHz)						
On-chip clock, cost/performance	150	200	300	400	500	625
On-chip clock, high performance	300	450	600	800	1000	1100
Chip-to-board speed, high performance	150	200	250	300	375	475
Chip size (mm^2)						
DRAM	190	280	420	640	960	1400
Microprocessor	250	300	360	430	520	620
ASIC	450	660	750	900	1100	1400
Maximum number wiring levels (logic)						
On-chip	4–5	5	5–6	6	6–7	7–8

Adapted from [69].

727

of 632.8 nm was translated to a wavelength of 850 nm. This class of optical-to-optical conversion could also be applied to telecommunication wavelengths such as 1.3 or 1.55 μm and would allow the actual processing system to operate at 850 nm, providing higher density smart pixel arrays.

A major advantage of using longer wavelengths for three-dimensional memory and optical interconnect applications is that silicon is transparent to light above approximately 1.0 μm. Vertical interconnections using 1.3-μm light have been demonstrated by emitting light from the bottom wafer through an upper silicon wafer to detectors on the top surface [52]. In this demonstration, thin-film InGaAsP/InP-based emitters and detectors were bonded to the respective silicon substrates using an ELO intergation process to allow operation at the longer wavelength.

Despite decreasing transistor feature size, the minimum dimensions of an individual smart pixel will be constrained by the size of the optical input and output devices. If the input is an array of light beams from an image, a diffractive optical element, or a spatial light modulator, each pixel must have the ability to detect and process the incoming information. The required photoreceptor will be much larger than the minimum feature size of the electronic circuits in order to accommodate the diffraction-limited spot size of the optical input. Currently, GaAs MSM detectors and silicon photodetectors operate efficiently in the 700- to 900-nm range. Low- and intermediate-growth temperature MSM devices that detect in the 850-nm range are also capable of very fast response times.

Linewidth is another important specification that is closely related to wavelength selection. Most passive modulators to date are based on MQW technology and, as a result, rely on the shift of narrow exciton peaks. These modulators require narrow linewidths for optimum device performance, which requires a stable, single-frequency source. In contrast, ferroelectric liquid crystals in conjunction with VLSI offer the designer some flexibility in wavelength selection.

20.4.3. INDIVIDUAL DEVICE SPEED

A typical smart pixel architecture includes an input optical device and circuitry capable of receiving optical information, an electronic processing circuit, and output driver circuitry and an optical output device. The input may be from an off-chip source or an on-chip emitter such as a VCSEL or LED. The output may be an emitter or modulator driven by the processing circuit. When considering operating speed, each element in the smart pixel system contributes to the overall performance. Many different types of devices are capable of converting the optical input into an electronic signal. Monolithic integration using compound semiconductors allows for MSM detectors at the front end of the smart pixels. These devices can be fabricated to operate at very high speeds with a wide range of efficiencies. Typically, however, process adjustments necessary to fabricate acceptable electronic devices will result in less than optimal detectors in a monolithic process. Specifically, conventional silicon CMOS fabrication processes are far from optimal for photodetector fabrication, resulting in CMOS detectors that are slow and inefficient in comparison with GaAs devices. However, the designer still must decide between slower, more highly efficient silicon phototransistors and faster, less efficient photodiodes. Liquid crystal on silicon systems are slowed by the requirement for these silicon photoreceptor devices as well as the switching speed of the liquid crystal material itself. Smart pixels using LCOS technology typically operate with microsecond switching speeds.

Bandwidth will often be limited by voltage or current amplifier circuits that serve as an interface between the detector and the processing circuit. Gain must usually be sacrificed to obtain higher speed amplifiers. Wider bandwidth amplifiers may be obtained by cascading a series of high-speed stages, each having lower gain. Similar constraints apply to the smart pixel output circuit and optical device. Trade-offs between gain, speed, and power must be carefully balanced when selecting the output emitter and designing the necessary driver circuit. On-chip power consumption and heat removal are other important considerations in the selection of the output optical device. Active emitters produce more on-chip heat than passive modulators that derive their optical energy from off-chip sources. Currently, off-chip heat removal techniques can be managed more easily at the system level.

Finally, the processing circuitry is unique to each smart pixel application. Bandwidth is a function of the number of devices necessary to satisfy the required functionality. As anticipated improvements in silicon and compound semiconductor processing become reality, the electronic processing circuits will be capable of progressively higher bandwidths, placing more emphasis on the limitations of the input and output circuits.

20.4.4. ARRAY SIZE

One of the fundamental reasons for the development of smart pixel technology is that large arrays of optical devices provide the mechanism to leverage the advantages associated with optics. To date, 1024 × 1024 arrays of SEED modulators and 256 × 256 arrays of LCOS devices have been demonstrated. Even larger arrays are necessary for high-capacity telecommunications switching fabrics and massively parallel image processing applications. As array sizes increase, smart pixels will experience the challenges associated with high-density circuits. These constraints include power dissipation, yield, alignment, and addressing. Monolithic integration and LCOS processes may offer better yields due to relaxed requirements for electrical continuity at the bonding points and spatial alignment of massive numbers of contacts. Although LCOS is not a monolithic process, alignment issues are minimized as a result of the fabrication of reflective electrodes as part of the normal silicon fabrication process [71]. Hybrid integration has the advantage of well-established, on-chip addressing techniques routinely used in CMOS for large device arrays.

A prime consideration in array size is the size of each pixel. The diffraction-limited spot size of incident light determines the size of the photoreceptor within a pixel and the pitch of the array or pixel spacing. Reduction of optical cross-talk in high-density arrays may drive the system designer to consider using a VCSEL or laser input instead of LEDs. The large divergence angle of LEDs limits the potential role of this device in high-density smart pixel applications. Resonant cavity enhanced (RCE) LEDs have improved linewidth and divergence [72] and can serve as an alternative to VCSELs when conditions permit.

20.4.5. SYSTEM INTEGRATION ISSUES

System integration issues are numerous and often specific to the selected architecture. Once the fabrication process has been determined, the designer must evaluate the type of optical input (VCSEL, laser diode, RCE LED, or LED) and the input medium (free space, fiber, light modulator, or diffractive/refractive element). The input photoreceptor is selected simultaneously with amplifier requirements while considering associated trade-offs in noise, gain, speed, and overall bandwidth. Processing circuitry requires interfacing with both the input and output circuits through buffers, impedance-matching amplifiers, or transconductance circuits. Output circuit considerations are similar to input requirements in terms of gain–speed, gain–bandwidth, and noise-resistance trade-offs. A compatible output device and output driver are required for systems having optical outputs. The overall system must then be evaluated for speed/bandwidth, power consumption and dissipation, power supply, and current limits.

Finally, different approaches to packaging and testing are available to the designer of smart pixel systems. Pixel arrays may be packaged on a single integrated circuit chip using monolithic or hybrid processing, designed as individual elements for use on either a standard optical bench or a custom bench, such as slotted plate technology, or integrated into a three-dimensional system using technologies such as multichip modules [73]. In the latter case, implementation alternatives include the use diffractive, refractive, or reflective optical components in conjunction with microlens and/or macrolens hardware [74]. If the system architecture relies on mono-chromaticity, then the system designed is forced to use laser sources, either on- or off-chip. Because packaging and testing can have a major impact on system performance [75] and cost, these two requirements are normally a part of the original system design. Optomechanical considerations are also important to consider in the context of system design. As the array size of smart pixels continues to grow, the susceptibility of the system to vibration and misalignment increases. As a result, smart pixel system designers have recently focused on improvements to the optomechanical system in an attempt to minimize these effects [76].

20.4.6. *VLSI CIRCUIT DESIGN CONSIDERATIONS*

Hybrid integration on silicon takes advantage of the maturity of CMOS fabrication processes and VLSI technology in general. As a result of smaller feature sizes, complex smart pixel circuitry will require less physical area but be susceptible to photon-induced charge transfer between devices. This phenomena occurs when optical energy, intended for a detector or modulator, falls instead on the surrounding electronic circuitry and creates charge carriers that are directly injected into the circuitry. Physical limitations on optical beam diameter require emphasis on controlling cross-talk within and between pixels. Metal shielding, guard rings, and other isolation procedures must be carefully considered and incorporated into the electronic circuit design.

With improvements in processing techniques also comes an increase in the number of metalization layers available to the silicon circuit designer. Hybrid, direct epitaxy, and LCOS processes, however, can be negatively affected by the addition of these metallization layers. Additional processing steps may be required to ensure adequate planarization of the silicon surface prior to the integration of the optical devices. This additional processing potentially adds to the complexity of bonding GaAs circuits to silicon or adding a liquid crystal layer to the chip surface. These additional processing steps could potentially add to the rising costs associated with retooling plants to meet expectations of the silicon road map.

20.4.7. SUMMARY

Table 20.2 summarizes some of the design considerations discussed in this section. This table is not meant to be all-inclusive, but instead is meant to provide a quick reference for system designers considering the use of smart pixel technology.

20.5. Applications

Applications of smart pixels to date have focused primarily on digital applications with a heavy emphasis toward addressing electronic interconnection limitations. These applications span from back-plane interconnects, which support printed circuit board (PCB)-to-PCB communications, to high-speed, wide-bandwidth switching and routing for telecommunication applications. Some work has also been done in the area of signal and image processing applications requiring massive parallelism such as neural networks and image halftoning. In this section, we introduce a few representative applications of smart pixel technology including back-plane interconnects, high-speed switching and routing, analog-to-digital (A/D) conversion, and digital image halftoning.

20.5.1. OPTICAL INTERCONNECTIONS

When considering interconnection issues, one can broadly categorize the different potential architectures to be interconnected in a hierarchical fashion from the highest to lowest levels as machine-to-machine, processor-to-processor, board-to-board, chip-to-chip, and finally intrachip interconnections. Optical techniques have already been successfully applied to machine-to-machine interconnections in the form of optical fiber communication links. Recently, considerable research has been directed toward bringing optics to the next lower levels of processor-to-processor, board-to-board, and chip-to-chip communications.

There exist several fundamental reasons for this interest in optics as an interconnect technology [77]. First, optical interconnects offer freedom from mutual coupling not afforded by electrical interconnects. This particular advantage becomes increasingly significant with increased bandwidth because the mutual coupling associated with electrical connections is proportional to the frequency of the signal propagating on the line. Increased routing flexibility is another advantage of optical interconnections. Electrical interconnect paths cannot cross and are constrained to reside close to a ground plane. In contrast, optical interconnections can be routed through one another and are not constrained to be physically located in the vicinity of a ground plane. Another potential advantage is freedom from capacitive loading effects. With electrical interconnects, this comes in the form of line-charging in which, as the interconnect line becomes longer, an increased number of electrons are employed to charge the capacitance of the line. No such line-charging effects are present with optical interconnects. Finally, at the quantum level there exists the potential advantage of impedance transformation that matches the high impedances of small devices to the low impedances associated with electromagnetic field propagation [78]. This impedance matching results in more energy-efficient communications using optical interconnects for all except the shortest intrachip interconnects.

Smart pixel technology provides additional advantages in terms of interconnect applications. Much of the work on interconnects at the board-to-board and chip-to-chip level is focused on reducing communication bottlenecks associated with high-performance processing systems that are predominately silicon based. Because the majority of smart pixels rely on silicon technology for increased functionality and programmability, optical interconnect applications of smart pixels provide a unique opportunity for producing compact systems with large aggregate input and output capacity. Another attractive feature is that most smart pixel interconnect architectures employ optical input and output that are normal to the surface of the chip, allowing two-dimensional arrays of interconnections to be formed and thereby increase the potential throughput of the interconnect system.

20.5.1.1. Back-Plane Interconnections

Future digital systems, such as asynchronous transfer mode (ATM) switching systems and massively parallel processing systems, will require large PCB-to-PCB connectivity to support the large aggregate throughput demands being placed on such systems. Projections from the Semiconductor Industry Association indicate that by the Year 2001, silicon CMOS feature size will be on the order of 0.18 μm, transistor density will be 13 mil-

Table 20.2 Design Considerations for Smart Pixel Technology

Smart pixel technology	Switching speed[a]	Array size[b]	System issues	Principal applications
Monolithic integration	Potentially the fastest	Limited	Reduced flexibility for optimization of electronic and optical technologies; process uniformity is critical for arrays	High-speed, single-channel, and small array applications
Direct epitaxy	Medium	Limited	Limited lifetime and reliability resulting from lattice mismatch and thermal expansion coefficient differences; process uniformity is critical for arrays	Predominately research demonstrations, to date
Hybrid integration				
Flip-chip bonding	Fast	Medium–large		High speed, high-density free-space interconnection applications
Epitaxial liftoff	Medium–fast	Medium	Detectors and RCE LEDs, to date	Through-chip and fiber optic interconnection applications
Liquid crystal on silicon	Slow	Large	Active DC balancing required to prevent liquid crystal degradation; no active emitters	Image processing and displays

Passive modulators				
MQW	Fast	Large	Precise, narrow linewidth optical sources required; low contrast ratio	Off-chip sources provide improved heat dissipation; differential signaling required to overcome limited contrast ratio
LCOS	Slow	Medium–large	Active DC balancing required to prevent long-term liquid crystal degradation	
Active emitters				
VCSELs		Small	Increased heat generation requires additional attention to heat removal techniques; device characteristics across arrays not yet uniform	
RCE LEDs		Medium	Incoherent	Architectures cannot employ diffractive components; short, through wafer 3-D interconnections, and multimode fiber applications
LEDs		Medium	Incoherent	Architectures cannot employ diffractive components; low-cost, multimode fiber applications

[a] Slow, μsec; fast; ps.
[b] Small, $<8 \times 8$; large, $>256 \times 256$.

735

lion transistors per 360 mm^2, and on-chip clock rates will be approaching 600 MHz with off-chip rates in excess of 250 MHz. These requirements provide the motivation to using photonics-based approaches to back-plane designs that overcome issues such as ground bounce, cross-talk, and reflections that result in signal distortion when using parallel electrical approaches at these frequencies.

One approach to solving the interconnection problem created by such high-performance systems is to exploit the temporal and spatial bandwidth of optical interconnections. At the system back-plane, free-space optical technology can be employed that supports a large number of digital optical communication channels, each of which is composed of simple optical connections between arrays of smart pixel devices on successive PCBs. This type of optical back-plane can potentially provide in excess of 10^3 high-performance connections, while the electronic circuitry associated with each smart pixel can provide the intelligence necessary to provide reconfigurable interconnections.

One recent approach to building such a free-space optical back-plane that implements a Hyperplane-based [79] ATM switching fabric is shown in Fig. 20.13 [60]. Here, electronic PCBs connect to a rigid back-plane that contains both electrical and free-space optical interconnection channels.

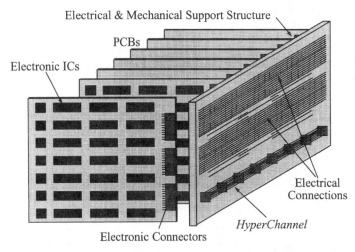

Fig. 20.13 Example of the application of smart pixel technology to back-plane interconnects [60].

The HyperChannel is composed of several smart pixel arrays that provide high-speed back-plane interconnections to and from the individual PCBs.

The smart pixels in this application integrate SEED modulators with CMOS silicon circuitry using the hybrid, flip-chip bonding approach. Arrays of differential reflection-mode MQW modulators and detectors are used as a method of improving noise immunity and overcoming the low contrast ratio limitation of the SEED modulators. The proximity of the optoelectronic devices with the CMOS circuitry allows decisions on the incoming optical data to be made local to the optical inputs and outputs and results in parallel whole-word processing and reduced requirements for driving long capacitive traces on the chip [80].

A photomicrograph of a buffered Hyperplane CMOS–SEED smart pixel is shown in Fig. 20.14. The core of this smart pixel array contains individual smart pixels that are organized into node channels. Control logic provides arbitration of data flow into and out of ATM cell queues while each of three queues buffer an entire ATM cell. A queue-addressing block places segments of the ATM cell into the queue at the appropriate location. An output multiplexer controls which of the three queues writes output data to the electrical pinouts of the chip.

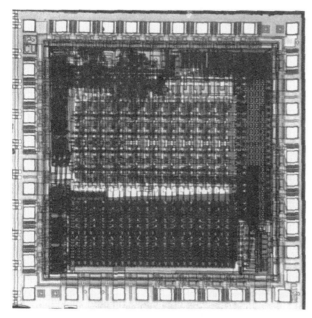

- Buffered *HyperPlane* Architecture

- 4 × 9 Smart Pixel Array

- ~60 Transistors

- ~20,600 Transistors

- 144 Optical I/O

- *Cross Out* Fabric

- Address Recognition

- 3 ATM Cell Buffers

- ATM Priority Bits

Fig. 20.14   Buffered Hyperplane CMOS–SEED smart pixel [60].

The physical size of this particular smart pixel chip is 2×2 mm. The electronic circuitry was designed using a 0.8-μm CMOS silicon process and provided a total transistor count in excess of 20,000 transistors with approximately 60 transistors per smart pixel. A 10×20 array of GaAs/AlGaAs SEED modulators was subsequently flip-chip bonded onto the processed silicon circuitry. The MQW modulator array provided the potential for up to 200 optical inputs or outputs, each of which had an optical window size of 18×18 μm.

The functionality of this smart pixel array, which includes packet injection, address recognition, packet extraction, and transparency, was experimentally verified to 10 Mb/s. A detailed system analysis predicts scalability to 2060 interconnections for a 1×1-cm chip area at a back-plane clock rate of 125 MHz. This provides a maximum aggregate optical throughput of 4.5 Gb/s, which is sufficient to provide four users each with a 622 Mb/s SONET STS-12 link [80].

The natural extension to this approach is to replace all the electrical interconnections on the back-plane with these high-speed HyperChannels and to add additional capability by incorporating fiber interfaces at other locations on the individual PCBs.

20.5.1.2. High-Speed Switching and Routing

As the demand for telecommunication switching capabilities continues to grow, the need for switching fabrics that are capable of switching large bandwidths of data becomes increasingly important. Examples of applications that are currently pushing the limits of electronic switching technology include ATM and video switching fabrics, pattern recognition and image processing systems, and high-performance computing systems. To meet the demands of these and future digital electronic systems, system designers are turning to optical interconnects to provide a potential solution to these massive interconnect applications.

Since about 1988, researchers at AT&T Bell Laboratories have been investigating the use of optical interconnects in high-speed switching applications. Figure 20.15 shows a representative block diagram of a switching fabric based on smart pixel technology [64].

AT&T Bell Laboratories has enjoyed tremendous success with various high-speed switching fabrics based on the SEED technology. Early demonstrations incorporated individual SEEDs and S-SEEDs in basic switching

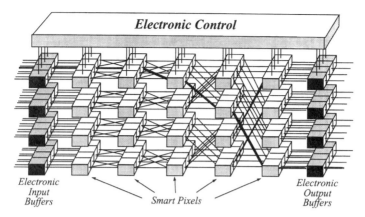

Fig. 20.15 Example of a high-speed switching fabric that incorporates electronic input and output buffers, electronic control, and smart pixels [64].

and logic demonstrations. In 1993, their first smart pixel switching demonstration used monolithically integrated FET–SEED technology. Recent advances use hybrid integration and the CMOS–SEED technology. Table 20.3 provides an overview of AT&T Bell Laboratories free-space photonic switching system demonstrations [81] to date and describes some of the technical details of each. These demonstrations have clearly led the community in the development of massively parallel interconnect and switching systems, custom optomechanical hardware, novel packaging and heat removal techniques, and other contributions that are critical to the successful evolution of a new technology such as smart pixels.

Recently, in an attempt to provide higher efficiency interfaces between the optical devices and the electronic digital-processing elements, new optical receiver designs have been investigated. Clocked-sense amplifier-based optical receivers are synchronous optical receivers that provide the potential for low-power, compact digital amplification. In Ref. [82], receivers of this type experimentally demonstrated operation in excess of 750 Mb/s within a layout area of 44×22 μm with a bias-dependent estimated power dissipation of 1–2 mW. A two-beam transimpedance smart pixel optical receiver using a CMOS–SEED design demonstrated 1 Gb/s performance within the same physical layout area [83]. Both of the receiver circuits used in these demonstrations were designed in a 0.8-μm CMOS foundry process. These demonstrations clearly show the potential for high-speed operation of receiver circuitry in smart pixel designs.

Table 20.3 **AT&T Bell Laboratories Free-Space Photonic Switching System Demonstrations**

				System			
System parameter	System$_1$ (1988)	System$_2$ (1989)	System$_3$ (1990)	System$_4$ (1991)	System$_5$ (1993)	System$_6$ (1995)	System$_7$[a] (1996)
Channels	2	32	64	1,024	32	256	1024
Bit rate	10 Kb/s	30 Kb/s	100 Kb/s	1 Mb/s	155 Mb/s	208 Mb/s	622 Mb/s
Technology	S-SEED	S-SEED	S-SEED	S-SEED	FET–SEED	CMOS–SEED	CMOS–SEED
Chip/system I/O	12/20	256/1024	1024/3072	10,240/61,440	192/960	4352/65,536	4096/1,048,576
Hardware							
Optics	Catalog	Catalog	Catalog	Custom	Custom	Custom	Custom
Mechanics	Catalog	Hybrid	Custom	Custom	Custom	Custom	Custom
System area (ft^2)	32	16	6	0.78	1.16	0.29	0.18
Reference	61	23	62	63	64, 65	5	6

[a] System figures are goals. To date, the System$_7$ smart pixel has demonstrated 625 Mb/s operation with 1024 differential optical inputs and 1024 differential optical

740

20.5.2. SIGNAL PROCESSING

A significantly smaller subset of the smart pixel community is focusing on analog applications of smart pixels for signal processing. In the following sections, we introduce two specific applications of smart pixel technology to a mixed technology implementation of error diffusion coding: over-sampled A/D conversion and digital image halftoning.

20.5.2.1. Analog-to-Digital Conversion

Because the majority of signals encountered in nature are continuous in both time and amplitude, the A/D interface is generally considered to be the most critical part of any overall signal acquisition and processing system. Because of the difficulty in achieving high-resolution and high-speed A/D converters, this A/D interface has been and continues to be a barrier to the realization of high-speed, high-throughput systems.

A technique that has become popular in the audio industry is that of oversampled A/D conversion or sigma–delta ($\Sigma\Delta$) modulation [84, 85]. Here, a low-resolution quantizer is embedded in a negative feedback architecture and subsequent digital signal processing techniques employed to improve the overall system signal-to-quantization noise ratio (SQNR). Figure 20.16 shows the block diagram for one realization of an oversampled A/D converter using a recursive error diffusion modulator and a digital postprocessor. In this architecture, the error introduced by the quantizer is physically computed and temporally redistributed to subsequent samples according to the feedback filter $H(z)$ in an effort to influence future quantization decisions. An analysis of this process in the frequency domain shows

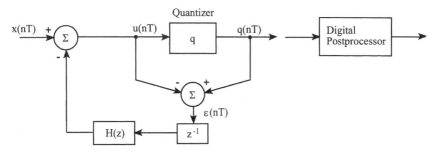

Fig. 20.16 Block diagram of an oversampled A/D converter consisting of a recursive error diffusion modulator and a digital postprocessor.

that the quantization error is spectrally shaped and redistributed to frequencies above the Nyquist frequency of the original sampled signal and then subsequently removed by low-pass filtering and decimation provided by the digital postprocessor.

The improvement in the performance of an oversampled A/D converter over that of a conventional Nyquist-rate converter can be seen by examining the SQNR. The maximum SQNR for a Nyquist rate A/D converter and an oversampled A/D converter, respectively, can be quantitatively described by

$$SQNR_{max}(b) = 3 \cdot 2^{2b-1}$$

and

$$SQNR_{max}(M, N) = \frac{3}{2} \cdot \left[\frac{2N + 1}{\pi^{2N}} \right] \cdot M^{2N+1},$$

where b is the number of bits resolution, N is the order of the modulator defined as the order of the feedback filter, and M is the oversampling ratio, which is the ratio of the sampling frequency to the Nyquist frequency of the sampled signal. Second- and third-order modulators are common practice in the audio industry, providing in excess of 16-bits of resolution at audio frequency bandwidths.

In an effort to extend the advantages of oversampled A/D conversion beyond audio frequency bandwidths, optoelectronic implementations have been investigated [24]. In this demonstration, discrete MQW modulators were used to provide the functionality necessary to implement a first-order oversampled modulator. Recently, extensions to this development have included smart pixel technology. Figure 20.17 shows the circuitry necessary to implement a first-order error diffusion modulator using an ELO-type smart pixel technology. Here, the intensity of the optical input represents the continuous amplitude signal to be A/D converted. The numbers represent electrical pinouts on the final integrated circuit chip. A silicon photodetector converts the light intensity to a photocurrent $x(nT)$, which is then buffered and amplified by the first set of current mirrors. The feed-forward section provides a scaled replica of the state variable $u(nT)$ for use in the quantizer differencing node and as the input to the comparator. The comparator provides the one-bit quantization necessary for A/D conversion, producing the ouput state variable $y(nT)$. The driver circuitry provides the necessary current gain to drive the active emitter, which in this demonstration is a RCE LED. The delay circuit is driven by a two-phase, nonover-

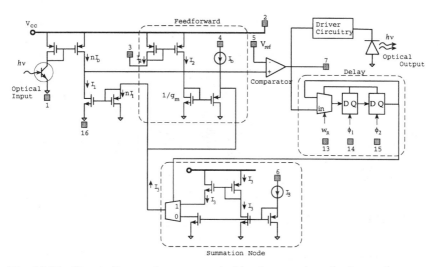

Fig. 20.17 Circuit diagram for a smart pixel implementation of a first-order error diffusion modulator based on an ELO-type smart pixel architecture.

lapping clock and provides a one-bit delay for the first-order modulator. The summation node circuitry provides voltage-to-current conversion for the feedback state variable $y(nT)$.

The RCE LEDs [86] provide several distinct advantages over conventional LEDs. The RCE LED is fabricated by placing the gain structure of a conventional LED into a Fabry–Perot cavity. The resonant cavity enhances the LED output spectrum, reducing the linewidth and providing increased directionality in the output. RCE LED's therefore offer a compromise between conventional LEDs and VCSELs. These active emitters can also be fabricated as thin-film structures and are therefore ideal candidates for ELO-type hybrid integration in smart pixel applications. Recently, RCE thin-film AlGaAs/GaAs/AlGaAs LEDs with metal mirrors experimentally demonstrated turn-on voltages of 1.3 V, linewidths of 10.4 nm, dispersion half-angles of 23.7°, and stable output over more than 1700 h of operation [86].

Figure 20.18 shows a photomicrograph of a smart pixel implementation of a first-order error diffusion modulator. The electronic circuitry was fabricated using a 2.0-μm CMOS process and was integrated with thin-film RCE LEDs using the hybrid ELO approach described previously. Two RCE LEDs and associated drivers are prominently visible in the center of the picture.

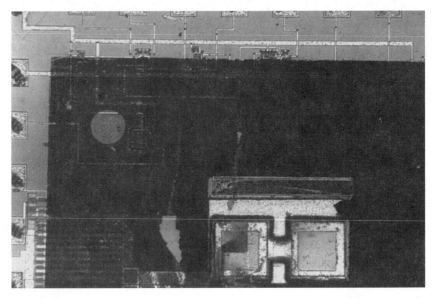

Fig. 20.18   Photomicrograph of an ELO-type smart pixel implementation of a first-order error diffusion modulator.

20.5.2.2.   Digital Image Halftoning

Digital image halftoning is an important class of A/D conversion within the context of image processing. Halftoning can be thought of as an image compression technique whereby a continuous-tone, gray-scale image is printed or displayed using only binary-valued pixels. Error diffusion is one method of achieving digital halftoning in which the error associated with a nonlinear quantization process is diffused within a local region and subsequent filtering methods employed in an effort to improve some performance metric such as signal-to-noise ratio. Classical error diffusion [87, 88] is a one-dimensional, serial technique in which the algorithm raster scans the image from upper-left to lower-right and, as a result, introduces visual artifacts directly attributable to the halftoning algorithm itself.

Artificial neural networks and their application to image halftoning have received considerable attention lately [89, 90]. This popularity lies in their ability to minimize a particular metric associated with a highly nonlinear system of equations. Specifically, the problem of creating a halftoned image can be cast in terms of a nonlinear quadratic optimization problem in which the performance metric to be minimized is the difference between the original and the halftoned images.

Figure 20.19 shows the error diffusion architecture and an electronic implementation of a four-neuron error diffusion-type neural network. Here, the individual neurons are represented as amplifiers (standard and inverting) and the synapses by the physical connections between the input and output of the amplifiers. Resistors are typically used to make these connections. The energy function of this error diffusion neural network can be described by

$$E(x, y) = y^T Ay - 2y^T Ax + x^T Ax = \underbrace{[B(y - x)]^T}_{\text{error}} \underbrace{[B(y - x)]}_{\text{error}},$$

where $y \in \{-1,1\}$ is a vector of quantized states with one element per pixel, $A = (I + W)^{-1}$, where W is derived from the original error diffusion filter weights through the relationship $W(i,j) = -w[(j - i) \text{ div } N, (j - i) \text{ mod } N]$, and $A = B^T B$. From this equation it is clear that as the neural network converges and the energy function is minimized, so too is the error between the output halftoned image and the input gray-scale image.

Figures 20.20a and 20.20b show 348 × 348 halftoned images of the Cadet Chapel at West Point using pixel-by-pixel thresholding and this error diffusion neural network, respectively.

Recently, a smart pixel approach to digital image halftoning has been investigated [7]. The smart pixel implementation provides the advantage that all pixel quantization decisions are computed in parallel and therefore the error diffusion process becomes two-dimensional and symmetric. Visual artifacts attributable to the halftoning algorithm are eleminated and overall

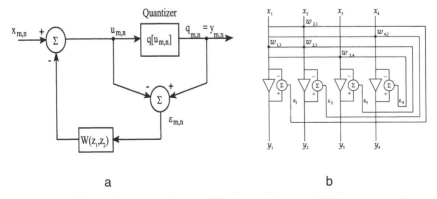

a   b

Fig. 20.19 (a) Two-dimensional error diffusion architecture; (b) four-neuron electronic implementation.

a   b

Fig. 20.20   Halftoned images of the Cadet Chapel at West Point (a) using pixel-by-pixel thresholding and (b) using an error diffusion neural network.

halftoned image quality is significantly improved. Also, the inherent parallelism associated with optical processing reduces the computational requirements while decreasing the total convergence time of the halftoning process. Figure 20.21 shows the circuitry necessary to implement a single neuron of this error diffusion neural network using a CMOS–SEED-type smart pixel.

All state variables in each circuit are represented as currents. Beginning in the upper-left circuit and proceeding clockwise, the input optical signal is continuous in intensity and represents the individual analog pixel intensity. The input SEED at each neuron converts the optical signal to a photocurrent, and subsequently current mirrors are used to buffer the input. The physical dimensions of the metal oxide semiconductor field-effect transistors (MOSFETs) determine the current gain that is used to amplify the photocurrent generated by the SEED. The first circuit produces two output state variables; $u(m,n)$ and $-u(m,n)$, which are used as the input to the quantizer and the feedback differencing node, respectively. The function of the quantizer is to provide a smooth, continuous sigmoidal function. This second electronic circuit produces a hyperbolic tangent sigmoidal function [81]. The third circuit takes as its input the state variable $y(m,n)$, produces a replica of the original signal, and drives an output optical SEED. In this case, the output signal $y(m,n)$ is a binary quantity represented

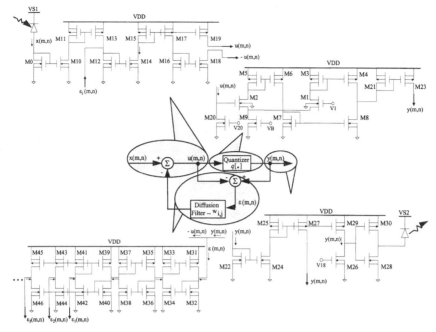

Fig. 20.21 Circuit diagram of a smart pixel implementation of a single neuron of a 5 × 5 error diffusion neural network based on a CMOS–SEED-type smart pixel architecture.

as the presence or absence of light. Here, the photoemission process is electroluminescence, which results from the SEED being forward biased. The last circuit implements the error weighting and distribution function of the error diffusion filter. The input error signal $\varepsilon(m,n)$ is weighted and distributed to adjacent pixels according to the coefficients of the error-diffusion filter. Current mirrors are used to replicate the error current and the physical dimensions of the MOSFETs are used to scale the currents according to the error diffusion filter design.

Figure 20.22 shows a photomicrograph of a 5 × 5 error diffusion neural network that was fabricated using CMOS–SEED smart pixels. The CMOS circuitry was produced using a 0.8-μm silicon process and the SEED modulators were subsequently flip-chip bonded to silicon circuitry using the same integration process described in Section 20.3.1.3.1. This smart pixel implementation resulted in a system with approximately 90 transistors per smart pixel and a total transistor count of almost 1800. A total of 50 optical

Fig. 20.22 Photomicrograph of a smart pixel implementation of a 5 × 5 error diffusion neural network for digital image halftoning.

input/output channels are provided in this implementation. Figure 20.23 shows a photomicrograph of a single neuron of the 5 × 5 neural network. The rectangular features are the MQW modulators, whereas the silicon circuits are visible between and beneath the modulators. The MQW modulators are approximately 70 × 30 μm and have optical windows that are 18 × 18 μm. The goal of this research is to produce a fast, massively parallel hardware implementation of the error diffusion algorithm that provides the capability of producing real-time halftoned images.

20.6. Future Trends and Directions

Although smart pixels are fairly young in terms of technological maturity, they have received tremendous focus and have emerged as a potential solution to many interconnection and massively parallel processing challenges. Individual smart pixel device development has been replaced by

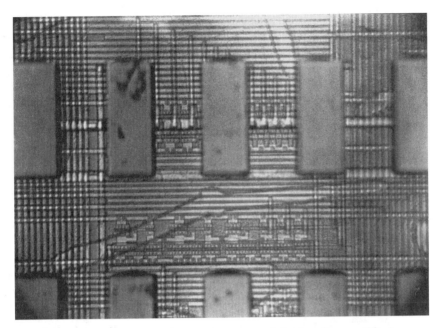

Fig. 20.23 Photomicrograph of a single neuron of the 5 × 5 error diffusion neural network.

system demonstrations and architectural development. At a recent conference dedicated to smart pixel technology [92], a majority of the papers were dedicated to issues such as optical power budget, efficient architectural designs, optimization of smart pixel receivers, and system demonstrations. Each of these topics relates to engineering issues, reflecting a maturing of the basic technology. As both silicon and compound semiconductor processing techniques mature, so too will smart pixel devices. Larger arrays of optical devices will provide increased aggregate capacity, whereas smaller feature sizes will lead to higher density electronic circuits that in turn, will provide increased complexity. Systems that provide aggregate capacity in excess of 10^{12} b/s can now be envisioned. Video-on-demand is one application that could immediately use this capability. The various integration techniques available with smart pixels provide the potential to integrate multispectral detectors on a common processing substrate. Large-scale and multispectral imagery applications in both the defense and commercial sectors could leverage some of the features of smart pixel technology.

Another potential area of application includes human perception and visualization. Portable smart pixel displays that project images the size of the human retina could provide processed information better suited to the reception and processing of the human brain. An interesting extension to the smart pixels discussed here is that of programmable smart pixels. Some recent work on field programmable smart pixels [93, 94] demonstrated the capability to dynamically program the electronic circuitry and therefore change the functionality of the smart pixel system. Although restricted to digital electronic circuitry, this approach to smart pixels holds promise for quickly reprogrammable interconnections for circuit switching and telecommunication applications. Another technology area closely related to and one that could have a significant impact on future smart pixel applications is that of microelectromechanical (MEM) device technology. Here, miromachining and nanofabrication techniques are used to create submicrometer mechanical devices in silicon and compound semiconductor materials. Optical applications of MEMs that would compliment and extend the applications of smart pixel technology include mechanically tunable VCSELs [95], optical scanning mirrors [96], and tunable optical filters [97]. Other uses include microsensors and microactuators that could potentially be integrated on a common substrate with the smart pixels providing the capability for mechanical and optical sensing, processing, and mechanical actuating from the same integrated device.

The integration of electronics with optics has produced hybrid devices that are fully capable of leveraging the advantages of each individual technology and providing total system performance well beyond the capability of either individual technology. This technology could provide solutions to many of the electrical interconnection problems currently facing many high-speed optoelectronic data communication, telecommunication, and signal and image processing applications.

References

1. Jenkins, B. K., A. A. Sawchuk, T. C. Strand, R. Forchheimber, and B. H. Soffer. 1984. Sequential optical logic implementation. *Appl. Opt.* 23:3455–3464.
2. Abu-Mostafa, Y. S., and D. Psaltis. 1987. Optical neural computers. *Sci. Am.*, 88–95.
3. Huang, A. 1984. Architectural considerations involved in the design of an optical digital computer. *Proc. IEEE* 72:780–786.

4. Hinton, H. S. 1988. Architectural considerations for photonic switching networks. *IEEE J. Select. Areas Commun.* 6:1209–1226.

5. Hinterlong, S. J., A. L. Lentine, D. J. Reiley, J. M. Sasian, R. L. Morrison, R. A. Novotny, M. G. Beckman, D. B. Buchholz, T. J. Cloonan, and G. W. Richards. 1996. An ATM switching system demonstration using a 40 Gb/s throughput smart pixel optoelectronic VLSI chip. Proceedings of the 1996 IEEE/LEOS Summer Topical Meeting on Smart Pixels, Keystone, CO, 47–48.

6. Lentine, A. L., K. W. Goossen, J. A. Walker, J. E. Cunningham, W. Y. Jan, T. K. Woodward, A. V. Krishnamoorthy, B. J. Tseng, S. P. Hui, R. E. Leibenguth, L. M. F. Chirovsky, R. A. Novotny, D. B. Buchholz, and R. L. Morrison. 1996. Optoelectronic switching chip with greater than 1 terabit per second potential optical I/O bandwidth. Proceedings of the 1996 IEEE/LEOS Summer Topical Meeting on Smart Pixels, Keystone, CO, Postdeadline Paper 001.

7. Sayles, A. H., B. L. Shoop, and E. K. Ressler. 1996. A novel smart pixel network for signal processing applications. Proceedings of the 1996 IEEE/LEOS Summer Topical Meeting on Smart Pixels, Keystone, CO, 86–87.

8. Shoop, B. L., and E. K. Ressler. 1994. Optimal error diffusion for digital halftoning using an optical neural network. Proceedings of the First International Conference on Image Processing, Institute of Electrical and Electronics Engineers, Austin, TX, 1036–1040.

9. McCormick, F. B., F. A. P. Tooley, T. J. Cloonan, J. L. Brubaker, A. L. Lentine, R. L. Morrison, S. J. Hinterlong, M. J. Herron, S. L. Walker, and J. M. Sasian. 1991. S-SEED-based photonic switching network demonstration. OSA Proceedings on Photonics in Switching, eds. H. S. Hinton and J. W. Goodman, 48–55.

10. Cloonan, T. J., G. W. Richards, A. L. Lentine, F. B. McCormick, and J. R. Erickson. 1992. Free-space photonic switching architectures based on extended generalized shuffle networks. *Appl. Opt.* 31:7471–7492.

11. Woodward, T. K., A. L. Lentine, L. M. F. Chirovsky, M. W. Focht, J. M. Freund, G. D. Guth, R. E. Leibenguth, and L. E. Smith. 1993. GaAs/AlGaAs FET-SEED receiver/transmitters. OSA Proceedings on Photonics in Switching, eds. J. W. Goodman and R. C. Alferness, 81–84.

12. Miller, D. A. B., D. S. Chemla, T. C. Damen, A. C. Gossard, W. Wiegmann, T. H. Wood, and C. A. Burrus. 1984. Band-edge electroabsorption in quantum well structures: The quantum-confined stark effect. *Phys. Rev. Lett.* 53:2173–2177.

13. Miller, D. A. B. 1984. Optical bistability and differential gain resulting from absorption increasing with excitation. *J. Opt. Soc. Am. B* 1:857–864.

14. Miller, D. A. B., D. S. Chemla, T. C. Damen, T. H. Wood, C. A. Burrus, A. C. Gossard, and W. Wiegmann. 1985. The quantum well self-electro-optic effect device: Optoelectronic bistability and oscillation, and self-linearized modulation. *IEEE J. Quantum Electron.* QE-21:1462–1476.

15. Boyd, G. D., A. M. Fox, D. A. B. Miller, L. M. F. Chirovsky, L. A. D'Asaro, J. M. Kuo, R. F. Kopf, and A. L. Lentine. 1990. 33 ps optical switching of symmetric self electro-optic effect devices. *Appl. Phys. Lett.* 57:1843–1845.

16. Lentine, A. L., F. B. Mccormick, R. A. Novotny, L. M. F. Chirovsky, L. A. D'Asaro, R. F. Kopf, J. M. Kuo, and G. D. Boyd. 1990. A 2-kbit array of symmetric self-electrooptic effect devices. *IEEE Photon. Tech. Lett.* 2:51–53.

17. Whitehead, M., A. Rivers, G. Parry, J. S. Roberts, and C. Button. 1989. Low-voltage multiple quantum well modulators with on:off ratios > 100:1. *IEEE Photon. Tech. Lett.* 25:984–986.

18. Yan, R. H., R. J. Simes, and L. A. Coldren. 1989. Electroabsorptive Fabry–Perot reflection modulators with asymmetric mirrors. *IEEE Photon. Tech. Lett.* 1:273–275.

19. Pezeshki, B., D. Thomas, and J. S. Harris, Jr. 1990. Optimization of modulation ratio and insertion loss in reflective electroabsorption modulators. *Appl. Phys. Lett.* 57:1491–1493.

20. Law, K.-K., R. H. Yan, L. A. Coldren, and J. L. Merz. 1990. Self-electro-optic device based on superlattice asymmetric Fabry–Perot modulator with an on:off ratio >100. *Appl. Phys. Lett.* 57:1345–1347.

21. Shoop, B. L., B. Pezeshki, J. W. Goodman, and J. S. Harris, Jr. 1992. Noninterferometric optical subtraction using reflection-electroabsorption modulators. *Opt. Lett.* 17:58–60.

22. Lentine, A. L., H. S. Hinton, D. A. B. Miller, J. E. Henry, J. E. Cunningham, and L. M. F. Chirovsky. 1989. Symmetric self-electrooptic effect device: Optical set–reset latch, differential logic gate, and differential modulator/detector. *IEEE J. Quantum Electron.* 25:1928–1936.

23. Cloonan, T. J., M. H. Herron, F. A. P. Tooley, G. W. Richards, F. B. McCormick, E. Kerbis, J. L. Brubaker, and A. L. Lentine. 1990. An all optical implementation of a 3D crossover network. *IEEE Photon. Tech. Lett.* 2:438–440.

24. Shoop, B. L., and J. W. Goodman. 1992. Optical oversampled analog-to-digital conversion. *Appl. Opt.* 31:5654–5660.

25. DeSouza, E. A., L. Carraresi, G. D. Boyd, and D. A. B. Miller. 1994. Self-linearized analog differential self-electro-optic-effect device. *Appl. Opt.* 33:1492–1497.

26. Shoop, B. L., B. Pezeshki, J. W. Goodman, and J. S. Harris, Jr. 1992. Laser power stabilization using a quantum well modulator. *IEEE Photon. Tech. Lett.* 4:136–139.

27. Miller, D. A. B. 1993. Novel analog self-electrooptic-effect devices. *IEEE J. Quantum Electron.* 29:678–698.

28. Woodward, T. K., L. M. F. Chirovsky, A. L. Lentine, L. A. D'Asaro, E. J. Laskowski, M. Focht, G. Guth, S. S. Pei, F. Ren, G. J. Przybylek, L. E. Smith, R. E. Leibenguth, M. T. Asom, R. F. Kopf, J. M. Kuo, and M. D. Feuer. 1992. Operation of a fully-integrated GaAs-Al$_x$Ga$_{1-x}$As FET-SEED: A basic optically addressed integrated circuit. *IEEE Photon. Tech. Lett.* 4:614–617.

29. Grot, A. C., D. Psaltis, K. V. Shenoy, and C. G. Fonstad, Jr. 1994. Integration of LED's and GaAs circuits by MBE regrowth. *IEEE Photon. Tech. Lett.* 6:819–821.

30. Shenoy, K. V., C. G. Fonstad, Jr., A. C. Grot, and D. Psaltis. 1995. Monolithic optoelectronic circuit design and fabrication by epitaxial growth on commercial VLSI GaAs MESFETs. *IEEE Photon. Tech. Lett.* 7:508–510.

31. Cheng, J., P. Zhou, S. Z. Sun, S. Hersee, D. R. Myers, J. Zolper, and G. A. Vawter. 1993. Surface-emitting laser-based smart pixels for two-dimensional optical logic and reconfigurable optical interconnections. *IEEE J. Quantum Electron.* QE-2:741–756.

32. Matsuo, S., T. Nakahara, Y. Kohama, Y. Ohiso, S. Fukushima, and T. Kurokawa. 1995. Monolithically integrated photonic switching device using an MSM PD, MESFET's, and a VCSEL. *IEEE Photon. Tech. Lett.* 7:1165–1167.

33. Zhou, P., J. Cheng, C. F. Schaus, S. Z. Sun, K. Zheng, E. Armour, W. Hsin, D. R. Myers, and G. A. Vawter. 1991. Low series resistance high efficiency GaAs/AlGaAs vertical-cavity surface-emitting lasers with continuously-graded mirrors grown by MOCVD. *IEEE Photon. Tech. Lett.* 3:591–593.

34. Goossen, K. W., G. D. Boyd, J. E. Cunningham, W. Y. Jan, D. A. B. Miller, D. S. Chemla, and R. M. Lum. 1989. GaAs-AlGaAs multiquantum well reflection modulators grown on GaAs and silicon substrates. *IEEE Photon. Tech. Lett.* 1:304–306.

35. Weiland, J., H. Melchior, M. Q. Kearley, C. Morris, A. J. Moseley, M. G. Goodwin, and R. C. Goodfellow. 1991. Optical receiver array in silicon bipolar technology with self-aligned, low parasitic III/V detectors for DC-1 Gbit/s parallel links. *Electron. Lett.* 27:2211–2213.

36. Camperi-Ginestet, C., M. Hargis, N. Jokerst, and M. Allen. 1991. Alignable epitaxial liftoff of GaAs materials with selective deposition using polyimide diaphragms. *IEEE Photon. Tech. Lett.* 3:1123–1125.

37. Yoffe, G. W., and J. M. Dell. 1991. Multiple-quantum-well reflection modulator using a lift-off GaAs-AlGaAs film bonded to gold on silicon. *Electron. Lett.* 27:557–559.

38. Yanagisawa, M., H. Terui, K. Shuto, T. Miya, and M. Kobayashi. 1992. Film-level hybrid integration of AlGaAs laser diode with glass waveguide on silicon substrate. *IEEE Photon. Tech. Lett.* 4:21–23.

39. Goossen, K. W., J. E. Cunningham, and W. Y. Jan. 1993. GaAs 850 nm modulators solder-bonded to silicon. *IEEE Photon. Tech. Lett.* 5:776–778.

40. Goossen, K. W., J. A. Walker, L. A. D'Asaro, S. P. Hui, B. Tseng, R. Leibenguth, D. Kossives, D. D. Bacon, D. Dahringer, L. M. F. Chirovsky, A. L. Lentine, and D. A. B. Miller. 1995. GaAs MQW modulators integrated with silicon CMOS. *IEEE Photon. Tech. Lett.* 7:360–362.

41. Lentine, A. L., K. W. Goossen, J. A. Walker, L. M. F. Chirovsky, L. A. D'Asaro, S. P. Hui, B. T. Tseng, R. E. Leibenguth, D. P. Kossives, D. W. Dahringer,

and D. A. B. Miller. 1995. 8 × 8 array of optoelectronic switching nodes comprised of flip-chip-solder-bonded MQW modulators on silicon CMOS circuitry. *Photon. Switching* 12 (OSA Technical Digest Series): 13–15.

42. Lentine, A. L., K. W. Goossen, J. A. Walker, L. M. F. Chirovsky, L. A. D'Asaro, S. P. Hui, B. T. Tseng, R. E. Leibenguth, J. E. Cunningham, D. W. Dahringer, D. P. Kossives, D. D. Bacon, R. L. Morrison, R. A. Novotny, and D. B. Buchholz. 1995. High speed optoelectronic VLSI switching chip with greater than 4000 optical I/O based on flip chip bonding of MQW modulators and detectors to silicon CMOS. 1995 Conference on Lasers and Electro-Optics, Postdeadline Paper.

43. Mosely, A. J., M. Q. Kearley, R. C. Morris, J. Urquhart, M. J. Goodwin, and G. Harris. 1991. 8 × 8 flipchip assembled InGaAs detector arrays for optical interconnect. *Electron. Lett.* 27:1566–1567.

44. Morgan, R. A., K. C. Robinson, L. M. F. Chirovsky, M. W. Focht, G. D. Guth, R. E. Leibenguth, K. G. Glogovsky, G. J. Przybylek, and L. E. Smith. 1991. Uniform 64 × 1 arrays of individually-addressed vertical cavity top surface emitting lasers. *Electron. Lett.* 27:1400–1401.

45. Chang-Hasnain, C. J., J. P. Harbison, C.-E. Zah, M. W. Maeda, L. T. Florez, N. G. Stoffel, and T.-P. Lee. 1991. Multiple wavelength tunable surface-emitting laser arrays. *IEEE J. Quantum Electron.* 27:1368–1376.

46. Banwell, T. C., A. C. Von Lehmen, and R. R. Cordell. 1993. VCSE laser transmitters for parallel data links. *IEEE J. Quantum Electron.* 29:635–644.

47. Yablonovitch, E., T. Gmitter, J. P. Harbison, and R. Bhat. 1987. Extreme selectivity in the liftoff of epitaxial GaAs films. *Appl. Phys. Lett.* 51:2222–2224.

48. Yablonovitch, E., E. Kapon, T. J. Gmitter, C. P. Yun, and R. Bhat. 1989. Double heterostructure GaAs/AlGaAs thin film diode lasers on glass substrates. *IEEE Photon. Tech. Lett.* 1:41–42.

49. Camperi-Gineset, C., M. Hargis, N. Jokerst, and M. Allen. 1991. Alignable epitaxial liftoff of GaAs materials with selective deposition using polyimide diaphragms. *IEEE Photon. Tech. Lett.* 3:1123–1126.

50. Jokerst, N. M. 1994. Parallel processing: Into the next dimension. *Opt. Photon. News* 5:8–14.

51. Yablonovitch, E., D. M. Hwang, T. J. Gmitter, L. T. Florez, and J. P. Harbison. 1990. Van der Waals bonding of GaAs epitaxial liftoff films onto arbitrary substrates. *Appl. Phys. Lett.* 56:2419–2421.

52. Calhoun, K. H., C. B. Camperi-Ginestet, and N. M. Jokerst. 1993. Vertical optical communication through stacked silicon wafers using hybrid monolithic thin film InGaAsp emitters and detectors. *IEEE Photon. Tech. Lett.* 5:254–257.

53. Wills, D. S. 1996. Smart pixel architectures for image processing. Proceedings of the 1996 IEEE/LEOS Summer Topical Meeting on Smart Pixels, Keystone, CO, 93–94.

54. Meyer, R. B., L. Liebert, J. Strzelecki, and P. Keller. 1975. Ferroelectric liquid crystals. *J. Phys. Lett.* 36:L69–L71.

55. Mao, C. C., and K. M. Johnson. 1993. Optoelectronic array that computes error and weight modification for a bipolar optical neural network. *Appl. Opt.* 32:1290–1296.

56. Jared, D. A., and K. M. Johnson. 1991. Optically addressed thresholding very-large-scale-integration/liquid-crystal spatial light modulators. *Opt. Lett.* 16: 767–769.

57. Kranzdorf, M., K. M. Johnson, J. Bigner, and L. Zhang. 1989. An optical connectionist machine with polarization-based bipolar weights. *Opt. Eng.* 28:844–848.

58. Wagner, K., and T. M. Slagle. 1993. Optical competitive learning with VLSI/liquid-crystal winner-take-all modulators. *Appl. Opt.* 32:1408–1435.

59. Johnson, K. M., D. J. McKnight, and I. Underwood. 1993. Smart spatial light modulators using liquid crystals on silicon. *IEEE J. Quantum Electron.* QE-2:699–714.

60. Hinton, H. S. 1995. (July 18–21). Progress and directions of smart pixel based systems. Paper presented at the AT&T/ARPA CO-OP Hybrid SEED Workshop, Fairfax, VA.

61. Kerbis, E., T. J. Kloonan, and F. B. McCormick. 1990. An all-optical realization of a 2 × 1 free-space switching node. *IEEE Photon. Tech. Lett.* 2:600–602.

62. McCormick, F. B., F. A. P. Tooley, T. J. Cloonan, J. L. Brubaker, A. L. Lentine, R. L. Morrison, S. J. Hinterlong, M. J. Herron, S. L. Walker, and J. M. Sasian. 1991. S-SEED-based photonic switching network demonstration. OSA Proceedings on Photonics in Switching, eds. H. S. Hinton and J. W. Goodman, 48–55.

63. McCormick, F. B., T. J. Cloonan, F. A. P. Tooley, A. L. Lentine, J. M. Sasian, R. L. Morrison, S. L. Walker, R. J. Crisci, R. A. Novotny, S. L. Hinterlong, H. S. Hinton, and E. Kerbis. 1993. Six-stage digital free-space optical switching network using symmetric self-electro-optic-effect devices. *Appl. Opt.* 32:5153–5171.

64. McCormick, F. B., T. J. Cloonan, A. L. Lentine, J. M. Sasian, R. L. Morrison, M. G. Beckman, S. L. Walker, M. J. Wojcik, S. J. Hinterlong, R. J. Crisci, R. A. Novotny, and H. S. Hinton. 1993. 5-Stage embedded-control EGS network using FET-SEED smart pixel arrays. OSA Proceedings on Photonics in Switching, eds. J. W. Goodman and R. C. Alferness, 81–84.

65. McCormick, F. B., T. J. Cloonan, A. L. Lentine, J. M. Sasian, R. L. Morrison, M. G. Beckman, S. L. Walker, M. J. Wojcik, S. J. Hinterlong, R. J. Crisci, R. A. Novotny, and H. S. Hinton. 1994. Five-stage free-space optical switching network with field-effect transistor self-electro-optic-effect-device smart-pixel arrays. *Appl. Opt.* 33:1601–1618.

66. Plant, D. V., A. Z. Shang, M. R. Otazo, B. Robertson, and H. S. Hinton. 1994. Design and characterization of FET-SEED smart pixel transceiver arrays for optical backplanes. Proceedings of the 1994 IEEE/LEOS Summer Topical Meeting on Smart Pixels, Lake Tahoe, NV, 26–27.

67. Lentine, A. L., K. W. Goossen, J. A. Walker, L. M. F. Chirovsky, L. A. D'Asaro, S. P. Hui, B. T. Tseng, R. E. Leibenguth, D. P. Kossives, D. W. Dahringer, D. D. Bacon, T. K. Woodward, and D. A. B. Miller. 1995. 700Mb/s operation of optoelectronic switching nodes comprised of flip-chip-bonded GaAs/AlGaAs MQW modulators and detectors of silicon CMOS circuitry. Technical Digest of Conference on Lasers and Electro-Optics, Postdeadline Paper CPD11.

68. McKnight, D. J., K. M. Johnson, and R. A. Serati. 1994. A 256 × 256 liquid-crystal-on silicon spatial light modulator. *Appl. Opt.* 33:2775–2784.

69. *The national technology roadmap for semiconductors.* 1994. Semiconductor Industry Association, published and distributed by SEMATECH, Incorporated.

70. Fisher, J. J., and B. L. Shoop. 1995. Optical-to-optical modulator based on a multiple quantum well device. Proceedings of the 9th National Conference on Undergraduate Research, vol. II, 614–618.

71. Kazlas, P. T., D. J. McKnight, and K. M. Johnson. 1996. Integrated assembly of smart pixel arrays and fabrication of associated micro-optics. Proceedings of the 1996 IEEE/LEOS Summer Topical Meeting on Smart Pixels, Keystone, CO, 51–52.

72. Schubert, E., Y. Wang, A. Cho, L. Tu, and G. Zydzik. 1992. Resonant cavity light emitting diode. *Appl. Phys. Lett.* 51:921–923.

73. Twyford, E., J. Chen, N. M. Jokerst, and N. Hartman. 1995. Optical MCM interconnect using thin film device integration. OSA Annual Meeting, Portland, OR.

74. McCormick, F. B. 1996. Smart pixel optics and packaging. Proceedings of the 1996 IEEE/LEOS Summer Topical Meeting on Smart Pixels, Keystone, CO, 45–46.

75. Kabal, D. N., G. C. Boisset, D. R. Rolston, and D. V. Plant. 1996. Packaging of two-dimensional smart pixel arrays. Proceedings of the 1996 IEEE/LEOS Summer Topical Meeting on Smart Pixels, Keystone, CO, 53–54.

76. Derstine, M. W., S. Wakelin, F. B. McCormick, and F. A. P. Tooley. 1994. A gentle introduction to optomechanics for free space optical systems. Available via anonymous ftp from the /pub/optomech directory of ftp.optivision.com, 18 pp.

77. Goodman, J. W. 1989. Optics as an interconnect technology. In *Optical processing and computing,* eds. H. H. Arsenault, T. Szoplik, and B. Macukow, 1–32. New York: Academic Press.

78. Miller, D. A. B. 1989. Optics for low-energy communication inside digital processors: Quantum detectors, sources, and modulators as efficient impedance converters. *Opt. Lett.* 14:146–148.

79. Hinton, H. S., and T. H. Szymanski. 1995. Intelligent optical backplanes. Proceedings of the Conference on Massively Parallel Processing with Optical Interconnections, San Antonio, TX.

80. Devenport, K. E., H. S. Hinton, and D. J. Goodwill. 1996. A Hyperplane smart pixel array for packet based switching. Proceedings of the 1996 IEEE/LEOS Summer Topical Meeting on Smart Pixels, Keystone, CO, 32–33.

81. Lentine, A. L. 1993. Advances in SEED based free space switching systems. OSA Proceedings on Photonics in Switching, eds. J. W. Goodman and R. C. Alferness, 81–84.

82. Woodward, T. K., A. V. Krishnamoorthy, K. W. Goossen, J. A. Walker, J. E. Cunningham, W. Y. Jan, L. M. F. Chirovsky, S. P. Hui, B. Tseng, D. Kossives, D. Dahringer, D. Bacon, and R. E. Leibenguth. 1996. Clocked-sense-amplifier-based smart-pixel optical receivers. *IEEE Photon. Tech. Lett.* 8:1067–1069.

83. Woodward, T. K., A. V. Krishnamoorthy, A. L. Lentine, K. W. Goossen, J. A. Walker, J. E. Cunningham, W. Y. Jan, L. A. D'Asaro, L. M. F. Chirovsky, S. P. Hui, B. Tseng, D. Kossives, D. Dahringer, and R. E. Leibenguth. 1996. 1-Gb/s two-beam transimpedance smart-pixel optical receivers made from hybrid GaAs MQW modulators bonded to 0.8-μm silicon CMOS. *IEEE Photon. Tech. Lett.* 8:422–424.

84. Inose, H., and Y. Yasuda. 1963. A unity bit coding method by negative feedback. *Proc. IEEE* 51:1524–1535.

85. Candy, J. C. 1974. A use of limit cycle oscillations to obtain robust analog-to-digital converters. *IEEE Trans. Commun.* 22:298–305.

86. Wilkinson, S. T., N. M. Jokerst, and R. P. Leavitt. 1995. Resonant-cavity-enhanced thin-film AlGaAs/GaAs/AlGaAs LED's with metal mirrors. *Appl. Opt.* 34:8298–8302.

87. Floyd, R., and L. Steinberg. 1975. An adaptive algorithm for spatial gray scale. *SID 75 Digest* 36:35–36.

88. Jarvis, J. F., C. N. Judice, and W. J. Ninke. 1976. A survey of techniques for the display of continuous-tone pictures on bilevel displays. *Comput. Graphics Image Processing* 5:13–40.

89. Anastassiou, D. 1989. Error diffusion coding for A/D conversion. *IEEE Trans. Circ. Syst.* 36:1175–1186.

90. Crounse, K. R., T. Roska, and L. O. Chua. 1993. Image halftoning with cellular neural networks. *IEEE Trans. Circ. Syst.* 40:267–283.

91. Mead, C. 1989. *Analog VLSI and neural systems,* Chap. 5. Reading, PA: Addison-Wesley.

92. Hinton, H. S., and N. M. Jokerst (Conf. Chairs). 1996. Digest of IEEE/LEOS 1996 Summer Topical Meeting on Smart Pixels, Keystone, CO.

93. Szymanski, T. H., and H. S. Hinton. 1994. Architecture of a field programmable smart pixel array. Proceedings of the International Conference on Optical Computing, Edinburgh, Scotland, 497–500.

94. Sherif, S. S., T. H. Szymanski, and H. S. Hinton. 1996. Design and implementation of a field programmable smart pixel array. Proceedings of the 1996 IEEE/LEOS Summer Topical Meeting on Smart Pixels, Keystone, CO, 78–79.

95. Harris, J. S., Jr., M. C. Larson, and A. R. Massengale. 1996. Broad-range continuous wavelength tuning in microelectromechanical vertical-cavity surface-emitting lasers. Proceedings of the 1996 IEEE/LEOS Summer Topical Meeting on Optical MEMs and Their Applications, Keystone, CO, 31–32.

96. Goto, H. 1996. Si micromachined 2D optical scanning mirror and its application to scanning sensors. Proceedings of the 1996 IEEE/LEOS Summer Topical Meeting on Optical MEMs and Their Applications, Keystone, CO, 17–18.

97. Arch, D., T. Ohnstein, D. Zook, and H. Guckel. 1996. A MEMS-based tunable infrared filter for spectroscopy. Proceedings of the 1996 IEEE/LEOS Summer Topical Meeting on Optical MEMs and Their Applications, Keystone, CO, 21–22.

Chapter 21 | Emerging Technology for Fiber Optic Data Communication

Chung-Sheng Li
Frank Tong

IBM T. J. Watson Research Center, Yorktown Heights, New York 10598

21.1. Introduction

Due to the explosive growth of the number of the Internet users and the World Wide Web sites, there has been a significant increase in the demand for network bandwidth. These web sites have spawned many academic and commercial applications such as digital library, distance learning, electronic commerce, cybermall, and on-demand video/audio. Consequently, the network congestion is aggravated at both the global and regional levels.

Currently, most of the interconnections between Internet routers are implemented using T1 (1.544 Mb/s), T3 (45 Mb/s) or OC-3 (155 Mb/s) lines. This network infrastructure is already inadequate to support the current traffic. Needless to say, future Internet application such as digital library and distance learning will create even more pressure on the transmission and switching capabilities of the system as more bandwidth-hungry information (graphics, images, and video) will be distributed electronically. Therefore, it is natural to expect that the backbone of the future Internet will be based on faster data rate such as OC-12 or beyond and employ some form of optical switching to alleviate the bandwidth problem.

Asynchronous transfer mode (ATM) over synchronous optical network has already been adopted as the primary transport mechanism for carrying broadband traffic for the future. Currently, the broadband traffic is carried on single-mode fiber between major switching hubs for data rates up to OC-48 (2.5 Gb/s). However, the speed of each fiber cannot be increased indefinitely. When the bandwidth required is more than that which can be supported by a single OC-48 connection, additional multiplexing techniques have to be incorporated in order to advance the link capacity.

759

HANDBOOK OF FIBER OPTIC
DATA COMMUNICATION

Currently, there are two proposals for implementing 10 Gb/s transmission: time-division multiplexing, in which four OC-48 channels are time multiplexed together into a single OC-192 channel, and wavelength-division multiplexing (WDM), in which four OC-48 channels are multiplexed at different wavelengths. WDM has been regarded as a promising approach for utilizing the vast optical bandwidth offered by the single-mode optical fiber without requiring the electronics to be operated at the aggregated speed as in time-division multiplexing. However, it imposes a different set of constraints on the optoelectronic components.

In this chapter, we will survey a number of promising technologies for fiber optic data communications. The goal is to investigate the potentials and limitations of each technology. The organization of the rest of this chapter is as follows: Section 21.2 describes the architecture of all-optical networks including both broadcast-and-select networks and wavelength-routed networks. The device aspects of tunable transmitters and tunable receivers for WDM networks are discussed in Sections 21.3 and 21.4, respectively. Section 21.5 describes the optical amplifiers, which is by far the most important technology for increasing the distance between data regeneration. Wavelength (de)multiplexer technologies are described in Section 21.6, whereas wavelength router technologies are discussed in Section 21.7. Section 21.8 discusses the wavelength converters, and this chapter is briefly summarized in Section 21.9.

21.2. Architecture of All-Optical Network

21.2.1. BROADCAST-AND-SELECT NETWORKS

A broadcast-and-select network consists of nodes interconnected to each other via a star coupler, as shown in Fig. 21.1. An optical fiber link carries signals from each node to the star. The star combines the signals from all the nodes and distributes the resulting optical signal equally among all its outputs. Another optical fiber link carries the combined signal from an output of the star to each node. Examples of such networks are Lambdanet [1] and Rainbow [2].

21.2.2. WAVELENGTH-ROUTED NETWORKS

A wavelength-routed network is shown in Fig. 21.2. The network consists of static or reconfigurable wavelength routers interconnected by fiber links. Static routers provide a fixed, nonreconfigurable routing pattern. A

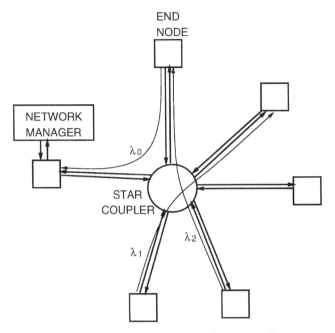

Fig. 21.1 A broadcast-and-select network.

reconfigurable router, on the other hand, allows the routing pattern to be changed dynamically. These routers provide static or reconfigurable light-paths between end nodes. A lightpath is a connection consisting of a path in the network between the two nodes and a wavelength assigned on the path. End nodes are attached to the wavelength routers. One or more controllers that perform the network management functions are attached to the end node(s).

21.2.3. CURRENT WDM/WDM ACCESS ACTIVITIES

The first field test of WDM system by the British Telecom in Europe dated back to 1991. This test was based on a five-node, three-wavelength OC-12 (622 Mb/s) ring around London with a total distance of 89 km. The ESPRIT program, funded by the European government since 1991, is a consortium funding multiple programs including OLIVES on optical interconnects and several WDM-related efforts. The RACE project, which is a joint university corporate program, has also included demonstrations of multiwavelength transport network. Since 1995, Advanced Communications Technologies

λ_i DATA WAVELENGTH

λ_0 CONTROL AND FAULT DETECTION WAVELENGTH

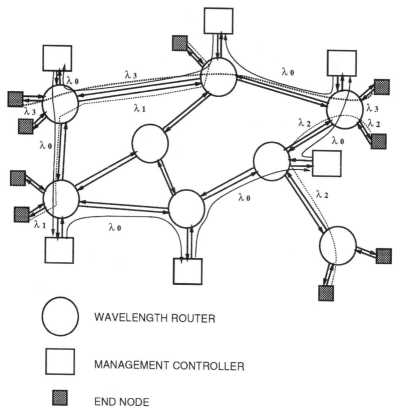

WAVELENGTH ROUTER

MANAGEMENT CONTROLLER

END NODE

Fig. 21.2   A wavelength-routed network.

and Services, which includes a total of 13 projects, started building a trans-European information infrastructure based on ATM and developing metropolitan optical network. In Japan, NTT is building a 16-channel photonic transport network with more than 320 Gb/s throughput. In the United States, ARPA/DARPA has funded a series of WDM/WDMA activities:

- All-Optical Network (AON) consortium, which includes ATT, DEC, and MIT, is focused on developing architectures and technologies that can support point-to-point or point-to-multipoint high-speed circuit-switched multigigabits per second digital or analog sessions.

- Optical Network Technology Consortium (ONTC) which includes Bellcore and Columbia University, is focused on scalable multiwavelength multihop optical network.
- Multiple Wavelength Optical Network (MONET), which is a consortium including Bell Labs, Bellcore, and three regional Bells, is chartered to develop WDM test beds and come up with commercial applications for the technology.

Currently, commercial WDM products include IBM (9729 Optical Wavelength Division Multiplexer) from the IBM Research Division. This is the first product that is developed solely for the computer industry. It allows the simultaneous transmission of up to 20 independent data streams at different wavelengths ranging from 1540 to 1560 nm in a single fiber. The channel spacing is 1 nm (or 120 GHz), and the maximum unrepeated distance is 50 km (or 30 miles) for data transmitting at 200 Mb/s. Ciena Corp. introduced MultiWave WDM terminal in March 1996 and demonstrated it at Supercomm in June 1996. This system can accommodate up to 16 OC-48 channels for long-haul carriers. A 40-channel system was due out toward the end of 1996. Pirelli demonstrated WaveMux 3200, which can handle up to eight OC-192 channels or 32 OC-48 channels, at Supercomm 96. A prototype that can handle 32 OC-192 channels has also been demonstrated.

21.3. Tunable Transmitter

The tunable transmitter is used to select the correct wavelength for data transmission. As opposed to a fixed-tuned transmitter, the wavelength of a tunable transmitter can be selected by an externally controlled electrical signal.

Currently, wavelength tuning can be achieved by one of following mechanisms:

- External cavity tunable lasers: A typical external cavity tunable laser, as shown in Fig. 21.3, includes a frequency selective component in conjunction with a Fabry–Perot laser diode with one facet coated with antireflection coating. The frequency selective component can be a diffraction grating or any tunable filter whose transmission or reflection characteristics can be controlled externally. This structure usually enjoys wide tuning range but suffers slow tuning time when

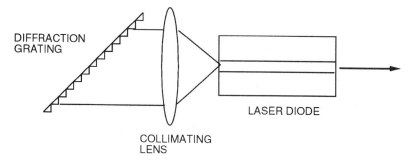

Fig. 21.3   Structure of an external cavity laser.

a mechanical tunable structure is employed. Alternatively, an electrooptic tunable structure can be employed to provide faster tuning time but narrower tuning range.

- Two-section tunable distributed Bragg reflector (DBR) tunable laser diodes: In a two-section device, as shown in Fig. 21.4, separate electrodes carry separate injection current: one is for the active area, whereas the other one is for controlling the index seen by the Bragg mirror. This type of device usually has a small continuous tuning range. For example, the tuning range of the device reported in Ref. [3] is limited to ≈5.8 nm (720 GHz at 1.55 μm).

- Three-section DBR tunable laser diodes: The major drawback of the two-section DBR device is the big gap in the available tuning range. This problem can be solved by adding a phase-shift section. With this additional section, the phase of the wave incident on the Bragg mirror section can be varied and matched, thus avoiding the gap of the tuning range.

- One- or two-dimensional laser diode array [4] or a multichannel grating cavity laser [5]: both allow only a few signaling channels. At another extreme, the wavelengths covering the range of interests can be reached by individual lasers in a one-dimensional or a two-dimensional laser array [4] with each lasing element emitting at a different wavelength. Single dimensional laser array, with each lasing element emitting at single transverse and longitudinal mode, can be fabricated. One possible scheme of single-mode operation can be achieved by means of short cavity (with high-reflectivity coatings) where the neighboring cavity modes from the lasing

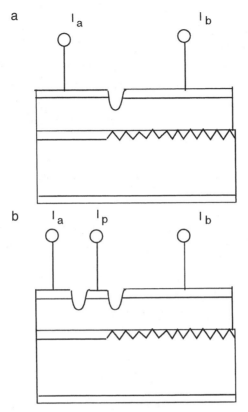

Fig. 21.4 Structure of a (a) two-section and (b) three-section laser diode.

modes are far from the peak gain. Different emission wavelengths can be achieved by tailoring the cavity length of each array element. For the two-dimensional laser array, such as the vertical surface-emitting laser array, each array element emits at a different wavelength by tailoring the length of the cavity [4]. It is possible to turn on more than one laser at any given time in both of these approaches. Heat dissipation, however, might limit the number of lasers that can be turned on.

The coupling of the laser emission from the one- or two-dimensional laser into a single fiber or waveguide can be achieved with gratings or a computer-generated holographic coupler. Coupling of four wavelengths from a vertical surface-emitting laser array has been demonstrated recently.

Holographic coupling has also been demonstrated to couple more than 100 wavelengths with two.

Due to the cross-talk and the limited bandwidth of the electronic switch, external modulator might be required to modulate the laser beam for higher bit rate. The light signals can be modulated by using a directional coupler-type modulator, a Mach–Zehnder type modulator [7], or a quantum well modulator [8]. The operation of these devices is required to be wavelength independent over the entire tuning range of the tunable transmitter.

21.4. Tunable Receiver

The tunable receiver is used to select the correct wavelength for data reception. As opposed to a fixed-tuned receiver, the wavelength of a tunable receiver can be selected by an externally controlled electrical signal.

Ideally, each tunable receiver needs a tuning range that covers the entire transmission bandwidth with high resolution and can be tuned from any channel to any other channel within a short period of time. Tunable receiver structures that have been investigated include:

- Single-cavity Fabry–Perot interferometer: The simplest form of a tunable filter is a tunable Fabry–Perot interferometer, which consists of a movable mirror to form a tunable resonant cavity. The electric field at the output side of the FP filter in the frequency domain is given by

$$S_{out}(f) = H_{FP}(f)S_{in}(f), \tag{21.1}$$

where H_{FP} is the frequency domain transfer function given by

$$H_{FP}(f) = \frac{T}{1 - Re^{j2\pi\frac{f-f_c}{FSR}}}, \tag{21.2}$$

where R is the power transmission coefficient and the power reflection coefficient of the filter, respectively. The parameter f_c is the center frequency of the filter. The parameter FSR is the free spectral range at which the transmission peaks are repeated and can be defined as FSR $= c/2\ \mu L$, where L is the FP cavity length and μ is the refractive index of the medium bounded by the FP cavity mirror. The 3-dB transmission bandwidth full width at half maximum (FWHM) of the FP filter is related to FSR and F by

$$\text{FWHM} = \frac{\text{FSR}}{F}, \tag{21.3}$$

where the reflectivity finesse F of the FP filter is defined as

$$F = \frac{\pi\sqrt{R}}{1 - R}. \tag{21.4}$$

Based on this principle, both fiber Fabry–Perot (as shown in Fig. 21.5) and liquid crystal Fabry–Perot tunable filters have been realized. The tuning time of these devices is usually on the order of milliseconds because of the use of electromechanical devices.

- Cascaded multiple Fabry–Perot filters [9]: The resolution of Fabry–Perot filters can be increased by cascading multiple Fabry–Perot filters by using either vernier or coarse-fine principle.
- Cascaded Mach–Zehnder tunable filter [10]: The structure of a Mach–Zehnder interferometer is shown in Fig. 21.6a. The light is first split by a 3-dB coupler at the input, then goes through two branches with a phase-shift difference and is then combined by another 3-dB coupler. The path length difference between two arms causes constructive and destructive interference depending on the input wavelength, resulting in a wavelength-selective device. To tune each Mach–Zehnder, it is necessary only to vary the differential path length by $\lambda/2$. Successive Mach–Zehnder filters can be cascaded together, as shown in Fig. 21.6b. In order to tune to a specific channel, filters with different periodic ranges must be centered at

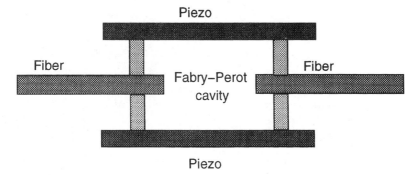

Fig. 21.5 Structure of a fiber Fabry–Perot filter.

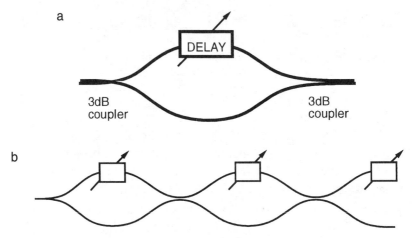

Fig. 21.6   (a) Single-stage and (b) multistage Mach–Zehnder tunable filters.

the same location. This can be accomplished by tuning the differential path length of individual filter.

- Acoustooptic tunable filter [10]: The structure of an acoustooptic tunable filter based on the surface acoustic wave principle is shown in Fig. 21.7. The incoming beam goes through a polarization splitter, separating horizontally polarized beam from the vertically polarized beam. Both of these beams travel down the waveguide with a grating established by the surface acoustic wave generated by a transducer. The resonant structure established by the grating rotates the polarization of the selected wavelength while leaving the other wavelength unchanged. Another polarization beam splitter at the output collects the signals with rotated polarization to ouput 1 while the rest is passed to output 2.

- Switchable grating [5]: The monolithic grating spectrometer is a planar waveguide device in which the grating and the input/output waveguide channels are integrated as shown in Fig. 21.8. A polarization-independent, 78-channel (channel separation of 1 nm) device has recently been demonstrated with cross-talk ≤ -20 dB [5]. For a 256-channel system, the detector array can be grouped into 64 groups with four detector array elements in each bar from which a preamplifier is connected to [11]. The power for the metal–semiconductor–metal (MSM) detectors is provided by a 2-bit control as shown. The outputs

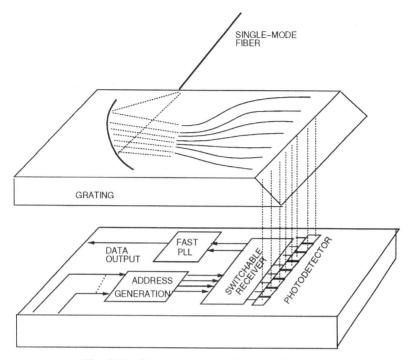

Fig. 21.7 Structure of a grating demultiplexer.

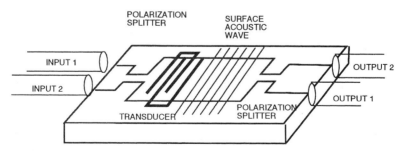

Fig. 21.8 Structure of optoelectronic tunable filter using grating demultiplexer and photodetector array.

from the preamplifiers are controlled by gates through a 3-bit control such that every eight outputs from the gates are fed into a postamplifier. The outputs from the postamplifiers are further controlled by another 3-bit controller. In this way, any channel from the 256 channels can be selected. The switching speed could be very fast, mostly due to the power-up time required by the MSM detectors (the total capacitance need to be driven is ~100 fF × 64).

21.5. Optical Amplifier

In order to achieve all-optical metropolitan/wide area networks, optical amplification is required to compensate for various losses such as fiber attenuation, coupling and splitting loss in the star couplers, and coupling loss in the wavelength routers. Both rare-earth ion-doped fiber amplifiers [12, 13] and semiconductor laser amplifiers can be used to provide amplification of the optical signals.

21.5.1. SEMICONDUCTOR OPTICAL AMPLIFIER

A semiconductor amplifier is basically a laser diode that operates below lasing threshold. There exist two basic types of semiconductor amplifiers: Fabry–Perot amplifier (FPA) and traveling wave amplifiers (TWA). In FPA structures, two cleaved facets act as partial reflective mirrors that form a Fabry–Perot cavity. The natural reflectivity for air-semiconductor facets is 32% but can be modified through a wide range by using antireflection coating or high-reflection coating. FPA is less desirable in many applications because of the nonuniform gain across the spectrum. TWA has the same structure as FPA except that antireflection coating is applied to both facets to minimize internal feedback.

The maximum available signal gain of both the FPA and the TWA is limited by gain saturation. The TWA gain, G_s, as a function of the input power $P_{in} = P_{out}/G_s$ is given by the following equation:

$$G_s = 1 + \frac{P_{sat}}{P_{in}} \ln \frac{G_o}{G_s}, \tag{21.5}$$

where G_o is the maximum amplifier gain, corresponding to the single pass gain in the absence of input light. It is easy to observe that G_s monotonically decreases to 1 as the input signal power increases, resulting the gain saturation effect.

Cross-talk occurs when multiple optical signals or channels are amplified simultaneously. In this circumstance, the signal gain for one channel is affected by the intensity levels of other channels as a result of the gain saturation. This effect depends on the carrier lifetime, which is on the order of 1 ns. Therefore, cross-talk among different channels is most pronounced when the data rate is comparable to the reciprocal of the carrier lifetime.

Another limit to the amplifier gain is due to the amplifier spontaneous emission. The amplification of spontaneous emission is triggered by the spontaneous recombination of electrons and holes in the amplifier medium. This noise beats with the signal at the photodetector, causing signal-spontaneous beat noise and spontaneous–spontaneous beat noise.

21.5.2. DOPED-FIBER AMPLIFIER

When optical fibers are doped with rare-earth ions, such as erbium, neodymium, or praseodymium, the loss spectrum of the fiber can be drastically modified. During the absorption process, the photons from the optical pump at wavelength λ_p are absorbed by the outer orbital electrons of the rare-earth ions and these electrons are raised to higher energy levels. The deexcitation of these high energy levels to the ground state might occur either radiatively or nonradiatively. If there is an intermediate level, additional deexcitation can be stimulated by the signal photon, providing that the bandgap between the intermediate state and the ground state corresponds to the energy of the signal photons. The result would be an amplification of the optical signal at wavelength λ_s.

The main difference between doped-fiber amplifier and semiconductor amplifiers is that the amplifier gain of doped-fiber amplifier is provided by means of optical pumping as opposed to electrical pumping.

Figure 21.9 shows a typical fiber amplifier system. Currently, the most popular doped-fiber amplifiers are based on erbium doping. Similar to

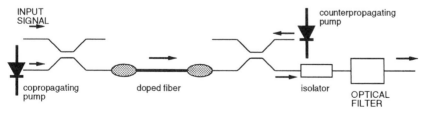

Fig. 21.9 A typical doped-fiber amplifer system with either copropagation pump or counterpropagation pump.

semiconductor amplifier, the gain of erbium-doped fiber amplifier also saturates. However, the cross-talk effect is much reduced thanks to the long fluorescence lifetime.

21.5.3. GAIN EQUALIZATION

The gain spectra of these optical amplifiers are nonflat over the fiber transmission windows at 1.3 and 1.55 μm, resulting in nonuniform amplification of the signals. Together with the near-far effect resulting from optical signals originating from various nodes at locations separated by large distances, there exists a wide dynamic range among various signals arriving at the receivers. The best dynamic range of lightwave receivers with high sensitivity reported thus far is limited to less than $\sim$20 dB at 2.4 Gb/s [14] and less than $\sim$30 dB at 1 Gb/s [15]. In addition, the signal with high average optical power saturates the gain of the optical amplifiers placed along the path of propagation. This limits the available gain for the remaining wavelength channels. Thus, signal power equalization among different wavelength channels is required.

Most of the existing studies on gain equalization have been focused on either statically or dynamically equalizing the nonflat gain spectra of the optical amplifiers, but they do not address the near-far effect. For static gain equalization, schemes including grating embedded in the Er^{3+} fiber amplifier [12], cooling the amplifiers to low temperatures [16], or a notch filter [17–19] were proposed previously to flatten the gain spectra. An algorithm is proposed to adjust the optical signal power at different transmitters to achieve equalization [20]. In Ref. [21], gain equalization is achieved by placing a set of attenuators in the arms of the back-to-back grating multiplexers to compensate for nonflat gain spectra of the fiber amplifier. For dynamic gain equalization, a two-stage fiber amplifier with offset gain peaks was proposed in Ref. [22] to equalize the optical signal power among different WDM channels by adjusting the pump power. This scheme, however, has a very limited equalized bandwidth of $\sim$2.5 nm. Dynamic gain equalization can also be achieved through controlling the transmission spectra of tunable optical filters. Using this scheme, three-stage (for 29 WDM channels) [23] and six-stage (for 100 WDM channels) [24] Er^{3+}-doped-fiber amplifier systems with equalized gain spectra were demonstrated using a multistage Mach–Zehnder interferometric filter. Acoustooptic tunable filter has also been used to equalize gain spectra for a very wide transmission window [25]. The combination of these schemes can, in principle, solve the near-far problem in the networks.

21.6.   Wavelength Multiplexer/Demultiplexer

Wavelength multiplexers and demultiplexers are the essential components for constructing wavelength routers. They can also be used for building tunable receivers and transmitters as described in the previous sections.

Two types of wavelength multiplexers/demultiplexers are most widely used: grating demultiplexers and phase arrays.

Figure 21.10 shows an etched-grating demultiplexer with N output waveguides and its cross-sectional view, respectively. The reflective grating uses the Rowland circle configuration in which the grating lies along a circle, whereas the focal line lies along a circle of half the diameter.

Phase array wavelength multiplexer/demultiplexers have been shown to be the superior WDM demultiplexers for systems with a small number of channels. A phase array demultiplexer consists of a dispersive waveguide array connected to input and output waveguides through two radiative couplers as shown in Fig. 21.11. Light from an input waveguide diverging in the first star coupler is collected by the array waveguides, which are designed in such a way that the optical path length difference between adjacent waveguides equals an integer multiple of the central design wavelength of the demultiplexer. This results in the phase and intensity distribution of the collected light being reproduced at the start of the second star coupler, causing the light to converge and focus on the

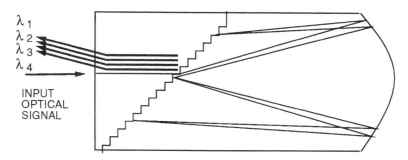

Fig. 21.10   Structure of a grating demultiplexer.

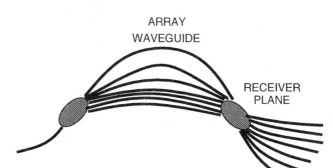

Fig. 21.11   Structure of phase array wavelength demultiplexer.

receiver plane. Due to the path length difference, the reproduced phase front will tilt with varying wavelength, thus sweeping the focal spot across different output waveguides.

21.7.   Wavelength Router

Wavelength routing for all-optical networks using WDMA has received increasing attention [26–29]. In a wavelength routing network, wavelength-selective elements are used to route different wavelengths to their corresponding destinations. Compared to a network using only star couplers, a network with wavelength routing capability can avoid the splitting loss incurred by the broadcasting nature of a star coupler [30]. Furthermore, the same wavelength can be used simultaneously on different links of the same network and reduce the total number of required wavelengths [26].

The routing mechanism in a wavelength router can either be static, in which the wavelengths are routed using a fixed configuration [31], or dynamic, in which the wavelength paths can be reconfigured [32]. The common feature of these multiport devices is that different wavelengths from each individual input port are spatially resolved and permuted before they are recombined with wavelengths from other input ports. These wavelength routers, however, have imperfections and nonideal filtering characteristics that give rise to signal distortion and cross-talk.

Figure 21.12 shows the structure of a static wavelength router that consists of K optical demultiplexers and multiplexers. Each input fiber to an optical demultiplexer is assumed to contain up to M different wavelengths,

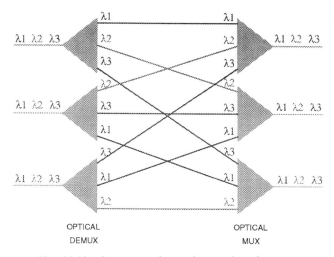

Fig. 21.12 Structure of a static wavelength router.

where $M \le K$. However, we only consider the case in which $M \le K$. The optical demultiplexer spatially separates the incoming wavelengths into M paths. Each of these paths is then combined at an optical multiplexer with the outputs from the other $M - 1$ optical demultiplexers.

The wavelength routing configuration in Fig. 21.12 is fixed permanently. The optical data at wavelength λ_j entering the ith demultiplexer exit at the $[(j - i)\mathrm{mod}M]^{\mathrm{th}}$ output of that demultiplexer. That output is connected to the ith input of the $[(j - i)\mathrm{mod}M)^{\mathrm{th}}$ multiplexer.

Because of the imperfections and nonideal filtering characteristics of the optical multiplexers and demultiplexers, cross-talks occur in the wavelength routers. On the demultiplexer side, each output contains both the signals from the desired wavelength and those from the other $M - 1$ cross-talk wavelengths. From reciprocity, both the desired wavelength and the cross-talk signals exit at the output on the multiplexer side. Thus, each wavelength at every multiplexer contains $M - 1$ cross-talk signals originating from all demultiplexers.

Cross-talk phenomena in wavelength routers have previously been studied [33–36]. It was shown in Ref. [33] that the maximum allowable cross-talk in each grating (grating as optical demultiplexers and multiplexers in the wavelength router) is -15 dB in an all-optical network with moderate size (e.g., 20 wavelengths and 10 routers in cascade). The results are based on using a 1-dB power penalty criterion and only considering the power addition effect of the cross-talk. Cross-talk can also arise from beating between the data signal and the leakage signal (from imperfect filtering) at the same output channel. The beating of these uncorrelated signals converts the phase noise of the laser sources into the amplitude noise and corrupts the received signals [37] when the linewidths of the laser sources are smaller than the electrical bandwidth of the receiver. Coherent beating, in which the data signal beats with itself, can occur as a result of the beatings among the signals from multiple paths or loops caused by the leakage in the wavelength routers in the system. It was shown in Ref. [35] that the component cross-talk has to be less than -20, -30, and -40 dB in order to achieve satisfactory performance for a system consisting of a single, 10, and 100 leakage sources, respectively.

Figure 21.13 shows the structure of the a dynamic wavelength routing device. This device consists of a total of N optical demultiplexers and N optical multiplexers. Each of the input fibers to an optical demultiplexer contains M different wavelengths. The optical demultiplexer spatially separates the incoming wavelengths into M paths. These paths pass through a photonic switch before they are combined with the outputs from the other $M - 1$ optical switches. When the cross-talk of the optical multiplexer/demultiplexer/switch is considered, each wavelength channel at each input optical demux can reach any of the output optical mux via M different paths.

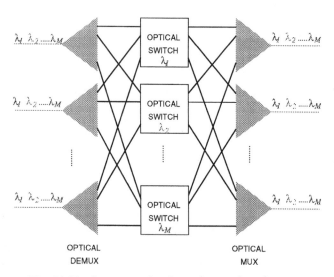

Fig. 21.13   Structure of a dynamic wavelength router.

21.8.   Wavelength Converter

The network capacity of WDM networks is determined by the number of independent light paths. One way to increase the number of nodes that can be supported by the network is to use wavelength router to enable spatial reuse of the wavelengths, as described in the previous section. The second method is to convert signals from one wavelength to another. Wavelength conversion also allows distributing the network control and management into smaller subnetworks and allows flexible wavelength assignments within the subnetworks.

There are three basic mechanisms for wavelength conversion:

1. Optoelectronic conversion: The most straightforward mechanism for wavelength conversion is to convert each individual wavelength to electronical signals, and then retransmit by lasers at the appropriate wavelength. A nonblocking cross-point switch can be embedded within the O/E and E/O conversion such that any wavelength can be converted to any other wavelength (as shown in Fig. 21.14a). Alternatively, a tunable laser can be used instead of a fixed tuned laser to achieve the same wavelength conversion capability (as shown in Fig. 21.14b). This mechanism requires only mature technology. The protocol transparency is completely lost if full data regeneration (which includes retiming, reshaping, and reclocking) is performed within the wavelength converter. On the other hand, limited transparency can be achieved by incorporating only analog amplification in the conversion process. In this case, other information associated with the signals, including phase, frequency, and analog amplitude, is still lost.

2. Optical gating wavelength conversion: This type of wavelength converter, as shown in Fig. 21.15, accepts an input signal at wavelength λ_1 which contains the information and a continuous wave (CW) probe signal at wavelength λ_2. The probe signal, which is at the target wavelength, is then modulated by the input signal through one of the following mechanisms:

    • Saturable absorber: In this mechanism, the input signal saturates the absorption and allows the probe beam to transmit. Due to carrier recombinations, the bandwidth is usually limited to less than 1 GHz.

    • Cross-gain modulation: the gain of a semiconductor optical amplifier saturates as the optical level increases. Therefore, it is

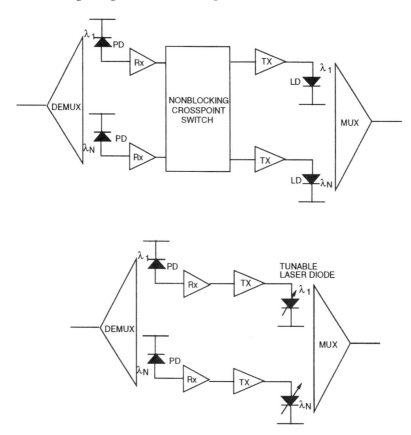

Fig. 21.14 (a) Structure of an optoelectronic wavelength converter using electronic cross-point switch. (b) Structure of an optoelectronic wavelength converter using tunable laser.

possible to modulate the amplifier gain with an input signal and encode the gain modulation on a separate CW probe signal.

- Cross-phase modulation: Optical signals traveling through semi-conductor optical amplifiers undergo a relatively large phase modulation compared to the gain modulation. The cross-phase modulation effect is utilized in an interferometer configuration such as in a Mach–Zehnder interferometer. The interferometric nature of the device converts this phase modulation to an amplitude modulation in the probe signal. The interferometer can operate in two different modes: a noninverting mode, in which an

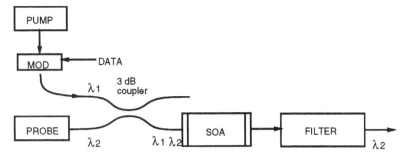

Fig. 21.15 Structure of an optical gating wavelength converter using semiconductor optical amplifier.

increase in input signal power causes a decrease in probe power, and an inverting mode, in which an increase in input signal power causes a decrease in probe power.

3. Wave-mixing wavelength conversion: Wavelength mixing, such as three-wave and four-wave mixing, arises from nonlinear optical response when more than one wave is present. The phase and frequency of the generated waves are linear combinations of the interacting waves. This is the only method that preserves both phase and frequency information of the input signals and is thus completely transparent. It is also the only method that can simultaneously convert multiple frequencies. Furthermore, it has the potential to accommodate signals at extremely high bit rates. Wave-mixing mechanisms can occur in either passive waveguides or semiconductor optical amplifiers.

21.9.   Summary

In this chapter, we have surveyed a number of promising technologies for fiber optic data communication systems. In particular, we have focused on technologies that can support gigabit all-optical WDM networks. Although most of these technologies are still far from being mature, they nevertheless hold the promise of dramatically improving the network capacity of existing fiber optical networks.

References

1. Goodman, M. S., H. Kobrinski, M. Vecchi, R. M. Bulley, and J. M. Gimlett. 1990 (August). The LAMBDANET multiwavelength network: Architecture, applications and demonstrations. *IEEE J. Select. Areas Commun.* 8(6):995–1004.

2. Janniello, F. J., R. Ramaswami, and D. G. Steinberg. 1993 (May/June). A prototype circuit-switched multi-wavelength optical metropolitan-area network. *IEEE/OSA J. Lightwave Tech.* 11:777–782.

3. Murata, S., I. Mito, and K. Kobayashi. 1987 (April). Over 720GHz (5.8nm) frequency tuning by a 1.5 μm DBR laser with phase and Bragg wavelength control regions. *Electron. Lett.* 23(8):403–405.

4. Maeda, M. W., C. J. Chang-Hasnain, J. S. Patel, H. A. Johnson, J. A. Walker, and C. Lin. 1991. Two dimensional multiwavelength surface emitting laser array in a four-channel wavelength-division-multiplexed system experiment. *OFC 91 Digest,* 73.

5. Kirkby, P. A. 1990. Multichannel grating demultiplexer receivers for high density wavelength systems. *IEEE J. Lightwave Tech.* 8:204–211.

6. Nyairo, K. O., C. J. Armistead, and P. A. Kirkby. 1991. Crosstalk compensated WDM signal generation using a multichannel grating cavity laser. *ECOC Digest,* 689.

7. Alferness, R. C. 1982 (August). Waveguide electrooptic modulators. *IEEE Trans. Microwave Theor. Techniques* 30(8):1121–1137.

8. Miller, D. A. B., D. S. Chemla, T. C. Damen, T. H. Wood, C. A. Burrus, Jr., A. C. Gossard, and W. Wiegmann. 1985 (September). The quantum well self-electrooptic effect device: Optoelectronic bistability and oscillation and self-linearized modulation. *IEEE J. Quantum Electron.* 21(9):1462–1476.

9. Hamdy, W. M. and P. A. Humblet. 1993. Sensitivity analysis of direct detection optical FDMA networks with OOK modulation. *IEEE J. Lightwave Tech.* 11(5/6):783–794.

10. DeCusats, C. and P. Das. 1991. *Acousto-optic signal processing fundamentals and applications.* Boston: Artech House.

11. Chang, G. K., W. P. Hong, R. Bhat, C. K. Nguyen, J. L. Gimlett, C. Lin, and J. R. Hayes. 1991. Novel electronically switched multichannel receiver for wavelength division multiplexed systems. *Proc. OFC91,* 6.

12. Tachibana, M., R. I. Laming, P. R. Morkel, and D. N. Payne. 1990. Gain-shaped erbium-doped fiber amplifier (EDFA) with broad spectral bandwidth. Topical Meeting on Optical Amplifier Application, MD1.

13. Desurvire, E., C. R. Giles, J. L. Zyskind, J. R. Simpson, P. C. Becker, and N. A. Olsson. 1990. Recent advances in erbium-doped fiber amplifiers at 1.5 μm. Proceedings of the Optical Fiber Communication Conference, San Francisco, CA.

14. Blaser, M., and H. Melchior. 1992 (November). High performance monolithically integrated $In_{0.53}Ga_{0.47}As/InP$ PIN/JFET optical receiver front end with adaptive feedback control. *IEEE Photon. Tech. Lett.* 4(11).

15. Mikamura, Y., H. Oyabu, S. Inano, E. Tsumura, and T. Suzuki. 1991. GaAs IC chip set for compact optical module of giga bit rates. Proceedings IECON'91 1991 International Conference on Industrial Electronics, Control and Instrumentation.

16. Goldstein, E. L., V. da Silva, L. Eskildsen, M. Andrejco, and Y. Silberberg. 1993 (February). Inhomogeneously broadened fiber-amplifier cascade for wavelength-multiplexed systems. *Proc. OFC'93.*

17. Tachibana, M., R. I. Laming, P. R. Morkel, and D. N. Payne. 1991 (February). Erbium-doped fiber amplifier with flattened gain spectra. *IEEE Photon. Tech. Lett.* 3(2):118–120.

18. Wilinson, M., A. Bebbington, S. A. Cassidy, and P. Mckee. 1992. D-fiber filter for erbium gain flattening. *Electron. Lett.* 28:131.

19. Willner, A. E., and S.-M. Hwang. 1993 (September). Passive equalization of nonuniform EDFA gain by optical filtering for megameter transmission of 20 WDM channels through a cascade of EDFA's. *IEEE Photon. Tech. Lett.* 5(9):1023–1026.

20. Chraplyvy, A. R., J. A. Nagel, and R. W. Tkach. 1992 (August). Equalization in amplified WDM lightwave transmission systems. *IEEE Photon. Tech. Lett.* 4(8):920–922.

21. Elrefaie, A. F., E. L. Goldstein, S. Zaidi, and N. Jackman. 1993 (September). Fiber-amplifier cascades with gain equalization in multiwavelength unidirectional inter-office ring network. *IEEE Photon. Tech. Lett.* 5(9):1026–1031.

22. Giles, C. R., and D. J. Giovanni. 1990 (December). Dynamic gain equalization in two-stage fiber amplifiers. *IEEE Photon. Tech. Lett.* 2(12):866–868.

23. Inoue, K., T. Kominato, and H. Toba. 1991. Tunable gain equalization using a Mach–Zehnder optical filter in multistage fiber amplifiers. *IEEE Photon. Tech. Lett.* 3(8):718–720.

24. Toba, H., K. Takemoto, T. Nakanishi, and J. Nakano. 1993 (February). A 100-channel optical FDM six-stage in-line amplifier system employing tunable gain equalizer. *IEEE Photon. Tech. Lett.,* 248–250.

25. Su, S. F., R. Olshansky, G. Joyce, D. A. Smith, and J. E. Baran. 1992. Use of acoustooptic tunable filters as equalizers in WDM lightwave systems. *OFC Proc.,* 203–204.

26. Brackett, C. A. 1994. The principle of scalability and modularity in multiwavelength optical networks. *Proc. OFC: Access Network,* 44.

27. Alexander, S. B. *et al.* 1993. A precompetitive consortium on wide-band all-optical network. *IEEE J. Lightwave Tech.* 11(5/6):714–735.

28. Chlamtac, I., A. Ganz, and G. Karmi. 1992 (July). Lightpath communications: An approach to high-bandwidth optical WAN's. *IEEE Trans. Commun.* 40(7):1171–1182.

29. Hill, G. R. 1988. A wavelength routing approach to optical communication networks. *Proc. INFOCOM,* 354–362.

30. Ramaswami, R. 1993 (February). Multiwavelength lightwave networks for computer communication. *IEEE Commun. Mag.* 31(2):78–88.

31. Zirngibl, M., C. H. Joyner, and B. Glance. 1994. Digitally tunable channel dropping filter/equalizer based on waveguide grating router and optical amplifier integration. *IEEE Photon. Tech. Lett.* 6(4):513–515.

32. d'Alessandro, A., D. A. Smith, and J. E. Baran. 1994 (March). Multichannel operation of an integrated acousto-optic wavelength routing switch for WDM systems. *IEEE Photon. Tech. Lett.* 6(3):390–393.

33. Li, C.-S., F. Tong, and C. J. Georgiou. 1993. Crosstalk penalty in an all-optical network using static wavelength routers. *Proc. LEOS Annu. Meeting.*

34. Li, C.-S., and F. Tong. 1994. Crosstalk penalty in an all-optical network using dynamic wavelength routers. *Proc. OFC'94.*

35. Goldstein, E. L., L. Eskildsen, and A. F. Elrefaie. 1994 (May). Performance implications of component crosstalk in transparent lightwave networks. *IEEE Photon. Tech. Lett.* 6(5):657–660.

36. Goldstein, E. L., and L. Eskildsen. 1995 (January). Scaling limitations in transparent optical networks due to low-level crosstalk. *IEEE Photon. Tech. Lett.* 7(1):93–94.

37. Gimlett, J. and N. K. Cheung. 1989. Effects of phase-to-intensity noise conversion by multiple reflections on gigabit-per-second DFB laser transmission systems. *IEEE J. Lightwave Tech.* 7(6):888–895.

Chapter 22 | Manufacturing Challenges

Eric Maass

Motorola, Chandler, Arizona 85248

22.1. Customer Requirements — Trends

The manufacturing challenges faced by optoelectronics components and systems are the direct effect of the technology challenges posed, which are in turn the indirect result of customer needs and expectations. These needs include performance (principally data rate and distance), reliability, service, and price.

A. PERFORMANCE

With the increasing speeds of hardware, such as microprocessors, and the increasing expectations of the users to include realistic video and user friendliness in the software or information communicated comes a need for increasing data rates on the interconnects from, to, and within the system. Some have called this the need to "feed the beast" with increasing supplies of its favorite "food" — data.

The need for higher data rates is driven by the increasing data rates required by applications that incorporate video or multimedia in general and the global sharing of information, including video conferencing, in which information in graphs, figures, spreadsheets, or other forms needs to be shared in real time. This trend for higher data rates is not expected to end — as the higher data rates are achieved, increasing expectations for better resolution, more interactiveness, and new capabilities not previously exploited because of system bottlenecks will drive new requirements for higher data rates. The trade-off in how to achieve the data rates will be based on the cost advantages and limitations of various approaches.

HANDBOOK OF FIBER OPTIC
DATA COMMUNICATION

B. RELIABILITY

Historically, optoelectronics applications have required high reliability primarily due to the early adoption of optoelectronics for telecommunications applications. Any expectation that data communication applications will involve lower reliability requirements and expectations will likely be short-lived. There are several reasons for this, including the rising expectations of customers in general and business customers in particular. However, the overriding trend will be the merging of data communications, telecommunications, and entertainment.

With the possible use of optoelectronics close to the home, the need for reliability can relate to "lifetime" uses, much as the telephone is now. Although it would be nice to believe that products will be driven toward increased reliability by the need for customer satisfaction and altruism, it may well prove that the avoidance of deep-pocket lawsuits will be an even greater driving force.

C. DELIVERY

The trend toward higher expectations will also include service in general, with on-time delivery as a subset. Customers increasingly expect promises to be kept. Companies that meet their delivery promises are rewarded with further business.

A possible strategy for achieving high expectations for service is illustrated in Fig. 22.1.

D. PRICE

Finally "the bottom line is the bottom line." Assuming that the component or system meets the performance and reliability requirements, the customer will generally purchase the least expensive product.

22.2. Manufacturing Requirements — Trends

The performance, reliability, delivery, and price expectations of the customers drive technology and manufacturing challenges. The manufacturing challenges involve anticipating and developing manufacturing strategies to deal with these customer requirements.

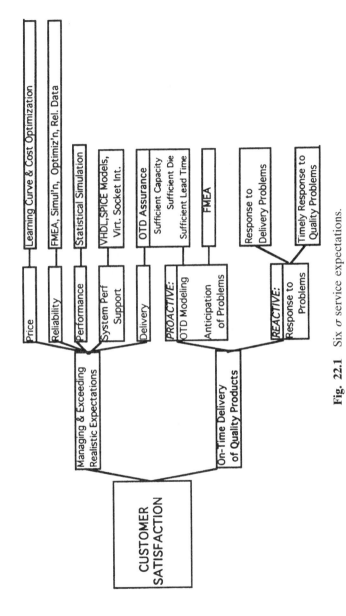

Fig. 22.1 Six σ service expectations.

A. *PERFORMANCE*

The data communication customers' rising expectations for performance—here defined as a high data rate, paired with a low error rate — will need to be met by combinations of system parameters, as illustrated in Figs. 4.2a and 4.2b in Chapter 4. Most of these system parameters involve tight requirements and tradeoffs, which pose challenges to the manufacturing sites.

Some of the key performance parameters from Figs. 4.2a and 4.2b are listed in Table 22.1, in which the parameters are separated into the separate components of the block diagram, as shown in Fig. 22.2, reproduced from Fig. 4.1 of Chapter 4.

1. Performance: Statistical Process Control, Reduction of Variability, Test, and Evaluation

In Fig. 4.2, the customer requirements are related to the performance parameters. Each of these performance parameters needs to be placed under statistical process control (SPC). SPC involves:

Control charts
    Appropriate control limits
    Corrective actions based on out-of-control condition
Preventative actions
    Anticipation of potential problems
    Poka Yoke/mistake proofing
    Feedback and control mechanisms within automated equipment
Reduction of variability

Control charting involves determining the natural variation of the performance parameter and developing control limits based on this natural variation. The control charts detect if the process varies beyond the natural variation so that appropriate corrective actions can be executed. These corrective actions avoid sending marginal product to the customer and correct the problem that caused the "out of control" condition. If the natural variation of some performance parameters exceeds the limits that are acceptable to the customers (specification limits), then design and process development efforts are required to reduce the variability.

Although control charting is an effective feedback system, it has been argued that control charting of the performance parameters is "after the fact": The product that is measured is an output of the process that needs to be controlled. It has been argued that it would be much preferred to control the inputs to the process that needs to be controlled rather than the output. (S. Shingo, *Zero quality control: Source inspection and the Poka-Yoke system*). Poka Yoke, or mistake proofing, involves anticipating and preventing the possibility of an out-of-control situation by methods such as

Checks and double checks prior to processing

Assistance to the operator

Source inspections

Source inspections detect nascent problems on the inputs and process motions or steps of the processing equipment and correct them, warn the operator, or shut down the processing before the product is affected.

The overall manufacturing process for an optoelectronic system is illustrated in Fig. 22.3. Within the integrated circuit (IC) processing areas, the controls for the ICs would be consistent with the requirements for other high-frequency integrated circuits, which would presumably be manufactured on the same processes. Nonetheless, for the sake of completeness, some of those requirements are listed below:

Complementary Metal Oxide Semiconductor

Effective gate length, including photolithography and etch at the gate polysilicon definition step and the critical dimension measurements after develop and after etch

Threshold voltages, both n and p channels

Gate oxide thickness, in that it affects the transconductance, the threshold voltages, and reliability

Capacitances, such as from gate to drain and drain to substrate/well

Isolation, especially radio frequency (RF) isolation, for the phase-locked loop (PLL) at higher frequencies

Bipolar

Current gain (beta) and the correlated $V\beta$

Capacitances, especially collector-base capacitance base resistance

Isolation, especially RF isolation, for the PLL at higher frequencies.

For all integrated circuits, the metal lines pose performance issues, contributing parasitic capacitances, inductances, and perhaps even act as undesired antennas. Note that a quarter wavelength at 1 GHz is about the length of some of the longer metal lines on larger ICs.

Table 22.1 Key Component Performance Parameters Related to System Performance Requirements

	Transmit/ Receive IC	Lasers	Packaging	Fiber	Photodetectors	PLL
High data rate	F_{max}, F_t Bipolar: collector/ base capacitance MOS: left V_t	Capacitance Resistance Threshold current	Parasitics Lead inductance Lead Capacitance	Attenuation	Capacitance Resistance Recombination?	High Q oscillation Filter w/low parasitics Substance Isoin Matching/ differential Sharp edges
Long distance	Noise figure Parasitics Adaptive threshold	Wavelength match w/fiber	Opto coupling	Loss Graded index		
Clean Data (Eye Opening)	Sharp edges Current to laser when off I/O levels		Parasitics			High Q oscillation Filter w/low parasitics Substance Isoin Matching/ differential Sharp edges

Low cost	Process cost/ complexity		OEI coupling	Material (plastic?)
Reliability	Dielectric Integration Interface charges/ traps Electromigration/ metal integration Contact integration Corrosion ESD Latchup	Crystal defects Point defects	Laser thermal Heat sink	

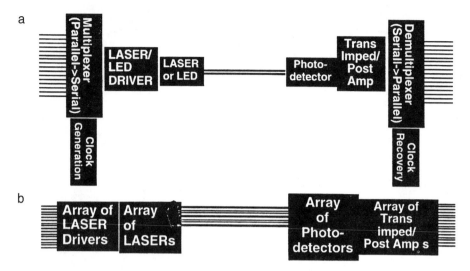

Fig. 22.2 Block diagrams of serial and parallel optoelectronic data communication systems.

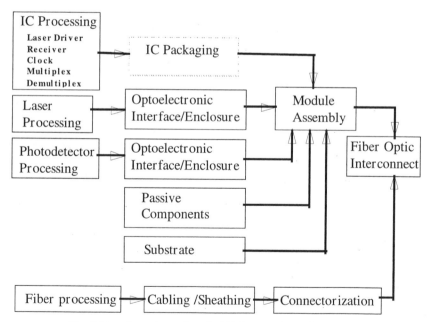

Fig. 22.3 Overview of optical link manufacturing process.

Within the assembly areas, the controls on die bonding, wire bonding, and molding operations would be required to prevent shorts and opens. For optoelectronics, tighter controls are required for all the parasitic inductances and capacitances associated with the packaging, especially the leads. The lead lengths should be minimized: Ideally, leads should be eliminated altogether.

Within the optoelectronic assembly area, the controls include the standard controls implemented for standard integrated circuits and new controls to meet the tolerances for good coupling between the light-emitting device and the packaging or between the photodetector and the packaging. This may require active alignment, which could use a feedback mechanism involving sensing the light output from the light-emitting device through the package to ensure that the alignment requirements are met. Although an active alignment system can be placed under controls that detect and prevent errors (Poka Yoke or mistake-proofing mechanisms), it can be an expensive system to implement.

Alternatively, passive alignment systems use the individual mechanical tolerances, determined by the manufacturing capabilities inherent in each of the components and piece parts as well as the assembly process. These tolerances determine where the light-emitting and light-detecting devices can be placed in the package and ensure that these tolerances are met during the assembly process without requiring activation during the assembly process. Passive alignment, if feasible, would be expected to have cost and cycle time benefits compared to active alignment of the light-emitting and photodetector devices during the assembly/alignment process.

Within the module assembly area, controls are required in the manufacture of the substrate (whether printed circuit or multichip module (MCM) substrate) and during the final assembly. The substrate manufacturing will generally require controls on the drilling (placement and diameter of holes), lamination (epoxy curing), photolithography, and etch processing. Final assembly will require controls on the insertion or placement of the components and on the solder joining process, including time, temperature, and ambient.

2. Meeting Customer Requirements: Design for Reliability, and Reliability Testing

Data communication customers will increasingly require high levels of reliability, here defined as the need to have the parts continue to meet the performance requirements over time. This requires that reliability be designed in and built in to every component, as illustrated in Fig. 22.4.

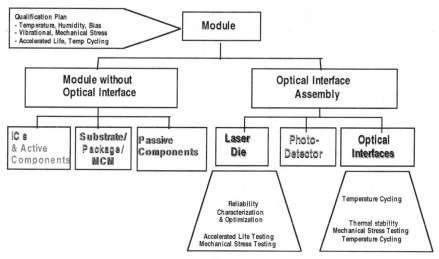

Fig. 22.4   Reliability requirements for optoelectronic systems, branched to individual component reliabilities.

Although designing and building in reliability is critical, customer satisfaction will require that testing of the integrated circuits also ensures reliability. Because most reliability testing, historically, has involved destructive testing, which can only be performed on a sample of the parts, this reliability testing will necessarily involve some innovative testing, such as the use of acceleration of known reliability degradation mechanisms. In integrated circuits, current is an acceleration factor for electromigration and contact integrity, voltage for dielectric integrity, and temperature for several mechanisms: low temperatures for hot carrier effects and high temperatures for corrosion. Some of the reliability degradation mechanisms for laser diodes are illustrated in Table 22.2 along with acceleration factors and some possible test structures for those reliability mechanisms.

3. Meeting Customer Requirements: Delivery, Optimizing On-Time Delivery, and Lead/Cycle Time

Figure 22.1 summarized customer service expectations. Service means many things, including the on-time delivery of quality products, providing the application support that the customers expect, and keeping the customer informed of progress in resolving issues that the customer expects to see

Table 22.2 **Laser Diode Reliability Mechanisms, Related to Acceleration Factors and Possible Test Structures to Ensure Reliability Is Built in**

Laser Diode Mechanism	*Acceleration Factors*	*Test Structures*
Crystal defect migration Dark line/area propagation Point defect formation	Current, temperature, Optical power, stress (packaging stress)	Standard laser
Contact integrity	Current, temperature	Laser, test pattern
Die bond integrity Corrosion Electromigration	 Temperature, humidity Current, temperature	Test patterns
Optical waveguiding Stress- or thermal-induced change in refractive index of III–V material	Stress, Temperature (?)	Laser
Facet defect migration (not important for VCELs) Facet erosion	Optical power, current	Laser

resolved. The immediate expectation is that the customer will receive reliable, high-quality products delivered on time.

On-time delivery of optical links means meeting the customer's requirements on the quantity delivered and when it is delivered. A related issue is the lead time — how long of a delay the customers must experience in receiving product after placing an order. Often, the lead time is not an issue: the customer may place an order for product that is not needed immediately. At other times, the delay in receiving the optical links could result in lost sales for the customer.

The overall delivery issue, then, involves three issues: optimizing the percentage of the product that is delivered on time, minimizing the lead time, and minimizing the inventory. The latter, inventory, is primarily a financial trade-off. The customers would not mind if the supplier kept a great deal of excess inventory available, as safety stock, and responded immediately to orders placed by shipping from this safety stock — as long as the customers do not experience higher prices as a result. Unfortunately, holding product in inventory can be expensive and generally has a substantial impact on the bottom line.

The percentage on-time delivery can be optimized in the overall manufacturing flow by treating it as the product of several probabilities:

$$Pr(OTD) = Pr(\text{sufficent amount}) \times Pr(\text{sufficient lead time}). \quad (22.1)$$

The probability that a sufficient amount of product, optoelectronic links, are produced to meet the customers' orders is obtained by determining the material that must be started, allowing for the yields of each of the components of the optoelectronic link, and ensuring that there is sufficient capacity available in the manufacturing line, described by

$$Pr(\text{sufficient amount}) = Pr(\text{sufficient capacity}) \times Pr(\text{sufficient material started}) \times Pr(\text{sufficient cumulative yield}). \quad (22.2)$$

The probability that sufficient capacity exists, $Pr(\text{sufficient capacity})$, is a matter of balancing the capacity development and expansion with the orders expected. Because anticipation of future orders is an inexact science, this can be a difficult issue, which is explored further under Section 2.B.1.

However, once the capacity is established and capacity expansions are planned and known, the key to $Pr(\text{sufficient capacity})$ is in order acceptance: Treating each order accepted as a promise that must be kept, for integrity reasons if not for customer satisfaction, allows the capacity to be matched to the orders. Accepting orders greater than maximum capacity, or accepting additional orders once all the capacity has been allocated, results in unpleasant and diversive trade-off issues on the level of "which customers are more important to us." From these perspectives, the probability that sufficient capacity exists is treated as a digital, 0 or 100% probability, that can reach 100% if sufficient integrity is maintained in the order-acceptance process.

The probability that sufficient material is started, $Pr(\text{sufficient material started})$, also encompasses the allowance for upsides in the demand from the customers. Generally, customers must project future sales when placing orders: This is an inexact science at best, and the customers will often underestimate or overestimate the sales. If the customers overestimate sales, the suppliers or the customer can keep the excess inventory, incurring a financial cost but the customer is generally satisfied. If the customers have underestimated sales, the customers are disappointing their customers and losing potential sales.

If the supplier provides sufficient product to meet the actual orders received but insufficient to meet the upside, the supplier may be on strong legal grounds, but the customer is still losing out. If the financial impact is not overwhelming, it might be preferable for the supplier to allow for some upside in the orders, allowing the potential to "delight the customer" by supplying the additional product in the happy case that the customers' sales exceed expectations: A delighted customer experiencing vigorous sales is a very favorable condition indeed!

Within the overall manufacturing flow, each of the steps has an associated yield — process, probe, assembly, and final test yields in the case of integrated circuit processing. Each of these yields follows a type of nonnormal distribution referred to as a beta distribution, which is a basic statistical distribution that is limited to the range of 0–100%. The beta distribution for MOS integrated circuits is illustrated in Fig. 22.5.

By determining the cumulative yield distribution associated with each of the components of the manufacturing flow illustrated in Fig. 22.3, the appropriate amount of material can be started through each of the processes to meet the delivery requirements in terms of quantity. A spreadsheet incorporating the statistical distributions of yield and cycle time can be used to determine the amount of material to be started and, when it needs to be started, to ensure on-time delivery (perhaps with 95% confidence) (Fig. 22.6)

From Eq. (22.1), the final probability associated with meeting on-time delivery requirements is the probability that the parts will be delivered to the customer when promised, Pr(sufficient lead time).

Each of the steps in Fig. 22.3 has a cycle time associated with it. Cycle times tend to follow another nonnormal distribution, called a gamma distribution, which is obtained when there is a minimum but no maximum: The

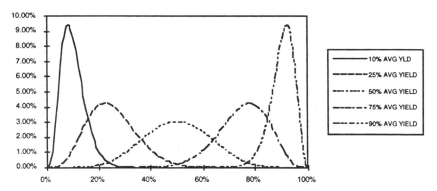

Fig. 22.5 Beta distributions for yields of MOS integrated circuits.

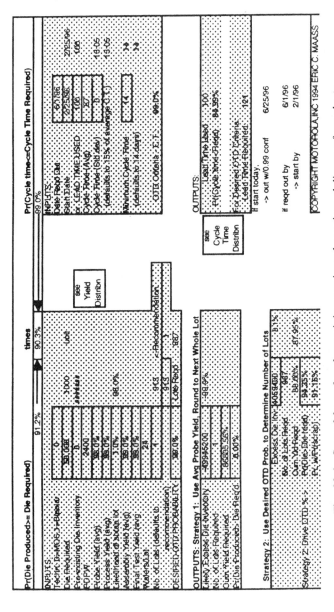

Fig. 22.6 Spreadsheet for determining projected on-time delivery of products based on yield, started material, and cycle time information.

cycle time for IC processing, for example, has a minimum theoretical cycle time including all the sequential times that wafers need to spend in furnaces, photolithographic and etch equipment, and so on. However, the cycle time for individual lots can exceed the theoretical cycle time, such that there is no maximum — the distribution is bounded by minimum theoretical cycle time and infinity. An example of a cycle time gamma distribution is shown in Fig. 22.7.

The overall cycle time is a complex combination of the cycle times for each of the component processes illustrated in Fig. 22.3. Because the manufacturing cycle time is one of the trade-offs involved in the decision making to allow for the manufacturing challenges in optoelectronic systems, it may be best to generate a hypothetical example of the overall cycle time for a hypothetical optoelectronic link. Although the example used here is not a real example, it will illustrate the process of optimizing cycle time and on-time delivery relative to the inventory and other associated costs for various alternative manufacturing strategies.

The nested histograms of cycle times displayed in Fig. 22.8 each have the same horizontal axis — cycle time in days. Proceedings from left to right in the manufacturing flow, the cycle time distributions are for the labeled steps, in each case inclusive of the cycle times of the preceding steps of the flow. In other words, the cycle time distribution displayed as a histogram for connectorization would be the entire cycle time experienced at that step, including the cycle time for fiber processing and for cabling/sheathing.

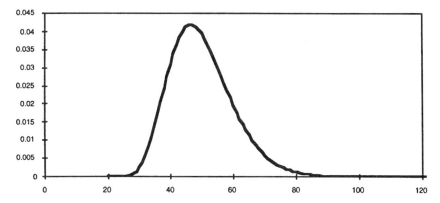

Fig. 22.7 Gamma model for cycle time, in which the average cycle time is 50 min, the standard deviation is 10 min, and the minimum theoretical cycle time is 20 min.

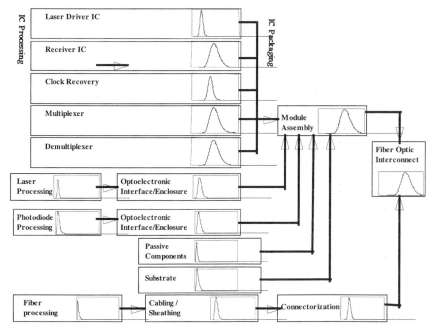

Fig. 22.8 Cycle time distributions associated with the manufacturing of an opto-electronic link.

Figure 22.9 illustrates the lead time and inventory trade-offs considering the overall fabrication and shipping of a complete link as a whole system. With no inventory, the total lead time is at its maximum, but the inventory cost is naturally at a minimum.

In this situation, the total lead time is also called the "visible horizon lead time." The visible horizon is the planning horizon — how far in advance the manufacturing company must plan starts. The manufacturer can quote a lead time substantially less than this total lead time to customers if the company is willing to start the processing early, in anticipation of orders.

The manufacturing company can reduce this visible horizon lead time by storing inventory (perhaps called "safety stock") for critical, longer cycle time processes. In Fig. 22.8, the progression in adding inventory is in the order of maximum favorable impact on lead time; that is, the cycle time in the longest step of the critical path (or "rate determining step," for those with a more chemical bent) is reduced using inventory. The lead time is

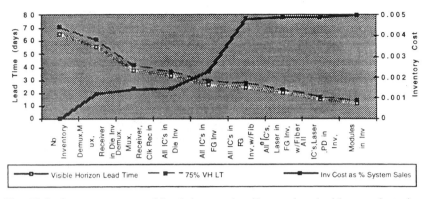

Fig. 22.9 Inventory cost and lead time trade-offs associated with manufacturing an optoelectronic link.

reduced, and the inventory cost increases. Because the jumps are rather sharp in places, the manufacturing company that constructs such a diagram can easily see reasonable trade-offs: In this example, the manufacturing company may be willing to store all integrated circuits in finished goods inventory but use just-in-time techniques for the remainder of the items required for manufacturing the optoelectronic link.

4. Meeting Customer Requirements: Price and Cost Trade-Offs

Optoelectronic data communication systems can be expected to be continuously under strong price pressure. Data communication systems will not only experience the learning curve expectations associated with computer systems, in general, and the price pressure from competitors but also experience price pressures from competing solutions, such as cable or other "copper wire" electronic solutions.

The learning curve expectations associated with computer systems in general, which can be expected to be applied to the associated data communication systems, have historically been what is called a "70% learning curve." This means that, for every doubling of cumulative volume shipped to customers, the price is expected to drop by 30% (Fig. 22.10); alternatively, the price expected at any cumulative volume is

New price expectation = (previous price) × .7∧ (Ln (new cum.

volume/previous cum. volume)/Ln(2))   (22.3)

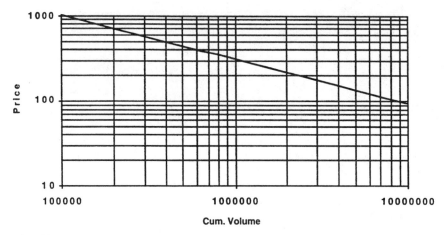

Fig. 22.10 Learning curve, with 70% historical expectations: reduction of price expected for increasing cumulative volume.

The competitive pressures that will influence pricing of optoelectronic systems come not only from competing optoelectronic manufacturers and solutions but also from nonoptoelectric technologies. Cable and other copper wire-based solutions can be expected to put competitive pressure on optoelectronic solutions. Fortunately for optoelectronic solutions, cable and copper wire solutions can be expected to encounter barriers that hinder high-speed operation, such as capacitative and inductive coupling that impacts noise immunity, as well as legal requirements such as electromagnetic interference. However, optoelectronic solutions will compete with cable and copper wire solutions at lower speeds. Also, it may be wise to consider the possibility that innovations in cable and wire solutions will surmount the high-speed barriers.

The competitive pressures and customer expectations influencing pricing will need to be addressed by minimizing the manufacturing costs associated with optoelectronic data communication systems. These include materials, direct and indirect labor, equipment and other fixed asset costs or depreciation, utilities, chemicals, and other consumables.

The materials, chemicals, and other consumables are considered variable costs: They increase directly with volume. Equipment, building, and similar asset costs (or depreciation) are considered fixed costs; They will not vary with the volume of product sold. Direct and indirect labor and utilities generally will behave as if they have both a fixed and a variable cost component. This is because running a manufacturing line requires a minimal level of personnel and utilities, such as electrical power, regardless of the volume of product; however, more people and power consumption will be needed for increasing volume.

The cost per unit behaves as shown in Fig. 22.11; the volume impact can be quite substantial, particularly for semiconductor processing portions of the flow such as manufacture of integrated circuits, laser diodes, and photodetectors. In fact, it would be expected that the volume impact is dominant: The greater the volume of optoelectronic systems sold, the lower the cost due to the allocation of fixed costs over a larger number of products sold. It would be very fortunate if this cost reduction with increasing volume met or exceeded the customers' expectations of lower prices, as perhaps represented by the aforementioned 70% learning curve.

The manufacturing costs can be addressed by optimizing the factors associated with the total cost. Optimizing the fixed-cost impact means maximizing the volume — to a point. If the factory is planned and set up in a modular fashion, it may be that the volume impact is totally benevolent. In many cases, however, there are natural break points in volume: The factory can be effectively run at a certain percentage of maximum available capacity, perhaps approximately 85%. Increasing beyond this percentage utilization may involve cycle time impacts that cause customer dissatisfaction that will impact future sales. Increasing beyond this point may involve purchase of additional fixed assets. In some cases, the purchase of additional

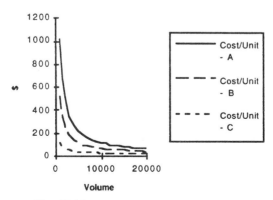

Fig. 22.11 Manufacturing cost per unit.

equipment may more than offset the cost reduction expected from increased volume of sales. This would be especially true if the capacity increase requires the construction of an additional factory. Hence, the up-front planning of the factory may be critical in allowing the increase in capacity in a controlled, perhaps modular, manner.

Optimization of direct and indirect labor costs involves trade-offs of labor rates versus location and automation. For example, labor costs in various countries may be lower, but the impact on cycle time may offset this: The most desirable cycle time would likely be achieved if all the factories are essentially co-located. Also, building manufacturing sites in countries with lower labor rates can involve other risks, such as political instability that can impact production or even lead to the loss of the factory, and the potential that the presence of the factory in the country will result in dissemination of intellectual knowledge that will ultimately lead to the creation of a competitor.

Reduction in material costs often involves negotiation with suppliers or process optimization that reduces the amount of wasted material. Reduction in material costs by using less material than required for quality and reliability is very unwise. Another means of reducing material costs can be by using newer technologies to miniaturize components — for example, using a new integrated circuit process with smaller geometries to increase the potential die per wafer, minimizing the cost per die (where a die is an unpackaged integrated circuit). In this case, the miniaturization generally improves performance because the device performance/speed of operation generally increases an integrated circuit geometries are reduced.

Reduction in utilities, chemicals, and other consumable costs is also largely a matter of negotiation with suppliers and process optimization that minimizes waste.

In summary, a cost reduction plan should be developed in order to meet the customers' expectations of price reductions with increased volumes. This cost reduction plan should include such items as maximizing utilization of the factories, process optimization, negotiation with suppliers, minimization of waste, and miniaturization of components. The projected cost savings should be calculated, with the results (if believable) prioritized in order of maximum favorable impact, allowing for negative side effects as risk factors. The cost savings plan should be highly visible, and people should be assigned direct responsibility to see that the cost reduction plan is successfully executed.

B. CHALLENGES

1. "Chicken and Egg" Syndrome

One key issue, which is especially relevant in emerging technologies, is the timing and risk of the investment in the technology. This issue can be called the chicken and the egg syndrome.

Customers do not want to commit to using an emerging technology without some assurance that it will be fully supported in a manufacturing sense. In the case of optoelectronic communications equipment, the customer would be extremely uncomfortable to have its product success be dependent on the production of something for which there is no factory.

On the other hand, the supplier does not want to commit to putting the emerging technology into production without having a customer that is committed to using the technology.

What often follows is a seemingly endless cycle that delays the implementation of the new technology—the supplier trying to get customers to commit to the new product/technology without a factory, and the customers showing interest but backing off gently when pressed for a commitment.

There appear to be three ways to resolve this chicken and egg syndrome:

1. A brave customer can commit, despite the risk
2. A brave supplier can commit and build the factory without a committed customer
3. The customer and supplier can share the risk: build the factory under a financial agreement that spreads the risk between the customer and supplier
4. An external agent can assume the risk for both parties—perhaps a venture capitalist, a government agency, or a potential partner of the supplier.

2. Diversity of Technologies

The technologies involved in optical fiber for communications are very diverse, as illustrated by Table 22.3.

The diversity of technologies generally results in using different manufacturing facilities for the different processing and assembly manufacturing involved in developing a complete fiber optics system. The integrated circuits may be manufactured in semiconductor "wafer fab" lines in the United

Table 22.3   **Technologies Involved in Optoelectronics for Data Communications**

Technology	Example in optoelectronics	Manufacturing step
Mechanical engineering	Fiber/laser alignment	Last step
Polymer chemistry	Laser package/ waveguide	Laser assembly
Digital electronics	Multiplexer IC	IC manufacturing
Analog electronics	Receiver IC	IC manufacturing
Materials science	Laser diode	Laser manufacturing
Optics	Photodiode, waveguide	Photodiode manufacturing Waveguide manufacturing
Thermodynamics	Laser diode packaging	Laser assembly

States, the lasers may be manufactured in Japan, the assembly of the components may be accomplished in Asia, and the fiber itself may be drawn, sheathed, and connectorized in three different locations around the globe.

The implications of the diversity of manufacturing locations include effects on cycle time, quality, and cost. Although some of the trade-offs are obvious, the issue of colocation of the key steps of manufacturing, or even the entire manufacturing process, needs to be explored up-front.

3. Colocation?

The most capital-intensive parts of the overall manufacturing process are the IC processing, followed by the laser diode processing, the photodiode processing, and the fiber processing. Complete colocation would therefore require building the new manufacturing near an existing wafer fabrication facility or else building an entirely new set of manufacturing facilities, including wafer fabrication, in the selected colocation site. Due to the very high capital costs of building a new wafer fabrication area (or several areas, if the processes for laser drivers, receivers, and multiplexers are not compatible), the former alternative seems more reasonable if colocation is desired.

The benefits of colocation can include measurable improvements, such as total cycle time, and less tangible improvements in areas such as communication between the parties involved and speed in resolving problems.

The benefits of colocation on cycle time can be estimated from the example used earlier. Without colocation, the cycle time includes the time to prepare the wafers, die, components, fibers, or similar item for shipping, the time required for transportation, and the time to unpack the components. For manufacturing sites within the same continent, this cycle time would be on the order of a few days — within the process flow for an optoelectronic data communication system illustrated in Figs. 22.8 and 22.9, the total cycle impact would be approximately 1 to 3 weeks, or approximately a 10–30% reduction.

The cycle time impact on the customer would be minimized by setting up storage facilities at the outgoing or the incoming manufacturing sites or both. The components must sit in storage until needed—this has minimal effect on cycle time, but affects costs through the costs of space and people to control this inventory.

Colocation also improves communication, which translates into more rapid detection and resolution of problems and improved yield and reliability. It also improves the development time of new generations of optoelectronic products and minimizes the unused, unsold inventory of components that may be rendered obsolete by the next generation of product.

Effectively, the colocation decision is a cost decision: If the labor costs in a geographically separate facility are significantly lower than the costs of maintaining and expediting inventories of components, or if the capital costs for building a manufacturing facility for some components are prohibitive so that the components should be purchased from an outside supplier in a distant location, then colocation may not be feasible.

Generally, colocation may make sense at some level in the overall manufacturing flow but not for the complete manufacture of the system. For example, semiconductor manufacturing facilities are extremely capital intensive, whereas assembly facilities are much less capital intensive.

Consequently, it may make sense to obtain semiconductor devices such as integrated circuits from a vendor — and similarly purchase the laser and photodetector devices and the fiber — setting up storage facilities for these components, but colocating the facilities for assembling the module as well as the optoelectronic interfaces and enclosures.

4. Integration

Integration seems to be the natural direction of electronics. The historical trend of electronic components has been from discrete devices (e.g., diodes and transistors) to small-scale integration (SSI; e.g., simple gates), to medium-scale integration (MSI; e.g., arithmetic logic unit), to large-scale integration, (e.g., microprocessor), to very large-scale integration (e.g., microcomputer chip).

However, the path toward increasing scales of integration and complexity has always been aligned with the directions of increased speed of operation and decreased cost. In fact, these latter two directions seem to be the driving force behind integration of circuitry.

In digital systems, the alignment of cost, speed, and complexity derives from shrinking geometries due to improved manufacturing processes, especially photolithography and etch processing.

Smaller geometry MOS transistors perform at higher speed owing to the shorter distance from source to drain that must be transversed by the charge carriers as well as reduced parasitic capacitances with the smaller transistors.

Lower cost is achieved by the smaller devices taking up less area on the wafer (commonly referred to as "real estate") so that more integrated circuits can be produced on the same-size wafer. This lower cost is complicated by the costs for equipment capable of resolving the small geometries, resulting in higher wafer costs, the trend to larger wafers, and the minimum die size set by the requirements of assembly (pad sizes and number of pads that must be wire bonded).

Smaller geometry devices also increase the level of complexity or integration that can be achieved. More devices can be integrated in the same area, allowing more functions to be integrated and eliminating the requirement to package some of the integrated circuits that were previously separate. The reduction in package count can reduce the total cost to the customer for populating the boards.

Although the directions of reduced cost, increased speed, and increased complexity are aligned very well in the fully digital applications, the directions are not so clearly aligned in the mixed-signal world associated with data communication.

Mixed-signal applications, in which there are both analog and digital functions, impose additional requirements on the semiconductor technology. On the receive side of a data communication system, the transimpedance and postamplifier functions are analog functions, the clock recovery function generally involves a phase-locked loop, a feedback circuit that in turn involves the analog function of a voltage-controlled oscillator (VCO), the digital function of a frequency divider (generally based on digital flip flops), and phase detector, and the analog function of an operational amplifier or charge pump to convert the digital signal from the phase detector into a voltage to control the VCO frequency.

An analog function such as a transimpedance amplifier requires high noise immunity and high transconductance. The necessity of handling a range of signals, dependent on the distance the light has travelled through the fiber or other medium, imposes requirements on noise/stray signal immunity that conflict with the need to handle an increasing number of signals on the same chip as the integration is increased.

The MOS transistors that have been so effective for digital applications experience difficulties in achieving these noise immunity and transconductance requirements. Integration of these functions may require a bipolar or BiCMOS process with excellent isolation — a more complex and expensive process compared to standard CMOS processes.

An alternative integration path is to integrate the transimpedance amplifier with the photodetector. This has the advantage of minimizing the parasitic inductances and capacitance experienced by the signal as it is conducted from the photodetector to the transimpedance amplifier, allowing higher frequency performance.

The analog functions, including the transimpedance amplifier, the post-amplifier, the VCO, and the charge pump or operational amplifier, also require passive elements such as capacitors, varactors, and inductors. These passive elements need high impedance per unit area and generally high Q values (where Q can be thought of as the ratio of how much the passive inductor behaves as an ideal inductor to how much it act as a capacitor and resistor). The optimization of these passive elements adds to the process complexity and, therefore, to the cost.

However, there may be a performance advantage of integrating passive elements — the requirement to go off-chip to connect to external capacitors and inductors adds the parasitic capacitances and inductances of bond wires, for example, into the total capacitance and inductance of the passive element. These parasitic capacitances and inductances generally degrade the effective Q of the passive element.

The added complexity required to integrate mixed-signal functions will generally increase the wafer cost, and the integration of multiple functions handling multiple signals may degrade the performance. As a consequence, the trend toward higher levels of integration is not as straightforward as it is in the digital arena.

The impact of integration on cycle time is also less favorable than might be thought. The integrated circuits are produced in parallel; therefore, the effective cycle time is approximately the same as the cycle time of the longest integrated circuit process. Integrating the functions onto one complex integrated circuit process would either keep the cycle time the same or possibly increase the cycle time, if the complex, mixed-signal integrated circuit process requires significantly more cycle time.

5. Assembly of Components

Assembly processes are involved in at least two stages of the overall manufacturing flow for optoelectronic systems. The individual components, such as integrated circuits, are assembled into packages, such as small outline integrated circuit or quad flat pack, and then the packaged components are assembled into a module such as a printed circuit board.

The purposes of the integrated circuit packages are to protect the integrated circuits and to facilitate the handling and attachment of the integrated circuits to the printed circuit boards.

For optoelectronic systems, however, the packaging of integrated circuits actually has some detrimental side effects on cycle time, performance, and cost. The cycle time required to assemble integrated circuits into packages is in series with the integrated circuit process time, adding to the total cycle time. This cycle time impact is generally on the order of a few days; however, because the package assembly sites are often in different countries (in order to minimize the associated labor costs), the cycle time impact may be on the order of a few weeks, allowing for transportation and customs inspections requirements.

The integrated circuit packages add parasitic capacitances, resistances, and inductances that impact high-frequency performance. Although these parasitic properties can be modeled, the Q of the capacitances and inductances may be such that the performance requirement of low jitter may be impacted by the phase noise of the PLL due to the parasitics, and the signal detection and low bit error rate may be impacted by the noise generated by the parasitic, low Q capacitances and inductances. The signal may be further degraded at high frequencies if, for example, the bond wires act as quarter-wavelength antennas.

Naturally, the assembly of integrated circuits into packages involves costs — at the SSI and even the MSI levels, these costs can dominate the costs of the complete integrated circuit, despite the high capital costs associated with wafer processing.

A reasonable manufacturing trend, therefore, might be to eliminate the packaging of the integrated circuits from the total manufacturing flow. The cost, cycle time, and performance advantages of this direction are obvious from the previous discussion; however, there are three issues raised by this elimination of unnecessary steps: the impact on handling and attachment during module assembly, the protection of the integrated circuits (now in die or wafer form rather than packaged form), and the thoroughness of testing of the integrated circuits.

The elimination of packaging of integrated circuits changes the module assembly from the simple process of attaching packaged parts onto a printed circuit board to a process that includes die attach and wire bonding or solder bumping of integrated circuits in die form. This migration is effectively a change from printed circuit boards to MCMs. The MCM process can be considered a hybrid of the printed circuit board process and the integrated circuit packaging process: Die are attached to the MCM substrate rather than to the leadframe of an integrated circuit package, and the bond pads of the integrated circuits are wire bonded or soldered (solder bump) to the MCM connections rather than wire bonded to the leads of the integrated circuit package.

The die are protected in the MCM assembly operation much as they would be in the integrated circuit package assembly operation. The testing requirements in wafer form, however, are more intense: Whereas packaged integrated circuits can be tested in wafer form and packaged form, with the more difficult tests perhaps being accomplished on packaged integrated circuits, the MCM approach requires that all testing be in wafer form, prior to MCM assembly, in order to minimize the costs. Yield loss after MCM assembly increases costs to include the costs of all other components assembled onto the MCM as well as the MCM assembly itself. Thorough testing in wafer form of DC and functional requirements, and some or all AC performance, is referred to as the known good die (KGD) approach. Testing AC performance at high frequency in wafer form may involve some challenges — contacting the pads in wafer form sufficiently well for RF and AC measurements, without damaging the pads, is very difficult. The KGD approach may also involve efforts to avoid reliability problems by testing the integrated circuits in wafer form at high voltages, currents, and/or temperatures.

If, as expected, the MCM costs follow a respectable learning curve, the advantages of improved performance, reduced cycle time, and the improved costs by the elimination of process steps (the packaging of components such as integrated circuits) make the MCM approach very attractive and perhaps the preferred path.

6. Flexibility

There is often a trade-off between flexibility and cost in manufacturing. Flexibility refers to both flexibility in the production volume and flexibility in the product manufactured.

Generally, low cost can best be achieved by constant, high-volume manu-facturing of a standard product. The manufacturing process itself can be optimized for this standard process, and equipment can be purchased and optimized for the standard process.

Deviations from this low-cost ideal include variations in the volume and variations in this standard process. Variations in the volume affect total cost and unit cost as shown in Fig. 22.12. In Fig. 22.12 (right), the total manufacturing cost is depicted. The fixed costs, which include depreciation expenses for the equipment and the manufacturing facility, correspond to the y-intercepts for cases A–C. The variable costs, which include materials used in the manufacturing process, correspond to the slopes. In Fig. 22.12 (left), the manufacturing cost per unit, which affects the price that the customer can be quoted while maintaining profitability (a worthy goal!),

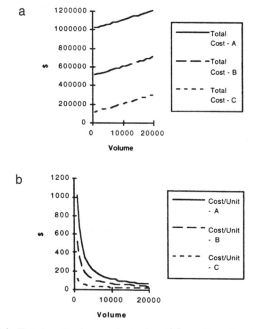

Fig. 22.12 (a), Total cost of manufacturing; (b), unit costs of manufacturing.

is depicted. Note that in Fig. 22.12 (left), the cost per unit rapidly increases at low volumes in case A, corresponding to the case of high fixed costs.

From the analysis of the cost impact of volume, the flexibility in production volume is constrained by upper and lower limits. The lower limit is set by the price acceptable to the customers, the profitability acceptable to the manufacturer, and the cost determined by a manufacturing cost per unit curve derived in Fig. 22.12 (left). The key determinant of the flatness of the cost per unit curve is the fixed costs — minimizing these fixed costs, therefore, is critical to flexibility. The upper limit is set by some natural break points in volume, as mentioned earlier: The factory can be effectively run at a certain percentage of maximum available capacity, perhaps approximately 85%. Increasing beyond this percentage utilization may involve cycle time impacts or the purchase of additional fixed assets.

Flexibility in terms of flexibility in the product manufactured involves the versatility of the manufacturing line both in running several different product types simultaneously and in terms of introducing new products, or product evolutions, into manufacturing. This flexibility also involves cost, primarily in terms of equipment and movement of material through the process but also in terms of planning and controlling the manufacturing process to minimize errors associated with the additional complexity introduced by maintaining multiple process flows.

a. "Job Shop" versus Flow Assembly Line

A manufacturing area that essentially runs only one product, efficiently, is often referred to as a "flow shop," whereas a manufacturing area that deals with a variety of products, each treated as a custom product, is often referred to as a job shop. These two extremes will generally have different manufacturing approaches, reflected most obviously in the factory layout.

A manufacturing area can be considered to consist of a material flow and an information flow. For any individual product type, the material flow can be drawn as a flowchart indicating the movement of material from workstation to workstation (a workstation generally being associated with each piece of equipment used in the process), with queues available between workstations.

In a flow shop, the layout of the manufacturing area will generally reflect the material flowchart in a cellular or modular layout. Equipment will be laid out to ease the movement of material between workstations in the sequence indicated by the flowchart. A simple example for an optoelectronic system is illustrated in Fig. 22.13. A modular layout that reflects the flowchart shortens the cycle time by minimizing the delay in moving material from workstation to workstation and by allowing operators at each workstation to readily see and respond to cycle time problems such as excessively large queues or equipment downtime at the next process. The flow layout also minimizes errors due to skipped processes and miscommunication between operators at adjacent work stations. The advantages of modular layout include

Reduced material handling
Reduced setup time
Reduced in-process inventory
Reduced need for expediting
Improved operator expertise
Improved communication, therefore human interactions

The disadvantages include

Reduced shop flexibility
Possible reduced machine utilization
Possible increased cycle times

In a job shop, the layout of the manufacturing area will often be a functional layout in which similar equipment or types of manufacturing processes are grouped together. Because the sequence of manufacturing operations varies from product to product type, the factory layout may concentrate on minimizing equipment downtime and maximizing equipment utilization and operator expertise in a particular type of equipment rather than reflecting the varying flowcharts. As simple example of this for an optoelectronic system is illustrated in Fig. 22.14.

For either job shops or flow shops, the movement of material between workstations can be accomplished by several means, including manual or automatic transport systems.

b.   Allowance for Product Evolution

The modular layout is optimal for one particular product type and the layout is derived from the material flowchart for one particular product. If all the products and anticipated new products will have this same flow, then the modular layout is clearly the preferred choice.

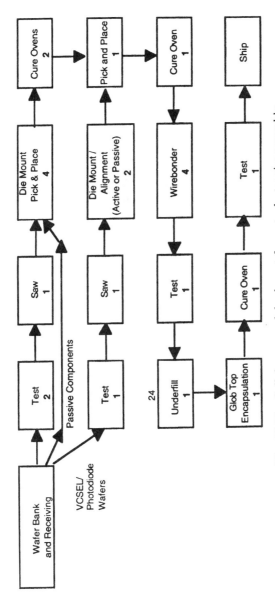

Fig. 22.13 Cellular or modular layout for optoelectronics assembly.

813

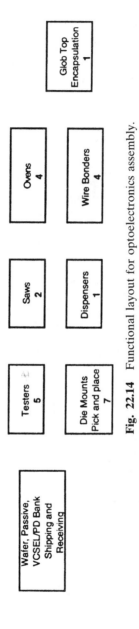

Fig. 22.14 Functional layout for optoelectronics assembly.

If several products and newer products are expected to differ from the material flowchart, the modular layout may experience problems. It is important to note that products that have material flowcharts that are nearly the same may actually involve worse problems than products with much different flowcharts because the very slight differences are likely to be missed by operators used to performing routine tasks the same way virtually every time.

The functional layout is not tied to any particular material flow — all flows are simply variations in which the material is transported between the groupings of similar equipment after each step in the process flow is completed.

Although the functional layout may appear to be the most flexible manufacturing layout, the cycle time and communication issues make this less than optimal, even for meeting the flexible needs associated with process evolution.

Instead, the optimal layout for a flexible optoelectronic manufacturing line is most likely a variation of the hybrid layout, in which some equipment is arranged in a modular layout (Fig. 22.13) and some equipment in a functional layout (Fig. 22.14). The intention is to anticipate the product evolution through development of a product road map, study the product road map to determine the portions of the material flow that will be common to all product types, and determine the portions of the material flow that will involve variations in the flowchart. The modular layout will be determined by the commonalities of all the flows, whereas the functional layout will reflect the steps in the material flow that will vary.

22.3. Manufacturing Alternatives

A. *AUTOMATION/ROBOTICS*

Automation can be applied to the manufacture of optoelectronic systems in several areas: automating the process operations at individual workstations such as attachment of the die to the package or MCM module, automating the transport of material between workstations, and automating the information flow (planning, control, and information gathering at each of the workstations).

Automation can provide advantages in terms of minimized cycle time (in terms of the expected cycle time or the variability of cycle time), minimized variability of processing, and perhaps reduced cost in complex labor-intensive operations. For example, the alignment of photonic devices such as light-emitting diodes (LEDs), or photodetectors to the fiber (either directly or indirectly through a lens or waveguide) can be a complex operation for an operator, whether the alignment is passive or active.

In the ideal situation, the entire manufacturing process would be automated. The information flow would plan and control the transport of the optoelectronic system from workstation to workstation, initiate the downloading of information to the workstation on the processing needed, provide for error prevention through the application of Poka Yoke (mistake proofing) approaches, initiate the processing of the material, provide feedback through the process consistent with error prevention, and initiate the uploading of information from the workstation regarding the processing or information on the results at a test operation. Finally, the information flow would initiate the transport of the material to the next workstation. After the last process is completed, the complete optoelectronic system would be transported to a staging area for delivery to the customer, with the information on the performance and processing readily available to the customer.

B. DESIGN FOR MANUFACTURABILITY

In designing an optoelectronic system for manufacturability, the first goal is to ensure that the processing variability does not impact meeting the customers' specifications. The second goal is to make the processing as simple as possible while still achieving the first goal.

Ensuring that the processing variability does not impact meeting the customers' requirements involves a combination of modeling and statistics: determining which processing variables affect which customer specifications, monitoring and obtaining measures of the variance of the relevant processing variables, obtaining an equation or simulation for the relationship between the processing variables and the specifications, and using this relationship combined with the variance of the processing variables to predict the impact of processing variables on the customers' requirements.

C. DESIGN FOR ASSEMBLY

To achieve the second goal, making the processing as simple as possible, the approach is to keep the first goal, achieving specifications allowing for process variability, and additionally determine the simplest assembly flow with the least amount of unique parts of processes. Achieving a simple assembly process for an optoelectronic system would involve the following:

1. Minimize the number of parts

   This could be achieved by integrating the ICs where appropriate; for example, the parallel-to-serial conversion, clock generation, and laser or LED driver could be integrated into one integrated circuit on an appropriate high-frequency semiconductor process (small geometry CMOS or BiCMOS, for example). On the receive side, the photodiode and transimpedance amplifier could be integrated into one part (perhaps in GaAs) and the post- or limiting amplifier, serial-to-parallel conversion, and clock recovery and decision circuitry on another integrated circuit (again in small geometry CMOS or BiCMOS).

   Integrating the passive components onto the ICs where possible, or otherwise minimizing the number of separately packaged passive or discrete components through system design or use of multiple passive elements packaged together could also help.

2. Maximize automatic or self-alignment and similar part handling.

   If possible, the alignment of the optical elements, lasers and photodiodes, should be a passive alignment. The optical elements can have alignment pins, notches, or other features molded in so that they can be automatically aligned with the complementary alignment structures on the module.

3. Minimize the number of separate attachment/fastening/connection processes

   The attachment of integrated circuits, optical elements, and passive and discrete components could be performed by robotic pick-and-place operations followed by a combined solder attachment heat cycle.

   Alternatively, in some MCM schemes, the integrated circuits can be processed to have solder bumps so that the pick-and-place operation is performed on the die upside down. A combined heat treatment performs the solder attachment, which is a combined die attachment and electrical connection to each of the pads.

   The MCM schemes effectively eliminate one set of assembly processes — the wire bonding operations associated with integrated circuit packaging.

D. FLEXIBLE MANUFACTURING APPROACHES

The "spine" approach shows promise for flexible manufacturing. The spine approach organizes the material flow and the information flow in a parallel fashion, analogous in some ways to the human spine, or to some topologies in information networking.

The spine is a linear axis, the main route of material movement, with branches for individual operations. The storage for materials and finished goods is also placed in the spine. Material handling equipment, such as conveyors or automated guided vehicle systems, can be incorporated into the spine.

The material handling equipment can be interfaced with automatic identification equipment, such as bar code reader systems, that can ensure appropriate movement of the materials and prevent errors. The automatic identification equipment can also be interfaced with a network communication system along the spine, which controls the movement of materials and obtains information on the procesesing of the equipment.

The modular nature of the spine flow is consistent with the cellular or modular flow discussed earlier and organizes the associated information

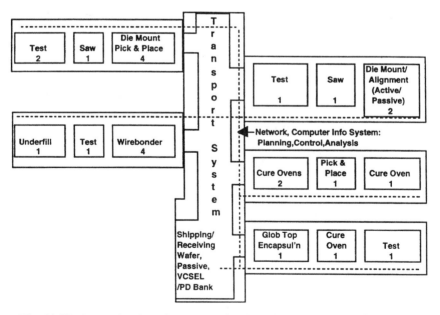

Fig. 22.15 Example of a spine approach adapted for optoelectronic systems.

flow of planning, control, and data acquisition along the spine. The modular nature allows for flexibility in the system and simplifies addition of modules to the spine for process evolution. Figure. 22.15 shows an example of a spine approach adapted for optoelectronic systems.

Appendix A: Measurement Conversion Tables

English-to-Metric Conversion Table

English unit	Multiplied by	Equals metric unit
Inches (in.)	2.54	Centimeters (cm)
Inches (in.)	25.4	Millimeters (mm)
Feet (ft)	0.305	Meters (m)
Miles (mi)	1.61	Kilometers (km)
Fahrenheit (F)	(°F − 32) × 0.556	Celsius (C)
Pounds (lb)	4.45	Newtons (N)

Metric-to-English Conversion Table

Metric unit	Multiplied by	Equals English unit
Centimeters (cm)	0.39	Inches (in.)
Millimeters (mm)	0.039	Inches (in.)
Meters (m)	3.28	Feet (ft)
Kilometers (km)	0.621	Miles (mi)
Celsius (C)	(°C × 1.8) + 32	Fahrenheit (F)
Newtons (N)	0.225	Pounds (lb)

Absolute Temperature Conversion

Kelvin (K) = Celsius + 273.15
Celsius = Kelvin − 273.15

Area Conversion

1 square meter = 10.76 square feet = 1550 square centimeters
1 square kilometer = 0.3861 square miles

HANDBOOK OF FIBER OPTIC
DATA COMMUNICATION

Copyright © 1998 by Academic Press.
All rights of reproduction in any form reserved.
ISBN: 0-12-437162-0

Metric Prefixes

zetta $= 10^{21}$
tera $= 10^{12}$
giga $= 10^9$
mega $= 10^6$
kilo $= 10^3$
hecto $= 10^2$
deca $= 10^1$
deci $= 10^{-1}$
centi $= 10^{-2}$
milli $= 10^{-3}$
micro $= 10^{-6}$
nano $= 10^{-9}$
pico $= 10^{-12}$
femto $= 10^{-15}$
atto $= 10^{-18}$
zepto $= 10^{-21}$

Appendix B: Physical Constants

Speed of light $= c = 2.99\ 792\ 458 \times 10^8$ m/s

Boltzmann constant $= k = 1.3801 \times 10^{-23}$ J/K $= 8.620 \times 10^{-5}$ eV/K

Planck's constant $= h = 6.6262 \times 10^{-34}$ J/S

Stephan–Boltzmann constant $= \sigma = 5.6697 \times 10^{-8}$ W/m^2/K^4

Charge of an electron $= 1.6 \times 10^{-19}$ C

Permittivity of free space $= 8.849 \times 10^{-12}$ F/m

Permeability of free space $= 1.257 \times 10^{-6}$ H/m

Impedance of free space $= 120\ \pi$ ohms $= 377$ ohms

Electron volt $= 1.602 \times 10^{-19}$ J

HANDBOOK OF FIBER OPTIC
DATA COMMUNICATION

Appendix C: Index of
Professional Organizations

AIP, American Institute of Physics
ANSI, American National Standards Institute
APS, American Physical Society
ASTM, American Society for Test and Measurement
CCITT, International Telecommunications Standards Body (see ITU)
CDRH, U.S. Center for Devices and Radiological Health
DARPA, Defense Advanced Research Projects Association (also referred to as ARPA)
EIA, Electronics Industry Association
FCA, Fibre Channel Association
FDA, U.S. Food and Drug Administration
HSPN, High Speed Plastic Network Consortium
IEC, International Electrotechnical Commission
IEEE, Institute of Electrical and Electronics Engineers
IETF, Internet Engineering Task Force
IrDA, Infrared Datacom Association
ISO, International Standards Organization
ITU, International Telecommunications Union (formerly CCITT)
NBS, National Bureau of Standards
NFPA, National Fire Protection Association
NIST, National Institute of Standards and Technology
OIDA, Optoelectronics Industry Development Association
OSA, Optical Society of America
OSHA, U.S. Occupational Health and Safety Administration
PTT, Postal, Telephone, and Telegraph Authority
SI, System International (International System of Units)
SIA, Semiconductor Industry Association
SPIE, Society of Photooptic Instrumentation Engineers
TIA, Telecommunications Industry Association
UL, Underwriters Laboratories

823

HANDBOOK OF FIBER OPTIC
DATA COMMUNICATION

Copyright © 1998 by Academic Press.
All rights of reproduction in any form reserved.
ISBN: 0-12-437162-0

Appendix D: OSI Model

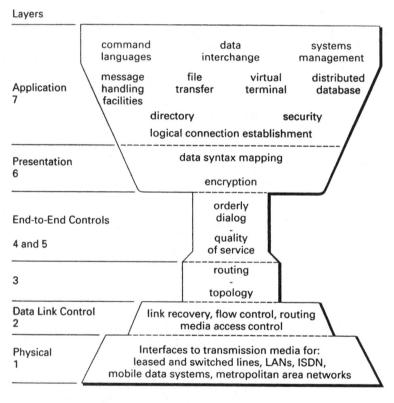

The wineglass of layered functions.

HANDBOOK OF FIBER OPTIC
DATA COMMUNICATION

Appendix E: Network Standards and Documents

The IEEE defines a common path control (802.1) and data link layer (802.2) for all the following LAN standards; although FDDI is handled by ANSI, it is also intended to fall under the logical link control of IEEE 802.2 (the relevant ISO standard is 8802-2).

IEEE LAN Standards

802.3-code sense multiple access/collision detection (CSMA/CD) also known as Ethernet. A variant of this is Fast Ethernet (100BaseX). 802.3z is the emerging gigabit Ethernet standard.

802.4-Token bus (TB)

802.5-Token ring (TR)

802.6-Metropolitan area network (MAN) [also sometimes called switched multimegabit data service (SMDS) to which it is related]

802.9-Integrated services digital network (ISDN) to LAN interconnect

802.11-Wireless services up to 5 Mb/s

802.12-100VG AnyLAN standard

802.14-100BaseX (version of Fast Ethernet)

ANSI Standards

Fiber distributed data interface (FDDI): ANSI X3T9.5 (the relevant ISO standards are IS 9314/1 2 and DIS 9314/3)

Physical layer (PHY)

Physical media dependent (PMD)

Media access control (MAC)

Station management (SMT)

Because the FDDI specification is defined at the physical and data link layers, additional specifications have been approved by ANSI subcommittees to allow for FDDI over single-mode fiber (SMF-PMD), FDDI over copper wire or CDDI, and FDDI over low-cost optics (LC FDDI). A time-division multiplexing approach known as FDDI-II has also been considered.

825

HANDBOOK OF FIBER OPTIC
DATA COMMUNICATION

Serial byte command code set architecture (SBCON): ANSI standard X3T11/95-469 (rev.2.2., 1996); follows IBM's enterprise systems connectivity (ESCON) standard as defined in IBM documents SA23-0394 and SA22-7202 (IBM Corporation, Mechanicsburg, PA)

Fibre channel standard (FCS)

ANSI X3.230-1994 rev. 4.3, physical and signaling protocol (FC-PH)

ANSI X3.272-199x rev. 4.5 (June 1995) Fibre Channel arbitrated loop (FC-AL)

High-performance parallel interface (HIPPI)

ANSI X3.183-mechanical, electrical, and signaling protocol (PH)

ANSI X3.210-framing protocol (FP)

ANSI X3.218-encapsulation of ISO 8802-2 (IEEE 802.2) logical link protocol (LE)

ANSI X3.222-physical switch control (SC)

Serial HIPPI has not been sanctioned as a standard, although various products are available.

HIPPI 6400 is currently under development.

Synchronous optical network (SONET): originally proposed by Bellcore and later standardized by ANSI and ITU (formerly CCITT) as ITU-T recommendations G.707, G.708, and G.709.

Other standards

Asynchronous transfer mode (ATM): controlled by the ATM Forum

Appendix F: Data Network Rates

The "fundamental rate" of 64 kb/s derives from taking a 4-kHz voice signal (telecom), sampling into 8-bit wide bytes (32 kb/s), and doubling to allow for a full-duplex channel (64 kb/s). In other words, this is the minimum data rate required to reproduce a two-way voice conversation over a telephone line. All of the subsequent data rates are standardized as multiples of this basic rate.

DS0	64 kb/s	
T1 = DS1	1.544 Mb/s	24 × DS0
DS1C	3.152 Mb/s	48 × DS0
T2 = DS2	6.312 Mb/s	96 × DS0
T3 = DS3	44.736 Mb/s	672 × DS0
DS4	274.176 Mb/s	4032 × DS0

Note: framing bit overhead accounts for the bit rates not being exact multiples of DS0).

STS/OC is the SONET physical layer ANSI standard. STS refers to the electrical signals and OC (optical carrier) to the optical equivalents. Synchronous digital hierarchy (SDH) is the worldwide standard defined by CCITT; formerly known as synchronous transport mode (STM). Sometimes the notation STS-XC is used, where X is the number (1, 3, etc.) and C denotes the frame is concatinated from smaller frames. For example, three STS-1 frames at 51.84 Mb/s each can be combined to form one STS-3C frame at 155.52 Mb/s. Outside the United States, SDH may be called plesiochronous digital hierarchy (PDH).

827

HANDBOOK OF FIBER OPTIC
DATA COMMUNICATION

Copyright © 1998 by Academic Press.
All rights of reproduction in any form reserved.
ISBN: 0-12-437162-0

STS-1 and OC-1	51.840 Mb/s	
STS-3 and OC-3	155.52 Mb/s	Same as SDH-1
STS-12 and OC-12	622.08 Mb/s	Same as SDH-4
STS-24 and OC-24	1244.16 Mb/s	Same as SDH-8
STS-48 and OC-48	2488.32 Mb/s	Same as SDH-16

The approach used in Europe and elsewhere:

E0	64 kb/s
E1	2.048 Mb/s
E2	8.448 Mb/s
E3	34.364 Mb/s
E4	139.264 Mb/s

The standardized network speeds are as follows:

Ethernet	10 Mb/s
Fast Ethernet (100BaseX)	100 Mb/s
Gigabit Ethernet	1000 Mb/s
100VG AnyLan	100 Mb/s
FDDI	100 Mb/s (effective data rate with 4B/5B encoding; basic rate is 125 Mb/s)
Token ring	4 or 16 Mb/s
MAN or SMDS	62 kb/s; 1.3, 4, 10, 16, 25, and 33 Mb/s
ATM	1.54, 25, 50, 100, 155, and 622 Mb/s (plus higher)
ESCON/SBCON	200 Mb/s (with 8B/10B encoding, max. 17 Mb/s)
Fibre channel (FCS)	See Chapter 12

Appendix G: Other Datacom Developments

Casimer DeCusatis

IBM Corporation
Poughkeepsie, New York

Because the field of fiber optic data communication is expanding so rapidly, inevitably new technologies and applications will emerge while this book is being prepared for print. In addition, there will be some recent developments that have not been incorporated into the previous chapters. This appendix is an attempt to address these changes by including a partial list of related datacom devices and standards for reference purposes and to include brief descriptions of several emerging datacom technologies that the reader may encounter in the literature (product names and terminology used in this appendix may be copyrighted by the companies that developed them).

Fibre Channel Standard

The ANSI standards body is currently investigating different connector options to reduce the size required by the current subscriber connector (SC) duplex interface. The new connector would be about half the size of an SC duplex and would allow more ports on large hubs, switches, and rack-mounted equipment as well as smaller panel openings to reduce electromagnetic noise emissions. There are several options for this new connector. One is the "mini-MT," a connector based on the MT ferrule concept with metal alignment pins but with only two fibers spaced 750 μm apart. The other option is the "Galaxy" connector, a radically different connector design that is the size of an RJ-45 jack and does not use ferrules; instead, a pair of fibers is cantilevered in free space and mated into a V groove assembly, where physical contact with another fiber pair causes a slight

HANDBOOK OF FIBER OPTIC
DATA COMMUNICATION

bend in the fibers and serves to hold them in place. At this writing, the cost/performance trade-offs of the two technologies remain undecided.

Fiber Cables

As an alternative to zero halogen fiber cables, some companies now offer a riser-rated cable with low-smoke attributes. The cable carries a UL rating of OFNR-LS and typically has less than 0.3% halogen content.

IBM has recently announced a license program for the enterprise system connection (ESCON) logo (ESCON is a registered trademark of IBM Corporation, 1991). Only cable vendors whose products have been tested and certified as IBM fiber cable suppliers may carry the ESCON logo. Currently, this logo is used by two cable vendors, Amp and ComputerCrafts (a third vendor, Siemens, is a qualified IBM cable supplier but has not licensed the ESCON logo as of this writing).

Free-Space Optical Links

Recent advances have made free-space optical data links practical in conditions when the weather is not a factor in attenuating the signal. Although bit error rates are typically on the order of 10^{-6} to 10^{-9} at 155 Mb/s, this is adequate for some applications such as voice and video transmission; recent experiments have shown that error rates as low as 10^{-12} can be obtained at 200 Mb/s using ESCON protocols on free-space optical links. Some examples of this technology and its applications include the following:

AstroTerra Corp. currently offers the TerraLink system, a tripod-mounted line-of-sight communication link with an 8-in. telescope aperture capable of 155 Mb/s transmission over 8 km in clear weather.

ThermoTrex Corp. in cooperation with SDL Inc. and funded by the Ballistic Missile Defense Organization, has demonstrated 1.2 Gb/s free-space communication over 150 km. This technology may form the basis of an aircraft-to-aircraft communications system and is scheduled to fly aboard the Space Technology Research Vehicle in 1998 to demonstrate communications among satellites.

Portable free-space laser communications are being developed by companies such as Leica Technologies, which has demonstrated a prototype binocular communications system for military applications at 100 kb/s over 3–5 km.

Gigabit Ethernet

Recent progress by the standards body has resulted in the following draft specification for the gigabit Ethernet transceiver, compatible with draft standard IEEE 802-3Z; a unique feature of this specification is the ability to use both multimode and single-mode fiber in the same transceiver.

Connector: SC duplex (possible alternative connectors include the Galaxy or Mini-MT)

Distance: 2 km using single-mode fiber, 550 m using 62.5-μm multimode fiber

Source: longwave laser (1270–1355 nm), RMS spectral width = 4 nm optical output power into a 62.5-μm fiber, NA = 0.275: -3 to -13 dBm

Receiver sensitivity: -3 to -20 dBm average

Optical Connectors

In addition to the Galaxy and Mini-MT described previously as potential connector options for Fibre Channel Standard, several other types of fiber optic connectors have recently begun to find applications. These include the following:

The mini-SC, a connector with the SC form factor but half the size (using a nonstandard ferrule): Several of these can be combined into a multifiber back-plane connector, called the MU connector, suitable for optical back planes and zero-insertion force applications, which provide 16 optical connections in a width of 100 mm (similar in form factor to the conventional DS-type back-plane connector). This connector was developed for the telecom market and is not yet well established for datacom applications. A separate tool is required to remove this connector from its receptacle. The MU connector is available from NTT International and Alcoa-Fujikura.

The Hi-PER Link connector: A multifiber ribbon connector for 2–12 fibers, based on the MT ferrule, and also used on parallel optical interconnects such as Motorola Optobus. The Hi-PER Link connector is available from Alcoa Fujikura Ltd.

The Opti-Jack connector (also known as Fiber-Jack): A duplex connector using conventional ferrules inside an RJ-45 style housing. It has been accepted into part of the ATM Forum baseline text and has the TIA/EIA standards designation FOCIS-6. It is available from Panduit and Berg.

The LC connector, developed by Lucent for the telecommunications market, may also find datacom applications including gigabit Ethernet links.

Previous sections of this book have already mentioned the emerging DC connector, developed by Siecor Corp. This connector uses multiple fibers in a single standard 2.5-mm ferrule and is available in both two-fiber and four-fiber versions; the external housing is identical to a simplex SC connection.

The LightRay MPX connector, developed by Amp for its family of parallel optical interconnect products, is a hermaphroditic 2- to 12-element multifiber connector based on the MT ferrule but not compatible with the MTP/MPO connectors used in some other datacom applications.

Specialty connectors with metal or polymer ferrules are also available from several major connector suppliers and may be encountered in some datacom links. Ferrules with a slight angle polish, typically between 6 and 8°, may be found in applications that require extremely low back reflection from the connector interface. Older connector types, such as the optical SMA connector, still resist being replaced and may be found in some installations. Various types of specialty optical fiber are also available for applications requiring nonstandard fiber diameters or doping for additional fiber strength.

There are many options available for multifiber optical connectors, most of which are mutually incompatible. For example, the MTP was recently adopted by the HIPPI 6400 standards body. Other versions, such as the MPO or MPX, are available from various sources. Another option for parallel optical links is a revised version of the AT&T MAC multifiber connector. The improved MAC-II connector still provides 12 fiber or more attachments but does so in a small package size. Currently Bellcore approved, another option is the mini-MAC, available from Berg as a back-plane connector option based on the MT ferrule.

There are many alternatives available for parallel optical connectors. Several recent papers in the literature have proposed stacking up to 6 MT ferrules in a single optical connector, providing a 72-fiber plug; one such connector has been proposed as part of the SIPAC system for large telecommunication switching. One of the largest single-element connectors proposed so far is the so-called super-MT, a 144-fiber connector using polymer optical waveguides. Developed by the POINT consortium (Amp, Allied Signal, Honeywell, and University of Columbia), this connector would be intended for optical back-plane applications.

Plastic Optical Fiber

At the Optical Fiber Conference 1997 in Dallas, Texas, plastic fiber applications were demonstrated for 155-Mb ATM and 100 Mb/s Ethernet LANs. This technology was developed by the High Speed Plastic Network Consortium (HSPN), a DARPA-funded venture consisting of Boeing, CEL, Boston Optical Fiber, Honeywell, Packard Hughes, and Lucent. Graded-index plastic fiber made from materials such as PMMA now offers bandwidth >3 Gb/s to 100 m and may be an alternative media for premises wiring. HSPN has submitted a proposal to the Optoelectronics Micro Networks OMNET program to continue development of plastic fibers.

Although some manufacturers have presented data indicating that plastic fiber will be unable to meet industry standard TIA/EIA-568A at 100 m, a number of companies are already developing plastic fiber applications. Some companies currently marketing plastic fiber or components including the following:

Amp Inc. offers plastic fiber connectors and components for airplanes and automobiles.

CEL (California Eastern Laboratories), a distributor for NEC Japan, offers the HiSpot transceiver with a visible red LED over step-index plastic fiber at 155 Mb/s using an F07 connector; applications include Fast Ethernet, ATM, and IEEE 1394 (Firewire).

Spec Trans Specialty Optics (distributor for Toray Industries of Japan) offers 200-μm step-index fiber; they have proposed to the ATM Forum a standard for 200-μm hard clad silica at 155 Mb/s, 100 m and plastic fiber for 50 Mb/s, 50 m.

Transceivers

Although discussion over the next generation of Fibre Channel continues, some companies are now offering an alternative, the gigabit interface connector (GBIC). This is a gigabit module that offers the convenience of being able to unplug a copper interface and replace it with an optical interface at a later time for increased distance and bandwidth. It features a parallel rather than a serial interface; the copper and optical modules plug into the same electrical connection point, which includes a pair of rails to guide the modules during plugging. Found mainly in workstation applications, the GBIC is available from several sources including IBM Rochester, Minnesota.

Low-cost bidirectional transceivers for fiber in the loop and similar applications (fiber to the curb or home) are also being developed. For example, a recent device reported by Amp in collaboration with Lasertron, Digital Optics Corp., GTE Labs, BroadBand Technologies, and the University of Colorado offers a connectorized, single-mode bidirectional transceiver with a source, modulator, and detector integrated into a single package. Each module includes an InP diode laser, silicon CMOS modulator/demodulator electronics, a PIN photodiode, an integrated beam splitter, and a built-in SC for ferrule-based fiber attachment. The beam splitter is used for bidirectional transmission at 1.55-μm wavelength and reception at 1.3-μm wavelength. Data rates up to 1.2 Gb/s is at 1.3 μm and a burst mode, 50 Mb/s receiver at 1.55 μm, are available in a package measuring 85 $\times$ 17 $\times$ 10 mm.

Other emerging areas include high-speed transceivers for OC-48 speeds (2.5 Gb/s). These products can typically reach 150 m using 850-nm lasers on 62.5-μm fiber or up to 350 m using 1300-nm lasers on 62.5-μm fiber. Many efforts are under way to develop high-speed optical transceivers compatible with ATM or Fibre Channel.

There are a number of trends in the future development of datacom transceiver technology. Among these is the migration to lower power supply voltages, following the trend of digital logic circuits. Most digital logic has migrated from 5 to 3.3 V and is on a path toward 2.5-V operation; optical transceivers are emerging that will be able to operate at 1 Gb/s speeds using a single 3.3-V power supply, although many sources will remain at 5 V for some time, especially VCSELs.

Although the telecommunications market continues to have distinct requirements from the datacommunications market, there is certainly some merging of requirements between these two areas as well. The telecom market is currently driving development of analog optical links for cable television applications, including bidirectional wavelength multiplexed links that transmit in one direction at 1300 nm and in the other at 1550 nm. This is achieved by incorporating some form of optical beam splitter in the transceiver package. Many of these devices incorporate optical fiber pigtails rather than connectorized transceiver assemblies. Packaging is another evolving area; existing datacom industry standards, such as the 1×9 pin serial transceiver or 2×9 pin transceiver with integrated clock recovery, may give way to surface-mount technology in the near future. In the telecom industry, the standard optical transceiver package has been the dual in-line pin (DIP) package (14 pins and two rows); recent developments in the so-called mini-DIP (8 pins and two rows) may develop into a new packaging standard for the industry. Finally, integrated optoelectronic integrated circuits are advancing to the point where they may replace the standard TO can and ball lens assembly for optical device packaging and alignment in the next 5 years.

VCSELs hold a great deal of promise as low-cost optical sources for data communications, especially for optical array interconnections. Research continues on VCSELs that can operate at long wavelengths (1300 nm); although some devices have been demonstrated under laboratory conditions, the technology is probably several years away from commercial applications.

Parallel optical transceivers will face new challenges in the area of laser safety. Although many parallel transceivers will be able to meet U.S. class 1 safety standards, it is more difficult to meet the international class 1 limits specified by IEC 825. One possible alternative is international certification as a class 3A product, which would require a warning label on the transceiver; it is unclear whether the IEC will allow short-wavelength lasers to hold this classification or whether the market will accept such products instead of class 1 sources.

Index

837